H.-P. Schultheiß (Ed.)

New Concepts in Viral Heart Disease

Virology, Immunology and Clinical Management

With 190 Figures and 58 Tables

Springer-Verlag
Berlin Heidelberg GmbH

PD Dr. med. Heinz-Peter Schultheiß
Med. Klinik I
Klinikum Großhadern
Marchioninistraße 15
8000 München 70

ISBN 978-3-642-73612-4 ISBN 978-3-642-73610-0 (eBook)
DOI 10.1007/978-3-642-73610-0

Softcover reprint of the hardcover 1st edition 1988

2119/3140/543210

Preface

The recent rapid advances in our knowledge of immunological and virological mechanisms involved in the pathogenesis of viral heart disease makes it difficult for everybody working in this field to keep up with the latest developments. However, much of what we know is still circumstantial and only vaguely substained. Interdisciplinary understanding and cooperation thus seems necessary to get a better insight into the mechanisms by which viruses may initiate immunological organ-specific tissue injury and disease!

This volume evolved out of an international symposium by the same title held on May 25.–28., 1988 in Tegernsee, near Munich, of which a wide spectrum of immunological, virological, diagnostical and clinical problems was covered.

Both review articles and new experimental and clinical data are included in this volume to give the reader an up-to-date information about current concepts and future aspects.

Chapter I serves as an excellent introduction to the epidemiology and natural history of dilated cardiomyopathy / viral heart disease. Although a definite viral etiology in myocarditis and dilated cardiomyopathy is often difficult to establish, epidemiological and serological data incriminate a viral etiology underlying many cases of "dilated cardiomyopathy". Chapters II and III describe the current thinking on virological and immunological mechanisms involved in the pathogenesis of viral heart disease. Among others virus topism, virus persistance, possible mechanisms and genetic basis of post-infection autoimmunity, and the virus-interaction with the immune system are discussed. Chapter IV provides an overview on humoral and cellular cardiac-specific autoimmunity in patients with myocarditis and dilated cardiomyopathy. Autoantibodies to several membrane proteins – such as the beta-adrenoreceptor, the calcium channel, the connexon and the ADP/ATP carrier – are described. The difficulties and future directions in the diagnosis of virus-myocarditis are reviewed in chapter V. Recent data demonstrating enterovirus RNA in myocardial tissue from patients with myocarditis and dilated cardiomyopathy and the meaning of this finding for the pathogenesis of viral heart disease are discussed, furthermore the signification of immunohistochemical-, in-situ hybridization- and immunoscintigraphic techniques. In the final chapters new aspects of the clinical management – specially the hemodynamic and metabolic effects of ACE-inhibitors and the current status of immunomodulating therapy – are critically reviewed.

The numerous suggestions and unrestrained support of Professor G. Riecker, Department of Internal Medicine, University of Munich, is greatly appreciated.

As organizer, it is a pleasure to thank the members of the scientific committee: Professor P. H. Hofschneider and Professor R. Kandolf, Department of Virology, Max-Planck-Institute for Biochemistry, Martinsried, Professor M. Klingenberg, Institute of Physical Biochemistry, University of Munich, and Professor G. Riethmüller, Institute of Immunology, University of Munich, for the fruitful and lively discussions, to all those at the Department of Internal Medicine who did so much to ensure the success of the symposium, and to Dr. C. Osthoff, Springer-Verlag, for publishing the book so rapidly.

Lastly but by no means least, I would like to thank Hoechst for their full support without which it would not have been possible to organize this meeting.

I hope that the presentations contained in this book convey the excitement and very productive scientific atmosphere of the conference, led to a better understanding of the basic aspects of viral heart disease and to a genuine exchange of concepts and experimental approaches between basic scientists and clinically oriented researchers.

HEINZ-PETER SCHULTHEISS

Table of Contents

I. Epidemiology and Natural History

II. Concepts in Viral Pathogenesis

VII. New Aspects of Clinical Management

I. Epidemiology and Natural History

Etiology, Pathogenesis, and Pathophysiology of Dilated Cardiomyopathies

W. H. ABELMANN

The Charles A. Dana Research Institute and the Harvard-Thorndike Laboratory of Beth Israel Hospital, Department of Medicine, Beth Israel Hospital and Harvard Medical School, Boston, MA 02215, USA

Introduction and Definitions

The term "cardiomyopathy" was introduced by Brigden in 1957 [18a] and defined as "non-coronary disease of heart muscle of unknown etiology." The first definition of cardiomyopathy by the World Health Organization (WHO) in 1968 [38] stressed cardiomegaly and heart failure as dominant factors and excluded myocardial dysfunction associated with valvular or coronary disease, or with systemic or pulmonary vascular disease. Myocardial disease of unknown etiology was labeled "primary cardiomyopathy," and that associated with a known single cause, systemic disease, or syndrome was called "secondary cardiomyopathy." In the 1980 Report of the WHO/ISFC Task Force [122], the term "cardiomyopathy" was redefined as including only heart muscle disease of unknown etiology; heart muscle disease of known etiology — previously known as "secondary cardiomyopathy" — became "specific heart muscle disease." Cardiomyopathy was further subclassified by pathophysiologic criteria into "dilated," "hypertrophic," and "restrictive/obliterative cardiomyopathy."

Dilated cardiomyopathy, by far the most frequent form of cardiomyopathy, is characterized by dilatation of one or generally both ventricles, associated with impaired pump function, low cardiac output, increased filling pressures, and decreased ejection fraction.

This chapter will address dilated cardiomyopathy defined functionally, i.e., including specific heart muscle diseases presenting as dilated (congestive) cardiomyopathy.

Etiology of Dilated Cardiomyopathy

The number of disease processes which may involve the heart is very large, and many comprehensive lists have been published [2, 40, 57, 60, 120, 125]. Table 1 presents specific causes of heart muscle disease which may present clinically as dilated cardiomyopathy. These causes include infections; granulomata; hematologic, metabolic, endocrine and nutritional diseases; neoplasms and neuromuscular diseases; as well as toxins and drugs. Cardiomyopathy may be acute or chronic, it often affects only a small percentage of the patients with the diseases listed, and may become manifest only late in the course of the disease.

H.-P. Schultheiß (Ed.)
New Concepts in Viral Heart Disease

Table 1. Etiologic classification of dilated cardiomyopathy

Infections	
Bacterial	Viral (cont.)
Diphtheria	Varicella
Tuberculosis	Epstein-Barr
Typhoid fever	Coxsackievirus
Scarlet fever	Echovirus
Meningococcosis	Cytomegalovirus
	Hepatitis
Spirochetal	Rabies
Leptospirosis	Mycoplasma
Lime fever	Psittacosis
	Arboviruses
Rickettsial	
Rocky Mountain spotted fever	Mycotic
	Histoplasmosis
Q fever	Coccidiomycosis
	Cryptococcosis
Viral	Protozoal
Poliomyelitis	South American trypanosomiasis
Influenza	African trypanosomiasis
Mumps	Toxoplasmosis
Rubella	Helminthic
Rubeola	Trichiniasis
Variola	Schistosomiasis
Metabolic	
Endocrine	Nutritional
Acromegaly	Beriberi
Thyrotoxicosis	Kwashiorkor
Hypothyroidism	Selenium deficiency (Keshan's disease)
Pheochromocytoma	
Diabetes mellitus	
Familial storage diseases	Others
Glycogen storage diseases	Hypokalemia
Fabry's disease	Carnitine deficiency
Gaucher's disease	Uremia
Sandhoff's disease	
Mucopolysaccharidosis	
Deposits	Endomyocardial diseases
Hemochromatosis	Endomyocardial fibrosis
	Loeffler's disease
Granulomatous	Endocardial fibroelastosis
Sarcoidosis	
Wegener's granulomatosis	Toxins and drugs
Granulomatous myocarditis	Ethyl alcohol
	Arsenic
Connective tissue diseases	Phosphorus
Lupus erythematosus	Amphetamine
Scleroderma	Cobalt
Dermatomyositis	Adriamycin
	Catecholamines

Table 1. Continued

Hematologic disorders	Cocaine
Leukemia	Radiation
Anemia	
Sickle cell anemia	
Henoch-Schönlein purpura	
Heredofamilial Neurological and neuromuscular diseases	
Progressive muscular dystrophy (Duchenne)	
Limb-girdle muscular dystrophy (Erb)	
Friedreich's ataxia	
Myotonia atrophica (Steinert)	

In well over 90% of patients who present with dilated cardiomyopathy, no specific etiology can be identified. It is highly unlikely that the causes of dilated cardiomyopathy are as numerous as the individual diseases with which this syndrome is associated. It is much more likely, especially since heart disease is generally not predictably associated and is often a late manifestation, that we are dealing with one or more final common pathogenetic mechanisms and with confluences of several etiologic and enhancing factors. Indeed, increasing evidence is accumulating for the concept of dilated cardiomyopathy as a multifactorial or pluricausal disorder [1, 47].

This presentation will consider the major processes and causes to which primary pathogenetic roles have been attributed, as well as secondary or modifying factors which have been found to affect the expression of heart muscle disease. Evidence for the mechanisms to be considered will be drawn from experimental models of cardiomyopathy as well as from the clinic.

Pathogenetic Mechanisms

Table 2 lists the major pathogenetic mechanisms to which a role in the causation of dilated cardiomyopathy has been attributed. They are not listed in order of perceived importance or frequency of incidence. Some pathogenetic mechanisms,

Table 2. Pathogenetic mechanisms

Cacium overload	Free radicals
Endogenous catecholamines	Infection
Microvascular spasm	Immune processes
Small vessel disease	Hypoxia
Altered myocardial metabolism	Heredity

which appear to be primarily contributory, i. e., affecting the expression of the disease, will be discussed separately.

Genetic

In 1955, Michelson et al. [84], described a hereditary myopathy in the mouse with involvement of the myocardium in 3% of the animals. A hereditary dystrophy-like myopathy in Syrian hamsters, described by Homburger et al. in 1962 [53], however, manifested 100% involvement of the heart and has become one of the most widely studied animal models of heart disease. The BIO 14.6 strain of hamsters was characterized as a reproducible model of cardiomyopathy which developed congestive heart failure in more than 90% of animals after 100 days; it was shown to be transmitted by an autosomal recessive gene [10].

In genetically diabetic mice, myocardial lesions preceding vascular lesions have been demonstrated [45]. Heredity was also thought to be the determinant of a generally fatal dilated cardiomyopathy seen in newborn calves in Australia [89].

The susceptibility to virus-induced myocarditis and its late sequel, cardiomyopathy, is also determined by genetic factors. This is best seen in the wide spectrum of expression of the infection with a specific strain of cardiotropic virus in different strains of inbred mice. Studies with coxsackievirus B3 [50], as well as with encephalomyocarditis virus [66], have shown correlation of severity and course of the myocardial disease with major histocompatibility genes. Furthermore, the host's genetic background has been shown to determine the immunologic mechanisms involved: Huber and Lodge [55] found that in BALB/c mice infected with coxsackievirus B3, cytolytic T lymphocytes were involved in the production of myocardial injury, whereas in DBA/2 mice inoculated with the same virus, heart-reactive antibodies were responsible.

In man also, dilated cardiomyopathy may have a genetic basis. Thus, a retrospective analysis of 169 patients seen at the Mayo Clinic yielded a positive family history in 6.5% of cases [83]. X-linked and autosomal dominant dilated cardiomyopathy has also been reported [11, 42], as well as autosomal recessive inheritance [46]. Many hereditary disorders are associated with myocardial involvement. This is seen in the glycogen storage diseases: Fabry's disease, mucopolysaccharidosis, muscular dystrophies, and many others [120].

Nutritional Deficiencies

Thiamine Deficiency

Vitamin B_1, or thiamine, is an essential co-enzyme in the decarboxylation of alpha-keto acids, pyruvate, and alpha-ketoglutarate, and in the utilization of pentose in the hexose-monophosphate shunt. Its absence constitutes a defect in myocardial energy production: interference with both aerobic and anaerobic glycolysis leads to impaired oxidative phosphorylation. Although many of the clinical cardiovascular manifestations, such as the high cardiac output state, are secondary to peripheral vasodilatation, largely attributable to the accumulation of intermediate carbohydrate metabolites, decreases in the duration of myocardial contraction and in the rate of development of tension have been reported in experimental thiamine deficiency [23], as has depression of left ventricular function in thiamine-deficient hamsters [52]. These changes resemble what is found under conditions of hypoxia. Both the experimental disease and thiamine deficiency in man, known as "beriberi heart disease", are readily reversible with administration of thiamine, without sequelae. Thus, this is one of the few completely curable (as well as preventable) dilated cardiomyopathies.

Protein-Calorie Malnutrition

Protein-calorie malnutrition is the basis for kwashiorkor, as well as marasmus, primarily seen in children. These syndromes may be associated with dilated cardiomyopathy, often in association with vitamin and electrolyte deficiencies [98]. Heart failure is seen especially under conditions of increased load, such as rehydration. Similar myocardial lesions have been produced in rhesus monkeys with pure protein deficiency by Deo et al. [28].

Selenium Deficiency

A dilated cardiomyopathy known as "Keshan disease" has been prevalent in northeast China, where the selenium content of the soil is low [127]. Patients with Keshan disease were found to have low blood selenium levels, as well as decreased activity of glutathione peroxides, an enzyme dependent on selenium. In the absence of selenium, lipid peroxides and free radicals are increased and are potentially toxic. Treatment and prevention with sodium selenite are said to have eradicated Keshan disease in China. Occasionally, clinical cardiomyopathy secondary to selenium deficiency has been reported from the West [59].

Taurine Deficiency

A form of dilated cardiomyopathy in cats depleted of taurine, which responded to taurine therapy, has been reported recently [97]. Taurine, ethanylamine sulfonic acid, is an essential nutrient for cats. This appears to be a species-specific problem.

The question of taurine deficiency playing a role in the cardiomyopathic hamster has been raised by the observation of a deficiency in atrial natriuretic factor [22] and its partial reversal by administration of taurine [30]. Taurine has also been noted to reduce the myocardial lesions in this model [118].

Carnitine Deficiency

Carnitine is a cofactor which is essential for the transport of long-chain fatty acids into mitochondria for oxidation. Depletion of carnitine interferes with oxidation of fatty acids, leading to accumulation of lipids in the cytoplasm [29]. Carnitine deficiency usually presents as dilated cardiomyopathy, may be familial, and may be associated with endocardial fibroelastosis [111]. The diagnosis is made by finding a low serum carnitine level and confirmed by endomyocardial biopsy. Treatment with L-carnitine orally is effective.

The possibility of a pathogenetic role for carnitine deficiency in the cardiomyopathic hamster has been raised by the finding of decreased myocardial levels of carnitine in this cardiomyopathy [128]. Indeed, treatment of animals with carnitine improved both the mechanical and hemodynamic performance of the isolated perfused heart [121].

A beneficial effect of carnitine therapy upon both myocardial histopathologic lesions and left ventricular function has also been observed in rats exposed to chronic adriamycin toxicity [82].

Vasoactive Substances

Catecholamines

Predominantly epinephrine-secreting adrenal medullary tumors or pheochromocytomas may be associated with dilated cardiomyopathy, which may be reversed, at least partially, by removal of the tumor, as well as by adrenergic blocking agents such as propranolol [58, 113]. Inhibition of biosynthesis of catecholamines by alpha-methyltyrosine has also been effective [16]. This form of cardiomyopathy has been attributed to the prolonged action of catecholamines, associated with increased preload and afterload due to tachycardia and hypertension.

Elevation of catecholamines in the plasma of patients with dilated cardiomyopathy, along with depletion of catecholamines in endomyocardial biopsy tissue has been reported [63].

Extensive studies of catecholamine cardiotoxicity have been carried out, but have focused largely upon the acute cardiac effects. The multifocal and often extensive myocardial necrotic lesions have been variously attributed to microvascular spasm and hypoxia (histopathologically, the lesions resemble ischemic lesions), calcium overload, and altered membrane permeability [104]. When rats were treated acutely with isoproterenol and then followed up, some animals developed cardiomegaly, aneurysms, and, at times, congestive heart failure [105].

Catecholamines, as well as other vasoactive substances, have also been incriminated as playing a role in the cardiotoxicity secondary to administration of anthracyclic antitumor agents such as adriamycin [20]. Rabbits can be partially protected from chronic anthracycline cardiomyopathy by administration of adrenergic blocking agents and antihistamines. Similarly, mice have been protected from chronic adriamycin toxicity by the administration of theophylline and disodium cromoglycate, which inhibits adriamycin-induced release of histamine [67]. Rats have been protected from late adriamycin toxicity by the administration of inosine, a potent coronary vasodilator [25]. The dilated cardiomyopathy produced in turkeys by the feed additive furazolidone, which can be prevented completely by the beta-adrenergic blocking agent propranolol, has also been attributed to increased production of, and sensitivity to, catecholamines [49].

Evidence of excessive circulatory catecholamines has also been considered of pathogenetic significance in diabetic cardiomyopathy [41]. Both epinephrine [10] and isoproterenol [73] have been reported to enhance the severity of myocardial necrosis in the hereditary cardiomyopathy of the Syrian hamster.

Microvascular Spasm

The resemblance of the myocardial lesions in several experimental models of dilated cardiomyopathy to those associated with reperfusion necrosis and their enhancement by isoproterenol have led Factor et al. [33–35] to investigate the microcirculation in several small models. By means of perfusing living rodent hearts with a solution of silicone rubber, they were able to demonstrate regions of narrowing and constriction in the cardiomyopathic hamster and attributed these to microvascular spasm [33]. The lesions could be prevented by verapamil. Similar lesions have also been noted in the hypertensive diabetic rat [34] and in the chagasic mouse [35].

Toxins

Many chemicals have been shown to have cardiotoxicity. However, the agents which have been associated with dilated cardiomyopathy are few.

Ethanol

The high prevalence of chronic alcoholism in patients with dilated cardiomyopathy has long been recognized [116, 123]. Bollinger [114] coined the term "Münchner Bierherz" (Munich beer heart) in 1884, when he reported marked dilatation and hypertrophy of the heart in patients with history of prolonged intake of large quantities of beer. The term "alcoholic heart disease" was first introduced by Mackenzie [75], who reported four patients with chronic alcoholism who died in congestive heart failure. In the early twentieth century, congestive heart failure in the alcoholic patient tended to be attributed to nutritional deficiencies rather than to the effects of alcohol per se, especially to thiamine deficiency [117]. In the second half of this century, however, it has become increasingly evident that most of the heart disease seen in chronic alcoholic patients is of the low output type and does not respond to thiamine, as noted by Brigden and Robinson [18] who reintroduced the label of "alcoholic heart disease" in 1957. Currently, alcohol is onsidered the major cause of dilated cardiomyopathy by many [115].

In an early hemodynamic study of myocardial function in chronic alcoholic patients with fatty liver by means of increasing afterload, Regan et al. [101] demonstrated impaired left ventricular function which was confirmed noninvasively in apparently healthy chronic alcoholic subjects by Spodick et al. [108].

Although the acute myocardial toxicity of alcohol has been well established in man as well as in experimental animals [51, 100], and chronic administration of ethanol was also shown to result in impaired function of ventricular muscle [65], the fact that chronic congestive heart failure could not be reproduced by administration of ethanol to animals raised major questions as to the etiologic role of alcohol itself in human cardiomyopathy. Many investigators felt that genetic, metabolic, nutritional, infectious, and other toxic factors might be needed for full expression of the clinical syndrome of alcoholic heart disease.

The pathogenetic mechanisms by which alcohol, administered chronically, affects myocardial function remain to be fully elucidated [99]. In early studies, Bing et al. [14] exposed dogs to alcohol for several months and reported significant reductions of calcium-binding as well as uptake by mitochondria and sarcoplasmic reticulum, suggesting interference with excitation-contraction coupling. Garrett et al. [43], after feeding alcohol to Syrian hamsters for many weeks, found depression of left ventricular function as well as of high-energy phosphates and adenosine. These effects were prevented by administration of verapamil. Thus, altered handling of calcium may be one mechanism by which alcohol depresses myocardial function.

Cobalt

Early experimental work by Kaufmann and Fleckenstein [61] had recognized that cobalt ions depressed myocardial function by interfering with the action of calcium. In 1966, several outbreaks of fulminant congestive heart failure in beer drinkers were reported from Canada [88], and later also from the United States and Belgium, which were subsequently attributed to cobalt sulfate, an additive to beer used to stabilize the foam. Since discontinuation of the use of cobalt, only rare cases related to industrial exposure have been reported. However, these outbreaks led to the production of an experimental model of dilated cardiomyopathy with heart failure in dogs given daily injections of cobalt over a period of 10 weeks [106].

Anthracyclines

Doxorubicin and daunorubicin, anthracycline antibiotics used widely in the chemotherapy of neoplastic disease, have acute as well as chronic cardiotoxicity. The latter comprises a dilated cardiomyopathy which is dependent upon the cumulative dose and is characterized by a relentless course and high mortality. The pathogenesis is still poorly understood. Studies in the rat revealed a decreased utilization of the radio-iodinated fatty acid analogue 15-I-131-iodophenylpentadecanoic acid, in association with decreased myocardial carnitine [72]. Ogawa et al. [92] found increased levels of peroxide and phospholipase activity as well as inhibition of mitochondrial respiration. Generation of superoxide has also been reported [109]. Release of histamine has also been incriminated [19].

Free Radicals

The formation of free radicals has been reported to occur in a number of experimental cardiomyopathies, including adriamycin toxicity [90]. Thus, the free radical scavenger α-tocopherol (vitamin A) exerts a protective action in the acute stage of anthracycline toxicity but is not effective in the chronic cardiomyopathy in rabbits [17]. Treatment of rabbits exposed to daunorubicin with a chelate of iron (ICRF-187), which blocks the generation of free radicals, also protected these animals from chronic cardiomyopathy. An increase in free radicals has also been demonstrated in cardiac mitochondria of cardiomyopathic hamsters [68].

Calcium Overload

The cardinal pathophysiologic abnormality of dilated cardiomyopathy is impaired systolic function of the heart. In view of the central role of Ca^{2+} in excitation-contraction coupling, a good deal of attention has been paid to the handling of calcium in both clinical and experimental dilated cardiomyopathies.

In the cardiomyopathic hamster, myocardial calcium content was found to be increased [74], and calcium levels were also elevated in mitochondria [124a]. Treatment of cardiomyopathic hamsters in the congestive heart failure stage with the calcium-blocking agent verapamil decreased congestion and mortality, in association with decreases of adenosine nucleotides, cyclic adenosine monophosphate (AMP), high-energy phosphates, as well as increases in the ratio of adenosine triphosphate (ATP) to total adenosine diphosphate (ADP), in comparison to untreated animals. Using nuclear magnetic resonance spectroscopy, verapamil has also been demonstrated to improve left ventricular hemodynamics as well as mitochondrial function [79].

Calcium overload has also been implicated as a pathogenetic mechanism of anthracycline toxicity [9]. The finding of depressed high-energy phosphates and adenosine, along with depressed left ventricular function, in Syrian hamsters fed ethanol for prolonged periods led to postulation of altered calcium handling here also: administration of verapamil, concomitantly with ethanol, prevented these effects [43].

Rats with streptozotocin-induced diabetes, on the other hand, showed evidence of depression of the calcium pump which could be reversed by insulin [78].

An interesting acute cardiomyopathy is that associated with seafood poisoning in which maitotoxin is the substance incriminated. Guinea pig atria exposed to this toxin manifest increased resting tension associated with increased uptake and content of calcium [69]. In rat cardiac myocytes, maitotoxin induced reversible contracture, which was suppressed by verapamil.

Infection

Acute myocarditis has been seen in association with almost all types of infections, be they bacterial, rickettsial, viral, mycotic, protozoal, or helminthic. In most but not all of these infections, acute or subacute cardiomyopathy has been described [2, 120]. Chronic dilated cardiomyopathy, in association with chronic myocarditis, is seen in toxoplasmosis, occasionally in mycoses, and rarely in tuberculosis. Chronic dilated cardiomyopathy as a late sequel of myocarditis, however, has been largely limited to viral infections and South American trypanosomiasis or Chagas' disease; clinical studies as well as studies of experimental models have yielded ample evidence for chronic dilated cardiomyopathy occurring as a late sequel [3, 103]. In endemic areas – primarily Central and South America – infection with *Trypanosoma cruzi* is now well accepted as the initiator of the majority of cases of chronic dilated cardiomyopathy on the basis of serologic and epidemiologic evidence, as well as certain distinguishing characteristics, such as conduction disturbances, apical aneurysms, and evidence of autonomic denervation of the heart and other organs [5]. Furthermore, careful examination of serial sections of myocardium may reveal a few persistent parasites [70]. It is clear, however, that the myocardial disease progresses unrelated to the occasional persistence of the parasite. In the case of viral myocarditis, the period of myocardial replication of virus is so brief – as a rule less than 2 weeks – that the chronic disease cannot be attributed to the virus itself, but has been attributed to immune/autoimmune processes (see below). Notwithstanding the increasing evidence for immune/autoimmune processes playing a pathogenetic role, even in acute and certainly in subacute myocarditis, dealt with extensively in other chapters in this volume, the initial cytopathic event in the myocardium remains the penetration of virus into the myocyte and its replication therein.

Immune/Autoimmune Mechanisms

In clinical as well as experimental heart disease due to *Trypanosoma cruzi,* the major pathogenetic role of the immune system has long been recognized. Comprehensive reviews are available [4, 5, 107].

Early evidence for the immune/autoimmune pathogenesis of chronic dilated cardiomyopathy as a late complication of viral myocarditis, both in experimental models and in man, has been summarized in a number of reviews and monographs [15, 27, 44, 56, 62, 76, 103, 124].

Inasmuch as many chapters in this volume deal with this topic, suffice it to say here that initial myocyte damage may stimulate either cellular or humoral immune responses or both, and that the responses may differ in the course of the disease and also between different hosts *(see above)*. In experimental models as well as in man, heart reactive antibodies have been demonstrated, and T lymphocytes have been found to have cytotoxic effects.

Enhancing Factors

Most of the etiologic and pathogenetic factors which have been discussed so far may also act as enhancing factors, and some of the evidence for this has already been cited above. This section will consider additional enhancing factors, listed in Table 3, which by themselves are not likely to produce dilated cardiomyopathy, but which in conjunction with other etiologic factors enhance or bring out the manifestations of dilated cardiomyopathy.

In classical studies, Pierce [96] demonstrated that *hypoxia* increased the severity of experimental myocarditis. Exposure to *hyperbaric oxygen* has been shown to increase the myocardial replication of coxsackievirus B1 in mice, associated with increased mortality [93].

Table 3. Enhancing Factors

Hypoxia	
Hyperbaric oxygen	
Myocardial hypoperfusion	
Nutritional deficiencies	
Ethanol	
Corticosteroids	
Nonsteroidal anti-inflammatory agents	
Radiation	
Increased afterload	Hypertension
Increased preload	Anemia
Pregnancy	
Exercise	
Tachycardia	

Pregnancy may be an enhancing factor: Farber and Glasgow [36] reported increased myocardial replication of virus in the heart of pregnant mice, associated with myocarditis, hepatic congestion, and increased mortality, not seen in infected, nonpregnant animals.

Deficiencies of vitamins of the B group, and especially of B1 or *thiamine,* have been reported to enhance experimental Chagas' disease in the rat [126].

The effects of prolonged ingestion of *ethanol* upon experimental trypanosomal myocarditis in mice was studied by Miller and Abelmann [85], who found that ethanol enhanced the effects of acute chagasic myocarditis and led to congestive heart failure not seen in control infected animals.

In this form of experimental myocarditis, severity and mortality were also found to be increased by forced *exercise* [31]. In a model of experimental autoimmune myocarditis in guinea pigs, the manifestations of dilated cardiomyopathy were enhanced after exercise [54].

Anthracycline cardiomyopathy is well known to be enhanced by concurrent administration of *cyclophosphamide* [12] or *radiation* therapy [13, 32]. Microvascular damage has been incriminated as a pathogenetic mechanism [8].

Autonomic imbalance, long known to play a role in Chagas' disease [70], may also be a contributory factor in idiopathic dilated cardiomyopathy: histopathologic studies by Amorim and Olsen [6] have demonstrated a significant reduction in the number of cardiac ganglia. A relationship to arrhythmia and sudden death has been postulated. Sinus tachycardia, which increases myocardial oxygen and blood flow requirements, as well as reduces diastolic filling and coronary perfusion times, may also be an enhancing factor. Sinus tachycardia has even been proposed as a primary cause of dilated cardiomyopathy [95].

Increased afterload is also an enhancing factor. Thus, whereas the myocardial lesions associated with experimental diabetes in the mouse are not accompanied by congestive heart failure, in the presence of *hypertension,* severe myocardial necrosis and subsequent fibrosis are associated with congestive heart failure [37].

The infectious-immune theory of pathogenesis of dilated cardiomyopathy has led to the expectation that immunosuppression might suppress the development of dilated cardiomyopathy after an acute viral myocarditis. Although uncontrolled preliminary studies in man appear to lend some support to this hypothesis [26, 80], studies in experimental models have shown consistently deleterious effects of immunosuppressive therapy. This has been demonstrated for corticosteroids in coxsackievirus disease [64] and in murine encephalomyocarditis [110], for cyclosporine in coxsackievirus disease [91] and in encephalomyocarditis [86], and for cyclophosphamide in experimental Chagas' disease [71]. Furthermore, even nonsteroidal anti-inflammatory agents have been reported to enhance myocardial coxsackievirus disease [24, 102].

Pathophysiology

Physiologically, dilated cardiomyopathy is characterized by biventricular dilatation associated with impaired pump function, low cardiac output, decreased ejection fraction, and increased ventricular filling pressures. Left ventricular dysfunction may dominate over right ventricular dysfunction, but occasionally the reverse is true [39]. As a rule, ventricular dysfunction is global, i. e., involving the ventricles diffusely, but there may be regional dominance [129]. While systolic ventricular dysfunction dominates, diastolic dysfunction may also be present, especially if ventricular compliance is reduced by heavy fibrosis or hypertrophy, resulting in impairment of ventricular relaxation during diastole [48, 81, 87]. Although occasionally, especially if the disease is detected in its earlier stage, ventricular dilatation may be slight, dilated cardiomyopathy is normally accompanied by progressive dilatation of the left ventricle, by the manifestations of low-output heart failure, and eventually by congestive heart failure. In the primary as well as compensatory manifestations, dilated cardiomyopathy may be difficult to distinguish from congestive heart failure of other causes, e. g., ischemic heart disease or valvular heart disease.

The cardiomyopathic process may result in partial or complete heart block, and it may give rise to almost any kind of cardiac arrhythmia. Ventricular ectopy is especially frequent and may reduce ventricular function further. Sudden death ist frequent [7].

The dilatation of one or both ventricles may result in insufficiency of the atrioventricular valves, resulting in an increase in both pre- and afterload and in acceleration of the course of the disease. Both mitral and tricuspid insufficiency are especially sensitive to increased afterload, i. e., systemic and pulmonary hypertension, respectively, which increase the regurgitant volumes.

The low-output state and the dilated cardiac chambers favor the formation of intracardiac thrombi, often resulting in pulmonary and/or systemic emboli.

Myofiber hypertrophy is the rule in dilated cardiomyopathy, although it tends to be not as severe or uniform as in hypertrophic cardiomyopathy. As a result, heart weight is always increased. Generally, hypertrophy does not keep pace with dilatation, however, and wall tension increases. There is some evidence that, when dilatation is mild and hypertrophy results in maintenance of normal wall thickness or even an increase, i. e., when the increase in systolic wall stress is limited, the prognosis is more favorable.

While dilated cardiomyopathy is associated with increased myocardial mass and hence with overall increased myocardial oxygen demand, myocardial oxygen demand and hence the demand for coronary blood flow per unit of myocardial mass is also increased because of the increased wall stress. At the same time, the coronary perfusion pressure (systemic arterial pressure) is often decreased in these patients, and the right atrial (coronary sinus) pressure is generally increased, adversely affecting coronary perfusion. This effect may be enhanced by compensatory tachycardia shortening the total diastolic time. It is thus not surprising that coronary blood flow [77], as well as coronary reserve [94] habe been found to be decreased in patients with dilated cardiomyopathy; this effect is especially significant for the subendocardium [112]. Myocardial ischemia may be one cause of chest pain not infrequently reported by patients with dilated cardiomyopathy [21].

Compensatory mechanisms operative in dilated cardiomyopathy, such as alterations in the autonomic nervous system, in catecholamine levels, and in the renin-angiotension system, are not distinctive in dilated cardiomyopathy when compared with other forms of congestive heart failure and hence will not be considered here.

In the assessment and staging of dilated cardiomyopathy, as well as in the evaluation of the course and effects of therapy, both noninvasive and invasive diagnostic methods are of value. While, initially, cardiac catheterization, coronary angiography, left ventricular angiography, and endomyocardial biopsy may yield valuable and unique diagnostic and quantitative information, noninvasive imaging is generally sufficient for ongoing monitoring of the clinical course, with echocardiography and gated blood pool scans being the most frequently used methods. Care must be taken, however, to carry out such studies under comparable loading conditions to permit comparisons.

Summary and Conclusions

"Dilated cardiomyopathy" may be defined as a chronic form of heart disease, characterized by left and right ventricular dilatation and impaired systolic function, with or without chronic congestive heart failure, often but not necessarily associated with arrhythmias or disturbances of conduction, not associated with primary disease of valves, coronary arteries, or lungs.

A review of clinical as well as experimental studies, with emphasis upon lessons lerned from animal models, reaffirms the concept that dilated cardiomyopathies are multifactorial or pluricausal. The major pathogenetic processes include heredity, calcium overload, endogenous catecholamines, microvascular spasm, altered myocardial metabolism, free radicals, infections, and immune processes. Enhancing factors include genetic predisposition, nutritional deficiency, toxins, drugs, ischemia, immunosuppression, radiation, exercise, increased preload or afterload.

In many animal models, as well as patients, several pathogenetic mechanisms and enhancing factors may be operative. Their recognition is an essential prerequisite to therapeutic interventions and to secondary prevention.

The pathophysiology of dilated cardiomyopathy in general does not differ from that of congestive heart failure of other etiologies. Systolic pump failure is the cardinal manifestation, but impaired diastolic function may also play a contributory role. The myocardial disease may lead to arrhythmias and disturbances of conduction, accounting for a high incidence of sudden death. Thromboembolic complications are not infrequent.

References

1. Abelmann WH (1971) Virus and the heart. Circulation 44: 950–956
2. Abelmann WH (1984) Classification and natural history of primary myocardial disease. Prog Cardiovasc Dis 27: 73–94
3. Abelmann WH (1988) Myocarditis as a cause of dilated cardiomyopathy. In: Engelmeier RS, O'Connell JG (eds) Drug therapy in dilated cardiomyopathy and myocarditis. Dekker, New York, pp 221–232
4. Acosta A, Santos-Buch CA (1985) Autoimmune myocarditis induced by Trypanosoma cruzi. Circulation 71: 1255–1261
5. Amorim DS (1979) Chagas' disease. Prog Cardiol 8: 235–279
6. Amorim DS, Olsen EGJ (1982) Assessment of heart neurons in dilated (congestive) cardiomyopathy. Br Heart J 47: 11–18
7. Anderson KP, Freedman RA, Mason JW (1987) Sudden death in idiopathic dilated cardiomyopathy. Ann Intern Med 107: 104–106
8. Arom KV, Bishop VS, Grover FL, Trinkle JK (1979) Effect of therapeutic-dose irradiation on left ventricular function in conscious dogs. Ann Thorac Surg 28: 166–175
9. Azuma J, Sperelakis N, Hasegawa H, Tanimoto T, Vogel S, Ogura K, Awata N, Sawamura A, Harada H, Ishiyama T, Morita Y, Yamamura Y (1981) Adriamycin cardiotoxicity: possible pathogenetic mechanisms. J Mol Cell Cardiol 13: 381–397
10. Bajusz E, Homburger F, Baker JR, Opie LH (1966) The heart muscle in muscular dystrophy with special reference to involvement of the cardiovascular system in the hereditary myopathy of the hamster. Ann NY Acad Sci 138: 213–229
11. Berko BA, Swift M (1987) X-linked dilated cardiomyopathy. N Engl J Med 316: 1186–1191
12. Billingham ME (1979) Some advances in cardiac pathology. Hum Pathol 10: 367–386
13. Billingham ME, Bristow MR, Glatstein E, Mason JE, Masek MA, Daniels JR (1977) Adriamycin cardiotoxicity: endomyocardial biopsy evidence of enhancement by irradiation. Am J Surg Pathol 1: 17–23
14. Bing RJ, Tillmanns H, Fauvel J-M, Seeler K, Mao JL (1974) The effects of prolonged alcohol administration on calcium transport in the heart muscle of the dog. Circ Res 35: 33–38
15. Bolte H-D (ed) (1984) Viral heart disease. Springer, Berlin Heidelberg New York, p 248
16. Bravo EL, Gifford RW (1984) Pheochomocytoma: diagnosis, localization and management. N Engl J Med 311: 1298–1303
17. Breed JGS, Zimmerman ANE, Dormans JAMA, Pinedo HM (1980) Failure of the antioxidant vitamin E to protect against adriamycin-induced cardiotoxicity in the rabbit. Cancer Res 40: 2033–2038
18. Brigden W, Robinson J (1964) Alcoholic heart disease. Br Med J 2: 1283–1289
18a. Brigden W (1957) Uncommon myocardial diseases. The noncoronary cardiomyopathies. Lancet 2: 1243–1249
19. Bristow MR, Kantrowitz NE, Harrison WD, Minobe WA, Sageman WS, Billingham ME (1983) Mediation of subacute anthracycline cardiotoxicity in rabbits by cardiac histamine release. J Cardiovasc Pharmacol 5: 913–919
20. Bristow MR, Minobe WA, Billingham ME, Marmor JB, Johnson GA, Ishimoto BM, Sageman WS, Daniels JR (1981) Anthracycline-associated cardiac and renal damage in rabbits. Evidence for mediation by vasoactive substances. Lab Invest 45: 157–168
21. Cannon RO, Cunnion RE, Parrillo JE, Palmeri ST, Tucker EE, Schenke WH, Epstein SE (1987) Dynamic limitation of coronary vasodilator reserve in patients with dilated cardiomyopathy and chest pain. J Am Coll Cardiol 10: 1190–2000

22. Chimoskey JE, Spielman WS, Brandt MA, Heidemann SR (1984) Cardiac atria of Bio 14.6 hamsters are deficient in natriuretic factor. Science 223: 820–822
23. Cohen EM, Abelmann WH, Messer JV, Bing OHL (1976) Mechanical properties of rat cardiac muscle during experimental thiamine deficiency. Am J Physiol 231: 1390–1394
24. Costanzo-Nordin MR, Reap EA, O'Connell JB, Robinson JA, Scalon PJ (1985) A nonsteroid anti-inflammatory drug exacerbates coxsackie B3 murine myocarditis. J Am Coll Cardiol 6: 1078–1082
25. Czarnecki A, Hinek A (1986) The influence of inosine on adriamycin-induced cardiomyopathy in rats. Eur J Cancer Clin Oncol 22: 1357–1363
26. Daly K, Richardson PJ, Olsen EG, Morgan-Capner P, McSorley C, Jackson G, Jewitt DE (1984) Acute myocarditis. Role of histological and virological examination in the diagnosis and assessment of immunosuppresive treatment. Br Heart J 51: 30–35
27. Das SK, Cassidy JT (1987) Antiheart Antibodies in cardiomyopathy. In: Kawai C, Abelmann WH (eds) Pathogenesis of myocarditis and cardiomyopathy: recent experimental and clinical studies. Tokyo University Press, Tokyo, pp 269–275
28. Deo MG, Sood SK, Ramalingaswami V (1965) Experimental protein deficiency. Arch Pathol 80: 14–23
29. DiMauro S, Trevisan C, Hays A (1980) Disorders of lipid metabolism in muscle. Muscle Nerve 3: 369–388
30. Dlouha H, McBroom MJ (1986) Atrial natriuretic factor in taurine-treated normal and cardiomyopathic hamsters. Proc Soc Exp Biol Med 181: 411–415
31. Elson SH, Abelmann WH (1965) Effects of muscular activity upon the acute myocarditis of C3H mice infected with Trypanosoma cruzi. Am Heart J 69: 629–636
32. Eltringham JR, Fajardo LF, Stewart JR (1975) Adiamycin cardiomyopathy: enhanced cardiac damage in rabbits with combined drug and cardiac irradiation. Radiology 115: 471–472
33. Factor SM, Minase T, Cho S, Dominitz R, Sonnenblick EH (1982) Microvascular spasm in the cardiomyopathic hamster: a preventable cause of focal myocardial necrosis. Circulation 66: 342–354
34. Factor SM, Minase T, Cho S, Fein F, Capasso JM, Sonnenblick EH (1984) Coronary microvascular abnormalities in the hypertensive-diabetic rat. A primary cause of cardiomyopathy? Am J Pathol 116: 9–20
35. Factor SM, Cho S, Wittner M, Tanowitz H (1985) Abnormalities of the coronary microcirculation in the acute murine Chagas' disease. Am J Trop Med Hyg 34: 246–253
36. Farber PA, Glasgow LA (1970) Viral myocarditis during pregnancy: encephalomyocarditis virus infection in mice. Am Heart J 80: 96–102
37. Fein FS, Capasso JM, Aronson RS, Cho S, Nordin C, Miller-Green B, Sonnenblick EH, Factor SM (1984) Combined renovascular hypertension and diabetes in rats: a new preparation of congestive cardiomyopathy. Circulation 70: 318–320
38. Fejfar Z (ed) (1968) Accounts of international meetings. Idiopathic cardiomegaly. Bull WHO 38: 979–992
39. Fitchett DH, Sugrue DD, MacArthur CG, Oakley CM (1984) Right ventricular dilated cardiomyopathy. Br Heart J 51: 25–29
40. Fowler NO (1973): Classification and diagnosis of myocardial diseases. In: Fowler NO (ed) Myocardial diseases. Grune and Stratton, New York, pp 25–36
41. Ganguly PK, Pierce GN, Dhalla NS (1987) Diabetic cardiomyopathy: membrane dysfuncton and therapeutic strategies. J Appl Cardiol 2: 323–338
42. Gardner RJM, Hanson WW, Ionasescu HH, Ardinger HH, Skorton DJ, Mahoney LT, Rose HF, Smith WL, Florenine MS, Hinrichs RL (1987) Dominantly inherited dilated cardiomyopathy. Am J Med Genet 27: 61–73
43. Garrett JS, Wikman-Coffelt J, Sievers R, Finkbeiner WE, Parmley WW (1987) Verapamil prevents the development of alcoholic dysfunction in hamster myocardium. J Am Coll Cardiol 9: 1326–1331
44. Gauntt CJ (1987) Cellular and humoral immune responses in coxsackievirus myocarditis. In: Kawai C, Abelmann WH (eds) Pathogenesis of myocarditis and cardiomyopathy: recent experimental and clinical studies. Tokyo University Press, Tokyo, pp 49–61

45. Giacomelli F, Weiner J (1979) Primary myocardial disease in the diabetic mouse. An ultrastructural study. Lab Invest 40: 460–473
46. Goldblatt J, Melmed J, Rose AG (1987) Autosomal recessive inheritance of idiopathic dilated cardiomyopathy in a Madeira Portuguese kindred. Clin Genet 31: 249–254
47. Goodwin JF (1974) Prospects and predictions for the cardiomyopathies. Circulation 50: 210–219
48. Grossman W, McLaurin LP, Rolett EL (1979) Alterations in left ventricular relaxation and diastolic compliance in congestive cardiomyopathy. Cardiovasc Res 13: 514–522
49. Gwathmey JK (1987) Furazolidone-induced cardiomyopathy in turkey poults. In: Kawai C, Abelmann WH (eds) Pathogenesis of myocarditis and cardiomyopathy: recent experimental and clinical studies. Tokyo University Press, Tokyo, pp 107–120
50. Herskowitz A, Wolfgram LJ, Rose NR, Beisel KW (1987) Coxsackievirus B3 murine myocarditis: a pathologic spectrum of myocarditis in genetically defined inbred strains. J Am Coll Cardiol 9: 1311–1319
51. Hirota Y, Bing OHL, Abelmann WH (1976) Effect of ethanol on contraction and relaxation of isolated rat ventricular muscle. J Mol Cell Cardiol 8: 727–732
52. Hirota Y, Kane RL, Abelmann WH (1979) Cardiovascular effects of exercise in hamsters with experimental thiamine deficiency. Jpn Circ J 43: 99–106
53. Homburger F, Baker JR, Nixon CW, Wilgram G (1962) New hereditary disease of Syrian hamsters. Primary, generalized polymyopathy and cardiac necrosis. Arch Intern Med 110: 660–662
54. Hosenpud JD, Campbell SM, Niles NR, Lee J, Mendelson D, Hart MV (1987) Exercise induces augmentation of cellular and humoral autoimmunity associated with increased cardiac dilatation in experimental autoimmune myocarditis. Cardiovasc Res 21: 217–222
55. Huber SA, Lodge PA (1986) Coxsackievirus B3 myocarditis. Identification of different pathogenetic mechanisms in DBA/2 and BALB/c mice. Am J Pathol 122: 284–291
56. Huber SA, Lodge PA, Herzum M, Estrin M, Olszewski J (1987) The role of T lymphocytes in the pathogenesis of coxsackievirus B3 myocarditis. In: Kawai C, Abelmann WH (eds) Pathogenesis of myocarditis and cardiomyopathy: recent experimental and clinical studies. Tokyo University Press, Tokyo, pp 9–21
57. Hudson REB (1972) Pathology of cardiomyopathy. Cardiovasc Clin 4: 3–59
58. Imperato-McGinley J, Gautier T, Ehlers K, Zullo MA, Goldstein DS, Vaughan ED Jr (1987) Reversibility of catecholamine-induced dilated cardiomyopathy in a child with a pheochromocytoma. N Engl J Med 316: 793–797
59. Johnson RA, Baker SS, Fallon JT, Maynard EP, Ruskin JN, Wen Z, Ge K, Cohen HJ (1981) An occidental case of cardiomyopathy and selenium deficiency. N Engl J Med 304: 1210–1212
60. Johnson RA, Palacios I (1982) Dilated cardiomyopathies of the adult. N Engl J Med 307: 105–1058, 1119–1126
61. Kaufmann R, Fleckenstein A (1965) Ca^{++}-Kompetititve elektromechanische Entkoppelung durch Ni^{++}- und Ca^{++}-Ionen am Warmblüter Myokard. Pflügers Arch 282: 290–297
62. Kawai C, Abelmann WH (eds) (1987) Pathogenesis of myocarditis and cardiomyopathy: recent experimental and clinical studies. Tokyo University Press, Tokyo, p 312
63. Kawai C, Yui Y, Hoshino T, Sasayama S, Matsumori A (1983) Myocardial catecholamines in hypertensive and dilated (congestive) cardiomyopathy: a biopsy study. J Am Coll Cardiol 2: 834–840
64. Kilbourne ED, Wilson CB, Perrier D (1956) Induction of gross myocardial lesions by a coxsackie (pleurodynia) virus and cortisone. J Clin Invest 35: 362–370
65. Kino M, Thorp KA, Bing OHL, Abelmann WH (1981) Impaired myocardial performance and response to calcium in experimental alcoholic cardiomyopathy. J Mol Cell Cardiol 13: 981–989
66. Kishimoto C, Kawai C, Abelmann WH (1987) Immuno-genetic aspects of the pathogenesis of experimental viral myocarditis. In: Kawai C, Abelmann WH (eds) Pathogenesis of myocarditis and cardiomyopathy: recent experimental and clinical studies. Tokyo University Press, Tokyo, pp 3–7
67. Klugman FB, Decorte G, Candussio L, Grill V, Mallardi F, Baldini L (1986) Inhibitors of adriamycin-induced histamine release in vitro limit cardiotoxicity in vivo. Br J Cancer 54: 743–748

68. Kobayashi A, Yamashita T, Kaneko M, Nishiyama T, Hayashi H, Yamazaki N (1987) Effects of verapamil on esperimental cardiomyopathy in the Bio 14.6 Syrian hamster. J Am Coll Cardiol 10: 1128–1138
69. Kobayashi M, Kondo S, Yasumoto T, Ohizumi Y (1986) Cardiotoxic effects of maitotoxin, a principal toxin of seafood poisoning, on guinea pig and rat cardiac muscle. J Pharmacol Exp Ther 238: 1077–1083
70. Köberle F (1957) Die chronische Chagaskardiopathie. Virchows Arch [Pathol Anat] 330: 267–295
71. Kumar R, Kline IK, Abelmann WH (1970) Immunosuppression in experimental acute and subacute myocarditis. Am J Trop Med Hyg 19: 932–936
72. Lenzhofer R, Dudczak R (1986) Indication of doxorubicin cardiotoxicity by impairment of ^{131}I-pIPPA utilization. Eur J Med 12: S32–S33
73. Lossnitzer K (1975) Genetic induction of a cardiomyopathy. In: Schmier J, Eichler O (eds) Experimental production of diseases. Part 3. Heart and circulation. Springer, Berlin Heidelberg New York, p 309
74. Lossnitzer K, Bajusz E (1974) Water and electrolyte alterations during the lifecourse of the Bio 14.6 Syrian golden hamster. A disease model of a hereditary cardiomyopathy. J Mol Cell Cardiol 6: 163–177
75. Mackenzie J (1902) The study of the pulse. Y. J. Pentland, Edinburgh, p 237
76. Maisch B (1987) Immune regulation, humoral and cell-mediated immune reactions in myocarditis and dilated cardiomyopathy. In: Kawai C, Abelmann WH (eds) Pathogenesis of myocarditis and cardiomyopathy: recent experimental and clinical studies. Tokyo University Press, Tokyo, pp 246–267
77. Magorien RD, Unverferth DV, Brown SP, Leier CV (1983) Dobutamine and hydralazine: comparative influences of positive inotropy and vasodilation on coronary blood flow and myocardial energetics in nonischemic congestive heart failure. J Am Coll Cardiol 1: 499–505
78. Makino N, Dhalla KS, Elemban V, Dhalla NS (1987) Sarcolemmal Ca^{2+} transport in streptozotocin-induced diabetic cardiomyopathy in rats. Am J Physiol 253 [Endocrine Metab 16]: E202–E207
79. Markiewicz W, Wu S, Parmley WW, Higgins CB, Sievers R, James T, Wikman-Coffelt J, Jasmin G (1986) Evaluation of the hereditary Syrian hamster cardiomyopathy by ^{31}p nuclear magnetic resonance spectroscopy: improvement after acute verapamil therapy. Circ Res 59: 597–604
80. Mason JW, Billingham ME, Ricci DR (1980) Treatment of acute inflammatory myocarditis assisted by endomyocardial biopsy. Am J Cardiol 45: 1037–1044
81. Mass-Schwartz A, Weinstock RS, Wagner RL, Abelmann WH (1977) Right and left ventricular compliance in the hereditary cardiomyopathy of the Syrian hamster. Cardiovasc Res 11: 367–374
82. McFalls EO, Paulson DJ, Gilbert EF, Shug AL (1986) Carnitine protection against adriamycin-induced cardiomyopathy in rats. Life Sci 38: 497–505
83. Michels VV, Driscoll IJ, Miller FA Jr (1985) Familial aggregation of idiopathic cardiomyopathy. Am J Cardiol 55: 1232–1233
84. Michelson AM, Russell ES, Harman PJ (1955) Dystrophia muscularis: a hereditary primary myopathy in the house mouse. Proc Natl Acad Sci 41: 1079–1084
85. Miller H, Abelmann WH (1967) Effects of dietary ethanol upon experimental Trypanosomal (T. cruzi) myocarditis. Proc Soc Exp Biol Med 126: 193–198
86. Monrad ES, Matsumori A, Murphy JC, Fox JG, Crumpacker CS, Abelmann WH (1986) Therapy with cyclosporine in experimental murine myocarditis with encephalomyocarditis virus. Circulation 73: 1058–1064
87. Monrad ES, McKay RG, Baim DS, Colcucci WS, Fifer MA, Heller GV, Royal HD, Grossman W (1984) Improvement in indexes of diastolic performance in patients with congestive heart failure treated with milrinone. Circulation 70: 1030–1037
88. Morin YL, Foley AR, Martineau G, Roussel J (1967) Quebec beer-drinkers cardiomyopathy: forty-eight cases. Can Med Assoc J 97: 881–883
89. Morrow CJ, McOrist S (1985) Cardiomyopathy associated with curly hair coat in Poll Hereford calves in Australia. Vet Rec 117: 312–313

90. Myers CE, McGuire WP, Liss RH, Ifrim I, Grotzinger K, Young RC (1977) Adriamycin: the role of lipid peroxidation in cardiac toxicity and tumor response. Sciense 197: 165–167
91. O'Connell JB, Reap EA, Robinson JA (1986) The effects of cyclosporine on acute coxsackie B3 myocarditis. Circulation 73: 353–359
92. Ogawa Y, Kondo T, Sugiyama S, Ogawa K, Ozawa T (1987) Role of phospholipase in the genesis of doxorubicin-induced cardiomyopathy in rats. Cancer Res 47: 1239–1243
93. Orsi EV, Mancini R, Barriso J (1970) Hyperbaric enhancement of coxsackievirus infection in mice. Aeros Med 41: 1169–1172
94. Opherk D, Schwarz F, Mall G, Manthey J, Baller D, Kübler W (1983) Coronary dilatory capacity in idiopathic dilated cardiomyopathy: analysis of 16 patients. Am J Cardiol 51: 1657–1662
95. Packer DL, Bardy GH, Worley SJ, Smith MS, Cobb FR, Coleman RE, Gallagher JJ, German LD (1986) Tachycardia-induced cardiomyopathy: a reversible form of left ventricular dysfunction. Am J Cardiol 57: 563–570
96. Pierce JM (1960) Heart disease and filtrable viruses. Circulation 21: 448–455
97. Pion PD, Kittleson MD, Rogers QR, Morris JG (1987) Myocardial failure in cats associated with low plasma taurine: a reversible cardiomyopathy. Science 237: 764–768
98. Piza J, Troper L, Cespedes R, Miller JH, Berenson GS (1971) Myocardial lesions and heart failure in infantile nutrition. Am J Trop Hyg 20: 343–355
99. Regan TJ (1984) Alcoholic cardiomyopathy. Prog Cardiovasc Dis 27: 141–152
100. Regan TJ, Koroxenidis GT, Moschos CB, Oldewurtel HA, Lehan PH, Hellems HK (1966) The acute metabolic and hemodynamic responses of the left ventricle to ethanol. J Clin Invest 45: 270–280
101. Regan TJ, Levinson GE, Oldewurtel HA, Frank MJ, Weisse AB, Moschos CB (1969) Ventricular function in noncardiacs with alcoholic fatty liver: the role of ethanol in the production of cardiomyopathy. J Clin Invest 48: 397–407
102. Rezkalla S, Khatib G, Khatib R (1986) Coxsackievirus B3 murine myocarditis: deleterious effects of non-steroidal anti-inflammatory agents. J Lab Clin Med 107: 393–395
103. Robinson JA, O'Connell JB (eds) (1983) Myocarditis: precursor of cardiomyopathy. Collamore, Lexington
104. Rona G (1985) Catecholamine cardiotoxicity. J Mol Cell Cardiol 17: 291–306
105. Rona G, Kahn DS, Chappel CI (1963) Studies on infarct-like myocardial necrosis produced by isoproterenol: a review. Rev Can Biol 22: 241–55
106. Sandusky GE, Crawford MP, Roberts ED (1981) Experimental cobalt cardiomyopathy in the dog: a model for cardiomyopathy in dogs and man. Toxicol Appl Pharmacol 60: 263–278
107. Santos-Buch CA (1979) American trypanosomiasis: Chagas' disease. Int Rev Exp Pathol 19: 63–100
108. Spodick DH, Pigott VM, Chirife R (1972) Preclinical cardiac malfunction in chronic alcoholism. Comparison with matched normal controls and with alcoholic cardiomyopathy. N Engl J Med 287: 677–680
109. Thomas CE, Aust SD (1986) Release of iron from ferritin by cardiotoxic anthracycline antibodies. Arch Biochem Biophys 248: 684–689
110. Tomioka N, Kishimoto C, Matsumori A, Kawai C (1986) Effects of prednisolone on acute viral myocarditis in mice. J Mol Coll Cardiol 7: 868-872
111. Tripp ME, Katcher ML, Peters HA, Gilbert E, Arya S, Hodach RJ, Shug AL (1981) Systematic carnitine deficiency presenting as familial endocardial fibroelastosis: a treatable cardiomyopathy. N Engl J Med 305: 385–390
112. Unverferth DV, Magorien RD, Lewis RP, Leier CV (1983) The role of subendocardial ischemia in perpetuating myocardial failure in patients with nonischemic congestive cardiomyopathy. Am Heart J 105: 176–179
113. Velasquez G, D'Souza VJ, Hackshaw BJ, Glass TA, Formanek AG (1984) Pheochromocytoma and cardiomyopathy. Br J Radiol 57: 89–92
114. Von Bollinger D (1884) Über die Häufigkeit und Ursachen der idiopathischen Herzhypertrophie in München. Dtsche Med Wochenschr 10: 180–181
115. Walsh TK, Vacek JL (1986) Ethanol and heart disease. An underestimated contributory factor. Postgrad Med 79: 60–75

116. Walshe WH (1873) Diseases of the heart and great vessels. Smith, Elder and Co, London, p 359
117. Weiss S, Wilkins RW (1937) The nature of the cardiovascular disturbances in nutritional deficiency states (beriberi). Am Int Med 11: 104–148
118. Welty MC, Welty JD, McBroom MJ (1982) Effect of isoproterenol and taurine on heart calcium in normal and cardiomyopathic hamsters. J Mol Cell Cardiol 14: 353–357
119. Wenger NK, Abelmann WH, Roberts WC (1986) Myocarditis. In: Hurst JW (ed) The heart, arteries and veins, 6th edn. McGraw-Hill, New York, pp 1158–1180
120. Wenger NK, Goodwin JF, Roberts WC (1986) Cardiomyopathy and myocardial involvement in systemic disease. In: Hurst JW (ed) The heart, arteries and veins, 6th edn. McGraw-Hill, New York, pp 1181–1248
121. Whitmer JT (1987) L-carnitine treatment improves cardiac performance and restores high-energy phosphate pools in cardiomyopathic Syrian hamster. Circ Res 61: 396–408
122. WHO/ISFC Task Force (1980) Report of the WHO/ISFC task force on the definition and classification of cardiomyopathies. Br Heart J 44: 672–673
123. Wood GB (1855) A treatise on the practice of medicine. Lippincott, Philadelphia, p 168
124. Woodruff JF (1980) Viral myocarditis: a review. Am J Pathol 101: 425–484
124a. Wrogemann K, Blanchaer M, Thakar JH, Mezon BJ (1975) On the role of mitochondria in the hereditary cardiomyopathy of the Syrian hamster. In: Fleckenstein A, Rona G (eds) Recent advances in studies on cardiac structure and metabolism Vol. 6: Pathophysiology and Morphology Cell Alterations. University Park Press, Baltimore, pp 231–241
125. Wynn J, Braunwald E (1988) The cardiomyopathies and myocarditides. In: Braunwald E (ed) Heart disease: a textbook of cardiovascular medicine, 3rd edn. Saunders, Philadelphia, pp 1410–1469
126. Yaeger RG, Miller DN (1960) Effect of malnutrition on susceptibility of rats to Trypanosoma cruzi. I. Thiamine deficiency. Exp Parasitol 9: 215–222
127. Yang GQ, Chen JS, Wen ZM, Ge KY, Zhu LZ, Chen XC, Chen XS (1984) The role of selenium in Keshan disease. Adv Nutr Res 6: 203–231
128. York CM, Cantrell CR, Borum PR (1983) Cardiac carnitine deficiency and altered carnitine transport in cardiomyopathic hamsters. Arch Biochem Biophys 221: 526–533
129. Zucchi R, Barsotti A, Mariotti R, Bradi O, Baldarini A, Mariani M (1987) Asynergy and left ventricular performance in dilative cardiomyopathy. Clin Cardiol 10: 153–158

Virological, Immunological and Clinical Features of Myocarditis and Dilated Cardiomyopathy

J. F. Goodwin

2 Pine Grove, Lake Road, Wimbledon, London SW 19 7HE, UK

For the purposes of this paper, myocarditis will be regarded as infective in origin. Virological features of myocarditis are essentially those of detection of virus in blood or other tissue fluids and detection of virus, virus sequences or particles in heart muscle. A rising titre of viral antibodies in the blood is good evidence of viral infection, but does not necessarily implicate the heart. Detection of virus products in the heart is essential for complete proof, and this is extremely difficult to achieve although new techniques of molecular biology, in which coxsackievirus type 3 (CVB 3) specific DNA hybridisation probes can be used to detect virus nucleic acid sequences in endomyocardial biopsy specimens from patients with a clinical diagnosis of myocarditis [1]. A further difficulty is that endomyocardial biopsy in acute virus myocarditis is rarely feasible although possible in subacute or chronic cases.

Histological features on endomyocardial biopsy specimens that are suggestive but not pathognommic of virus myocarditis are pronounced cellular infiltration, myocyte necrosis in the acute phase and replacement fibrosis of varying degree with minor cellular infiltration in the subacute or chronic stages. It is essential that histological evidence, based on the Dallas criteria [2] for a diagnosis of virus myocarditis should be agreed and meticulously followed, otherwise unwarranted diagnoses of virus myocarditis are likely to be made, and there has been a wide variability of the apparent incidence of virus myocarditis from different institutions in the past few years. It is also important to realise that healing or progressing myocarditis can only be diagnosed on repeated endomyocardial biopsy specimens.

The virological features of dilated cardiomyopathy are even more tennuous than those of myocarditis. In *restrictive cardiomyopathy* due to endomyocardial fibrosis there is an acute *inflammatory myocarditis*, which is both histologically and clinically evident, followed by a chronic fibrotic stage and clinical evidence of established heart diease. Pathological and clinical features are in harmony, but no virus or virus products have so far been identified. In *dilated cardiomyopathy*, in at least 50% or possibly more patients, the histological features in the myocardium are non-specific, but in some, particularly those with an acute onset and short history, there may be features suggestive of acute or subacute myocarditis on the biopsy, while in others, with a longer history, biopsy may show extensive fibrosis suggestive of chronic myocarditis.

A viral aetiology of dilated cardiomyopathy is strengthened if there is a history of viral illness preceding the development of heart failure. Evidence of virally re-

H.-P. Schultheiß (Ed.)
New Concepts in Viral Heart Disease

lated sequences in the myocardium in patients regarded as having dilated cardiomyopathy is strong supporting evidence of a viral aetiology of dilated cardiomyopathy in at least some patients with this diagnosis. It is apparent that the diagnosis of viral myocarditis must be confirmed by techniques of molecular biology and histology, while the diagnosis of dilated cardiomyopathy is made on clinical grounds supported by invasive and non-invasive haemodynamic studies of structure and function. Patients diagnosed on these grounds as having dilated cardiomyopathy and found to have extensive fibrosis of the myocarditium on biopsy or necropsy possibly may be given the diagnostic label of "post-infective (viral) dilated cardiomyopathy".

Immunological features of myocarditis have been much discussed. Reduced suppressor T cell activity is reported in myocarditis in animals [3] and also in man [4] though some workers have denied this in man. The sarcolemma and myolemma have been identified as antigens in patients with CVB myocarditis [5]. In animals, Huber et al. [6] have shown that myocardial damage in CVB3 viral infection is mainly mediated by T lymphocyte response to infection rather than by direct viral injury to the myocytes.

In dilated cardiomyopathy, defective suppressor T cell function has been reported [7, 8], but this finding is non-specific and has been denied by other workers [9].

Considering the evidence of immunological disturbance in viral infection in animals and the classic experiments of Woodruff and Woodruff [10] who showed that mice inoculated with CVB3 virus did not develop myocarditis if previously depleted of T cells, and the abnormal T lymphocyte response in dilated cardiomyopathy, it seems likely that immunological disorder is related to myocardial cell damage in both myocarditis and dilated cardiomyopathy. Woodruff and Woodruff suggested that myocardial damage in their mice infected with virus was due to a cellular immune reaction against antigens induced by the virus.

This evidence, together with the viral sequences found in the myocardium and the data suggesting progression of viral myocarditis to dilated cardiomyopathy [11–13] supports the contention that an autoimmune reaction set up by a preceding viral infection can progressively damage the myocardium, leading to dilated cardiomyopathy.

The clinical features of acute myocarditis are those of myocardial insufficiency of recent onset against a background of a viral-type illness. Severe peripheral muscle pains, tachycardia out of proportion to fever, gallop rhythm and raised jugular venous pressure are the principle aspects. A pericardial friction rub is not uncommon, and in severe cases there may be pulmonary oedema, hypotension and circulatory collapse. A weak thready pulse, peripheral cyanosis, sweating and anxiety are seen in such cases. Arrythmias of various types or even heart block may occur. Sudden death is well recognised though uncommon. Echocardiography reveals dilated ventricles, poor systolic function and low stroke volume and cardiac output. The electrocardiogram is usually non-specific, but ST elevation may occur if there is pericarditis. Presence of a third heart sound suggests either impaired left ventricular myocardial function or impending acute constrictive pericarditis. A fourth heart sound indicates an elevated left ventricular diastolic pressure.

Although myocarditis and dilated cardiomyopathy are usually global disorders, regional myocardial damage has been reported in scintigraphic studies in both conditions [14–15]. Regional myocarditis may produce a syndrome indistinguishable from myocardial infarction with severe ischaemic chest pain. Since the diagnosis of acute viral myocarditis is so often inferential rather than proven, I have suggested the use of a "points" system to assist diagnosis based on the clinical features and virology without the need for invasive investigation [16].

The clinical features of dilated cardiomyopathy are those of congestive heart failure of unknown cause. Early cases may be suspected by symptoms suggesting impending heart failure; dilatation of the left ventricle, gallop rythm and impaired left ventricular function on echocardiography. Dilated cardiomyopathy may occasionally involve only (or predominantly) the right ventricle [17] when it tends to be associated with more overt arrythmias than in the usual type.

Prognosis of acute myocarditis is usually excellent, probably over 90% of patients recover fully. A few die in the acute stage, a few progress rapidly to develop a clinical syndrome of dilated cardiomyopathy, and a few probably enter a latent asymptomatic phase and later develop typical dilated cardiomyopathy. By contrast, the prognosis of dilated cardiomyopathy is very poor, with a 5-year-survival rate of around 50%, the most important determinant being haemodynamic.

Time will tell whether effective antiviral therapy will become a reality, and whether immunosuppressive therapy will significantly improve the prognosis of myocarditis and dilated cardiomyopathy. It also remains to be seen whether immunosuppressive therapy will prove effective for dilated cardiomyopathy; whether β-adrenergic blockade will provide a breakthrough; and what may be expected from the newer drugs for the treatment of heart failure. Gilbert et al. [18] have recently published a case report on the treatment of myocarditis with OKT3 monoclonal antibody with encouraging results. Apart from this, the most effective method of treatment for intractable myocarditis and severe progressive dilated cardiomyopathy remains cardiac transplantation.

References

1. Bowles NE, Richardson PJ, Olsen EGJ, Archard LC (1986) Detection of coxsackie B virus-specific RNA sequences in myocardial biopsy from cases of myocarditis and cardiomyopathy. Lancet I: 1120–1123
2. Aretz HT, Billingham ME, Edwards WD, Factor SM, Fallon JT, Fenoglio JJ, Olsen EGJ, Schaen FJ (1986) Myocarditis: a histopathologic definition and classification. Am J Cardiovasc Pathol 1: 3–14
3. Gauntt CJ (1987) Cellular and humoral immune responses in coxcackie virus myocarditis. In: Kuwaii C, Abelmann WH (eds) Pathogenesis of myocarditis and cardiomyopathy. University of Tokyo Press, Tokyo, pp 49–61
4. Eckstein R, Mempel M, Heim M, Bolte HD (1984) The role of human leucocyte antigen genes and suppressor call activity in the pathogenesis of myocarditis and dilated cardiomyopathy. In: Bolte HD (ed) Viral heart disease. Springer, Berlin Heidelberg New York, pp 144–149
5. Maisch B, Schwab D, Bauer E, Sandhage K, Schmaltz AA (1987) Antimyolemmal antibodies in myocarditis in children. Eur Heart J 8 [Suppl J]: 167–173
6. Huber SA, Lodge TA, Job LP (1984) The role of virus and immune mediated cardiocyte injury in coxsackie B3 induced myocarditis. In: Bolte HD (ed) Viral heart disease. Springer, Berlin Heidelberg New York, pp 64–73
7. Anderson JL, Carlquist JF, Hammond EH (1982) Deficient natural killer cell activity in patients with idiopathic dilated cardiomyopathy. Lancet II: 1124–1127

8. Fowles RE (1984) Dilated cardiomyopathy: a consequence of defective suppressor cell function? In: Bolte HD (ed) Viral heart disease. Springer, Berlin Heidelberg New York, pp 144–149
9. Gentle TA, Baynham MID, Gammage JMD, Lowry PJ, Compston RA, Littler WA (1984) T-cell function in idiopathic congestive cardiomyopathy. In: Bolte HD (ed) Viral heart disease. Springer, Berlin Heidelberg New York, p 145
10. Woodruff JF, Woodruff JJ (1984) Involvement of T lymphocytes in the pathogenesis of coxsackie B3 heart disease. J Immunol 113: 1716–1734
11. Cambridge G, MacArthur CGC, Waterson AP, Goodwin JF, Oakley CM (1979) Antibodies to coxsackie B virus in congestive cardiomyopathy. Br Heart J 41: 693
12. Quigley PJ, Richardson PJ, Meany BT, Olsen EGJ, Monaghan MJ, Jackson G, Jewitt DE (1987) Long term follow-up of acute myocarditis. Correlation of ventricular function and outcome. Eur Heart J 8 [Suppl J] 39–42
13. Goodwin JF (1974) Prospects and predictions for the cardiomyopathies. Circulation 50: 210–219
14. Haber E, Yasuda T, Palacios IF, William G, Kwaw BA (1987)Scintigraphy in the diagnosis of acute myocarditis. In: Kawaii C, Abelmann W (eds) Pathogenesis of myocarditis and cardiomyopathy. University of Tokyo Press, Tokyo, pp 277–280
15. O'Connell JB, Mason JW (1987) The role of endomyocardial biopsy and gallium 67 scintigraphy in the assessment and treatment of active myocarditis. In: Kawaii C, Abelmann W (eds) Pathogenesis of Myocarditis Cardiomyopathy. University of Tokyo Press, Tokyo, pp 281–292
16. Goodwin JF (1984) Summary of workshop. In: Bolte HD (ed) Viral heard disease. Springer, Berlin Heidelberg New York, pp 234–239
17. Fitchett DH, Sugrue DD, MacArthur CGC, Oakley CM (1984) Right ventricular dilated cardiomyopathy. Br Heart J 51: 25
18. Gilbert EM, O'Connell JB, Hammond ME, Rowland DG, Watson FS, Bristow MR (1988) Treatment of myocarditis with OKT_3 monoclonal antibodies. Lancet 1:8588759

Epidemiology and Pathogenicity of Coxsackieviruses

N. R. GRIST

Communicable Diseases (Scotland) Unit, Ruchill Hospital, Glasgow G20 9NB, UK

Characteristics of Coxsackieviruses

The general properties and behaviour of coxsackieviruses (CV) have been the subject of recent reviews [1–3]. These members of the *Enterovirus* (EV) genus (Table 1) within the *picornavirus* family were originally discovered and then distinguished from polioviruses, and from the other EV found in later years, by their pathogenicity for suckling mice. Group A CV typically produce lethal, paralysing infections in newborn mice with severe, generalised myositis of skeletal muscles. Those classified in group B produce only scattered, focal myositis but more severe damage to the nervous system, brown fat, pancreas and other viscera – including the heart to a variable extent. The distinction from other viruses of the EV group became less absolute when later experience showed that

a) most group B (CVB) a few group A CV (CVA) can be grown in the same types of cell cultures used for the cultivation of polioviruses and echoviruses;
b) paralytic poliomyelitis can sometimes be caused by group B CV, by certain CVA (notably type A7), rarely by some echoviruses and also by the type 71 EV discovered later;
c) acute muscle disease (Bornholm disease; epidemic pleurodynia) is mainly caused by CVB, sometimes by CVA, but occasionally by echoviruses (especially type 6);
d) skin rash and oropharyngeal lesions can be caused by many CVA and CVB (especially types A4, A9, A16) and echoviruses (especially type 9);
e) all types of EV can cause acute, mainly upper respiratory infections;
f) the pathogenic properties for newborn mice of both CVA and CVB show considerable overlap and a few echoviruses can also cause myositis, notably type 9 (also known as CV type A23) and strains of type 6 [4]. Roberts and Boyd [5] have recently described their histopathological experience with routine diagnostic isolations in suckling mice from clinical specimens. Table 2 is based on their findings and shows a wide range of changes in newborn mice inoculated with materials containing typed EV. The effects of CV, especially those of group B, were much more severe than those of echoviruses and showed major impact on brown fat, nervous system, the heart and pancreas. Because of this overlap of properties between different groups of EV, recently discovered types have not been distinguished in the same way but are classified as higher-numbered "enteroviruses" (Table 1).

H.-P. Schultheiß (Ed.)
New Concepts in Viral Heart Disease

Table 1. Enteroviruses

Polioviruses	types 1–3
Coxsackieviruses	types A1–A24[a]
Coxsackieviruses	types B1–B6
Echoviruses	types 1–34[b]
Enteroviruses	types 68–72[c]

[a] Coxsackievirus A23 is also echovirus type 9
[b] Echovirus 9 is also coxsackievirus type A23;
former echovirus 10 is now reovirus type 1; former echovirus 28 is now rhinovirus 1A
[c] Enterovirus 72 is hepatitis A virus

General Epidemiology

Over several decades, knowledge of the epidemiology of CV and other EV has been built up from the application of diagnostic techniques (virus isolation; antibody studies) to clinical specimens from diseased patients and from surveys of various populations. Although biased by selection of cases for investigation, by availability of specimens for testing, and by the availability and application of various test methods, a reasonably clear picture has emerged and has been reviewed recently on the basis of global surveillance reports to the World Health Organisation [6, 7] and for the earlier period 1967–1974 [8]. EV are most prevalent in warm climates and in the warm season of temperate countries, in children and young persons, and in low socioeconomic conditions with poor hygiene and sanitation, and overcrowding. Infection spreads mainly by the faecal-oral route, but also by the respiratory route which is important when higher standards minimise faecal-oral transmission. Infection can spread within the familiy group, often asymptomatically, and nosocomially. Reinfections can occur but are generally transient and silent. Immunity is mainly type specific, and a broadening range of antibodies is built up during life from exposure to different EV which share various antigens.

The types of EV prevalent in a community vary from year to year, those types with the youngest age distribution tending to be relatively endemic in most years, while those with older age distribution appear in outbreaks or epidemics at longer intervals [9].

The history of poliomyelitis this century, before its control by vaccination, and of hepatitis A in recent decades in northwest Europe, illustrates the changing epidemiological pattern of EV infection. As improved living conditions slow the circulation of these viruses, the average age of first infections and of the acquisition of post-infective active immunity is postponed. The larger pool of non-immune persons, now including adults, allows outbreaks to occur periodically. With poliomyelitis and hepatitis A (EV type 72), and probably with other EV infections, the severity of illness is greater in older children and adults. Non-immune infants, especially as neonates, are, however, vulnerable to severe, life-threatening, systemic EV infection [10]. In populations where most adult women are immune,

Table 2. Histological findings in newborn mice inoculated with specimens containing enteroviruses[a]

Enterovirus group		Echovirus	Coxsackie A	Coxsackie B
Types represented		6, 9, 11, 25, 30	1–10, 14, 16, 20	1–5
Mice studied	*(n)*	94	321	42
Mice with listed features				
Myositis	*(n)*	38[c]	312	25
	(%)	40[c]	97	60
Encephalitis	*(n)*	4	52	16
	(%)	4	16	38
Poliomyelitis	*(n)*	8	52	11
	(%)	9	16	26
Carditis[b]	*(n)*	5	13	16
	(%)	5	4	38
Brown fat necrosis	*(n)*	0	29	26
	(%)		9	62
Pancreatitis	*(n)*	0	7	12
	(%)		2	29
Hepatitis	*(n)*	0	5	3
	(%)		2	7

[a] Commonest findings tabulated; additional changes listed by Roberts and Boyd [5]
[b] Endocarditis, myocarditis, pericarditis: one or more of these changes seen with echovirus 6, 9, 25; CVA1, 2, 7, 10, 16; CVB1–4
[c] Types 9 [35], 6 [2] and 11 [1]

passive transfer of antibodies prevents these severe neonatal infections. In more advanced countries and population groups, however, many pregnant women may not be immune, their infants may not receive passive immunity, and outbreaks of severe EV infection can affect maternity and neonatal units with acute myocarditis as a prominent feature [11, 12]. Efficient virological surveillance of communities may provide useful warning of when such outbreaks may occur [13].

In general, most EV infections are of minor severity, often silent, but they cause a wide range of illnesses including those affecting the respiratory, central nervous and cardiovascular systems, the muscles, skin and mucosae including eyes, the liver and occasionally other organs and tissues [1–3].

General Pathogenicity

Most information is available about the six group B CV because they can usually be isolated readily in routine cell cultures, whereas most CVA have required the inconvenient and expensive use of suckling mice and they also exist as at least 23 different antigenic types requiring differentiation. Serological diagnosis has mainly been done by neutralisation tests (easy in cell cultures but inconvenient in suckling mice), occasionally by haemagglutination inhibition (where applicable) or complement fixation, but interpretation is complicated by frequent cross-reactions and multiple antibody responses, especially in older persons. New methods are now becoming available for the diagnosis and investigation of individual patients by detection of specific viral antigen by immunofluorescence or enzyme-linked immunoabsorbent assay (ELISA) tests using polyclonal or monoclonal antibody, and by detection of virus-specific RNA sequences by hybridisation with cloned DNA probes (to be discussed in other chapters in this volume). Serological diagnosis is also becoming greatly improved by the development of tests for specific IgG or IgM by μ-capture ELISA tests [14–19].

Acute Infections

These have been referred to above and are discussed in greater detail in the recent reviews [1–3]. Those enteroviruses with the greatest propensity for damaging the heart and skeletal muscles of mice are also those most clearly associated with Bornholm disease and cardiac disease in man. Table 3 summarises the reports of enterovirus infections made to the World Health Organisation in the period 1975–1983 in relation to diagnoses of "cardiovascular" and "muscle-joint" disease. It shows the dominant role of CVB compared with that of CVA and other enteroviruses. Analysis by individual types did not show the significant role of certain CVA and echoviruses reported for earlier years [8, 20, 21]. Perhaps the virulence for muscles of circulating enterovirus strains can vary over the years.

Table 3. Enteroviral infections with diagnoses of cardiovascular and muscle-joint diseases

Infections reported		Clinical diagnosis			
Virus group	Reports	Cardiovascular		Muscle-joint	
	(n)	*(n)*	(%)	*(n)*	(%)
Coxsackievirus A	5781	57	0,98	35	0.61
Coxsackievirus B	14934	596	3.99	302	2.02
Echovirus, all types	38191	266	0.70	223	0.58
Enterovirus 68–71	217	0		0	
Poliovirus 1–3	43	1	2.33	0	

Data from Virus Unit of World Health Organisation: Global Surveillance Programme 1975–1983

Table 4 shows the correlation by individual CVB type. CVB5, which caused worldwide epidemics in 1965 with numerous cases of acute myopericarditis [22–26], shows the lowest proportion of cardiovascular diagnoses and the second lowest proportion with muscle-joint diseases. The high proportion of "untyped" CV with these diagnoses is probably due to excess use of serological diagnosis (which usually fails to indicate the type) selectively in older patients. Acute CV infection

Table 4. Coxsackie group B infections with diagnoses of cardiovascular and muscle-joint diseases

Infections reported		Clinical diagnosis			
CV type	Reports	Cardiovascular		Muscle-joint	
	(n)	*(n)*	(%)	*(n)*	(%)
B1	1940	65	3.35	49	2.53
B2	2891	130	4.50	57	1.97
B3	2846	96	3.37	56	1.97
B4	3690	196	5.31	70	1.90
B5	3010	77	2.56	48	1.59
B6	347	14	4.03	5	1.44
Untyped[a]	210	45	21.43	17	8.10

Data from World Health Organisationn: Global Surveillance Programme 1975–1983
[a] Probably serologically diagnosed — see text

of the heart has most impact on the myocardium in infants, although most of those who survive the initial illness generally recover completely without long-term sequelae [27, 28]. In later childhood and adult life pericarditis becomes the more prominent clinical feature although the myocardium is probably involved to some degree in all cases.

Chronic, Persistent and Recurrent CV Infections

Persistent or recurrent myopericardial illness associated with CV infections is a recognised problem [29–31]. Suggestive evidence of persistent CV infection of cardiovascular tissues was reported by Burch and Colcolough [32] and Burch et al. [33, 34] on the basis of mmunofluorescence tests for specific CV antigens in post mortem tissues. New and supportive evidence is now coming forward from the application of nucleic acid probes and monoclonal antibodies [35–37].

New models of persisting infection are suggested by the recent studies of post-viral fatigue syndrome (myalgic encephalomyelitis). Yousef et al. [38] reported the presence in such cases, studied up to 12 years after onset, of infectious virus detectable in faeces by acid dissociation from antibody and persisting for at least 1 year. Most of these patients also had IgM antibodies, usually with heterotypic specificity. Most of them also had circulating immune complexes of IgM with enteroviral group-specific protein. The field is now open to explore the possible significance of these findings and to determine the extent to which diseases of cardiac and other muscles and tissues may result from direct or indirect damage by such persisting infections or by immune processes stimulated and maintained by them. Since infections by CV are common, yet most people do not develop post-viral fatigue syndrome or heart disease as a result, it seems clear that constitutional host factors must be important determinants in those who do so.

Summary

CV and other EV are common agents of infection, spreading mainly by faecal-oral transmission. Rising living standards are associated with a shift in epidemiological pattern and postponement of infections to older ages, more often then in epidemics and with more serious clinical consequences. Viraemic spread of infection within the patient can cause a variety of disease syndromes. Acute disease of the heart with enterovirus infection shows the highest correlation with CVB, less with CVA and least with echoviruses. Evidence for chronic infections is increasing through the application of new techniques of investigation. Host factors may determine whether infection persists and what type of disease may result.

Acknowledgements. I am grateful to the late Dr. F. Assaad, late Head of the Division of Communicable Diseases, and to Mrs. K. Esteves, Epidemiology and Management Support Services, World Health Organisation, Geneva, Switzerland, for data from the Global Surveillance Programme. I thank Miss E. H. Simpson for her secretarial help with this text.

References

1. Moore M, Morens DM (1984) Enteroviruses, including polioviruses. In: Belshe RB (ed) Textbook of human virology. PSB, Littleton, pp 407–483
2. Grist NR (1987) Enteroviruses. In: Weatherall DJ, Ledingham JGG, Warrell DA (eds) Oxford textbook of medicine, 2nd edn. Oxford University Press, Oxford, pp 5.93–5.102
3. Friedman H, Bendinelli M (eds) (1988) Coxsackieviruses: a general update. Plenum, New York
4. Vasilenko S, Atsev S (1965) Experimental infection of mice with ECHO-6 virus. Acta Virol (Praha) 9: 541–545
5. Roberts GBS, Boyd JF (1987) The histopathology of enterovirus infections of new-born mice. J Infect 15: 45–56
6. Grist NR, Reid D (1988) General pathogenicity and epidemiology. In: Friedman H, Bendinelli M (eds) Coxsackieviruses: a general update. Plenum, New York, pp 221–239
7. Bell EJ, Assaad F, Esteves K (1988) Neurological disorders. In: Friedman H, Bendinelli M (eds) Coxsackieviruses: a general update. Plenum, New York, pp 319–337
8. Grist NR, Bell EJ, Assaad F (1978) Enteroviruses in human disease. Prog Med Virol 24: 114–157
9. Grist NR, Bell EJ, Reid D (1975) The epidemiology of enteroviruses. Scott Med J 20: 27–31
10. Kaplan MH (1988) Coxsackie infection in children under three months of age. In: Friedman H, Bendinelli M (eds) Coxsackieviruses: a general update. Plenum, New York, pp 241–251
11. Gear JHS (1958) Coxsackievirus infections in the newborn. Prog Med Virol 1: 106–121
12. Gear JHS, Measroch V (1973) Coxsackievirus infections of the newborn. Prog Med Virol 15: 42–61
13. Schoub BD, Johnson S, McAnerney JOM, Dos Santos IL, Klassen KIM (1985) Epidemic coxsackie B virus infection in Johannesburg, South Africa. J Hyg (Camb) 95: 447–455
14. Hannington G, Booth JC, Wiblin CN, Stern H (1983) Indirect enzyme-linked immunosorbent assay (ELISA) for detection of IgG antibodies against coxsackie B viruses. J Med Microbiol 16: 459–465
15. King ML, Shaikh A, Bidwell D, Voller A, Banatvala JE (1983) Coxsackie-B-virus-specific IgM responses in children with insulin-dependent (juvenile-onset; Type I) diabetes mellitus. Lancet I: 1397–1399

16. McCartney RA, Banatvala JE, Bell EJ (1986) Routine use of μ-antibody capture ELISA for the serological diagnosis of coxsackie B virus infections. J Med Virol 19: 205–212
17. Bell EJ, McCartney RA, Basquill D, Chaudhuri AKR (1986) μ-antibody capture ELISA for the rapid diagnosis of enterovirus infections in patients with aseptic meningitis. J Med Virol 19: 213–217
18. Dörries R, Ter Meulen V (1983) Specificity of IgM antibodies in acute human coxsackie B infections, analysed by direct solid phase enzyme immunoassay and immunoblot technique. J Gen Virol 64: 159–167
19. Morgan-Capner P, McSorley C (1983) Antibody capture radioimmunoassay (MACRIA) for coxsackievirus B4 and B5-specific IgM. J Hyg (Camb) 90: 333–349
20. Grist NR, Bell EJ (1969) Coxsackie viruses and the heart. Am Heart J 77: 295–300
21. Bell EJ, Grist NR (1971) ECHO viruses, carditis, and acute pleurodynia. Am Heart J 82: 133–135
22. Bennett NMcK (1966) Coxsackie B pericarditis. Med J Aust II: 178–179
23. Castro A, Mazzone O, Di Fazio S, Tempera G (1975) Coxsackie-induced cardiomyopathies: virological, clinical and epidemiological observations. G Mal Infett 27: 1101–1111
24. Helin M, Savola J, Lapinleimu K (1968) Cardiac manifestations during a coxsackie B5 epidemic. Br Med J III: 97–99
25. Public Health Laboratory Service (1967) Coxsackie B5 virus infections during 1965. Br Med J IV: 575–577
26. Sainani GS, Dekate MP, Rao CP (1975) Heart disease caused by coxsackie virus B infection. Medicine (Baltimore) 47: 133–147
27. Suckling PV, Vogelpoel L (1970) Coxsackie myocarditis of the newborn. Lancet II: 421
28. Rantakallio P, Saukkonen A, Krause U, Lapinleimu K (1970) Follow-up study of 17 cases of neonatal coxsackie B5 meningitis and one with suspected myocarditis. Scand J Infect Dis 2: 25–28
29. Soutar CA (1971) Unusual case of viral pericarditis. Lancet I: 498
30. Tilzey AJ, Signy M, Banatvala JE (1986) Persistent coxsackie B virus specific IgM response in patients with recurrent pericarditis. Lancet I: 1491–1492
31. Reyes PR, Lerner AM (1988) Myocarditis: clinical and experimental correlates. In: Friedman H, Bendinelli M (eds) Coxsackieviruses: a general update. Plenum, New York, pp 253–270
32. Burch GE, Colcolough HL (1969) Progressive coxsackie viral pancarditis and nephritis. Ann Intern Med 71: 963–970
33. Burch GE, Sun SC, Chu KC, Sohal RS, Colcolough HL (1968) Interstitial and coxsackievirus B myocarditis in infants and children. J Am Med Assoc 203: 1–8
34. Burch GE, Sun SC, Colcolough HL, Sohal RS, De Pasquale NP (1967) Coxsackie B viral myocarditis and valvulitis identified in routine autopsy specimens by immunofluorescent techniques. Am Heart J 74: 13–23
35. Bowles NE, Richardson PJ, Olsen EG, Archard LC (1986) Detection of coxsackie-B-virus-specific RNA sequences in myocardial biopsy samples from patients with myocarditis and dilated cardiomyopathy. Lancet I: 1120–1123
36. Easton AJ, Eglin RP (1988) The detection of coxsackievirus RNA in cardiac tissue by *in situ* hybridization. J Gen Virol 69: 285–291
37. Kandolf R (1988) The impact of recombinant DNA technology on the study of enterovirus heart disease. In: Friedman H, Bendinelli M (eds) Coxsackieviruses: a general update. Plenum, New York, pp 293–317
38. Yousef GE, Bell EJ, Mann GF, Murugesan V, Smith DG, McCartney RA, Mowbray JF (1988) Chronic enterovirus infection in patients with postviral fatigue syndrome. Lancet I: 146–150

Natural History of Acute Viral or Idiopathic Myocarditis: A Clinical and Endomyocardial Biopsy Follow-up

M. Sekiguchi[1], M. Hiroe[1], S. Hiramitusu[2], and T. Izumi[3]

[1] The Heart Institute of Japan and Department of Radiology, Tokyo Women's Medical College, 8-1 Kawada-cho, Shinjuku-ku, Tokyo 162, Japan
[2] Department of Internal Medicine, Fujita Gakuen University School of Medicine, Toyoake, 470-11 Japan
[3] Department of Internal Medicine, Niigata University School of Medicine, Niigata, 951 Japan

Introduction

In 1899, Fiedler [1–3] reported four acute fatal cases with an abrupt onset of fever and chills, dyspnea, vomiting, and convulsive seizure with unconsciousness. He presented autopsy and clinical findings and described the condition as acute interstitial myocarditis. According to the current understanding of the disease, three of Fiedler's four cases seemed to be viral myocarditis.

We have seen 30 cases of this disease over the past 14 years, and have performed serial endomyocardial biopsies in 14 cases. We have found this approach to be useful for diagnosing and understanding the natural course of the disease [5–22]. Through examination of the biopsy specimens [4], we were able to observe that cardiac myocytes can recover extremely well after undergoing the most acute stage of the disease. A number of the patients, however, showed ultrastructural changes which suggested advanced pathology and no possibility of recovery.

Detailed descriptions of the clinical and serial biopsy findings of both histopathological and ultrastructural observations were presented in our previous papers [5–22]. The present article summarizes our additional clinical observations of patients who have undergone endomyocardial biopsy or autopsy with a diagnosis of acute viral or idiopathic myocarditis.

Recently, the importance of immunosuppressive therapy has been stressed by various investigators, but some contradictory reports have also appeared. Systematic immunosuppressive therapy was not applied in the patients presented here. Therefore, our results may be considered to be those of the natural history of acute viral or idiopathic viral myocarditis.

A review of our observations and of those from clinical or experimental studies is also presented [23–65].

Patients and Methods

From 1971 to 1988, 25 biopsy cases and five autopsy cases were analyzed clinicopathologically. Findings of 20 of the 30 patients have been reported elsewhere [66]. In this presentation, 20 patients were from the Heart Institute of Japan, four

H.-P. Schultheiß (Ed.)
New Concepts in Viral Heart Disease

Table 1. Diagnostic criteria of acute viral or idiopathic myocarditis. These eriteria were devived from our clinical analysis of the patients who had undergone endomyocardial biopsy at the acute stage. From [8]

Major manifestation
1. Cardiac symptoms[a] appear within 10 days after onset of the precuding symptoms[b]
2. Electrocardiographic abnormalities (sercic arrhythmias such as atrioventricular block. ST-T changes, low R-wave, abnormal Q-wave, and intraventricular conduction disturbances), and an associated rise in cardiac sarcoplasmic enzymes in the serum

Minor manifestation
1. Positive results in neutralizing antibody, complement fixation, or hemagglutination inhibition titers
2. Positive histopathologic diagnosis employing endomyocardial biopsy[c]
3. No significant stenosis revealed by coronary angiography

The diagnosis is justified if both of the major plus two of the minor manifestations are present
[a] Cardiac symptoms: Adams-Stokes attack, congestive heart failure, cardiogenic shock
[b] Preceding symptoms: fever, influenza-like symptoms (cough, sore throat, general malaise, myalgia, arthralgia) gastrointestinal symptoms (nausea, vomiting, anorexia, abdominal pain), chest pain or discomfort
[c] Endomyocardial biopsy findings: interstitial cellular infiltration consisting of small round cells and large mononuclear cells with or without basophilic cytoplasm, degeneration, and lysis of myocytes, and fragmentation of muscle bundles

from Fujita-Gakuen Medical School Hospital, and six from Niigata University Hospital. Routine histopathological ($n = 30$) and ultrastructural ($n = 23$) analyses were performed. There were 15 males and 15 females, with an age range of 20–66 years, mean 47 years.

Our series of studies on this disease have been reported previously. These include early clinical features [6, 8], histopathological profiles [6, 15], ultrastructural findings [5, 15], long-term prognosis of up to 6 years [9], ventriculographic findings at the convalescent stage [10], a hemodynamic study in relation to treatment [7, 12], and long-term follow-up with a radionuclide study [13]. In the course of our experience, various vasodilators, catecholamines, and intra-aortic balloon pumps (IABP) have been applied. Although symptomatic treatment with steroids was sometimes given when the patients were in cardiogenic shock [6], programed immunosuppressive therapy was not applied in any of our cases.

Results of the long-term prognosis are based on a clinical follow-up of the patients who were discharged from the hospitals. The period of the follow-up ranged from 1 to 14 years, averaging 4.1 ± 4.4 (mean ± SD). Criteria for the clinical diagnosis of acute viral or idiopathic myocarditis are shown in Table 1 [8].

Clinical Features from Acute to Convalescent Stages

Preceding symptoms were observed in all patients. They included fever and chills (81%), upper respiratory infection syndrome (33%), and gastrointestinal symptoms such as vomiting and loss of appetite (62%), and chest pain or discomfort

(48%). The interval between the onset of the preceding symptoms and the occurence of cardiac symptoms and signs was 3.4 ± 1.4 days, range 1–8 days.

Results of routine laboratory examinations were characterized by increases in creatine phosphokinase, glutamic oxalacetic transaminase, glutamic pyruvic transaminase, and lactic acid dehydrogenase [8]. In some cases, the increase was not so marked, which may have been due to the delay between the onset of the disease and the time determinations were made. An increase in amylase, creatinine, and blood urea nitrogen was characteristically seen. Leukocytosis was also observed in all patients. In our viral study, one patients had a rise in the neutralizing antibody titer for herpes simplex, and in another the neutralizing antibody titer of coxsackievirus type B4 was prominent. A complement fixation test was positive in the above-mentioned patients, in one other patient with influenza, and another with coxsackie B3 virus. Therefore, positive results were obtained in four out of the 15 patients studied (27%) [8].

Time-Course Changes in Hemodynamic Findings

Bedside sequential hemodynamic studies were conducted in nine patients during the acute, subacute and convalescent stages of this disease, employing a Swan-Ganz catheter [12]. The mean right atrial pressure was 13.0 ± 3.8 mmHg, mean pulmonary arterial pressure 25.6 ± 4.5 mmHg, mean pulmonary capillary wedge pressure 22.3 ± 4.6 mmHg (showing a rise), and the mean systemic arterial pressure 75.9 ± 7.9 mmHg (showing a decrease). Cardiac index was 1.55 ± 0.3 liters min^{-1} m^{-2}, stroke work index 20.8 ± 5.1 mm $beat^{-1}$ m^2, left ventricular stroke work index 14.9 ± 3.1 mm $beat^{-1}$ m^{-2} (showing a decrease). Also noted was a systemic vascular resistance of 26.8 ± 9.3 resistance units, showing a rise. The above data suggest that the cardiac function at the acute stage of severe myocarditis may show an increase in preload and depressed myocardial contractile force with biventricular failure. However, all but two patient showed normalization of hemodynamic parameters 3–4 weeks after the initial examination, showing a notable difference from that observed in acute myocardial infarction.

In 14 patients, left ventriculographic findings at the convalescent stage (3–4 weeks) were analyzed. The ejection fraction (EF) was 57.4% ± 9.1%; four of the 14 patients revealed depressed EF of less than 50% and there were three cases where wall motion abnormalities were seen, showing that the residual signs of left ventricular dysfunction may become evident in some cases. There was no significant coronary artery stenosis in any of the patients, as demonstrated by coronary arteriography.

Biopsy Profiles at the Acute and Convalescent Stages

Of the 20 patients, 19 had so-called pernicious myocarditis. Serial endomyocardial biopsy findings in ten of these patients were analyzed (Fig. 1). The histopathological features were investigated at the acute (0–10 days after the onset), subacute (11–21 days), and convalescent stages (22–167 days). The incidence and

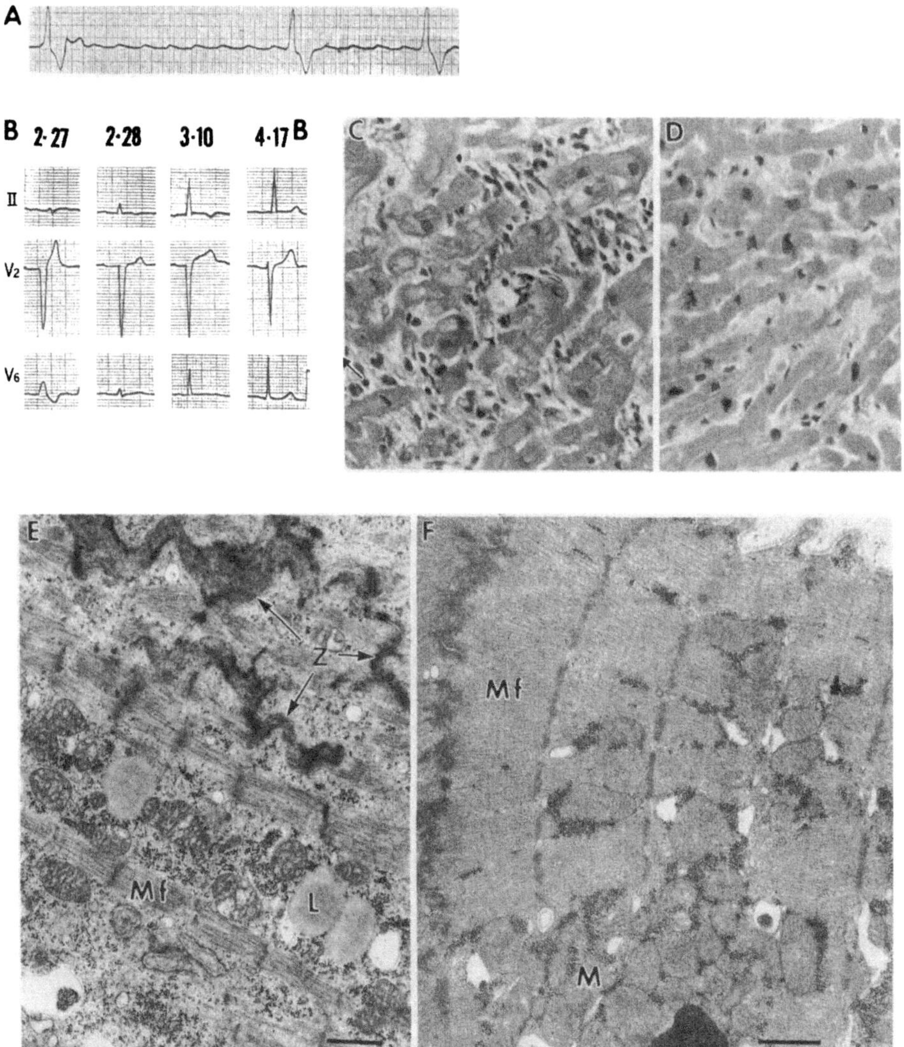

Fig. 1 A–F. An example of acute myocarditis in a 25-year-old female who underwent serial endomyocardial biopsy. **A** On February 25, 1975, the patient was admitted with complete atrio-ventricular block and cardiogenic shock. **B** ECG course after recovery of the atrio-ventricular conduction showing that the left bundle branch block pattern is improving to that of the normal QRS configuration during the course of disease. Note month and date. **C** Biopsy findings at the acute stage (15 days after the onset), showing interstitial cell infiltration and myocytolysis. **D** Second biopsy finding at the convalescent stage (56 days after the onset). **E** Ultrastructure of a myocyte, showing disruption and looseness of myofibrils *(Mf)* and swelling of mitochondria *(M)*. Increased Z-band material *(arrows)* is also noteworthy (15 days after the onset). *(Bars* in all of the following electron-microscopic pictures indicate 1 μm). **F** Ultrastructure of a myocyte at the second biopsy, showing almost complete normalization of the fine structure (56 days after the onset).

severity of various cardiac myocytes and interstitial changes were compared at each stage of the disease. At the acute stage, interstitial cell infiltrations were composed of fibroblasts, macrophages, and lymphocytes, in descending order of frequency. At the convalescent stage, interstitial cell infiltration showed a marked decrease and then became replaced by fibrocytes. At the subacute stage, transitional changes between the acute and convalescent stages were observable (Figs. 2–4). A control study in 21 patients with myocardial infarction, in whom interstitial cell infiltrations were compared at various stages, revealed that at the acute stage neutrophils were most prominent, and at the subacute and convalescent stages macrophages were most conspicuous and plasma cells were often observed.

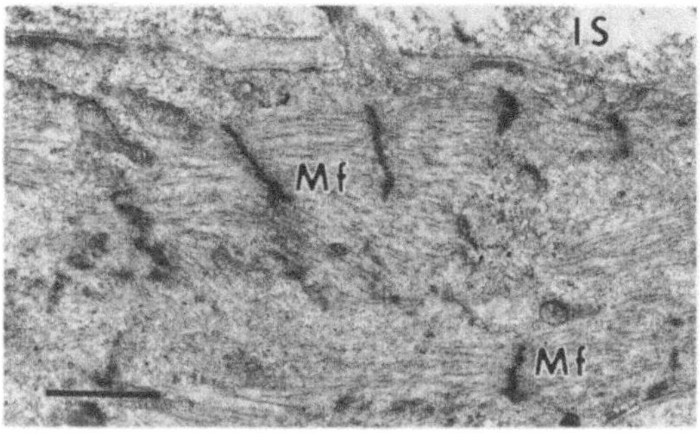

Fig. 2. A biopsy performed in a patient with subacute myocarditis (21 days after the onset) showing that the ultrastructure of the myocytes is still much deteriorated, with loose myofibrils *(Mf)* and deposition of glycogen particles, indicating that recovery has not occured sufficiently. This patient developed characteristics of dilated cardiomyopathy during the long-term follow up

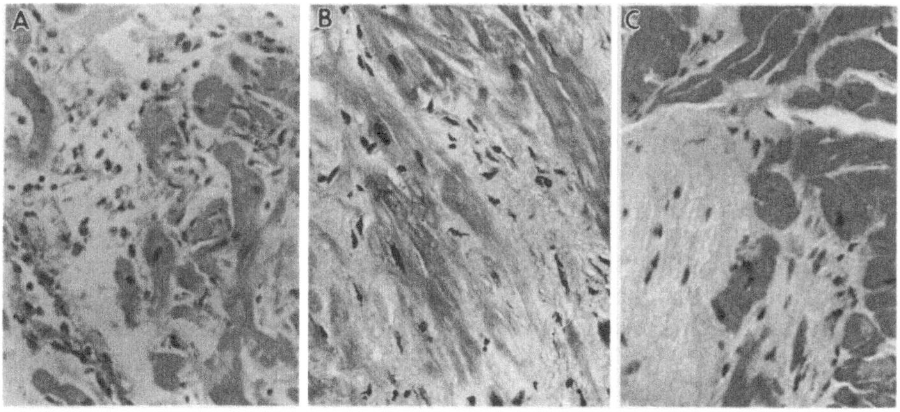

Fig. 3 A–C. Serial endomyocardial biopsy findings in a case with acute myocarditis. This patient had typical symptoms and signs of acute myocarditis at the age 42 years. **A** Acute stage (5th day after the onset) showing interstitial edema and inflammatory cell infiltration. Fragmentation of muscle bundles and myocytolysis are also prominent. At this stage, the patient had cardiogenic shock and complete atrio-ventricular block. **B** Convalescent stage (167 days after onset), showing fragmented myocytes and increased interstitial fibrosis. An increase in fibrocytes is conspicuous. **C** Nine years after the onset, this patient developed dyspnea on exertion and showed residual ECG changes (left axis deviation and complete right branch block). The third biopsy revealed marked interstitial fibrosis, showing a histology of healed myocarditis

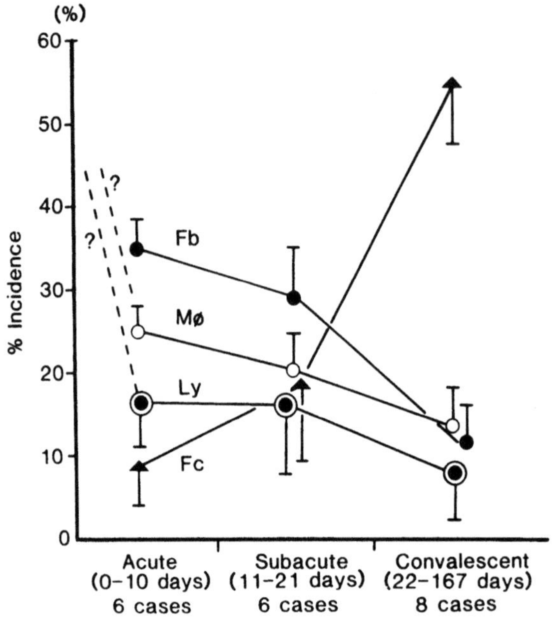

Fig. 4. Incidence of interstitial cellular constituents in cases with acute myocarditis. This figure is derived from our histopathological analysis in our serial biopsy study at the acute, subacute, and convalescent stages. Note that fibroblast *(Fb)* were most frequent in the cell population and lymphocytes were not prominent at every stage. Fibrocytes *(Fc)* became most frequent at the convalescent stage
M ∅, macrophage

The time course changes of the histopathological findings in acute myocarditis were summarized as follows [11]: at the acute stage, the following were seen: (a) interstitial cell infiltration composed of fibroblasts, macrophages and lymphocytes, (b) fragmentation of the muscle bundles, (c) myocytolytic changes, (d) swelling and scarcity of the cytoplasm and swelling of nuclei, (e) variation in size of the myocytes, (f) disarrangement of the muscle bundles, (g) interstitial edema, (h) increased glycogen deposition in the myocytes, (i) abnormal branching of the myocytes, and (j) interstitial fibrosis. At the convalescent stage, most of these findings could still be seen, except for myocytolytic change, swelling of myocytes, and interstitial edema. At the subacute and convalescent stages, an increase in abnormal branching and in the number of double nuclei in the myocytes, and nuclear degeneration became evident. A further control study comparing the change in the convalescent stage of myocarditis and the myocardial changes in 58 cases with chronic right ventricular overload revealed that in the former, fragmentation of the muscle bundles, abnormal branching, size variation, glycogen deposition, and large mononuclear cell infiltrations were significantly more frequent.

Ultrastructural observation of the myocytes [5, 15] revealed that they have a great potentiality for regeneration; myocytes showing severe myocytolytic changes may recover to regain an almost normal architecture (Fig. 1). Some cases, however, still showed deteriorated myocytes even at the convalescent change, and one case showed persisting left ventricular dilatation (Fig. 2).

Table 2. Breakdown of the 30 patients with acute myocarditis who underwent endomyocardial biopsy. See text for details

			(n)	(%)
1. Pump failure 1			27	90
a. Shock	12			
b. Congestive heart failure	25			
2. Electric failure			16	53
a. Third degree atrioventricular block	15			
b. Ventricular tachycardia	1			
3. Multiorgan failure			3	10
a. Renal	3			
b. Disseminated intravascular coagulation	2			
4. Thromboembolic			1	3
5. Death		6		20

Natural History

Clinical Manifestations and Treatments

Clinical profiles of the patients and the treatments we applied are summarized in Figs. 1–3 and Table 2.

After viral invasion, two major disease conditions occur, one being myocardial contractility failure, and the other, cardiac conduction abnormalities. Usually, the atrio-ventricular (A-V) conduction abnormalities disappeared within 14 days after the onset. There are many complications, however, which occur at the acute stage, such as renal failure, disseminated intravascular coagulation, encephalopathy, and pancreatitis. Death occurred at the acute stage (range 1–14 days) because of cardiogenic shock and/or refractory ventricular arrhythmias. A comparison of the clinical outcome before and after application of vasodilating agents and cathecholamines revealed a decrease in the death rate from 43 % to 17 %, showing a statistical difference ($P < 0.05$).

Electrocardiographic Change and Its Course

Electrocardiographic (ECG) abnormalities were seen in all of 21 patients analyzed at the time of admission. Most frequent were third-degree A-V Block in 14 patients (67 %) and then ST-T abnormalities in 13 patients (ST segment elevation in six, ST segment depression in four, and negative T in three). Other changes were poor R-wave progression in five patients (24 %), and low-voltage, second-decree A-V block (Mobitz type 2) and ventricular premature contraction in two patients (10 %). The fatal cases showed ventricular tachycardia, sinus tachycardia, third-degree A-V block, and low voltage. Take [9] conducted a long-term follow-up of up to 5.3 years with a mean follow-up period of 26 month in 13 of our patients with a definitive diagnosis of acute myocarditis and found that in seven

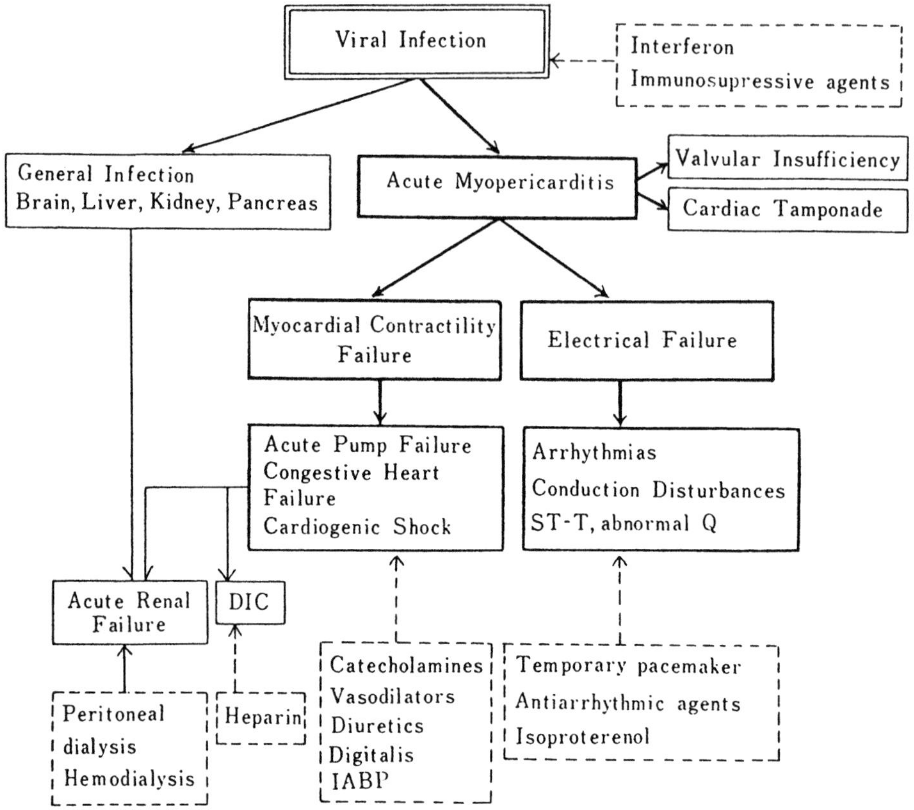

Fig. 5. Clinical manifestations and treatment of choice in acute viral myocarditis. *Dotted square* indicates treatment of choice

patients (54%) the ECG changes persisted. Among these, three patients showed third-degree AV block, and therefore permanent cardiac pacemakers were implanted. In addition, right bundle branch block plus left axis deviation was found in three patients, a combination of left axis deviation and abnormal Q waves in one patient, and T wave abnormalities in one patient, showing that conduction abnormalities were prominent.

Long-term Prognosis Assessed by a Radionuclide Study

Eleven patients with previous myocarditis were evaluated to determine the cardiac condition through thallium(Tl)-201 myocardial perfusion imaging and stress radionuclide ventriculography [13]. Compared with 11 sex- and age-matched controls, the exercise capacity employing an bicycle ergometer was reduced in myocarditis. The resting left ventricular ejection fraction (LVEF) of myocarditis was 57% ± 4%, similar to that of the controls. LVEF response to stress in myocarditis

Table 3. Result of a long-term follow-up of 20 patients who had acute myocarditis. See text for details

		(n)	(%)
1. No apperant residual signs		13	65
2. Residual ECG abnormalities		7	35
a. Right bundle branch block and/or left axis deviation	4		
b. Complete atrioventricular block (pacing)	3		
3. Congestive heart failure (dilated cardiomyopathy-like)		1	5
4. Death		1	5

was abnormal, with an increment of end-systolic volume, while in the controls the LVEF significantly increased under stress. Eight of the nine patients with abnormal EF responses revealed constant perfusion defects. This study indicates that latent myocardial dysfunction is present, and that nuclear cardiology is useful in a long-term follow-up.

Recent Long-term Prognosis of the Patients Who Underwent Endomyocardial Biopsy

Results of the clinical follow-up of our patients in this study are shown in Table 3. It is worthy of note that in 13 of the 20 patients (65 %) who left hospital, no apparent residual signs were revealed by conventional clinical observations. This contradicts our previous reports from the Heart Institute of Japan, where residual ECG changes were seen in 80 % of the cases, ventriculographically demonstrated depressed EF in 30 %, and abnormal EF ejection fraction response to ergometer exercise in 89 % (eight out of nine patients), which was well correlated with myocardial perfusion defect by ^{201}Tl scintigraphy. A case where a ^{201}Tl scintigraphy was performed is presented in Fig. 6. Regarding the discrepancy between the results of the present analysis and our previous results, it is worthy of note that a patient (case 5 of [6] and case 2 of [22], see Fig. 3) who was regarded as having dilated ventricle at convalescence (167 days after the onset of the illness) showed improvement 9 years later, as demonstrated by left ventriculography. Therefore, further recovery may occur after several years. Another reason for the high rate of absent residual signs in the present study may be that although the patients admitted to Fujita-Gakuen Medical School and Niigata University Hospital had grave cardiac conditions at the onset, they had potentiality for rapid recovery because they were admitted without delay.

Review of the Literature with Regard to the Natural History of Viral Myocarditis

Many reports appeared on this subject after 1965 [23–38], before modern methods such as echocardiography or radionuclide studies were introduced. Helin [27] reported that during an endemic manifestation of coxsackie B5 virus infec-

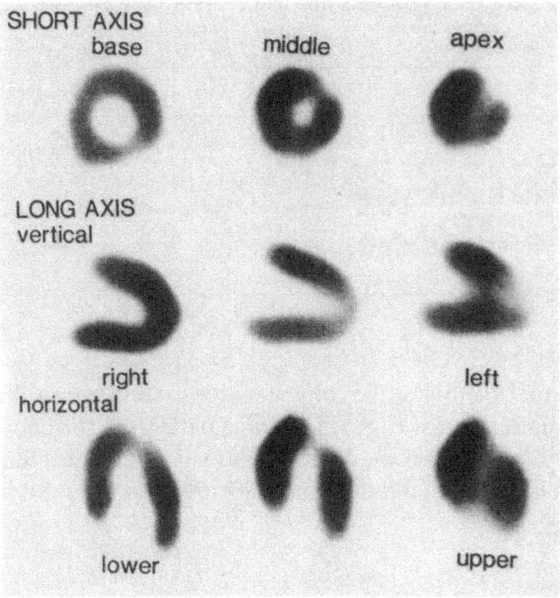

Fig. 6. A radionuclide study in a patient with prior myocarditis. This 48-year-old male patient had acute myocarditis and showed normalization of the ECG findings. However, these single photon emission computed tomography *(SPECT)* images show defects at the apical, anterior, and lateral portions of the left ventricle in a thallium scintigraphy

tion, 12 out of 18 patients with acute myocarditis were followed up for 7 months. No patients developed cardiomegaly, and residual ECG abnormalities were seen in only on patient.

In 1975, Sainani et al. [35] reported 19 cases of coxsackie B infected heart disease in which five cases were of acute myocarditis, five acute myopericarditis, three acute pericarditis, three unexplainable congestive heart failure, two pleuropericarditis, and one patient had postpartum heart failure which was associated with cardiogenic shock. Sainani et al. followed them up for 10 weeks after the disease onset. He was unable to conduct a long-term follow-up in five of the patients, but three patients developed heart failure, and cardiomegaly was observed in two of them. Another three patients showed typical chest pain, and the remaining eight showed no apparent subjective or objective signs of residual cardiac disease. However, four of the eight patients showed ECG abnormalities. They concluded that postmyocardic patients do not always show a benign prognosis, but a dilated cardiomyopathy-like condition may be a sequela of coxsackie B myocarditis.

Smith [31] reported 42 coxsackie B-infected Australian patients with acute myocarditis, experienced during 1962–1969. Of the 42 patients, 20 revealed associated pericarditis. Thirty-five patients showed signs of complete recovery, but 12 of them needed 3 month or longer for complete recovery. The remaining seven patients showed a recurrence of the myocarditis and two of them died of congestive heart failure. Six patients showed residual ECG changes, three with both cardiomegaly and ECG abnormalities. This suggests that this dilated cardiomyopathy-like condition may occur in post-coxsackie-virus infection.

The report of Levi et al. [37] described a rather favorable prognosis after coxsackievirus infection. They followed up ten myocarditis patients with coxsackie infection for 42–68 month after the onset, and observed a prolongation of the preejection period, cardiomegaly or slight ECG abnormalities, but there was no severe cardiac disability in the cases studied. Obeyesekere and Hermon [34] followed up Arvo virus-infected acute myocarditis cases and found that one of 35 patients died after the acute stage and two patients died within 6 month after the onset. Six patients showed signs of recovery, and two showed residual ECG changes. The remaining 26 patients revealed both the ECG abnormalities and cardiomegaly, three showed signs of dilated cardiomyopathy and one had signs of a restrictive cardiomyopathy-like condition.

Das et al. [38] followed-up ten patients with clinically diagnosed viral myocarditis and performed a radionuclide study with thallium scintigraphy and technetium ventriculography which was assessed by ergometer exercise. They found decreased exercise tolerance in all cases, dysfunction of the cardiac reserve in eight cases, and defect by thallium scintigraphy in eight cases.

Hayakawa et al. [49] followed-up 20 cases with viral or idiopathic myocarditis with Holter analysis and echocardiogram for 89.1 ± 39.3 months and found cardiac abnormalities in ten patients, in six of whom dilated cardiomyopathy-like left ventricular enlargement was observed. It is worthy of note that they observed asymmetric septal hypertrophy in three cases.

Miklozek et al. [53] followed up 16 patients for 6 months, six with echovirus and one with confirmed Epstein-Barr virus. They found cardiac abnormalities in 14 cases. According to their clinical evaluation using a scoring system, the condition of nine patients improved, six worsened, and one remained unchanged. They stated that most of the cases showed subclinical features, but one died because of sustained left ventricular failure.

Kawamura [63] followed-up 87 cases among 218 patients surveyed in a nationwide survey in Japan. According to a follow-up questionnaire in 67 cases, 13 patients showed a complete recovery, four had cardiac death, and 49 still revealed residual cardiac abnormalities, eight of whom showed a dilated cardiomyopathy-like condition and one a hypertrophic cardiomyopathy-like ventricular hypertrophy.

Experimentally Induced Viral Myocarditis with Regard to Natural History

Wilson et al. [28] injected ICR suckling mice with coxsackie-B3 virus (Nancy strain) and observed them for a period of up to 185 days. They found that, contrary to coxsackie A9 virus injection where complete recovery of the myocarditis occurred, coxsackie B-infected animals showed myocarditic change as well as a deposit of calcium in the myocardium and showed marked interstitial fibrosis, residual inflammatory change, and mild hypertrophy of the myocytes. Since this report, it has been recognized that coxsackie B-infected murine myocarditis has similarities to human viral myocarditis, and therefore it is considered a good model for research on the prognosis for viral myocarditis.

Kitaura [42] reported his long-term observation in coxsackie-B-infected hamster. He found that four of ten animals showed cardiomegaly 14 months after inoculation with the virus and also signs of congestion in the liver and lung. Therefore, myocarditis in hamsters infected with coxsackie B3 is also considered a good animal model, with similarities to human myocarditis without calcium deposition.

Matsumori and Kawai [46, 47] observed encephalomyocarditis (EMC) virus-injected mice for 3 months after inoculation and found cardiomegaly, myocardial hypertrophy, and myocardial fibrosis and noticed a dilated cardiomyopathy-like condition in the animals. In an additional study, he inoculated BALB/c mice with the EMC virus. This resulted in severe myocarditis and marked cardiomegaly and congestive heart failure at the acute and subacute stages of the disease was observed.

Aggravating Factors

Interestingly, gender is related with the occurrence of viral diseases [26]. Among coxsackie B1 virus-infected CD-1 mice, a higher death rate was revealed in male animals and the death rate decreased after castration. Also, among BALB/c mice which were infected with coxsackie B3 virus, an earlier death was seen in males than in females [41].

Pregnancy or delivery has also been shown to be an aggravating factor in EMC virus-infected mice in which the myocarditis was exacerbated during pregnancy, and also the viral concentration was higher in the myocardium [29].

Exercise is also considered in aggravating factor in coxsackie-injected mice [25, 30]. With forced swimming, the death rate increased from 5% to 50%. Also, there was an increase in necrosis of the cardiac muscle, and myocardial hypertrophy was also recognized. Nutrition is also a contributing factor [32, 33]. Mice injected with coxsackie B3 virus, and also those with coxsackie B3 induced myocarditis, showed a higher death rate when fed with a high-cholesterol diet [34]. Cold has also been proven to be an aggravating factor for coxsackie virus-injected myocarditis in mice [23].

Some Notes on the Usefulness of Immunosuppressive Therapy

There are differences of opinion regarding the application of anti-inflammtory agents such as steroids and other immunosuppressive agents [48, 52, 58, 60]. Generally, the use of corticosteroids appears to be controversial in myocarditis of viral etiology because of their suppressive effect on systemic defense and myocardial regeneration, as well as the suppression of interferon [24, 60]. Mason et al. [40] have recently suggested, on the basis of endomyocardial biopsy findings, that immunosuppressive therapy using prednisone and/or azathioprine was effective in eliminating myocardial inflammation and improving myocardial performance in ten patients with acute myocarditis. These patients, however, suffered from congestive heart failure for over 3 weeks following the onset of myocarditis, and therefore one should consider that some immune reaction is progressing in this type of patient from the start. Our experience of employing sequential endomyocardial biopsy in cases of acute myocarditis has shown that the healing of myocarditis can occur without the use of immunosuppressive agents [5–22].

Steroids may be beneficial in the prevention of cardiogenic shock, but not in the healing process of myocarditis. Therefore, the question of whether or not immunosuppressive therapy is necessary and effective during the acute stage of myocarditis remains unsolved.

Camerini's group [67] reported improvement of cardiac function and biopsy findings in their long-term follow-up of patients who received a programmed immunosuppressive therapy. However, their results do not seem to differ from those of our study, in which the immunosuppressive therapy was not applied. It is worthy of note that 52% of the patients (12 out of 23) who underwent immunosuppressive therapy developed characteristic features of dilated cardiomyopathy in a long-term follow-up [65].

On the Relationship Between Dilated Cardiomyopathy and Myocarditis

As described by many investigators [59, 64], histopathological findings, observed either by endomyocardial biopsy or autopsy in patients with dilated cardiomyopathy, do not always reveal myocarditis. The incidence of myocarditis in dilated cardiomyopathy described in the literature varies widely (10%–65%) from author to author [54, 55–57].

Problems exist in the histopathological diagnosis [44, 45, 51, 61, 62] which is derived from a single rather than a serial biopsy, thus causing confusion as to whether the histopathological findings are those of healing or healed myocarditis. Through our experience of performing serial endomyocardial biopsy in cases with myocarditis, we have constructed a diagnostic categorization system which classifies the biopsy findings into (a) definitely present, (b) highly suggestive, (c) slightly suggestive, (d) doubtful, and (e) not suggestive. Using this system, we are able to classify the biopsy findings from specimens taken during a single rather than a serial biopsy [22]. We were also able to divide the disease into acute, subacute, and convalescent stages. The incidence of highly suggestive postmyocarditis change in our study is shown in Fig. 7 [17].

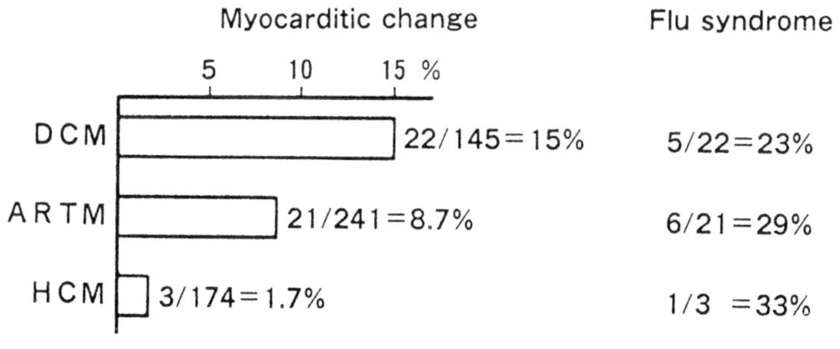

Fig. 7. Incidence of postmyocarditic change in biopsy specimens from three heart muscle diseases and relationship with the history of influenza syndrome. See text for details. *DCM*, dilated cardiomyopathy; *ARTM*, arrhythmias or conduction disturbances; *HCM*, hypertrophic cardiomyopathy

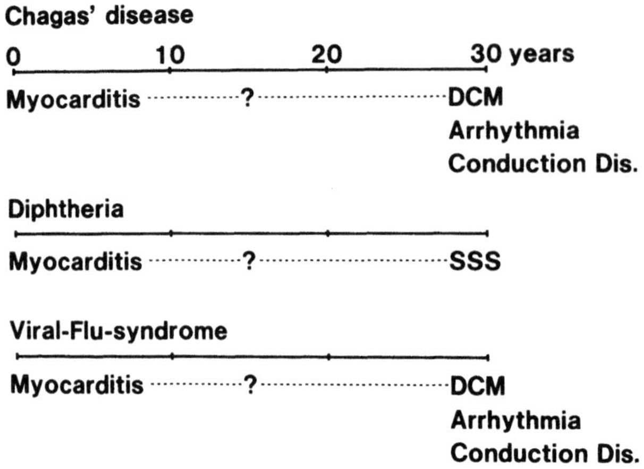

Fig. 8. Importance of past history in etiological considerations of cardiomyopathy. Note that in the three diseases, an interval of up to 20–30 years after the initial infection is presumed to exist in the development of various types of cardiac manifestations. *DCM,* dilated cardiomyopathy; *SSS,* sick sinus syndrome

A View of the Clinical Spectrum of Myocarditis and Its Course

The authors' view of the natural history of viral myocarditis is summarized in Fig. 8. As described by Amorin [68] and Rosenbaum [69], sick sinus syndrome, atrioventricular conduction abnormalities, and dilated cardiomyopathy-like ventricular dilatation may be seen decades after initial Chagasic myocarditis. Our analysis of cases with sick sinus syndrome showed that 24% of the patients (18 out of 74) had a history of diphteria in childhood although they had no concomitant cardiovascular disease at the time of the infection [21]. Patients who suffer from grave conditions, such as congestive heart failure and ventricular arrhythmia, presumably die at the acute stage and, therefore, only the patients who suffer from atrial disease survive. The sick sinus syndrome develops decades after the initial infection with diphteria.

These facts regarding Chagas' disease or diphteria may be applicable to viral myocarditis. In view of our experience, arrhythmias or conduction disturbance together with ventricular dilatation can be seen many years later. It is suggested, therefore, that the clinical spectrum of myocarditis is wide and shows electrical disturbance. One bout of viral infection may not be sufficient to cause dilated cardiomyopathy but repeated attacks or some chronic pathogenetic mechanism, such as age-associated progress of disease and/or an immune process [50], may take place leading to dilated cardiomyopathy.

Acknowledgements. This study was supported in part by a Research Grant for Intractable Diseases from the Ministry of Health and Welfare of Japan.

Summary

Fourteen years have passed since we started to perform endomyocardial biopsy in patients with acute viral or idiopathic myocarditis. A review of 30 patients in whom the diagnosis of acute myocarditis was made by autopsy or biopsy (including serial biopsy) is presented. In our initial experience five autopsied patients revealed cardiogenic shock, congestive heart failure, and malignant arrhythmias or conduction disturbance. Twenty-five patients showed a fairly good recovery (NYHA I-II) even though they had grave clinical features at the onset. Only conventional medical treatment without programed immunosuppressive therapy was applied in our cases. A serial biopsy analysis revealed a resolving or healing course of the myocarditis. A clinical follow-up for 1–14 years (mean 4.1 ± 4.4) was made in 20 patients who left hospital after recovery. Seven patients (35%) showed residual conduction abnormalities, such as advanced atrio-ventricular block, complete right bundle branch block, and/or left axis deviation. Three of them were fitted with permanent pacemaker. Seven patients showed normal ECG findings after recovery. It is worthy of note that only two patients (10%) showed clinical evidence of dilated cardiomyopathy. A review of the clinical and experimental aspects of myocarditis with regard to the natural course in the literature was also made.

References

1. Fiedler A (1899) Über akute intestitielle Myokarditis. Vortrage Gesellschaft Natur und Heilkunde Dresden 9.10.1897; ausführliche Festschrift zur Feier des fünfzigsten Bestehens des Stadtkrankenhauses zu Dresden-Friedrichstadt. Baensch, Dresden, pp 3–24
2. Jarco S (1973) Fiedler on acute interstitial myocarditis (1899) I. Am J Cardiol 32: 221–223
3. Jarco S (1973) Fiedler on acute interstitial myocarditis (1899) II. Am J Cardiol 32: 716–718
4. Sekiguchi M, Hiroe M, Ogasawara S, Nishikawa T (1981) Practical aspects of endomyocardial biopsy. Ann Acad Med Singapore 10, Suppl: 115–128
5. Sekiguchi M, Hiroe M, Take M (1979) On the potentiality of myocardial cell regeneration. Ultrastructural observation employing serial endomyocardial biopsy in 9 cases of acute myocarditis. J Moll Cell Cardiol, Suppl 3: 21 (abstract)
6. Sekiguchi M, Hiroe M, Take M, Hirosawa K (1980) Clinical and histopathological profile of sarcoidosis of the heart and acute myocarditis: II. Myocarditis, Jpn Circul J 44: 264–273
7. Sekiguchi M, Hiroe M, Take M, Akamatsu T, Takahashi S (1980) Vasodilator therapy in congestive cardiomyopathy and myocarditis. In: Sekiguchi M, Olsen EGJ (eds) Cardiomyopathy. Clinical, pathological and theroretical aspects, University of Tokyo Press, Tokyo and University Park Press, Baltimore, pp 257–275
8. Take M, Sekiguchi M, Hiroe M, Hirosawa K (1981) Early clinical profiles of histopathologically proven cases with acute idiopathic myocarditis and a proposal for diagnostic criteria. Jpn Circul J 45: 1415–1420
9. Take M, Sekiguchi M, Hiroe M, Hirosawa K (1982) Long-term follow-up of electocardiographic findings in patients with acute myocarditis proven by endomyocardial biopsy. Jpn Circul J 46: 1127–1234
10. Hasumi M, Sekiguchi M, Morimoto S, Take M, Hirosawa K (1983) Ventriculographic findings at the convalescent stage in eleven cases with acute myocarditis. Jpn Circul J 47: 1310–1316
11. Yu Z-X, Sekiguchi M, Hiroe M, Take M, Hirosawa K (1984) Histopathological findings of acute and convalescent myocarditis obtained by serial endomyocardial biopsy. Jpn Circul J 48: 1368–1374
12. Hiroe M, Sekiguchi M, Take M, Matsuda M, Hirosawa K (1985) Hemodynamic studies and response to a combined therapy of nitroglycerin ointment and dopamine in patients with acute myocarditis. In: Sekiguchi M, Olsen EGJ, Goodwin JF (eds) Myocarditis and related disorders. Heart and Vessels (Suppl 1): 180–186

13. Hiroe M, Sekiguchi M, Take M, Kusakabe K, Shigeta A, Hirosawa K (1985) Long follow-up study in patients prior myocarditis by radionuclide methods. In: Sekiguchi M, Olsen EGJ, Goodwin JF (eds) Myocarditis and related disorders. Heart and Vessels (Suppl 1): 199–203
14. Hiroe M, Sekiguchi M, Morimoto S, Take M, Hirosawa K (1985) Relation of hemodynamics in acute myocarditis to a quantitative histopathologic analysis employing endomyocardial biopsy. In: Sekiguchi M, Olsen EGJ, Goodwin JF (eds) Myocarditis and related disorders. Heart and Vessels (Suppl 1): 310 (abstract)
15. Sekiguchi M, Yu Z-X, Hasumi M, Hiroe M, Morimoto S, Nishikawa T (1985) Histopathologic and ultrastructural observations of acute and convalescent myocarditis. A serial endomyocardial biopsy study. Heart Vessels Suppl 1: 143–153
16. Sekiguchi M, Olsen EGJ, Goodwin JF (1985) Editor's comments on the biopsy diagnosis of myocarditis. In Sekiguchi M, Goodwin JF, Olsen EGJ (eds) Myocarditis and Related Disorders. Heart Vessels Suppl 1: 306–308
17. Take M, Sekiguchi M, Hiroe M, Hirosawa K (1985) A clinicopathologic study on a cause of idiopathic cardiomyopathy and arrhythmia and conduction disturbance employing endomyocardial biopsy. Heart Vessels Suppl 1: 159–164
18. Yu Z-X, Sekiguchi M, Hiroe M, Take M, Hirosawa K (1985) A comparative ultrastructural study on the nature of interstitial cell constituents in idiopathic myocarditis and myocardial infarction. Heart Vessels Suppl 1: 154–158
19. Yu Z-X, Sekiguchi M, Hiroe M, Hasumi M, Morimoto S, Hirosawa K (1985) On the interstitial fibrotic changes in acute and convalescent myocarditis obtained by serial endomyocardial biopsy. Jpn Circul J 49: 1270–1276
20. Hasumi M, Sekiguchi M, Yu Z-X, Hirosawa K, Hiroe M (1986) Analysis of histopathologic findings in cases with dilated cardiomyopathy with special reference to formulating diagnostic criteria on the possibility of postmyocarditic change. Jpn Circul J 50: 1280–1287
21. Sekiguchi M, Nishino H, Nishikawa T, Morimoto S, Hiroe M (1986) Age associated myocardial changes in various heart diseases. A clinocopathologic analysis in biopsied and autopsied myocardium. Jpn Circul J 50: 1023–1032
22. Sekiguchi M, Hiroe M, Yu Z-X, Hasumi M (1987) A serial endomyocardial biopsy study on myocarditis. In: Kawai C, Abelmann WH (eds) Pathogenesis of myocarditis and cardiomyopathy. Recent experimental and clinical studies. Univ. of Tokyo Press, Tokyo, pp 213–231
23. Boring WD, ZuRhein GM, Walker (1956) Factors influencing host-virus interactions: II. Alteration of coxsackie virus infection in adult mice by cold. Proc Soc Exp Biol Med 93: 273–277
24. Kilbourne E, Wilson CB, Perrier D (1956) The induction of gross myocardial lesions by a coxsackie (pleurodynia) virus and cortisone. J Clin Invest 35: 362–370
25. Tilles JG, Elson SH, Shaka JA, Abelmann WH, Lerner AM, Finland M (1964) Effects of exercise on coxsackie A-9 myocarditis in adult mice. Proc Soc Exp Biol Med 117: 777–782
26. Berkovich S, Ressel M (1967) Effect of sex on susceptibility of adult mice to coxsackie B1 virus infection. Arch Ges Virus Forsch 22: 246–251
27. Helin M, Savola J, Lapinleimu K (1968) Cardiac manifestation during a coxsackie B-5 epidemic. Br Med J 3: 97–99
28. Wilson FM, Miranda QR, Chason JL, Lerner AM (1969) Residual pathologic changes following murine coxsackie A and B myocarditis. Am J Path 55: 253–265
29. Farber PA, Glasgow LA (1970) Viral myocarditis during pregnancy: encephalomyocarditis virus infection in mice. Am Heart J 80: 96–102
30. Gatmaitan BG, Chason JL, Lerner AM (1970) Augmentation of the virulence of murine coxsackie virus B3 myocardiopathy by exercise. J Exp Med 131: 1121–1136
31. Smith WG (1970) Coxsackie B myopericarditis. Am Heart J 80: 34–46
32. Woodruff JF, Kilbourne ED (1970) The influence of quantitated post-weaning undernutrition on coxsackie virus B3 infection of adult mice. I. Viral persistence and increased severity of lesions. J Infect Dis 121: 137
33. Woodruff JF (1970) The influence of quantitated post-weaning undernutrition on coxsakkie virus B3 infection of adult mice. II. Alteration of host defense mechanisms. J Infect Dis 121: 164

34. Obeyesekere I, Hermon Y (1973) Arbovirus heart disease: Myocarditis and cardiomyopathy following dengue and chikungunya fever — a follow-up study. Am Heart J 85: 186–194
35. Sainani SG, Dekate MP, Rao CP (1975) Heart disease caused by coxsackie virus B infection. Br Heart J 37: 819–823
36. Loria RM, Kibrick S, Madge GE (1976) Infection of hypercholesterolemic mice with coxsackie virus B. J Infect Dis 133: 655–662
37. Levi GF, Proto C, Quadri A, Ratti S (1977) Coxsackie virus heart disease and cardiomyopathy. Am Heart J 93: 419–421
38. Das SK, Brady TJ, Thrall JN, Pitt B (1980) Cardiac function in patients with prior myocarditis. J Nucl Med 21: 689–693
39. Woodruff JF (1979) Lack of correlation between neutralizing antibody production and suppression of coxsackie virus B3 replication in target organs: Evidence for involvement of mononuclear inflammatory cells in host defense. J Immunol 123: 31–36
40. Mason JW, Billingham ME, Ricci DR (1980) Treatment of acute inflammatory myocarditis assisted by endomyocardial biopsy. Am J Cardiol 45: 1037–1044
41. Woodruff JF (1980) Viral myocarditis. A review. Am J Pathol 101: 427–479
42. Kitaura Y (1981) Experimental coxsackie B virus myocarditis in mice. 18 month histopathological and virological study. Jpn Circul J 45: 747–762
43. Morita H (1981) Experimental coxsackie B3 virus myocarditis in golden hamsters — light and electron microscopy findings in a long-term follow-up study. Jpn Circ J 45: 713–729
44. Baandrup U, Florio RA, Olsen EGJ (1982) Do endomyocardial biopsies represent the morphology of the rest of the myocardium? A quantitative light microscopic study of single v. multiple biopsies with King's bioptome. Eur Heart J 3: 171–178
45. Edwards WD, Holmes DR Jr, Reeder GS (1982) Diagnosis of active lymphocytic myocarditis by endomyocardial biopsy: Quantitative criteria for light microscopy. Mayo Clin Proc 57: 419–425
46. Matsumori A, Kawai C (1982a) An animal model of congestive (dilated) cardiomyopathy: Dilation and hypertrophy of the heart in the chronic stage in DBA/2 mice with myocarditis caused by encephalomyocarditis (EMC) virus. Circulation 66: 355–360
47. Matsumori A, Kawai C (1982b) An experimental model for congestive heart failure after encephalomyocarditis virus myocarditis in mice. Circulation 65: 1230–1235
48. Fenoglio JJ, Ursell PC, Kelogg CF, Pil M, Drusin RE, Weiss MB (1983) Diagnosis and classification of myocarditis by endomyocardial biopsy. N Engl J Med 308: 12–18
49. Hayakawa M, Inoh T, Yokota Y, Kawanishi H, Matsumoto K, Kumaki T, Fukuzaki H (1983) A long-term follow-up study of acute viral and idiopathic myocarditis. Jpn Circul J 47: 1304–1309
50. Maisch B, Deeg P, Liebau G, Kochsiek K (1983) Diagnostic relevance of humoral and cytotoxic immune reactions in primary and secondary dilated cardiomyopathy. Am J Cardiol 52: 1072–1078
51. Baandrup U, Mortensen SA (1984) Histopathological aspects of myocarditis with special reference to mumps, cytomegalovirus infection and the role of endomyocardial biopsy. In: Bolte HD (ed) Viral heart disease. Springer, Heidelberg New York, pp 13–25
52. Daly K, Richardson PJ, Olsen EGJ, Morgan-Capner P, McSorley C, Jackson G, Jewitt DE (1984) Acute myocarditis, role of histological and virological examination in the diagnosis and assessment of immunosuppressive treatment. Br Heart J 51: 30–35
53. Miklozek CL, Come PC, Royal HD, Crumpacker CS, Abelmann WH (1984) In: Bolte HD (ed) Viral heart disease. Springer, Berlin Heidelberg New York, pp 95–98
54. Parrillo JE, Aretz HT, Palacios I, Fallon JT, Block PC (1984) The results of transvenous endomyocardial biopsy can frequently be used to diagnose myocardial diseases in patients with idiopathic heart failure. Endomyocardial biopsies in 100 consecutive patients revealed a substantial incidence of myocarditis. Circulation 69: 93–101
55. Zee-Cheng C, Tsai CC, Palmer DC, Codd JE, Pennington DG, Williams GA (1984) High incidence of myocarditis by endomyocardial biopsy in patients with idiopathic congestive cardiomyopathy. J Am Coll Cardiol 3: 63–70
56. Dec GW Jr, Palacios IF, Fallon JT, Aretz HT, Mills J, Lee DC-S, Johnson RA (1985) Active myocarditis in the spectrum of acute dilated cardiomyopathies. Clinical features, histologic correlates, and clinical outcome. New Eng J Med 312: 885–890

57. Dec GW Jr, Palacios IF, Fallon JT (1985) Does histology on repeat right ventricular biopsy predict improved ventricular function in patients treated for myocarditis? Circulation 72 (Supp. III): III–109
58. Hosenpud JD, McAnulty JH, Niles NR (1985) Lack of objective improvement in ventricular systolic function in patients with myocarditis treated with azathioprine and predonisone. J Am Coll Cardiol 6: 797–801
59. Lowry PJ, Littler WA (1985) Dilated (congestive) cardiomyopathy — the evidence for and against a disorder of cellular immunity and infection. In: Goodwin JF (ed) Heart muscle disease. MTP, Lancaster, pp 263–271
60. Tomioka N, Kishimoto C, Matsumori A, Kawai C (1986) Effects of prednisolone on acute viral myocarditis in mice. J Am Coll Cardiol 7: 868–872
61. Arets HT, Billingham ME, Edwards WD, Factor SM, Fallon JT, Feroglio JJ Jr, Olsen EGJ, Schoen FJ (1987) Myocarditis. A pathologic definition and classification. Am J Cardiovasc Pathol 1: 3
62. Goodwin JF (1987) Myocarditis and perimyocarditis. Historical survey, epidemiology and clinical features. Eur Heart J 8: 7–9
63. Kawamura K (1987) Report of a committee on myocarditis. Annual Report for Intractable Diseases, Ministry of Health and Welfare of Japan, p. 13 (in Japanese)
64. Marboe CC, Schwartz RA, Adelson D, Escala EL, Weiss MB, Fenoglio JJ Jr (1987) Active myocarditis with fibrosis. Poor correlation of ejection fraction with histology of follow-up biopsy. Eur Heart J 8 (Suppl. J): 43–45
65. Quigley PJ, Richardson PJ, Meany BT, Olsen EGJ, Monaghan MJ, Jackson G, Jewitt DE (1987) Long-term follow-up of acute myocarditis. Correlation of ventricular function and outcome. Eur Heart J 8 (Suppl. J): 39–42
66. Sekiguchi M, Hiroe M, Take M, Kaneko M, Kusakabe K. Natural history of 20 patients with biopsy proven acute myocarditis: A 10 year follow-up. Circulation 72: Suppl III–109, 1985 (abstr.)
67. Salvi A, Hrovatin E, Dreas L, Silvestri F, Camerini F (1987) Changes in histology and left ventricular ejection fraction during immunosuppressive treatment in active myocarditis. Eur Heart J 8 (Suppl. J): 267–269
68. Amorim DS, Manco JC, Gallo L, Marin, Meto JM (1980) Chronic chagas' heart disease. A clinical and functional study. In: Sekiguchi M, Olsen EGJ (ed) Cardiomyopathy. Clinical, pathological and theoretical aspects. Univ. of Tokyo Press, Tokyo and Univ. Park Press, Baltimore, pp 317–335
69. Rosenbaum MB, Elizari MV, Lazzari JO, Nau GJ, Levi RJ, Halpern MS (1969) Intraventricular trifascicular blocks. The syndrome of right bundle branch block with intermittent left anterior and posterior hemiblock. Am Heart J 78: 306–317

many authors [8]. To discuss the natural history of myocarditis in dilated cardiomyopathy, we will present repeated measurements of left ventricular function and its correlation with histologic findings in patients not treated by immunosuppression.

Incidence of Myocarditis in Patients with Dilated Cardiomyopathy

The histologic diagnosis of myocarditis varies considerably with the definition used and with the investigator [1, 2, 9]. Our first experience in 187 patients, from 1978 to 1985, was made by using an internationally reputable classification resulting in an incidence of myocarditis of 27% [10]. From 1986, after the final publication of the Dallas criteria, and taking into account our own previous experience, our classification became slighthly modified [11], resulting in a lower incidence of myocarditis of about 10% (Fig. 1). This incidence of roughly 10% is now also found in larger international trials dealing with the assessment and the treatment of myocarditis in dilated cardiomyopathy [6, 7].

Comparison of Clinical, Electrocardiographic, Virologic, and Hemodynamic Features in Myocarditis and Dilated Cardiomyopathy

Myocarditis as an inflammatory disease is thought to be viral in most cases in Europe. It therefore seems obvious to screen for evidence of previous viral disease in the patients concerned. We compared in 187 consecutively studied patients, with

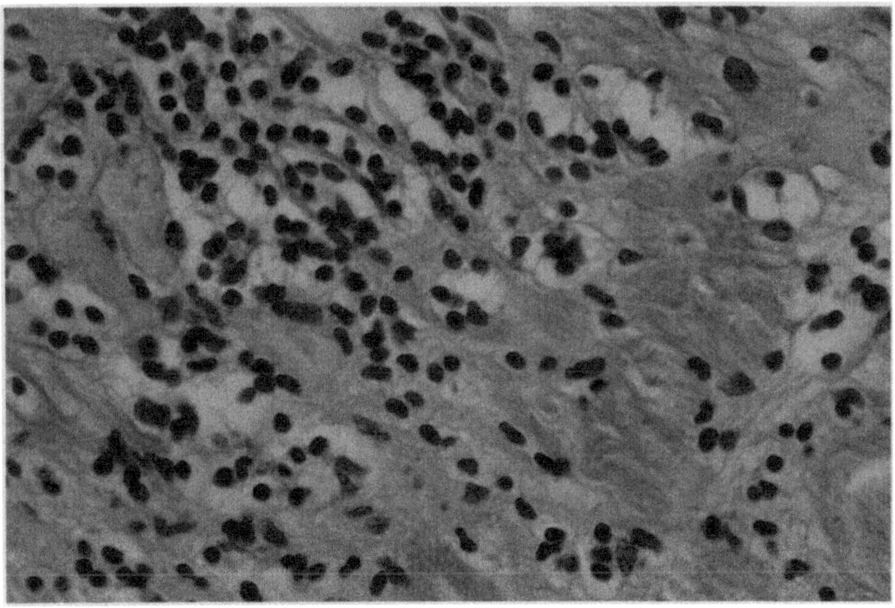

Fig. 1. Histologic example of active myocarditis, demonstrating lymphocytic infiltrates and myocyte necrosis

Myocarditis in Patients with Dilated Cardiomyopathy – Correlation with Clinical, Hemodynamic, and Biochemical Findings

V. REGITZ[1], R. STRASSER[1], G. CHMIELEWSKI[2], S. BLÜMCKE[2], and E. FLECK[1]

[1] Department of Cardiology, German Heart Institute Berlin, Augustenburger Platz 1, D-1000 Berlin 65, FRG
[2] Department of Pathology, Klinikum Charlottenburg, University of Berlin, Spandauer Damm 130, D-1000 Berlin 19, FRG

Introduction

The histologic features of myocarditis have been described in patients with dilated cardiomyopathy in varying incidences ranging from 1% to 66% [1, 2], the significant variability probably reflecting more the differing interpretations of histologic findings than seasonal and regional differences in myocarditis. As the patients frequently present without obvious clinical evidence of inflammatory disease, it seems questionable whether viral heart disease may be diagnosed based on histologic findings, whether patients with myocarditis represent a separate subgroup in dilated cardiomyopathy, and whether the histologic diagnosis can contribute valuable information for patient management or for the estimation of prognosis.

To characterize patients with histologically documented myocarditis as a distinct group with viral etiology in cardiomyopathy, discriminating parameters other than histology ought to be established. Differences are most likely to be found regarding the incidence of previous viral disease or regarding immunologic markers, for example, elevated circulating antiviral or autoantibodies [3]. Regarding hemodynamic features, in myocarditis, in comparison to patients with dilated cardiomyopathy, smaller end-diastolic volume indexes at comparable ejection fractions have been reported [4]. Recent biochemical findings postulate that autoantibodies against the adenosine tri-/di-phosphate (ATP/ADP) carrier occur in viral heart disease and, therefore, specific alterations in cellular energy metabolism could be attributed to myocarditis [5]. In order to differentiate myocarditis from dilated cardiomyopathy, it therefore appears reasonable to compare clinical, immunologic, and hemodynamic parameters as well as myocardial metabolism in patients from both groups.

So far there are no conclusive data available to predict the development of inflammatory infiltrates and their association with alterations in left ventricular function in patients with dilated cardiomyopathy and myocarditis. The success of immunosuppressive therapy is often estimated by the relief of inflammatory signs although it has yet to be determined, in untreated patients, if disappearance of infiltrates is associated with hemodynamic improvement and persistence of infiltrates with worsening of left ventricular function [6, 7]. Such a course would not fit with a transition from myocarditis into dilated cardiomyopathy as is postulated by

H.-P. Schultheiß (Ed.)
New Concepts in Viral Heart Disease

Table 1. Clinical and electrocardiographic findings in patients with dilated cardiomyopathy, myocarditis and a subgroup with active myocarditis

	DCM ($n = 137$)	MC ($n = 50$)	AMC ($n = 14$)
Duration of dyspnea (months)	21.8	23.8	18.8
Chest pain (%)	49.2	36.2	38.5
Palpitations (%)	57.6	52.2	41.7
Fever (%)	20.6	25.0	25.0
Atrial fibril. (%)	31.1	25	30.8
BBB (%)	47.2	32	23
T inversions (%)	32.8	35	50

DCM, dilated cardiomyopathy; MC, myocarditis; AMC, active moycarditis; BBB, bundle branch block

the hemodynamic features of dilated cardiomyopathy after complete hemodynamic and angiographic diagnostic procedures, the duration of dyspnea, the incidence of chest pain, of palpitations, and of febrile illness before diagnosis. According to histologic findings, 50 of these patients suffered from myocarditis and 137 from dilated cardiomyopathy. No significant differences were found between both patient groups and a subgroup with active myocarditis (Table 1).

Electrocardiographic parameters, particularly T wave inversions, bundle branch blocks, atrioventricular blocks, and ventricular arrhythmias have been reported as characterizing myocarditis [3]. In the above-mentioned 187 patients, the incidence of atrial fibrillation, bundle branch block, or T wave inversions in patients with the histologic features of myocarditis, a subgroup with active myocarditis, and those with dilated cardiomyopathy is almost identical (Table 1).

Antiviral antibody titers against the most important cardiotropic viruses, coxsackie-, influenza, adeno- and picornavirus, could provide another hint to the presence of viral infection or past viral disease. We therefore studied these titers in subgroups of our patients with myocarditis and dilated cardiomyopathy (Table 2). Low titers against cardiotropic viruses were found in a high percentage of patients in both groups, in agreement with the age-dependent occurrence of these antibodies in about 30%–50% of the general population [12]. High antibody titers were negligible in both groups. According to these findings, histologically documented myocarditis in the patients studied did not present as an inflammatory viral disease. Anamnestic and virologic data gave no evidence that present or past inflammation was more frequent in the myocarditis compared to the dilated cardiomyopathy group.

Smaller end diastolic volume indexes at comparable ejection fractions have been reported in myocarditis compared to dilated cardiomyopathy. These studies, however, may have been biased by the fact that only patients with dilated cardiomyopathy of recent onset, of less than 6 months, were included, possibly resulting in a special selection of patients [4]. We measured hemodynamic and angiographic parameters, including left ventricular end-diastolic pressure, left

Table 2. Antiviral antibody titers in subjets of patients with myocarditis and dilated cardiomyopathy

	Cardiomyopathy ($n = 48$) (% patients)	Myocarditis ($n = 24$) (% patients)
Influenza, adeno-, picornavirus		
1:8	48	71
1:256	8	2
Coxsackie virus B		
1:8	23	38
1:256	2	8

ventricular end-diastolic and end-systolic volume index, ejection fraction, stroke volume index, mass index, wall thickness, and cardiac index in the 187 patients already mentioned. Fifty patients with myocarditis, including a subgroup of 14 with active myocarditis, could not be distinguished from 137 with dilated cardiomyopathy. In all groups, left ventricular ejection fraction was comparably impaired, and the end-diastolic volume index was enlarged to the same degree. Wall thickness and cardiac index as the other parameters were almost identical. There were also no significant differences when only the 14 patients with active myocarditis, where the histologic diagnosis is made without any questions, were compared with the patients with dilated cardiomyopathy.

Comparable dilatation and hemodynamic alterations accompanied corresponding decreases in ejection fraction in myocarditis and dilated cardiomyopathy. This can be explained by the concept that ventricular dilatation and rise in end-diastolic pressure represent mechanisms compensating for and resulting from a long-standing decrease in systolic function. Exceptions to this rule, as described in five patients with end-stage mildly dilated cardiomyopathy by the Stanford group, are rare [13]. However, these patients had no histologic evidence of myocarditis and are more likely to represent a family-linked subgroup of dilated cardiomyopathy with a specific, yet unknown, defect [13].

Hemodynamic and Histologic Features at Follow-up in Myocarditis

Hypotheses derived from experimental models of myocarditis suggest that lymphocytic infiltrates reflect an active disease process leading to cellular necroses and disturbing myocardial function. However, experience with transplant rejection acquired during recent years has shown that cellular infiltrates with myocyte necroses, larger than found in most myocarditis cases, can be tolerated by the heart with only modest functional impairment, although fulminant diffuse infiltrates that are present in every biopsy fragment clearly lead to heart failure. This makes it questionable in patients with dilated cardiomyopathy or myocarditis to rely only on the presence or absence of inflammatory infiltrates as a guide to therapy or prognosis.

To evaluate if a correlation of hemodynamic and histologic parameters in myocarditis exists, follow-up studies have been undertaken by several authors, one of the earliest by ourselves [10]. Preliminary results from 12 patients can now be substantiated in a larger number with the tendency established in the first patients being confirmed: no correlation between histologic findings and hemodynamic changes could be established. We have so far accumulated repeated biopsies and catheterization data from 23 patients with myocarditis. All follow-up studies were done for at least 6 months (range 6–54 months, mean 19 months) after diagnosis. Myocarditis (active 1 patient: healing, 2 patients; and healed, 6 patients) was also found at the second study in nine patients (39%), whereas 14 patients with previously documented myocarditis showed histologic features of dilated cardiomyopathy at a second biopsy (61%). Hemodynamic improvement, as well as deterioration, was seen independently of changes in the histologic findings (Fig. 2). Improvement can be explained by the good response to conventional medical therapy as can be found in some patients with dilated cardiomyopathy, and must not be contributed to resolving myocarditis. Hemodynamic deterioration occurred independently of histologic changes: in one patient, where active myocarditis was

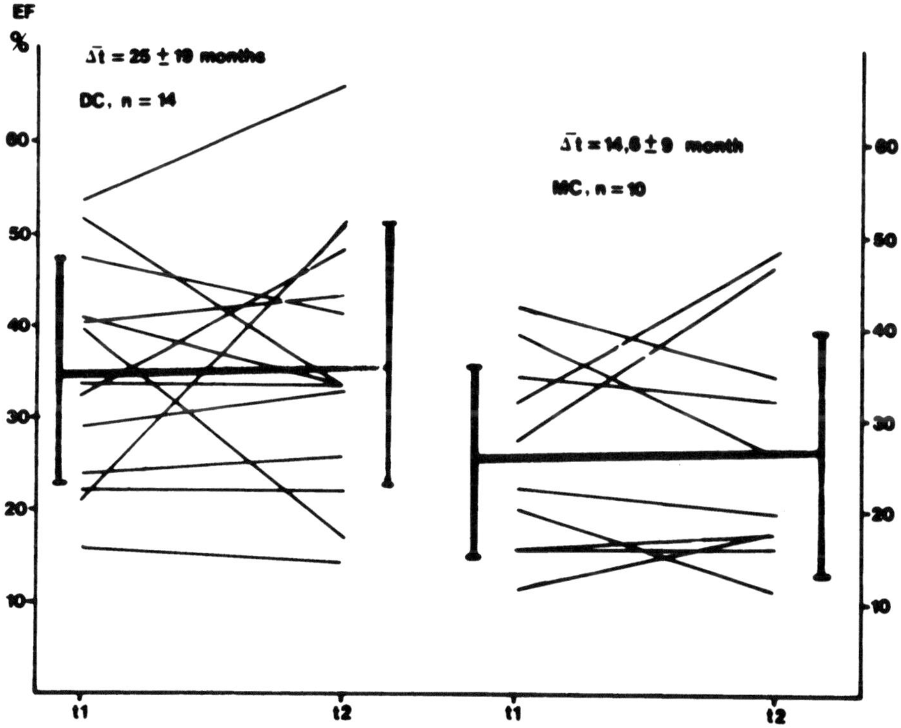

Fig. 2. Left ventricular ejection fraction at the first *(t1)* and at the second hemodynamic study *(t2)* in 23 single patients with myocarditis at initial biopsy. On the *left*, patients with dilated cardiomyopathy *(oc)* and on the *right*, patients with myocarditis *(mc)* at the second biopsy are shown. *Thick bars* represent mean values ± SD for both groups. Hemodynamic improvement and deterioration occurred in both groups

found in two repeated biopsies separated by 15 months, left ventricular ejection fraction decreased from 20% to 11%, and the patient died in cardiac failure. In another patient, in whom a fall of the left ventricular ejection fraction from 40% to 18% and death from cardiac failure, there was active myocarditis in the initial biopsy turned into dilated cardiomyopathy at a second biopsy (after 6 months) and at autopsy.

Sampling error is frequently discussed when the diagnosis of myocarditis cannot be repeated at a control biopsy. In the second example above, sampling error was excluded by a careful search at autopsy. The complete remittance of inflammatory changes together with a hemodynamic deterioration should be accepted in this patient, indicating that myocarditis can be the underlying disease in patients representing with the clinical and histologic features of dilated cardiomyopathy and that resolving infiltrates are not necessarily associated with hemodynamic improvement.

Since the publication of our first series of 12 patients, 1985, more data became available on this subject [10]. In a recent study, seven out of 14 patients with myocarditis showed dilated cardiomyopathy at a second biopsy [14]. The lack of correlation between hemodynamic and histologic findings has also been confirmed by others [15]. In combination, the results may indicate that the presence or absence of lymphocytic infiltrates does not determine the clinical course in dilated cardiomyopathy.

Myocardial Carnitine Levels, Enzyme Activities and Norepinephrine in Myocarditis and Dilated Cardiomyopathy

Recently published biochemical results [5] suggest that autoantibodies against the ATP/ADP carrier occur in patients with myocarditis and postmyocarditic dilated cardiomyopathy, inhibiting nucleotide transport and disturbing the balance between mitochondrial and extramitochondrial energy metabolism. Therefore differences in myocardial metabolism could be expected in patients with myocarditis and dilated cardiomyopathy. As myocardial carnitine controls mitochondrial fatty acid oxidation, delivering about 70% of energy in fully oxygenated hearts, and as myocardial lactate dehydrogenase (LDH) and hydroxybutyrate dehydrogenase (HBDH) activities reflect extramitochondrial substrate utilization, we measured myocardial carnitine content and LDH and HBDH activity in controls and failing hearts with histologic findings of myocarditis and dilated cardiomyopathy.

We described a significant loss of carnitine in endomyocardial biopsies and in explanted hearts from patients with myocarditis and dilated cardiomyopathy [16, 17]. In four patients with myocarditis and 16 with dilated cardiomyopathy, comparably reduced carnitine levels were found in endomyocardial biopsies (Fig. 3). Accordingly, a similar loss of myocardial carnitine occurred in explanted hearts with and without inflammatory infiltrates (Fig. 3). It remains to be determined whether reduced myocardial carnitine levels are caused by insufficient precursor synthesis, reduced cellular uptake, or whether they represent the consequences of regulatory mechanisms. In 23 of these explanted hearts, a significant increase in

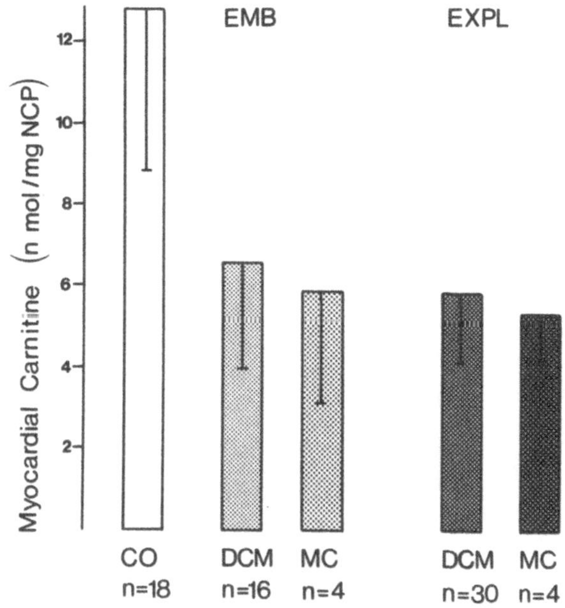

Fig. 3. Loss of myocardial carnitine in dilated cardiomyopathy *(DCM)* and histologically confirmed myocarditis *(MC)*, compared with controls *(CO)*. Results are shown in 20 endomyocardial biopsies *(EMB)* and 34 explanted hearts *(EXPL)*

LDH and HBDH activity was found, indicating increased glucose or lactate utilization (Fig. 4). Both findings were consistent in all areas measured: right atrium and ventricle, and left ventricle. LDH and HBDH activity in myocarditis and dilated cardiomyopathy were not significantly different. Both reduced carnitine levels and increased LDH/HBDH activity indicate altered energy metabolism in

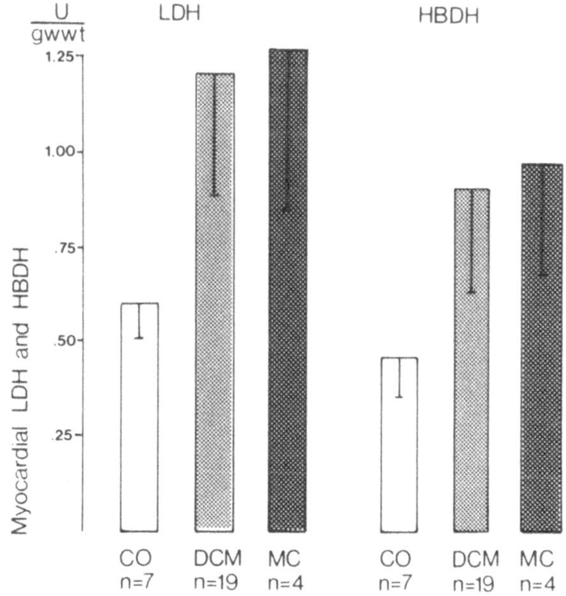

Fig. 4. Increase in myocardial LDH and HBDH activity in explanted hearts in dilated cardiomyopathy *(DCM)* and histologically confirmed myocarditis *(MC)* compared with controls *(CO)*

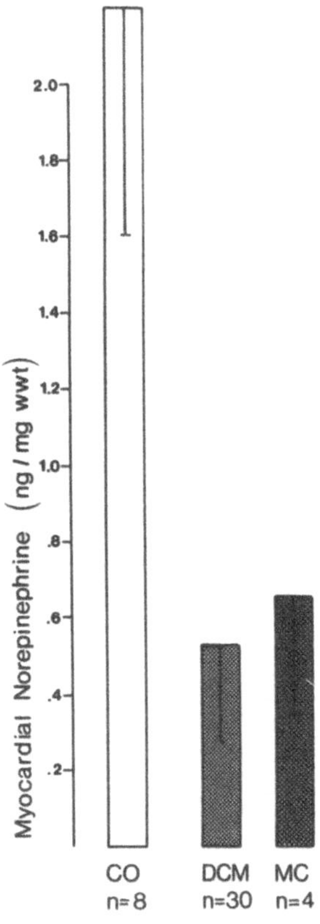

Fig. 5. Loss of myocardial norepinephrine in explanted hearts with dilated cardiomyopathy *(DCM)* and histologically confirmed myocarditis *(MC)* compared with controls *(CO)*

failing hearts. As in the hemodynamic studies, no differences were found in myocarditis and dilated cardiomyopathy.

A loss of myocardial norepinephrine and reduced β-receptor density has been found in failing hearts [18, 20]. This has generally been contributed to agonist-induced downregulation of the receptors. However, recent studies suggest that in patients with dilated cardiomyopathy autoantibodies against β-adrenergic receptors controlled by the major histocompatibility locus may play a role [21]. Therefore, disturbance of catecholamine metabolism may have different features in myocarditis and dilated cardiomyopathy. Myocardial norepinephrine levels in explanted hearts from patients with dilated cardiomyopathy and myocarditis compared with controls are represented in Fig. 5. As about the same loss of myocardial norepinephrine was found in both patient groups, we have no evidence for specific mechanisms in one group.

Summary and Conclusions

As the histologic features of myocarditis are frequently found in patients with dilated cardiomyopathy without clinical evidence of inflammatory disease, the question arises whether histologic findings alone permit the diagnosis of viral heart disease and so characterize a separate subgroup of patients in dilated cardiomyopathy.

To investigate this question, 237 patients with decreased left ventricular function of unknown etiology after complete clinical and hemodynamic investigation were studied. Serious efforts were made to establish differences between patients with and without histologically documented myocarditis regarding anamnestic, clinical, electrocardiographic, virologic, and complete hemodynamic data. Inspite of these efforts, no correlates of the histologic features of myocarditis were detected. Association of myocarditis with previous viral disease could not be substantiated by history or antiviral antibody titers. These data do not support the notion that patients with myocarditis represent a distinct subgroup in dilated cardiomyopathy.

Follow-up studies, including hemodynamic and biopsy studies in patients with myocarditis, demonstrated changes in histologic and hemodynamic parameters occurring independently of each other. This makes it unlikely that prognostic predictions can be based on histologic changes and causes difficulties in understanding how therapeutic decisions could be derived from these findings. The latter issue, hopefully, will be answered by the ongoing multicenter study concerning immunosuppressive therapy of myocarditis [6].

Although, based on recent hypotheses, myocardial energy metabolism is suspected to be specifically modified in myocarditis, the parameters measured were almost identical in patients with and without inflammatory infiltrates, indicating that the changes observed represent no specific metabolic alterations in myocarditis or dilated cardiomyopathy. In contrast, they probably reflect a nonspecific loss in myocardial metabolic function that accompanies a loss in contractile function, leaving it unproven whether metabolic alterations can also contribute to further functional impairment.

According to recent insights from experimental studies and from immunologic and molecular biology techniques, lymphocytic infiltrations and myocyte necrosis may reflect autoimmune mechanisms that are part of the disease process in some patients with dilated cardiomyopathy, possibly initiated by an earlier viral infection. Similar mechanisms, however, may be active in patients with dilated cardiomyopathy without inflammatory infiltrates [3]. Therefore other parameters characterizing viral etiology more specifically may in the future be useful to identify viral etiology and secondary immune mechanisms in dilated cardiomyopathy and to serve as a basis for the indication of immunosuppressive therapy.

Acknowledgements. Histological studies in the first 187 patients were done by Dr. E. G. J. Olsen, National Heart Hospital, Westmoreland Street, London W1M8BA, UK. We thank Prof. A. L. Shug, University of Wisconsin, USA, for many fruitful discussions and valuable advice regarding the carnitine measurements in the myocardium, and Prof. Hetzer, German Heart Institute, Berlin, for providing us with tissue samples from the explanted hearts. We gratefully acknowledge the qualified technical assistance of H. Hensel, D. Möske, M. Müller, S. Sasse, M. Wessel. This study was supported by Grant Re 662/1–1 from the Deutsche Forschungsgemeinschaft.

References

1. Kunkel B, Schneider M, Kober G, Bussmann WD, Hopf R, Kaltenbach M (1982) Morphological analysis and clinical implications. (Die Morphologie der Myocardbiopsie und ihre klinische Bedeutung) Z. Kardiol. 71: 787–794
2. Dec GW jr, Palacios IF, Fallon JT, Aretz HT, Mills J, Lee DC, Johnson RA (1985) Active Myocarditis in the spectrum of acute dilated Cardiomyopathies. The New England Journal of Medicine 312: 14, 885–890
3. Maisch B (1988) Myokarditis und dilatative Kardiomyopathie. Die internistische Welt 11: 48–63
4. Hosenpud JD (1988) Immunosuppressive therapy in myocarditis. In: "New concepts in viral heart disease", Schultheiss HP ed. Springer, Berlin Heidelberg New York
5. Schultheiss, HP (1987) The mitochondrium as an antigen in inflammatory heart disease. Eur Heart J., Suppl J: 203–210
6. Anderson JL, Fowles RE, Unverferth DV, Mason JW (1987) Immunosuppressive therapy of myocardial inflammatory disease. Eur Heart J., Suppl J: 263–266
7. Salvi A, Hrovatine E, Dreas L, Silvestri F, Camerini F (1987) Changes in histology and left ventricular ejection fraction during immunosuppressive treatment in active myocarditis. Eur Heart J., Suppl J: 267–270
8. Goodwin JF (1987) Myocarditis and perimyocarditis. Historical survey, epidemiology and clinical features. Eur Heart J., Suppl J: 7–10
9. Olsen EGJ (1985) The role of biopsy in the diagnosis of myocarditis. Herz 1: 21–26
10. Regitz V, Olsen EGJ, Rudolph W (1985) Histologisch nachweisbare Myokarditis bei Patienten mit eingeschränkter linksventrikulärer Funktion. Herz 1: 27–35
11. Billingham ME (1987) Acute myocarditis: a diagnostic dilemma. British Heart J 58: 6–8
12. Eggers Th Viral heart disease – views from a virologist. European Heart Journal, Suppl J, 8:
13. Keren P, Billingham ME, Weintraub D, Stinson RB, Popp RL (1985) Mildly dilated congestive cardiomyopathy. Circulation 72: 2, 302–309
14. Quigley PJ, Richardson PJ, Meany BT, Olsen EGJ, Monaghan MJ, Jackson G, Jewitt DE (1987) Long-term follow-up of acute myocarditis. Correlation of ventricular function and outcome. European Heart Journal, Suppl J, 8: 39–42
15. Marboe CC, Schwartz RA, Adelson D, Escala EL, Weiss MB, Fenoglio JJ jr (1987) Active myocarditis with fibrosis. Poor correlation of ejection fraction with histology of follow-up biopsy. European Heart Journal, Suppl J, 8: 43–45
16. Regitz V, Müller M, Schüler S, Yankah CA, Hetzer R, Shug AI, Fleck E (1987) Carnitinstoffwechsel – Veränderungen im Endstadium der dilatativen Kardiomyopathie und der ischämischen Herzmuskelerkrankung. Z. Kardiol. 76, Suppl. 5: 1–8
17. Regitz V, Sasse S, Schüler S, Yankah AC, Hetzer R, Fleck E Variabilität biochemischer Bestimmungen im menschlichen Myocard – Voraussetzung für Stoffwechseluntersuchungen in Endomyocardbiopsien (1988). Z. Kardiol. 77:56 (P190)
18. Pierpont GK, Francis GS, DeMaster EG, Levine TB, Bolman RM, Cohn JN (1988) Elevated left ventricular myocardial dopamine in preterminal idiopathic dilated cardiomyopathy. Amer. J. Cardiol. 52: 1033
19. Regitz V, Shug AI, Schüler S, Yankah AC, Hetzer R, Fleck E (1988) Herzinsuffizienz bei dilatativer Kardiomyopathie und koronarer Herzkrankheit. Dtsch. med. Wschr. 113: 781–786
20. Bristow MR, Kantrowitz N, Ginsburg R, Fowler MB (1985) Beta-Adrenergic function in heart muscle disease and heart failure. J. molec. cell. Cardiol. 17, 2: 42
21. Limas CJ, Limas C (1988) Autoantibodies against cardiac Beta-adrenoceptors in idiopathic dilated cardiomyopathy, in HP Schultheiss, ed, New concepts in viral heart disease, Springer Verlag, in press

Coxsackievirus Cardiomyopathy in Animal Models, Including the Augmenting Effect of Exercise During the Acute Phase

M. P. REYES, A. M. LERNER, and R. KHATIB

Wayne State University School of Medicine, Department of Internal Medicine,
4707 St. Antoine Blvd., Detroit, MI 48201, USA

In series of outbred and inbred mice we have studied coxsackievirus (CV) infections of the heart, both as acute diseases and in end point studies over time extending for 1 year. The disease may be benign (e.g., involving the interstitium of the heart including coronary arteries, veins, lymphatics, and pericardium) or virulent (involving muscle fibers). In benign disease, healing is complete, and myonecrosis and fibrosis are minimal. Virulent disease is destructive of muscle fibers with replacement by fibrous tissue. Recovery in the latter case is incomplete with residual scarring; chronic primary myocardial disease results [1].

The disease produced is CV type specific, HLA specific, and age dependent in the host. In an inbred model of BALB/c mice we produced benign pericardial disease predominantly of the right ventricle with CV prototype B3 (CVB3). Healing left only a pericardial scar. In outbred Swiss (ICR) mice CVA_9 produced benign interstitial disease. Mature adult mice were most susceptible [2].

Specific Models of Heart Muscle Disease

We have developed specific models of acute and chronic heart disease in our laboratory.

Acute Myocarditis

CVB3 is the most cardiotropic to weanling mice, whereas CVB4 causes severe necrotizing myocarditis in suckling mice [6].

CVB1 and CVB4 Murine Models

Suckling Mice (48 h Old). Intraperitoneal inoculation of CVB1 and CVB4 into Swiss Webster mice induces focal transmural necrotizing myocarditis [3]. Left ventricular aneurysms are found in surviving animals (Fig. 1).

Viremias appear on the 2nd day of infection and persist for 24 h. Mean peak titers of virus in blood occur on days 2 or 3 (CVB1, $10^{2.8}$ TCD_{50}; CVB4, $10^{3.1}$ TCD_{50}).

H.-P. Schultheiß (Ed.)
New Concepts in Viral Heart Disease

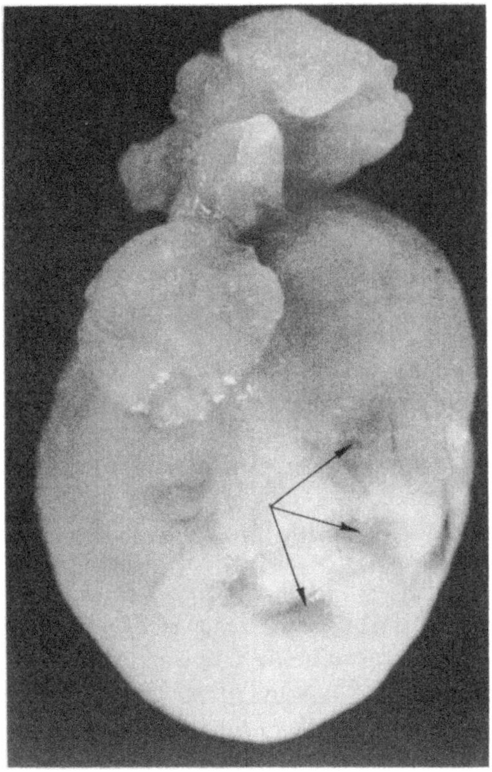

Fig. 1. The heart of a 19-day-old mouse on day 17 after an intraperitoneal inoculation of 10^4 TCD_{50} CVB1. The dilatation of the freshly autopsied specimen and dimpling *(arrows)* of the left ventricular wall due to the aneurysm after fixation is visible. (From El Khatib MR, Chason JI, Lerner AM (1979) Ventricular aneurysms complicating coxsackievirus B types 1 and 4 murine myocarditis. Circulation 59: 412–416)

Virus is first isolated from the myocardium on day 2 and reaches peak titers (mean peak CVB_1, $10^{2.2}$ TCD_{50}) on day 3 (Table 1).

Myocarditis occurs in 65% of mice infected with CVB1 and 78% of mice infected with CVB4. Histopathologic changes characterized by swollen foci of myocardial fibers and mononuclear infiltrates appear 2–3 days after infection. By days 5–7, necrotic fibers are calcified and fibroblasts appear in involved areas. By day 14, mononuclear infiltrates are sparse.

Ventricular aneurysms occur in 14% of mice. Microscopic findings consist of severe necrosis of myocardial fibers, mononuclear cell infiltration and early fibrosis [4].

Weanling Mice (14 Days Old). In contrast to suckling mice, myocarditis occurs only in 30% of CVB1-inoculated and 17.9% of CVB4-inoculated weanlings. Pathologic changes are also less severe.

Mice (3–5-Months Old) Myocarditis does not occur in this age group. Virus is not isolated from either blood or myocardium. No animals die after intraperitoneal infection.

Table 1. Clinical, virologic, and histopathologie findings in Swiss-Webster mice infected with group B coxsackieviruses types 1–4

Virus type	Inoculum (TCD_{50})	Age of mice at inoculation (days)	Viremia (Peak day)	Virus in myocardium (Peak day)	Histopathology No. with myocarditis/ No. examined (n)	(%)	Day of maximal pathology	Mortality rate (%)
1	10^3	2	2	3	63/96	65.6	3–5	46
		14	ND		3/10	30	ND	
		> 30	ND		0/16		None	
4	10^3	2	2	2	48/61	78.7	2–3	23
		14	2	2	7/39	17.9	7	
		> 30	ND	ND	0/36		None	
2	10^3	2	2–3	2–3	3/30	10	ND	23
		14	2	2	20/32	62.5	5–7	12
		> 30			0/24		None	
3	10^3	2	2–3	2	1/ 9	11.1	ND	43
		14	2	3–5	34/40	85	5–7	6
		> 30			11/13	84.6	ND	

ND, not determined

Pathogenesis

The mechanisms by which CVs induce cardiac injury are incompletely understood. Pathologic changes in the heart are synchronous with maximal cardiac titers of CVB1 and CVB4 viruses. Thus the pathogenesis of myofiber necrosis in these animal models may be the direct result of viral cytopathogenic effects.

CVB2 and CVB3 Murine Models

Weanling Mice (14 Days Old) In contrast to suckling mice, myocarditis is characteristic in weanling mice infected with CVB2 (62.5%) and CVB3 (85%). A virulent, diffuse, myonecrotic inflammatory lesion involving 25%–50% of the myocardium occurs (Fig.2). Deaths occur in 12% of CVB2-inoculated mice and 6% of CVB3-inoculated mice compared to none in CVB1- and CVB4-inoculated mice.

Viremias appear on the 2nd day of infection. Virus is isolated from myocardium on day 2 for CBV2 with a mean peak titer of $10^{2.8}$/g, and on days 3–5 for CVB3 with a mean peak titer of $10^{2.3}$/g.

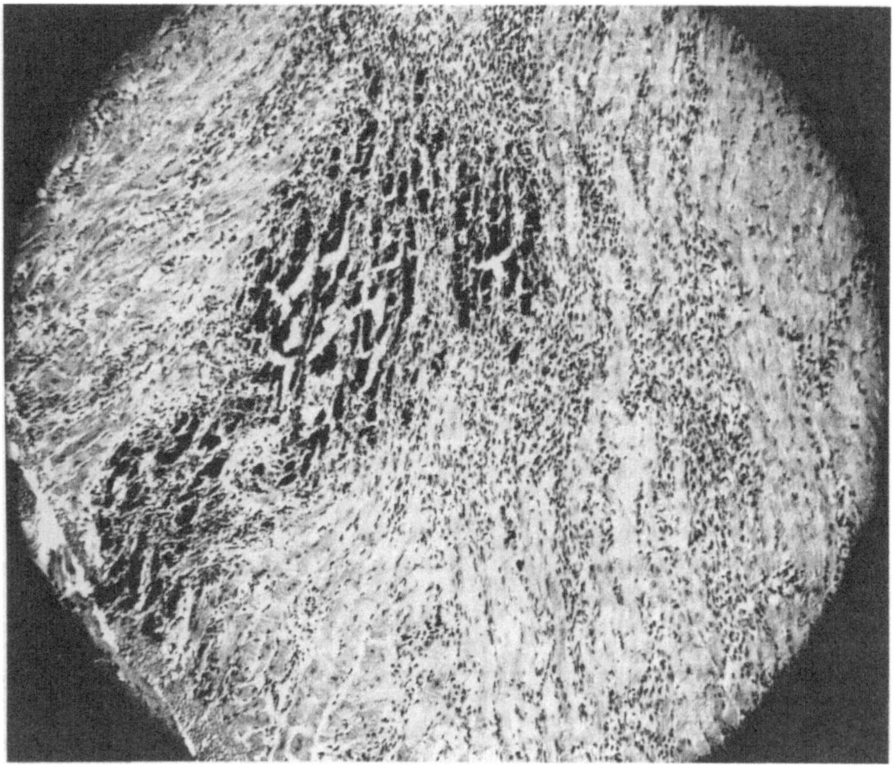

Fig. 2. Effect of swimming on pathologic findings in acute CVB3 myocardiopathy. The extensively involved myocardium of a 27-day-old mouse on the 13th day after infection is shown. (From Lerner AM, Wilson FM (1973) Virus myocardiopathy. Prog Med Viral 15: 63–91)

Mice (3–5 Months Old) Adult animals reveal inflammatory changes only after infection with CVB3 (85%).

Pathogenesis

In these models, maximal virus titers which occur on days 2–5 are asynchronous with maximal histopathologic changes which appear on days 5–7.

CVB2 and CVB3 may therefore be immunopathic myocardial viruses which induce myofiber necrosis via an immunopathologic mechanism. Direct experimental evidence for a T cell-mediated immunologic mechanism for CVB3 as well as identification of specific cytotoxic T lymphocyte subsets has now been provided by Woodruff, Huber and coworkers [5].

Chronic Heart Muscle Disease

We further followed the courses of the CVB4 and CVB3 murine models of necrotizing myocarditis in Swiss ICR mice over a 1-year period.

CVB4 Murine Model

Suckling mice that survive CVB4 infection develop myofiber replacement with fibrous tissue over a 12-month period [6]. A pattern of myocardial infarction with patent coronary arteries develops in 75% of animals. Transmural myocardial fibrosis with fibrous scar occurs in the left ventricle (41.3%), interventricular septum (56.5%), and right ventricle (37.7%). Ventricular aneurysms, which appear as early as day 5 after infection, are sequelae seen in the left ventricle (61.5%), right ventricle (19.2%), and interventricular septum (19.2%).

CVB3 Murine Model

In the murine model of CVB3 heart muscle disease, subjects forced to swim during the acute phase of infection and followed over a 15-month period develop changes similar to the findings in humans with congestive cardiomyopathy [7]. Myofiber loss is replaced with fibrous tissue and heavy deposits of calcium in the injured myocardium with atrial hypertrophy and thrombi therein. Pathologic findings are concentrated in the left ventricle, interventricular septum, and atrioventricular junction. These findings represent the changes of a dilated cardiomyopathy (Figs. 3–4). Intense scarring of the atrioventricular node with atrial thrombi suggests that arrhythmias and conduction disturbances occur in mice with chronic heart muscle disease. Residual pathologic changes are also seen in infected, nonexercised mice but they are much milder.

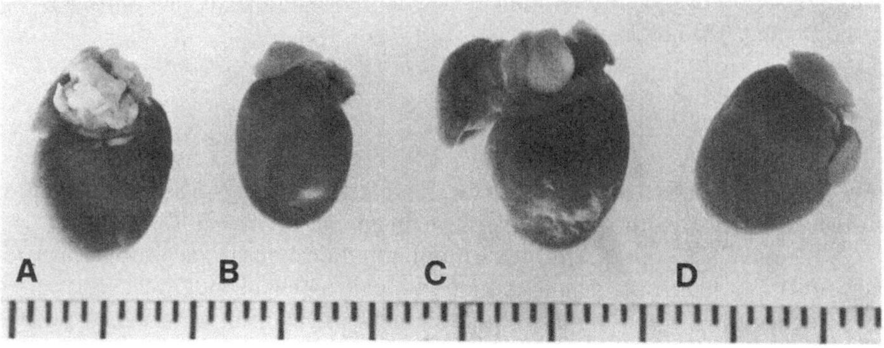

Fig. 3 A–D. Hearts from ICR Swiss mice with and without chronic cardiomyopathy (9 months' duration) caused by CVB3. **A** In group 2 (infected but not exercised), the heart is enlarged, and a gray fibrous scar is seen at the apex. **B** In group 1 (neither infected nor exercised), the heart appears normal. **C** In group 3 (infected and exercised) the heart is grossly enlarged, with obvious massive atrial and ventricular hypertrophy and many shiny, fibrous scars are diffusely present (seen here over the epicardium of the left ventricle). **D** In group 4 (not infected but exercised), the heart is enlarged but otherwise normal. Ratio of heart weight to body weight of the hearts shown: **A** group 2, 184.7 mg/43.5 g; **B** group 1, 137.0 mg/36.9 g; **C** group 3, 396.1 mg/49.9 g; **D** group 4, 187.0 mg/45.5 g. *Scale* is in millimeters. (From Reyes MP, Ho KL, Smith FE, Lerner AM (1981) A mouse model of dilated-type cardiomyopathy due to coxsackievirus B3. J Infect Dis 155: 232–236)

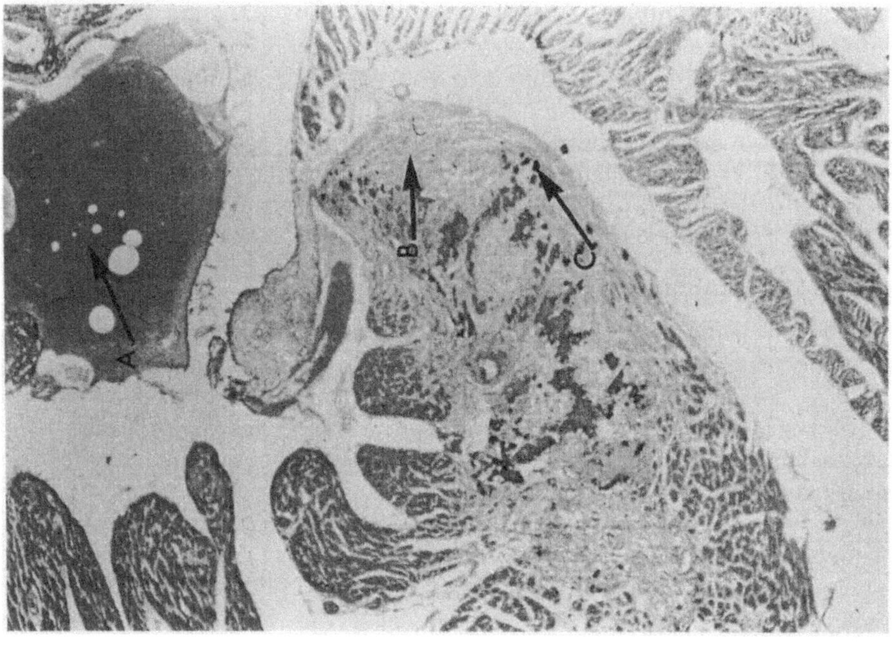

Fig. 4. Effect of swimming on pathologic findings in chronic CVB3 cardiomyopathy. The extensive scarring *(A)*, fibrous tissue *(B)*, and calcium deposition *(C)* involving the myocardium and atrioventricular septum are shown. (From Reyes MP, Lerner AM (1985) Coxsackievirus myocarditis with special reference to acute and chronic effects. Prog Cardiovaskular Dis XXVII: 373–394)

CVB3 Murine Model and Exercise

Striking augmentation in the virulence of CVB3 myocarditis by swimming during the acute phase of infection is seen [8]. Quantitative titers of virus increase 500 times in the blood on days 3–6 after infection, and in the heart on days 6–9 after infection. Mortality rate increases to 50% and the myocardium is converted into a necrotic, calcified mass. Interferon titers increase in the blood and heart; neutralizing antibody to CVB3 appears on day 3 and in unchanged amounts [9].

Absolute counts of mononuclear cells increase on days 1 and 2 after infection. Total WBC counts decrease on days 2 and 3. On days 2–4, mononuclear cells become CVB3 positive. Mononuclear cells in the thymus and spleen contain virus as well [10].

Erythrocytes and polymorphonuclear leukocytes remain virus free through day 6 after infection.

We then followed thymocyte circulation in infected and normal mice to explore further the mechanisms of the virulence-enhancing effects of exercise. Significant decreases in mean thymic indes occur on days 6 through 9 in healthy animals conditioned by swimming. At autopsy, thymuses are small, and their cortex is thin, but the density and microscopic appearance of retained thymocytes is normal [11]. Thymuses from infected and exercised mice are even smaller (Fig. 5). Forced swimming of healthy, uninfected mice may induce enhanced release of T lymphocytes from the thymus.

We also assayed levels of cyclic nucleotides, 3′, 5′-adenosine monophosphate (cAMP) and 3′, 5′-guanosine monophosphate (cGMP), and norepinephrine in isolated thymocytes from these mice. Norepinephrine and cGMP levels were increased in the thymocytes of infected and exercised subjects on day 9 after infection [12]. Elevations in intracellular thymocyte cGMP augment the effector functions of these cells.

During infections, sensitized T and B lymphocytes migrate toward the sites of infection. Migration of antigen-directed, thymus-derived lymphocytes was followed using sensitized and nonsensitized Cr-labeled T lymphocytes at days 5.5, 6.0, 7.0, and 9.0 Cardiac xeroradiographs were used to assess T cell-dependent in situ calcification. Significant rerouting of sensitized Cr-labeled T cells to the heart occurred on day 9 in CVB3-infected but not exercised subjects. Myocardial calcifi-

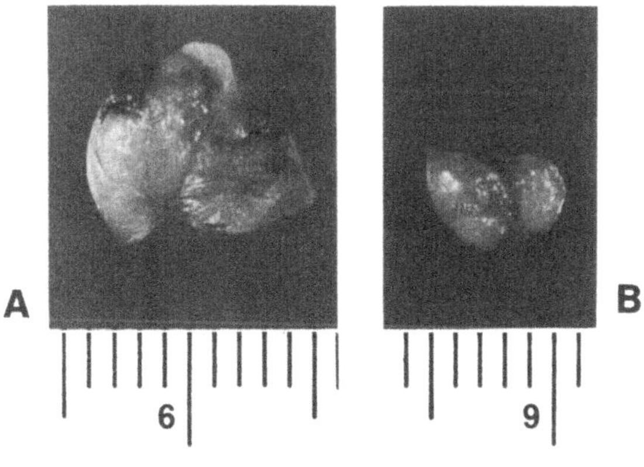

Fig. 5 A–B. Contraction of the thymus in 20-day-old albino ICR Swiss mice after a program of vigorous swimming. *Scale* is in millimeters. (From Reyes MP, Lerner AM (1987) Animal models of coxsackievirus myocarditis. In: Kawai C, Abelman WH (eds) Pathogenesis of myocarditis and cardiomyopathy: recent expermental and clinical studies, vol 1. University of Tokyo Press, Tokyo)

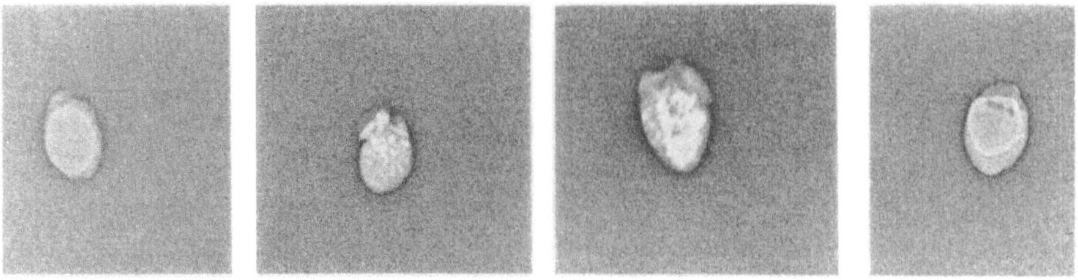

Fig. 6 A–D. Xeroradiograms of representative hearts from each of the four groups of 19-day-old mice given intravenous infusions of sensitized ^{51}Cr-labeled T lymphocytes on the 5th day after an intraperitoneal inoculation of 0.05 ml CVB3 containing 10^4 TCD_{50}/0.1 ml are shown. Autopsies were done 96 h (day 9) later. **A, D** Mice in group 1 (neither infected nor exercised) and group 4 (not infected but exercised) show no calcium deposition within the myocardium. The pools of intracardiac blood are accentuated as a rim lining the ventricular cavity. **B** Mice in group 2 (infected but not exercised) transfused with CVB3-sensitized, Cr-labeled T lymphocytes show only occasional minimal spotty calcium deposits on day 9. **C** Animals in group 3 (infected and exercised) given sensitized T cells show diffuse dense calcification on day 9. Hearts in this subcohort are also grossly enlarged. A dilated-type cardiomyopathy is present. (From Reyes MP, Smith FE, Lerner Am (1984) An enterovirus-induced murine model of an acute dilated-type cardiomyopathy. Intervirology 22: 146–155)

cation accompanied this increased T cell migration. In mice infected with CVB3 and concomitantly forced to swim, sensitized Cr-labeled T cells redirected to the heart increased several times [13]. Myocardial calcification was much more severe at xeroradiograph and it was not accompanied by cardiac dilatation (Fig. 6).

These experiments provide further insights toward understanding the pathogenesis of CVB3 murine myocarditis and the virulence-enhancing effects of exercise. At least three factors may contribute to this exercise-induced augmentation in virulence:

a) an increased presence of a CVB3 induced cardiac antigen;
b) increased T cell traffic, redirected to the heart; and
c) delay in turning off cell-mediated immune response reflected by an increased T cell cGMP on day 9 after infection [13].

CVB3 Murine Model: Anti-inflammatory Agents and Beta-Blockers

Anti-inflammatory agents (prostaglandin inhibitors) and beta-blockers also augment myocardial damage and increase mortality by apparently different mechanisms. Intraperitoneal administration of sodium salicylate (0.01 mg/g weight daily) and indomethacin (0.002 mg/g weight daily) during the first 10 days following CVB3-infection of 3-week old CD1 mice resulted in higher mortality (2/33 or 6%, 4/15 or 26%) and virus titers respectively by log 1 on day 3. Interferon levels were three fold lower in treated animals, whereas neutralizing antibody titers were comparable. Histopathologic changes were more extensive in the treated groups [14].

Beta-blockers show promise in the management of cardiomyopathies; however, their role in acute myocarditis is unknown. Metoprolol (MP) was administered i.p. at a dose of 0.031 mg/g at 07.00 h and 17.00 h and 0.0065 mg/g at 24.00 h daily for 10 days following CVB3 infection of CD1 and Swiss Webster mice. MP administration resulted in a significant increase in mortality (60% vs 0%), increased viral replication on day 10 of infection (50-fold) and more extensive and severe cardiac damage on days 17 and 30 after infection. The mechanism of this effect is not known.

Conclusion

Murine models have enabled investigators to understand better the pathogenesis and the acute and chronic effects of CV myocarditis. Future studies in these animal models should include development of newer and better methods for diagnosis as well as search for potential modalities of treatment.

References

1. Lerner AM, Wilson FM, Reyes MP (1975) Enteroviruses and the heart (with special emphasis on the probable role of coxsackieviruses, group B, types 1–5). II. Observations in humans. Mod Conc Cardiovasc Dis 44: 11–15
2. Lerner AM (1965) An experimental approach to virus myocarditis. Prog Med Virol 7: 97–115
3. Khatib R, Chason JL, Silberberg B, Lerner AM (1980) Age dependent pathogenicity of group B coxsackievirus in Swiss-Webster mice: infectivity for myocardium and pancreas. J Infect Dis 141: 394–403
4. El-Khatib MR, Chason JL, Lerner AM (1979) Ventricular aneurysms complicating coxsackievirus B types 1 and 4 murine myocarditis. Circulation 59: 412–416
5. Woodruff JF, Huber SA, Job LP (1980) Lysis of infected myofibers by coxsackieviruses B3 immune T lymphocytes. Am J Pathol 98: 681–694
6. Khatib R, Chason JL, Lerner AM (1982) A mouse model of transmural myocardial necrosis due to coxsackievirus B4: observations over twelve months. Intervirology 18: 197–202
7. Reyes MP, Ho KL, Smith F, Lerner AM (1981) A mouse model of dilated-type cardiomyopathy due to coxsackievirus B3. J Infect Dis 144: 232–236
8. Gatmaitan BG, Chason JL, Lerner AM (1970) Augmentation of the virulence of murine coxsackievirus B3 myocardiopathy by exercise. J Exp Med 131: 1121–1136
9. Reyes MP, Lerner AM (1976) Interferon and neutralizing antibody in sera of exercised mice with coxsackievirus B3 myocarditis. Proc Soc Exp Biol Med 151: 333–338
10. Gomez MP, Reyes MP, Smith F, Lerner, AM (1980) Coxsackievirus B3 positive mononuclear leukocytes in peripheral blood of Swiss and athymic mice during forced swimming. Proc Soc Exp Biol Med 165: 107–113
11. Reyes MP, Lerner AM, Ho KL (1981) Diminution in the size of the thymus in mice during forced swimming. J Infect Dis 143: 292
12. Reyes MP, Thomas JA, Ho KL, Smith F, Lerner AM (1982) Elevated thymocyte norepinephrine and cyclic guanosine 3′, 5′-monophosphate in sensitized T-lymphocytes from exercised mice with coxsackievirus B3 myocarditis. Biochem Biophys Res Commun 109: 704–708
13. Reyes, MP, Smith, FE, Lerner, AM (1982) An enterovirus-induced murine model of an acute dilated-type cardiomyopathy. Intervirology 18: 197–202
14. Rezkalla S, Khatib G, Khatib R (1986) Coxsackievirus B3 murine myocarditis: deleterious effects of nonsteroidal antiinflammatory agents. J Lab Clin Med 107: 393–395

II. Concepts in Viral Pathogenesis

Replication, Molecular Biology, and Pathogenesis of Coxsackieviruses

N. K. CHATTERJEE

New York State Department of Health, Wadsworth Center for Laboratories and Research, Empire State Plaza, Albany, NY 12201, USA

Introduction

The family Picornaviridae contains four genera: enteroviruses, cardioviruses, rhinoviruses, and aphthoviruses [26]. Enteroviruses that infect humans are polio, Coxsackie (CV) group A (CVA) and group B (CVB), echo, hepatitis A, and other enteroviruses. CV infections have been associated with a variety of human diseases, ranging from common cold to myocarditis, pericarditis, and juvenile onset (type I, insulin-dependent) diabetes.

All enteroviruses are acid stable; they inhabit the alimentary tract, and most of them except hepatitis A virus can infect the central nervous system. They multiply primarily in the pharynx and small intestine. *In vitro,* these viruses have been grown in egg yolk sacs; in monkey, hamster, or pig kidney cell cultures; and in other tissue culture cells. CVA are more difficult to grow than CVB. The RD cell line, derived from human rhabdomyosarcoma, supports growth of several CVA, but not of CVB which replicate poorly or not at all [30]. Different members of the genus possess similar physicochemical properties, such as structure, buoyant density in CsCl, molecular weight, and protein and RNA composition (for reviews see [3, 27, 36]).

Replication

The mechanism of replication of enteroviruses is very similar and can be subdivided into the following steps:

Cellular Entry, Uncoating, and RNA Release. Studies by several investigators [13, 37] have significantly contributed in this field. Cellular entry of the virus is via receptor-mediated endocytosis. Virus binding to cellular receptor is mediated by a protein called "virion attachment protein". Capsid protein VP2 of CVB3, which induces neutralizing antibodies, may function as the virion attachment protein [1]. Uncoating of the virus occurs in the cytoplasmic vesicles and/or lysosomes.

Viral Protein and RNA Synthesis. Protein synthesis occurs at the rough endoplasmic reticulum after the release of a protein called "VPg" from the genomic RNA. RNA synthesis occurs at the smooth endoplasmic reticulum in a structure called "replication complex" using enzymes poly(U) polymerase and viral replicase.

H.-P. Schultheiß (Ed.)
New Concepts in Viral Heart Disease

Viral Morphogenesis. The process of assembly of plus-stranded RNA with capsid proteins to produce progeny virus identical to the parent virions, occurs at the cytoplasmic membranes.
Release of Virions. The process of release of newly synthesized virions is poorly understood. A tight binding of a population of particles to cytoplasmic membranes has been demonstrated in several CVB [4, 9]. This population is separate from the virion population with a buoyant density of 1.34 that is usually released in the culture medium following a lytic infection.

Molecular Biology

Virions and Membrane-Bound Virions

Extraction of freshly harvested tissue culture cells infected with CVB2, CVB4, CVB5, and CVB6 with the detergent Nonidet P-40 releases, in addition to virions, a population of virus particles, which we called "membrane-bound virions" (MBV) [4, 9]. MBVs represent a separate and distinct population, have a pronounced polygonal appearance, and are slightly smaller than virions. The particles have a lower buoyant density (about 1.30) than virions (1.34) and sediment more slowly (at about 107S) compared to virions (at about 150S). A lower A260:280 ratio for MBVs compared to virions (1.21 versus 1.56) suggests the presence of relatively more protein than RNA in the former.

MBVs are 18 to 100 times lower in infectivity than virions. At about 24 h post infection in CVB5-infected cultures, the infectivity of the MBVs range from 7.6 x 10^7 to 2.2 x 10^9 total plaque-forming units, as compared with 13.5 x 10^8 to 2.3 x 10^{11} units for the virions. Cells infected with highly purified MBVs produce both types of progeny, virions and MBV. In coinfected cultures, the yield of progeny is lower than in cells infected with virions alone, suggesting interference by MBVs. No interconversion of the two types of particles has been observed in our studies. Particles of dissimilar densities, infectivities, and morphologies were described in swine vesicular disease virus and CVB5 [2, 16],rhinovirus 14 [14] and rhinovirus 2 [22].

Capsid Proteins

Like other enteroviruses , CVBs contain four capsid proteins, VPl-4, of molecular weights ranging from 37000 to 5000 [7]. MBV appears to contain only three capsid proteins plus at least seven additional proteins with apparent molecular weights of 92000 to 21000 [9]. Some of these additional proteins may be of host and/or viral origin. CVA13 was reported to contain five capsid proteins VP1, VP2a, VP2b, VP3, and VP4 [12]. Cleavage of VP2 precursor at two sites was adduced to explain the presence of both VP2a and VP2b.

Synthesis and Processing of Capsid Proteins

The exact pathway of processing of CV capsid proteins has not been determined. However, studies using poliovirus show that a polyprotein of molecular weight 247000 undergoes extensive processing to generate the capsid proteins and a series of noncapsid proteins (see [28]). Many of these intermediate proteins of molecular weights ranging from >97000 to about 7300 were detected in poliovirus-infected cells. Large protein precursors of over 150000, many intermediate cleavage products, and virus capsid proteins have been detected in several CVB-infected cells [7, 18, 23].

Protein Kinase

Alkali-dissociated CVB4 releases a cyclic nucleotide-independent protein kinase enzyme activity that phosphorylates several proteins of virions and MBV, and exogenous phosphate acceptor proteins, such as protamine sulfate [5]. Nearly 20-fold more enzyme activity was detected in MBV than in virions of the virus. The enzyme activity appears to be located internally in the virus; it may be host cell coded. The enzyme appears to be lacking in a strain of the virus that induces diabetes in susceptible mice.

Viral RNA and the Genome

The genomic RNA of CVB is a single-stranded molecule which is polyadenylated at the 3′ end [8]. The estimated molecular weight of CVB5 RNA is 2.53×10^6, which contains information for 2480 amino acids and can code for proteins of a total molecular weight of about 273000. The RNA directs synthesis of virus-specific proteins very efficiently in cell-free extracts. Inhibition of viral RNA-directed protein synthesis in the presence of the inhibitor 7-methylguanosine-5′-monophosphate suggested that the 5′ end of the RNA is uncapped. Later studies [4] identified a small protein VPg covalently linked to the 5′ end of the RNA. VPg contains two proteins of different charge as was described in foot-and-mouth disease viral RNA [21]. Furthermore, MBV VPg appears to be considerably smaller than the 5300-dalton virion VPg [4].

The genomic RNA of virions and MBV of CVB5 [9], and CVB4 [4, 11] are indistinguishable by velocity sedimentation and migration in denaturing gels. Hybridization experiments show that the two RNAs are almost completely homologous. However, T_1 oligonucleotide fingerprint analysis shows minor differences in the two RNAs. These minor differences between virion RNA and MBV RNA could have originated due to mutations and small deletions, which is not surprising considering the high (10^{-4}) mutation frequency in CBV [29].

The entire nucleotide sequences of CVB3 [24], CVB1 [19], and CVB4 [20] have been determined. These viruses exhibit gene organization quite similar to other enteroviruses. The genome is about 7400 nucleotides long and encodes a polyprotein of 2183–2185 amino acids. Tracy et al. [32] sequenced the 5′ end of CVB3

RNA to nucleotide 3822 and identified the noncoding region extending to nucleotide 738.

Hybridization data of Young [35] and Tracy [31] show about 5% genomic homology among all enteroviruses, about 20% between groups, and 30%–50% within groups. CVB5 shares about equivalent homology with the other CVB; but CVB1, CVB2, and CVB3 have diverged in one direction from CVB5. CVB4 and CVB6 also diverged from CVB5 but perhaps earlier in time or at a greater rate.

Pathogenic Potential

To analyze the disease potential of these two types of particles we have employed a particular strain E2 of CVB4 which induces a diabetes syndrome in susceptible mice resembling insulin-dependent diabetes in humans. The strain E2 was obtained by plaque purification of a human isolate, Edwards [17], originally isolated from myocardial tissue of an infant dying of generalized CV infection, focal necrosis, and inflammation of the pancreas [33]. Both the Edwards isolate and the E2 strain produce abnormalities of sugar metabolism and hyperglycemia in mice [10, 34]. Pancreatic beta cells from these mice synthesize significantly less protein and insulin [10] due to a shortage of insulin mRNA [6]. Increased expression of a molecular weight 64000 islet cell antigen in the infected mice prior to the development of hyperglycemia strongly suggests an autoimmune basis for the disease [15].

The E2 strain also contains virions and MBV, which differ in buoyant density, infectivity, and protein composition [11]. T_1 fingerprint analysis suggests some heterogeneity in virion RNA and MBV RNA. MBVs are 10–70 times less infective in tissue culture, yet they induce significantly higher glucose intolerance than virions in susceptible mice. Hyperglycemic mice contain viral RNA in the beta cells but not infectious virus [6], suggesting a role of persistent infection in inducing diabetes. Persistent infection in tissue culture cells by CVB has been reported [25].

Acknowledgements. I thank Nancy Miller for her expert assistance during the preparation of this manuscript. This work was supported by a Public Health Service Grant from the National Institutes of Health.

References

1. Beatrice ST, Katze MG, Zajac BA, Crowell RL (1980) Induction of neutralizing antibodies by the coxsackievirus B3 virion polypeptide, VP2. Virology 104: 426–438
2. Brown F, Talbot P, Burrows R (1973) Antigenic differences between isolates of swine vesicular disease and their relationship to coxsackie B5 virus. Nature (London) 245: 315–316
3. Chatterjee NK (1988) Replication. In: Bendinelli M, Friedman H (eds) Coxsackieviruses: a general update. Plenum, New York, p. 35
4. Chatterjee NK, Nejman C (1985) Membrane-bound virions of coxsackievirus B4: cellular localization, analysis of the genomic RNA, genome-linked protein, and effect on host macromolecular synthesis. Arch Virol 84: 105–118

5. Chatterjee NK, Nejman C (1986) Protein kinase in nondiabetogenic coxsackievirus B4. J Med Virol 19: 353–365
6. Chatterjee NK, Nejman C (1988) Insulin mRNA content in pancreatic beta cells of coxsackievirus B4-induced diabetic mice. Mol Cell Endocrinol 55: 193–202
7. Chatterjee NK, Tuchowski C (1981) Comparison of capsid polypeptides of group B coxsackieviruses and polypeptide synthesis in infected cells. Arch Virol 70: 255–269
8. Chatterjee NK, Tuchowski C (1981) Translation of coxsackievirus B RNAs in a rabbit reticulocyte lysate: characterization of the genome RNA, reaction conditions for translation, and analysis of the products. Arch Viral 70: 271–283
9. Chatterjee NK, Samsonoff WA, Tuchowski C (1983) Isolation and characterization of a membrane-bound population of group B coxsackieviruses. J Virol 45: 832–841
10. Chatterjee NK, Haley TM, Nejman C (1985) Functional alterations in pancreatic beta cells as a factor in virus-induced hyperglycemia in mice. J Biol Chem 260: 12786–12791
11. Chatterjee NK, Nejman C, Gerling I (1988) Purification and characterization of a strain of coxsackievirus B4 of human origin that induces diabetes in mice. J Med Virol (in press)
12. Cords CE, James CG, McLaren LC (1975) Alteration of capsid proteins of coxsackievirus A 13 by low ionic concentrations. J Virol 15: 244–252
13. Crowell RL, Landau BJ (1983) Receptors in the initiation of picornavirus infections. In: Fraenkel-Conrat H, Wagner RR (eds) Comprehensive virology, vol 18. Plenum, New York, p 1
14. Gauntt CJ, Griffith MM, Sauck JR, Upson HR, Carlson EC (1975) Properties and origins of infectious rhinovirus type 14 particles of different buoyant densities. J Virol 16: 1265–1272
15. Gerling I, Nejman C, Chatterjee NK (1988) Coxsackievirus B4 infection in mice increases the expression of the 64K autoantigen and glucose sensitivity of the islets prior to the development of hyperglycemia. Diabetes (in press)
16. Harris TJR, Doel TR, Brown F (1977) Molecular aspects of the antigenic variation of swine vesicular disease and coxsackie B5 viruses. J Gen Virol 35: 299–315
17. Hartig PC, Madge GE, Webb SR (1983) Diversity within a human isolate of coxsackie B4: relationship to virus-induced diabetes. J Med Virol 11: 23–30
18. Holland JJ, Kiehn ED (1968) Specific cleavage of viral proteins as steps in the synthesis and maturation of enteroviruses. Proc Natl Acad Sci USA 60: 1015–1022
19. Iizuka N, Kuge S, Nomoto A (1987) Complete nucleotide sequence of the genome of coxsackievirus B1. Virology 156: 64–73
20. Jenkins O, Booth JD, Minor PD, Almond JW (1987) The complete nucleotide sequence of coxsackievirus B4 and its comparison to other members of the picornaviridae. J Gen Virol 68: 1835–1848
21. King AMQ, Sanger DV, Harris TJR, Brown F (1980) Heterogeneity of the genome-linked protein of foot-and-mouth disease virus. J Virol 34: 627–634
22. Korant BD, Lonberg-Holm K, Yin FH, Noble-Harvey J (1975) Fractionation of biologically active and inactive populations of human rhinovirus type 2. Virology 63: 384–394
23. Kiehn ED, Holland JJ (1970) Synthesis and cleavage of enterovirus polypeptides in mammalian cells. J Virol 5: 358–367
24. Lindberg AM, Stalhandske POK, Pettersson U (1987) Genome of coxsackievirus B3. Virology 156: 50–63
25. Matteucci D, Paglianti M, Giangregorio AM, Capobianchi MR, Dianzani F, Bendinelli M (1985) Group B coxsackieviruses readily establish persistent infection in human lymphoid cell lines. J Virol 56: 651–656
26. Matthews REF (1982) Classification and nomenclature of viruses. Karger, Basel, p 129
27. Melnick JL (1985) Enteroviruses: polioviruses, coxsackieviruses, echoviruses, and newer enteroviruses. In: Fields BN (ed) Virology. Raven, New York, p 739
28. Pallansch MA, Kew OM, Semler BL, Omilianowski DR, Anderson CW, Wimmer E, Rueckert RR (1984) Protein processing map of poliovirus. J Virol 49: 873–880
29. Prabhakar BS, Haspel MV, McClintock PR, Notkins AL (1982) High frequency of antigenic variants among naturally occurring human coxsackie B4 virus isolates identified by monoclonal antibodies. Nature (London) 330: 374–376

30. Schmidt NJ, Ho HH, Lennette EH (1975) Propagation and isolation of group A coxsackieviruses in RD cells. J Clin Microbiol 2: 183–185
31. Tracy S (1985) Comparison of genomic homologies in the coxsackievirus B group by use of cDNA:RNA dot-blot hybridization. J Clin Microbiol 21: 371–374
32. Tracy S, Liu HL, Chapman NM (1985) Coxsackievirus B3: primary structure of the 5′ noncoding and capsid protein-coding regions of the genome. Virus Res 3: 263–270
33. Webb SR, Loria RM, Madge GE, Kibrick S (1976) Susceptibility of mice to group B coxsackie virus is influenced by the diabetic gene. J Exp Med 143: 1239–1248
34. Webb SR, Madge GE (1980) The role of host genetics in the pathogenesis of coxsackievirus infection in the pancreas of mice. J Infect Dis 141: 47–54
35. Young NA (1973) Polioviruses, coxsackieviruses, and echoviruses: comparison of the genomes by RNA hybridization. J Virol 11: 832–839
36. Zeichhardt H (1986) Enteroviruses. In: Specter S, Lancz GJ (eds) Clinical virology manual. Elsevier, New York, p 283
37. Zeichhardt H, Wetz K, Willingmann P, Habermehl KO (1985) Entry of poliovirus type 1 and mouse Elberfeld (ME) virus into HEp-2 cells: receptor-mediated endocytosis and endosomal or lysosomal uncoating. J Gen Virol 66: 483–492

A Murine Model for Coxsackievirus B3-Induced Acute Myocardial Necrosis for Study of Cellular Receptors as Determinants of Viral Tropism

R. L. CROWELL[1], S. D. FINKELSTEIN[2], K.-H. L. HSU[1], B. J. LANDAU[1], P. STALHANDSKE[1], and P. S. WHITTIER[1]

[1] Department of Microbiology and Immunology,
[2] Department of Pathology and Laboratory Medicine, Hahnemann University School of Medicine Broad and Vine, Philadelphia, PA 19102, USA

Receptor Hypotheses

A number of studies have contributed to the working hypothesis that specific cellular receptors, if present in sufficient quantity, serve as major determinants of both the host range and tissue tropism in the pathogenesis of human and animal picornavirus infections [14, 21, 28, 36, 42]. Though picornaviruses contain relatively small genomes, they produce a bewildering array of diseases. It is likely that one of the several factors contributing to disease diversity is due to the specific requirement of individual viruses for host cell receptors. For example, we have postulated [12a] that genetic changes in the virion attachment site could occur during virus replication in infected individuals. These changes could influence the course of the disease [9, 23, 51] by altering the virus receptor specificity. This would be of special importance if different receptors occurred on different tissues.

We also have hypothesized [12a] that receptors may be different (polymorphic) in genetically dissimilar individuals. If so, these differences may result in the selection and amplification of variant viruses and thereby cause a more or less severe disease than that produced by the parental virus in other individuals. Furthermore, we view cellular receptors as having a dual function, i.e., to bind virus and to facilitate disassembly of the capsid to release the viral genome [17]. In some cases virus may attach to specific receptors, but infection does not occur because of some limitation in the capacity of the receptors to eclipse virus infectivity [52]. Finally, recent results from our laboratory reveal that more than one type of specific receptor protein may exist on cell surfaces for binding a given virus and influence virus tropism [19, 30, 48]. Collectively, these several observations and hypotheses (Table 1) illustrate the complexity which can be encountered in investigation of cellular receptors for picornaviruses and the role of receptors as determinants of virus tropism.

Nevertheless, results of studies of other virus-host cell systems support the concept that receptors influence pathogenesis. For example, a strain of mice (SJL/J) resistant to mouse hepatitis virus (a coronavirus) is homozygous for an autosomal recessive mutant receptor gene and cannot attach virus, whereas, a normal mouse strain (BALB/c) which produces a specific receptor is susceptible to this virus infection [3]. Another example of receptor-mediated pathogenesis can be found for

H.-P. Schultheiß (Ed.)
New Concepts in Viral Heart Disease

Table 1. Hypotheses about cellular receptors for picornaviruses

1. Determinants of viral tropism in the pathogenesis of infection.
2. Receptor variation among individuals may account for selection of virus variants which may be more (or less) virulent than parental virus.
3. Receptors have dual functions: to attach virus and to initiate virus disassembly.
4. More than one type of specific receptor may serve to initiate virus infection.

reoviruses in a mouse model [54]. Reovirus T1 has a distinct tropism for ependymal cells of the brain, whereas reovirus T3 seeks out neurons. However, since these viruses share a receptor on mouse L cells [37], undoubtedly more than one receptor specificity accounts for the specific histotropisms of reoviruses in mice [8].

Despite continued progress in learning about cellular receptors for viruses in general [16] and picornaviruses in particular, little is known about their molecular structure, cellular function or role in viral pathogenesis. This chapter outlines some of our knowledge about receptors on cultured cells and presents some preliminary experimental results together with strategies for testing our several hypotheses about receptors for coxsackieviruses (CV) of group B (CVB) in a murine model.

Receptors on Cultured Cells for CVB

Receptor specificity for the different species of picornaviruses was first demonstrated by the binding competition between two virus serotypes which share the same receptor [11, 12, 18, 39, 47]. On HeLa cells, specific receptors were discovered for polioviruses, CVB, and for two separate families of human rhinoviruses, respectively. Studies from our laboratory have focused attention on the characterization of the HeLa cell receptor for the CVB. Our current knowledge of the properties of the receptor for the prototype CVB3 (Nancy strain) is summarized in Table 2. A receptor protein of approximately 50 kilodalton (kd) has been identified and purified from HeLa cells [40]. This protein is currently being sequenced (N-terminal amino acids), and oligonucleotide probes are being prepared to permit screening of a HeLa cell cDNA library for isolation of a receptor gene (collaboration with R. Colonno and J. Tomasinni, Merck, Sharp, and Dohme Research Laboratories, West Point, PA). Whether this glycoprotein [35] occurs in pentameric units to comprise a receptor site as suggested previously [13] remains to be determined. Additional studies are also needed to determine whether purified receptors reconstituted in liposomes possess the postulated dual functions of virus attachment and virus disassembly.

A second cellular receptor for a host range variant of CVB3, referred to as "CVB3-RD" following selection by serial passage in human rhabdomyosarcoma cells (RD), has been identified [48]. This receptor has been designated as "human receptor 2" (HR2) to distinguish it from the 50-kd receptor (HR1) described

Table 2. Properties of receptors from HeLa cells for CVB3 (Nancy)

	Reference
1. Chymotrypsin sensitive; trypsin resistant	61
2. Inactivated by periodate, B glucosidases, and mannosidase	35
3. Regeneration requires mRNA and protein synthesis	37 a
4. Under genetic control of cell	41
5. Integral plasma membrane protein approximately 50-kd monomere	40
6. Approximately 275-kd (Sepharose 4B) multimere	35
7. Approximately 10^5 sites/cell	12
8. Density on sucrose 1.06	35
9. Virus species specificity	18
10. Stable at pH 1, 2 °C, 10 min; 60 °C, 30 min	62
11. Monoclonal antibodies to receptor block	20
virus attachment	30

above. HR2 has been found on RD cells, which lack HR1, whereas HeLa cells possess both HR1 and HR2. The Buffalo green monkey kidney cell line possesses simian receptor one (SR1) which appears to be closely related to HR1 by assays with a receptor-specific monoclonal antibody, designated "RmcB" [30]. Another receptor-specific monoclonal antibody, designated "RmcA", has specificity for HR2 [20, 30]. These and other receptors for CVB which are found on cells of different species (mouse, rat) are listed in Table 3. Human erythrocytes also have a receptor (HR3) to which CVB3-RD and RmcA attach. Thus, more than one receptor for CVB may serve to initiate infection and each receptor will need to be characterized.

Table 3. Nomenclature for CVB cellular receptors defined by cells and probes[a]

Receptor	Origin		Probe		
Designation	Species	Cell	Virus	Rmc[d]	Other
HR1	Human	HeLa	CVB1–6	B	Ad2F[e]
HR2	Human	RD[b,c]	CVB3-RD	A	–
HR3	Human	Erythrocyte	CVB3-RD	A	–
MR1	Mouse	YAC-1	CVB1–6	–	Ad2F
RR1	Rat	L8	CVB3	–	–
SR1	Monkey	BGM	CVB1–6	B	Ad2F

[a] It is anticipated that additional receptors will be found that can be distinguished by use of specific probes
[b] Rhabdomyosarcoma cells
[c] HR2 are also found on HeLa cells
[d] Receptor-specific monoclonal antibody
[e] Adenovirus type 2 fiber protein (studies with K. Lonberg-Holm)

Table 4. Comparative number of binding sites of different cells for CVB3 and the monoclonal antibodies RmcA and RmcB. (From [30])

Ligand	Binding sites HeLa (*n*)	RD (*n*)	BGM (*n*)
CVB3	1.8×10^4	–[a]	2.7×10^3
CVB3-RD	5.6×10^5	3.7×10^3	3.3×10^3
RmcA	2.5×10^6	2.7×10^4	–
RmcB	5.0×10^4	–	3.6×10^4

[a] Less than 100 per cell as limit of detection, considered to be absent

Monoclonal antibodies have also been prepared against cellular receptors for polioviruses [44, 46] and rhinoviruses [10]. These reagents, which are very useful for probing the different cellular receptors, were first obtained by Campbell and Cords [6] for a receptor epitope from HeLa cells for some of the CVB. A different monoclonal antibody against another HeLa cell receptor epitope for CVB2, CVB4, and CVB6, but not reactive for CVB1, CVB3, and CVB5 has recently been described [26]. It is predicted that many additional epitope specificities, recognized by other monoclonal antibodies, will be identified if the search is continued.

The number of receptors per susceptible cell for most picornaviruses is relatively low, even though there may be as many as 10^5 receptor molecules per HeLa cell. This value is undoubtedly high for most normal cells in the intact animal. We have estimated the number of receptors for CVB3 and CVB3-RD on HeLa cells, RD cells, and BGM cells using radiolabeled virions and monoclonal antibodies (Table 4) [30]. The approximate number of specific binding sites per cell at ligand saturation was determined by extrapolation of the slope of the nonspecific binding to intercept the ordinate at zero input. The results of this type of assay are represented in Fig. 1, which depicts the titration of receptors for CVB3-RD on RD cells. It will be a challenge to conduct quantitative studies of receptors on cells in situ which have a low receptor density.

The relationship between patterns of pathogenesis and the quantitative distribution of specific virus receptors, however, must be examined in the whole animal since cultured cells may synthesize receptors which are not expressed in vivo [29]. In searching for receptors for CVB in the experimental mouse model, the effects of age and genetic strain of the animals will also be of critical interest. Molecular biological methods, in addition to the use of labeled receptor ligands, may permit more sensitive assays of receptors in mouse tissues. For example, receptor mRNA could be measured by hybridization with labeled DNA probes prepared from receptor genes. We soon hope to be able to apply some of the tools derived from molecular studies of the CV receptors on cultured cells to the ultimate question of how a virus causes a specific pattern of disease.

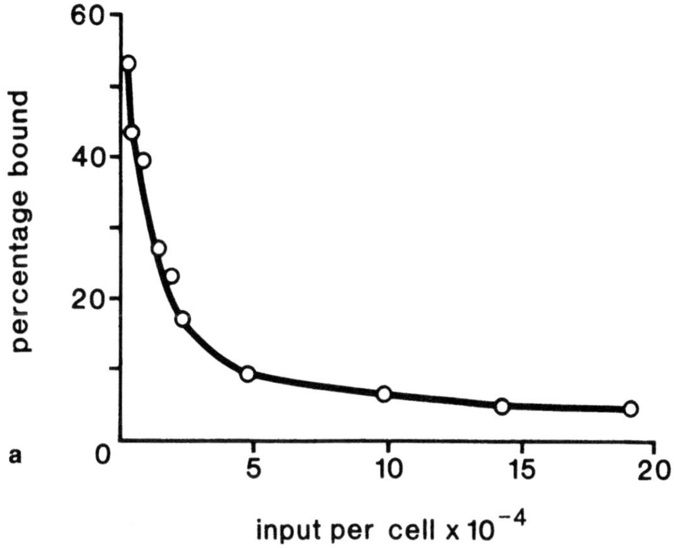

a

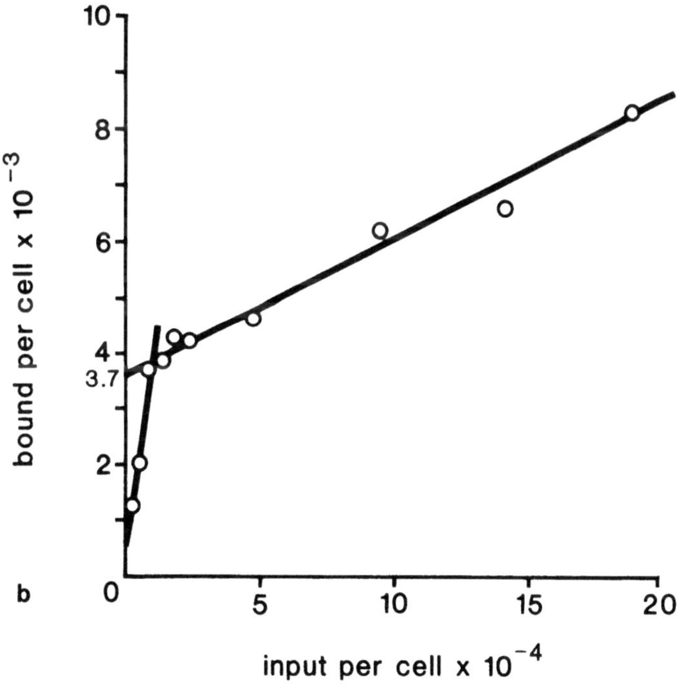

b

Fig. 1 a, b. Titration of receptors for CVB3-RD on RD cells. Dilutions of ^{35}S-CVB3-RD were incubated 90 min with 10^7 RD cells per milliliter at 24°C. The cells were washed twice and counted for radioactivity. **a)** The relationship of input virus multiplicity to percentage of cell-bound labeled virions. **b)** The results in **a** transformed to particles bound per cell input virus multiplicity. Approximately 3.7 x 10^3 virions were estimated to bind specifically per cell

CVB3-Induced Histopathology in Different Strains of Inbred Mice

We have compared the temporal sequence of tissue damage following the intraperitoneal inoculation of a myocarditic strain of CVB3 (Woodruff strain; CVB3-W) into 8-week-old mice of three inbred strains, BALB/C, SJL/J and C3H/HeJ. A strain of BALB/c mice has been shown to produce cytotoxic T lymphocytes which resulted in myocardial injury following CVB3 infection [27], although this reponse has not been uniformly detected [49]. The SJL/J mouse has a tendency to develop a diabetes-like disease following infection with CB4 or encephalomyocarditis virus [60], whereas the C3H/HeJ mouse has a partial defect in cell-mediated immunity with diminished macrophage stimulation to lipopolysacharide [58].

Although our initial objectives were to study the development of virus-induced heart disease in BALB/c mice, we found that these animals died following inoculation of only 5 x 10^4 plaque-forming units (PFU) of CVB3-W (Table 5). This observation prompted a histopathologic study of multiple tissues at intervals following CVB3-W infection. In general, three phases of disease were identified (Table 6). Phase I (pancreatic disease) occurred 2–3 days postinfection (p.i.) and was characterized by subtotal or total pancreatic acinar cell necrosis with sparing of islets. Phase II (hepatic disease) occurred 4–5 days p.i. and was marked by hepatocyte swelling and vacuolization leading to coagulative necrosis. We concluded that the BALB/c mice, inoculated with 5 x 10^4 PFU of virus, died from massive hepatocellular necrosis. If a lower amount of inoculum was used, the mice developed reversible changes in the liver and went on to Phase III (heart disease) which developed 7–10 days p.i. The heart disease was characterized by multifocal myocardial necrosis with calcification as shown in Fig. 2. Resolution of lesions occurred over a 28-day period. It is significant that there was no evidence of an inflammatory infiltrate in the early phase in the pancreas, the liver, or the heart; and the pathologic changes were associated with increasing titers of virus. Although peak virus titers preceded peak tissue damage by 1–3 days, only a mild inflammatory cell infiltrate composed of scattered lymphocytes and macrophages was observed secondary to coagulative necrosis. In each case inflammation was associated with removal of necrotic cellular debris caused by virus infection and not to immunopathologic mechanisms [25].

Table 5. Comparative susceptibility of BALB/c and C3H/HeJ mice to CVB3-W. (Modified from [19])

Mouse strain[a]	Virus inoculum	% Surviving at 14 days (%)
BALB/c	5 x 10^2	40
	5 x 10^4	0
C3H/HeJ	5 x 10^6	100

[a] Eight week-old males

Table 6. General pattern of disease following i. p. inoculation of CVB3-W into BALB/c mice

Disease phase	Primary target organ	Days of peak change	Description of pathology
I	Pancreas	2–3	Total acinar cell necrosis followed by cell drop-out Inflammation minimal after coagulative necrosis
II	Liver	4–5	Hepatocyte swelling, vacuolization and fatty change leading to coagulative necrosis Begins pericentrally
III	Heart	7–10	Multifocal myocardial necrosis with calcification Lesions resolve over time Inflammation follows viral-induced necrosis

[a] Leads to death if inoculum contains 5 x 10^4 PFU, whereas animals resolve liver disease and progress to heart disease with 5 x 10^2 PFU

The C3H/HeJ mice, infected with as much as 5 x 10^6 PFU of CVB3-W, neither died (Table 5), nor developed histologic evidence of liver disease (Table 7). However, these animals developed prominent multifocal myocardial necrosis with onset on day 7 p.i.. SJL/J mice developed only subliminal liver disease and some myocardial necrosis. However, a chronic inflammatory skeletal muscle disease was produced in these mice following infection with CVB3-WRD [53]. Thus, CVB infections of inbred mice provide a system in which different patterns of pathogenesis can be compared with possible differences in cellular receptors between strains of mice and between different organs.

Table 7. Comparative occurrence of pathologic changes in selected organs of mouse strains infected with CVB3–W

		Occurrence of pathologic changes[b]			
Mouse strain[a]	Inoculum (PFU)	Pancreas	Liver	Heart	Skeletal muscle
C3H/HeJ	5 x 10^6	+	–	+	–
SJL/J	5 x 10^6	+	+/–	+	+
BALB/c	5 x 10^2	+	+	+	–

[a] Eight-week-old males
[b] Fourteen days p. i.

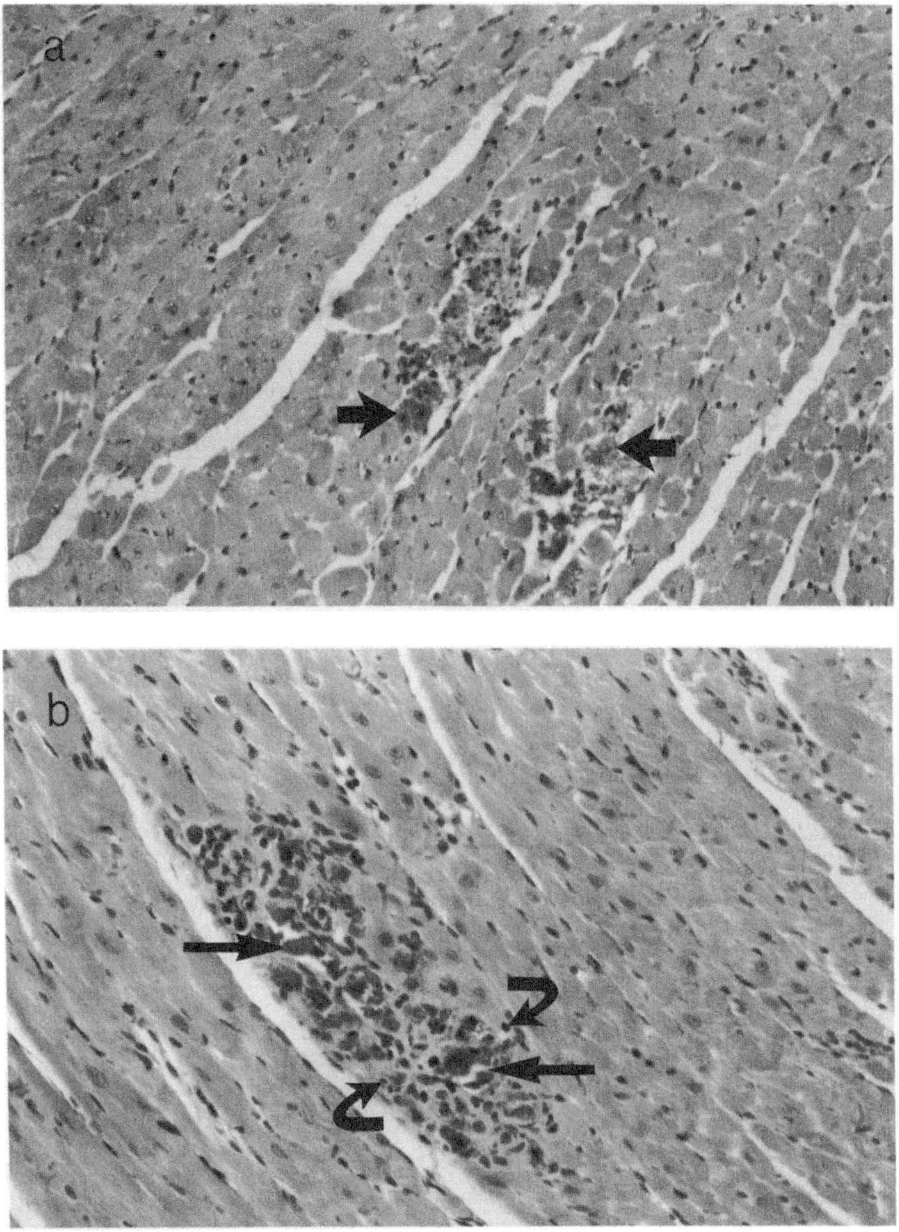

Fig. 2 a-d. Male C3H/HeJ mice 6–8 weeks old were infected i.p. with 5 x 10^6 CVB3-W. **a)** At day 7 p.i., heart shows focal areas of myocardial cell necrosis *(straight arrows)*. Remainder of heart is normal. **b)** At day 10 p.i., foci of necrotic myocardial cells show calcification *(straight arrows)*. Scattered lymphocytes have entered the lesion *(curved arrows)*. **c)** At day 14 p.i., lesions have attained maximal size and contain areas of dense calcification *(straight arrow)*. A mild to moderate infiltrate is seen *(curved arrows)*. Note the continual focal nature of the heart lesions which spare the remainder of the myocardium. **d)** At day 28 p.i., areas of focal necrosis with calcification *(straight arrows)* are undergoing resorption. Note reduced size of lesion. From [25] H & E, X 100

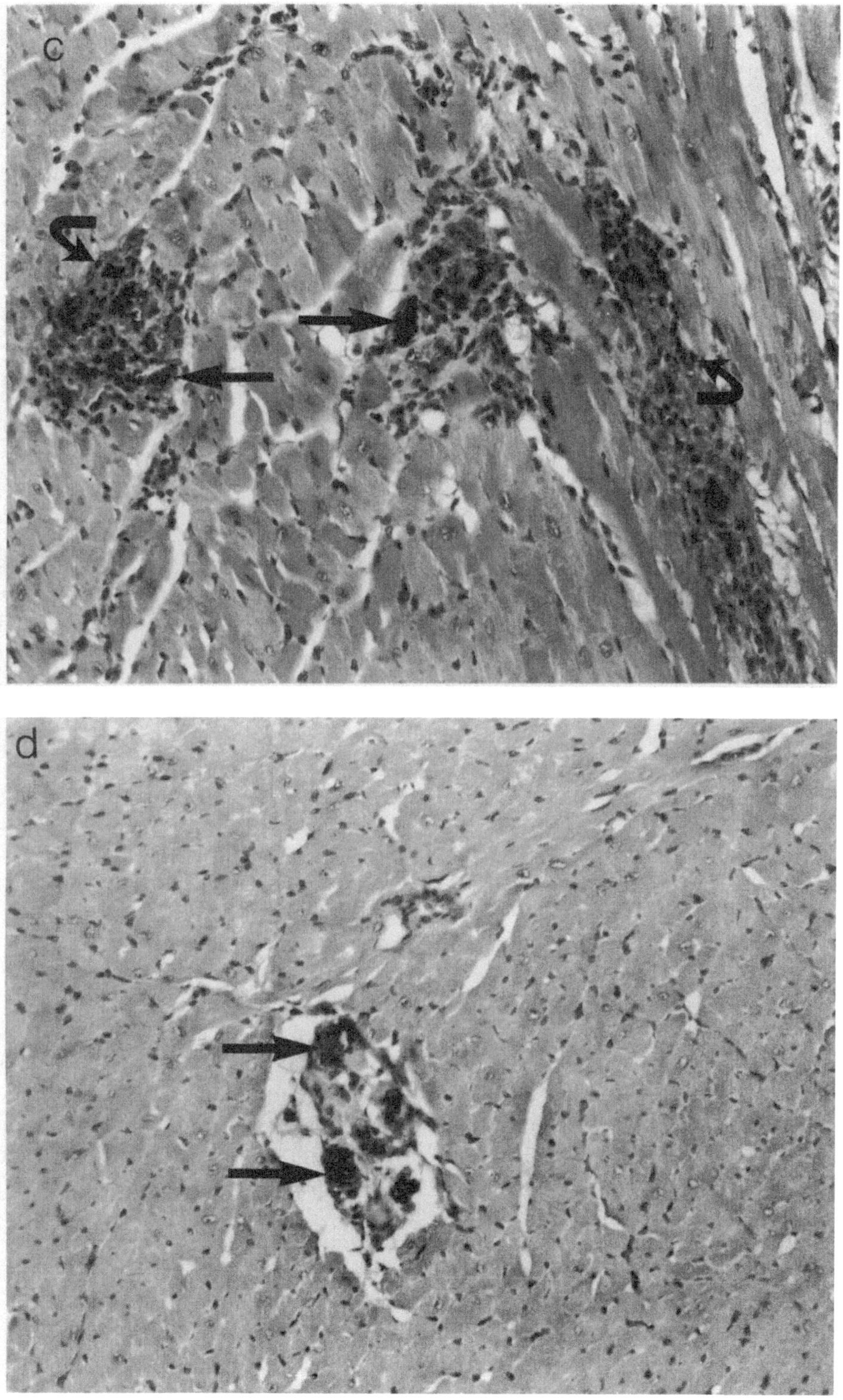
c
d

Strategies for Localization of Receptors, Receptor mRNA, and Virus in Mouse Tissues

It has been axiomatic to us that receptors contribute to the pattern of pathogenesis of the CVB. Yet there are presently only imprecise methods for measurement of the distribution of receptors in different organs and in different individuals or strains of experimental animals [36]. This deficiency is attributable to the low number of receptor molecules on the cell surface and to the fact that neither picornavirus virions, nor any virion substructure or polypeptide can be used to probe for the presence of receptors in situ [15, 50]. Earlier studies which attempted to quantitate receptors in tissue homogenates have been aided by use of methods to solubilize the receptor proteins prior to assay [34]. Currently we are developing molecular and monoclonal antibody probes that are better suited to determining if there is a correlation between receptor concentration and susceptibility to CVB infection with accompanying pathologic change.

Localization of Receptors

Receptors in detergent-solubilized extracts of different tissues can be measured by quantitative immunoblot assay using monoclonal and, as available, high titer polyclonal antibodies. Recently, we have also found that radiolabeled virions can be employed in dotblots and in viroblot assays (comparable to Western blot) [3, 45]. Solubilized receptor proteins can be adsorbed to nitrocellulose and detected with either radiolabeled antibody or virions. Bound radioactive probe is detected by radioautography and can be quantitated by densitometry [56] or by counting radioactivity. In addition, immunogold labeling techniques can be used to determine receptor density and distribution on cells by electron microscopy [55]. We seek to compare the presence and amount of receptors for CVB on different organs from BALB/c, C3H/HeJ, and SJL/J strains of mice which show different sensitivities to infection and differences in the ability of virus to produce liver necrosis (Table 7).

Localization of Receptor mRNA

There are two reasons to examine receptor mRNA as well as receptor protein for identifying the distribution of receptors in human and animal tissues. In some situations (e.g., when there is rapid turnover or down-regulation in infected animals or low numbers of receptor molecules) it may not be possible to demonstrate functional receptors by direct assay. More significantly, it may be possible to determine the distribution of structurally different but antigenically homologous receptor proteins by using specific probes for their mRNA [38].

Initially, we plan to use labeled copies of receptor specific cDNA for detection and quantitation of receptor mRNA. If it is found that there are different species of related mRNAs, probes with different specificities can be prepared from restriction fragments of their cDNA.

Localization of Virus

Viruses have been detected in infected tissues by plaque titration or by immunohistologic identification of viral antigen in sections. Viral RNA can also be detected by in situ hybridization [4, 5]. To evaluate fully the persistent state of infection, we will probe for both plus and minus strands of viral RNA [7]. Detection of viral RNA permits identification of latently or defectively infected cells which may not contain infectious particles [2, 22, 32]. We intend to quantitate viral nucleic acids in organs of infected mice by dot hybridization [31] and then to examine sections of major organs by both immunoperoxidase staining for viral antigens and by in situ hybridization using ^{35}S-labeled probes or peroxidase-labeled nucleic acid probes to identify viral RNA.

Summary

We hypothesize that specific cellular receptors serve as major determinants of CVB tropism in the pathogenesis of human and animal diseases. The CVB cause a wide spectrum of human diseases including: pleurodynia, meningitis, pancreatitis, myositis, hepatitis, pericarditis, and heart muscle disease [43]. They also produce a similar spectrum of diseases in inbred mice which provide useful animal models for laboratory investigation [1, 19, 24, 30a, 33, 57, 59]. An increasing amount of information about the nature of cellular receptors for the CVB has been obtained from studies of cultured cells; however, cultured cells may express receptors which are not expressed in vivo. The challenge remains to apply this information to determine receptor distribution, receptor abundance, and receptor specificities by in situ methods in human tissues and in animal models. Several strategies for the detection of receptors by monoclonal antibodies and molecular probes have been presented. The finding of more than one receptor species for prototype and variant viruses on different target cell types will make these determinations more complicated than previously anticipated.

Acknowledgements. Studies from our laboratory were supported by US Public Health Service Research Grant AI-03771 from the National Institute of Allergy and Infectious Diseases. The secretarial help of Mrs. Mary Ellen Bealor is gratefully acknowledged.

References

1. Bendinelli M, Friedman H (eds) (1988) Coxsackieviruses: a general update. Plenum, New York, pp 1–430
2. Bowles NE, Richardson PJ, Olsen EGJ, Archard LC (1986) Detection of coxsackie B virus-specific RNA sequences in myocardial biopsies from cases of myocarditis and dilated cardiomyopathy. Lancet I: 1120–1122
3. Boyle JF, Weismiller DG, Holmes KV (1987) Genetic resistance to mouse hepatitis virus correlates with absence of virus-binding activity on target tissues. J Virol 61: 185–189

4. Brahic M, Haase AT, Cash E (1984) Simultaneous in situ detection of viral RNA and antigens. Proc Natl Acad Sci USA 81: 5445–5448
5. Brigati DJ, Myerson D, Leary JJ, et al. (1983) Detection of viral genomes in cultured cells and paraffin-embedded tissue sections using biotin-labeled hybridization probes. Virology 126: 32–50
6. Campbell BA, Cords CE (1983) Monoclonal antibodies that inhibit attachment of group B coxsackieviruses. J Virol 48: 561–564
7. Cash E, Chamorro M, Brahic M (1988) Minus-strand RNA synthesis in the spinal cords of mice persistently infected with Theiler's virus. J Virol 62: 1824–1826
8. Choi AHC, Lee PWK (1988) Does the B-adrenergic receptor function as a reovirus receptor? Virology 163: 191–197
9. Choppin PW, Eggers HJ (1962) Heterogeneity of coxsackie B4 virus: two kinds of particles which differ in antibody sensitivity, growth rate, and plaque size. Virology 18: 470–476
10. Colonno RJ, Callahan PL, Wong WJ (1986) Isolation of a monoclonal antibody that blocks attachment of the major group of human rhinoviruses. J Virol 57: 7–12
11. Crowell RL (1963) Specific viral interference in HeLa cell cultures chronically infected with coxsackie B5 virus. J Bacteriol 86: 517–526
12. Crowell RL (1966) Specific cell-surface alteration by enteroviruses as reflected by viral attachment interference. J Bacteriol 91: 198–204
12a. Crowell RL (1976) Comparative generic characteristics of picornavirus-receptor interactions. In: Beers RF Tr and Bassett EG (eds) Cell Membbrane Receptors for Viruses, Antigens and Antibodies, Polypaptide Hormones and Small Molecules. Raven Press, New York pp 179–202
13. Crowell RL (1987) Cellular receptors in virus infections. ASM News 53: 422–425
14. Crowell RL, Landau BJ (1979) Receptors as determinants of cellular tropism in picornavirus infections. In: Bearn AG, Choppin PW (eds) Receptors and human diseases. Macy Foundation Press, New York, pp 1–33
15. Crowell RL, Landau BJ (1983) Receptors in the initiation of picornavirus infections. In: Fraenkel-Conrat H, Wagner RR (eds) Comprehensive virology, vol 18. Plenum, New York, pp 1–42
16. Crowell RL, Lonberg-Holm K (eds) (1986) Virus attachment and entry into cells. American Society for Microbiology, Washington, pp 1–216
17. Crowell RL, Siak J-S (1978) Receptor for group B coxsackieviruses: characterization and extraction from HeLa cell membranes. In: Pollard M (ed) Perspectives in virology, vol 10. Raven, New York, pp 39–53
18. Crowell RL, Syverton JT (1961) The mammalian cell-virus relationship. VI. Sustained infection of HeLa cells by coxsackie B3 virus and effect on superinfection. J Exp Med 113: 419–435
19. Crowell RL, Reagan KJ, Schultz M, Mapoles JE, Grun JB, Landau BJ (1985) Cellular receptors as determinants of viral tropism (Banbury Report 22). Genetically altered viruses and the environment. Cold Spring Harbor Laboratory, Cold Spring Harbor, pp 147–161
20. Crowell RL, Field AK, Schleif WA, Long WL, Colonno RJ, Mapoles JE, Emini EA (1986) Monoclonal antibody that inhibits infection of HeLa and Rhabdomyosarcoma cells by selected enteroviruses through receptor blockade. J Virol 57: 438–445
21. Crowell RL, Hsu K-HL, Schultz M, Landau BJ (1987) Cellular receptors in coxsackievirus infections. In: Brinton MA, Rueckert RR (eds) Positive strand RNA viruses. Liss, New York, pp 453–466
22. Easton AJ, Eglin RP (1988) The detection of coxsackievirus RNA in cardiac tissue by in situ hybridization. J Gen Virol 69: 285–291
23. Eggers HJ, Sabin AB (1959) Factors determining pathogenicity of variants of Echo 9 virus for newborn mice. J Exp Med 110: 951–967
24. Gauntt CJ (1988) The possible role of viral variants in pathogenesis. In: Bendinelli M, Friedman H (eds) Coxsackieviruses: a general update, Plenum, New York, pp 159–179
25. Grun JB, Schultz M, Finkelstein SD, Crowell RL, Landau BJ (1988) Pathogenesis of acute myocardial necrosis in inbred mice infected with coxsackievirus B3. Microbial Pathogenesis 4: 417–430
26. Gu F, Mu G, Lu H, Xu R (1988) Preparation and identification of monoclonal antibodies against cell receptors of coxsackievirus. Chin J Virol 4: 8–11

27. Guthrie M, Lodge PA, Huber SA (1984) Cardiac injury in myocarditis induced by coxsackievirus group B, type 3 in Balb/c mice is mediated by Lyt 2^+ cytolytic lymphocytes. Cell Immunol 88: 558–567
28. Holland JJ (1961) Receptor affinities as major determinants of enterovirus tissue tropisms in humans. Virology 15: 312–326
29. Holland JJ (1964) Enterovirus entrance into specific host cells and subsequent alterations of cell protein and nucleic acid synthesis. Bacteriol Rev 28: 3–13
30. Hsu K-HL, Lonberg-Holm K, Alstein B, Crowell RL (1988) A monoclonal antibody specific for the cellular receptor for the group B coxsackieviruses. J Virol 62: 1647–1652
30a. Huber SA, Lodge PA (1986) Coxsackievirus B3 myocarditis. Identification of different pathogenic mechanisms in DBA/2 and BALB/c mice. Am J Pathol 122: 284–291
31. Hyypiä TP, Stalhandske P, Vainionpaa R, Petterson U (1984) Detection of enteroviruses by spot hybridization. J Clin Microbiol 19: 436–438
32. Kandolf R, Ameis D, Kirschner P, Canu A, Hofschneider PH (1987) In situ detection of enteroviral genomes in myocardial cells by nucleic acid hybridization: an approach to the diagnosis of viral heart disease. Proc Natl Acad Sci USA 84: 6272–6276
33. Khatib R, Probert A, Reyes MP, Khatib G, Chason JL (1987) Mouse strain-related variation as a factor in the pathogenesis of coxsackievirus B3 murine myocarditis. J Gen Virol 68: 2981–2988
34. Krah DL, Crowell RL (1982) A solid-phase assay of solubilized HeLa cell membrane receptors for binding group B coxsackieviruses and polioviruses. Virology 118: 148–156
35. Krah DL, Crowell RL (1985) Properties of the deoxycholate-solubilized HeLa cell plasma membrane receptor for binding group B coxsackieviruses. J Virol 53: 867–870
36. Kunin CM, Halmagyi NE (1961) The relative abundance of viral receptors. An explanation of the differential susceptibility of suckling and adult mice to coxsackie B1 infection. J Clin Invest 40: 1055–1056
37. Lee PWK, Hayes EC, Joklik WK (1981) Protein sigma-1 is the reovirus cell attachment protein. Virology 108: 156–163
37a. Levitt NH, Crowell RL (1967) Comparative studies of the regeneration of HeLa cell receptors for poliovirus T1 and coxsackievirus B3. J Virol 1: 693–700
38. Lo CW (1986) Localization of low abundance DNA sequences in tissue sections by in situ hybridization. J Cell Sci 81: 143–162
39. Lonberg-Holm K, Crowell RL, Philipson L (1976) Unrelated animal viruses share receptors. Nature 259: 679–681
40. Mapoles JE, Krah DL, Crowell RL (1985) Purification of a HeLa cell receptor protein for the group B coxsackievirus. J Virol 55: 560–566
41. Medrano L, Green H (1973) Picornavirus receptors and picornavirus multiplication in human mouse hybrid cell lines. Virology 54: 515–524
42. McLaren LC, Holland JJ, Syverton JT (1959) The mammalian cell-virus relationship. I. Attachment of poliovirus to cultivated cells of primate and non-primate origin. J Exp Med 109: 475–485
43. Melnick JL (1985) Enteroviruses: polioviruses, coxsackieviruses, echoviruses, and newer enteroviruses. In: Fields BN (ed) Virology. Raven, New York, pp 739–794
44. Minor PD, Pipkin PA, Hockley D, Schild GC, Almond JW (1984) Monoclonal antibodies which block cellular receptors of poliovirus. Virus Res 1: 203–212
45. Mischak H, Neubauer C, Kuechler E, Blaas D (1988) Characteristics of the minor group receptor of human rhinoviruses. Virology 163: 19–25
46. Nobis P, Zibirre R, Meyer G, Kuhne J, Warnecke G, Koch G (1985) Production of a monoclonal antibody against an epitope on HeLa cells that is the functional poliovirus binding site. J Gen Virol 66: 2563–2569
47. Quersin-Thiry L, Nihoul E (1961) Interaction between cellular extracts and animal viruses. II. Evidence for the presence of different inactivators corresponding to different viruses. Acta Virol 5: 282–293
48. Reagan KJ, Goldberg B, Crowell RL (1984) Altered receptor specificity of coxsackievirus B3 after growth in rhabdomyosarcoma cells. J Virol 49: 635–640
49. Robinson JA, O'Connell JB, Roeges LM, Major EO, Gunnar RM (1981) Coxsackie B3 myocarditis in athymic mice. Proc Soc Exp Biol Med 166: 80–91

50. Rossmann MG, Arnold E, Erickson JW, Frankenberger EA, Griffith JP, Hecht HJ, Johnson JE, Kamer G, Luo M, Mosser AG, Rueckert RR, Sherry B, Vriend G (1985) Structure of a human common cold virus and functional relationship to other picornaviruses. Nature 317: 145–153
51. Rozhon EJ, Wilson AK, Jubelt B (1984) Characterization of genetic changes occurring in attenuated poliovirus 2 during persistent infection in mouse central nervous systems. J Virol 50: 137–144
52. Schultz M, Crowell RL (1983) Eclipse of coxsackievirus infectivity: the restrictive event for a non-fusing myogenic cell line. J Gen Virol 64: 1725–1734
53. Schultz M, Heiman-Patterson TD, Landau BJ, Crowell RL (1986) The development of chronic myopathy in SJL/J mice infected with a variant of coxsackievirus B3. Muscle Nerve 9: 219
54. Sharpe AH, Fields BN (1985) Pathogenesis of viral infections. Basic concepts derived from the reovirus model. N Engl J Med 312: 486–497
55. Stump RF, Pfeiffer JR, Seagrave J, Oliver JM (1988) Mapping goldlabelled IgE receptors on mast cells by scanning electron microscopy: receptor distributions revealed by silver enhancement, backscattered electron imaging, and digital image analysis. J Histochem Cytochem 36: 493–502
56. Thomas PS (1980) Hybridization of denatured RNA and small DNA fragments transferred to nitrocellulose. Proc Natl Acad Sci USA 77: 5201–5205
57. Toniolo A, Onodera T, Jordan G, Yoon J-W, Notkins AL (1982) Virus-induced diabetes mellitus, glucose abnormalities produced in mice by the six members of the coxsackie B virus group. Diabetes 31: 496–499
58. Vogel SN, Fersch D (1987) Macrophages from endotoxin-hyporesponsive (LSP) C3H/HeJ are permissive for vesicular stomatitis virus because of reduced levels of endogeneous interferon: possible mechanism for natural resistance to virus infection. J Virol 61: 812–818
59. Wolfgram LJ, Beisel KW, Rose NR (1985) Heart specific autoantibodies following murine coxsackievirus B3 myocarditis. J Exp Med 161: 1112–1121
60. Yoon J-W, Onodera T, Notkins AL (1978) Virus-induced diabetes mellitus. XV. Beta cell damage and insulin-dependent hyperglycemia in mice infected with coxsackievirus B4. J Exp Med 148: 1068–1080
61. Zajac I, Crowell RL (1965) Effect of enzymes on the interaction of enteroviruses with living HeLa cells. J Bacteriol 89: 574–587
62. Zajac I, Crowell RL (1969) Differential inhibition of attachment and eclipse activities of HeLa cells for enteroviruses. J Virol 3: 422–428

Induction of Soluble Mediators During Coxsackievirus B3 Replication In Vitro and In Vivo

C. J. GAUNTT[1], C. W. LUTTON[1], G. E. REVTYAK[2], A. J. VALENTE[3], A. N. FERNANDES[4], H. M. ARIZPE[1], and M. M. ROZEK[3]

Departments of Microbiology[1], Pathology[3], and Medicine[4]
The University of Texas Health Science Center, 7703 Floyd Curl Drive, San Antonio, TX 78284, USA
[2] Department of Medicine (Cardiovascular Division), College of Medicine, University of Iowa and VAMC, Iowa City, IA 52240, USA

Introduction

Intraperitoneal (i.p.) inoculation of several strains of mice with myocarditic strains of coxsackievirus B3 (CVB3) induces focal lesions in the myocardium [7, 9, 18, 41–43]. Within hours of virus inoculation, virus appears in the blood [24] and remains in the plasma phase over the next 2–3 days [1, 8]. Virus is also associated with circulating leukocytes [6, 13]. During this period, virus enters the myocardium and replicates in unidentified cells. Histologic examination of heart tissues during this prelesion period reveals single and multiple foci of necrotic myocytes [17]. Beginning on day 4 postinoculation (p.i.), lesion formation is detected in the myocardium of CVB3-inoculated CD-1 mice [11]. The initial cells detected in the nascent lesion were macrophages, T suppressor/cytotoxic lymphocytes, and natural killer cells. T helper lymphocytes were detected 1 day later. At no time during the 4–10 day (p.i.) period were cells producing viral antigens [11] or viral RNA (N. M. Chapman, S. M. Tracy, and C. J. Gauntt, unpublished data) detected in nascent or developing lesions. The data suggest that virus infection of a cell (or cells) in the myocardium initiates a series of events that induce a focal inflammatory response.

There are three major pathways through which inflammation could be induced in this animal model. One is through virus induction of a surface change (or changes) on infected cells which macrophages respond to and stimulate cell-mediated immune responses. Secondly, direct viral stimulation of an infected cell (or cells) may induce the production of pro-inflammatory mediators. The third may be by the release of pro-inflammatory viral or cellular products following virus-induced cell lysis. CVB3 infection of murine neonatal skin or adolescent heart fibroblasts in vitro results in induction of a cell surface change detectable by binding of two lectins [22]. This surface change can also be detected in foci of cells in heart tissues of CVB3-inoculated mice at the time of maximal virus replication, i.e., 3 days p.i. [11]. Studies with the CVB3 mouse model have detected virus-induced production of three soluble mediators. Mice challenged with CVB3 produce interferon which is detectable in sera by day 2 p.i. [10] and is of both gamma and beta types [6]. Earlier studies showed that CVB3-infected heart tissues contained an extractable soluble neoantigen (or neoantigens) which, during incubation with peritoneal leukocytes from CVB3-inoculated mice, resulted in production of a so-

H.-P. Schultheiß (Ed.)
New Concepts in Viral Heart Disease

luble cell migration inhibition factor which inhibited migration of macrophages in the population [9, 23, 28, 29].

The present study shows that infection of murine fibroblast cultures with a myocarditic CVB3 variant results in production of several new soluble mediators. Furthermore, heart tissues from mice inoculated with this virus also produce increased levels of several arachidonic acid metabolites. The role(s) played by these chemical mediators in CVB3-induced myocarditis is under study.

Materials and Methods

A myocarditic strain of CVB3 ($CVB3_m$) was used in these studies [9]. Virus was propagated and assayed by a plaque method in HeLa cells [9]. HeLa cells and neonatal skin or adolscent heart fibroblasts were cultured in complete minimal essential medium (MEM) supplemented with 10% fetal calf serum, 4 mM L-glutamine, penicillin (100 units/ml) and streptomycin (100 μg/ml). Preparation and culture of both types of fibrolasts have previously been described [8,22].

Breeding pairs of CD-1 mice were initially purchased from Charles River Breeding Laboratories, Inc. (Boston, Mass). Adolescent mice reared from these matings were maintained in the Laboratory Animal Resources facilities at this institution. All animals were given water and laboratory mouse chow ad libitum. Mice were inoculated by the i. p. route with 10^5 plaque-forming units (PFU) of purified virus [7] in 0.2 ml MEM. Control mice received 0.2 ml MEM only.

Monocyte/macrophage chemoattractant and prostaglandins were assayed in culture medium from fibroblast cultures using the following conditions. Fibroblasts were seeded to plastic Petri dishes. After 18–24 h in complete MEM, the cultures were rinsed once with MEM-1% bovine serum albumin (BSA), challenged with 100–200 PFU/cell of $CVB3_m$ in MEM-BSA or MEM alone and incubated for 1.5 h at 37 °C in a humidified CO_2 incubator. All cultures were rinsed three times with MEM-BSA and incubated in the latter medium. Zero time was the time of addition of virus to the cells. At various times p. i., culture fluids were assayed for monocyte/macrophage chemotactic activity using modified Boyden (blind well) chambers and 5μm-pore size polycarbonate filters. Unelicited mononuclear cells were obtained from adolescent CD-1 mice by peritoneal lavage, washed once in MEM-BSA and resuspended in the same medium. Conditioned medium from $CVB3_m$-infected or normal primary mouse fibroblast cultures was placed below the filter and 2.5 x 10^5 mononuclear cells above the filter. The chambers were incubated for 4 h at 37 °C. The filters were fixed, stained and chemotaxis was quantitated as described previously [39]. Fluids from similar cultures were assayed by radioimmune assay (RIA) for prostaglandins [2, 3].

Heart tissues from normal and $CVB3_m$-inoculated mice were assayed for prostaglandins as follows. Mice were sacrificed by cervical dislocation, and hearts were rapidly removed, washed in Dulbecco's phosphate-buffered saline, minced with scissors and fragments incubated in hydroxyethyl piperazine ethane sulphonic acid (HEPES)-buffered MEM containing 1 μm ^{14}C-arachidonic acid (1 μCi) for 30 min at 37 °C. The culture medium was acidified (pH 3) with glacial acetic acid and extracted with 3–5 v/v of ethylacetate: cyclohexane (1:1) to obtain prostaglandins and hydroxyeicosatetraenoic acids (HETEs). Radiolabeled metabolites were separated by high-pressure liquid chromatography (HPLC) using reverse phase C-18 or normal silica gel columns with selected elution programs and solvent systems [30]. The identity of specific metabolites was established by comigration of radioactive peaks eluted from an HPLC column with standard compounds coinjected or cochromatographed under identical conditions. RIA was used to quantitate levels of eicosanoids.

Thin layer chromatography (TLC) was used to quantify levels of HETEs. A portion of the acidified organic solvent-extracted material above was evaporated to dryness under a stream of N_2 and the residue resuspended in chloroform: methanol and applied to a multichanneled silica gel G thin layer plate. The plate was developed using the organic phase of ethyl acetate: 2,2,4-trimethylpentane:acetic acid:water. Standards were cochromatographed on the same plate and visualized in iodine vapor. Radioacitivity in selected areas of the silica gel was determined by liquid scintillation spectrometry.

Results

In Vitro Replication of Virus in Fibroblasts

Replication of $CVB3_m$ occurs for days in primary cultures of neonatal skin or adolscent heart fibroblasts [8, 12, 22]. Primary cultures refer to cells within passage levels 1–4 or 1–8, respectively, for adolescent heart or neonatal skin fibroblasts. Two types of replication curves were found for either type of cell: sustained replication of virus at moderate titers (100–300 PFU/cell) for 5–7 days as typified in Fig. 1 or peak replication of virus around 1–2 days and diminution in titers each day thereafter. The data in Fig. 1 illustrate two features of $CVB3_m$ replication in either type of fibroblast culture: a) immediate release of nascent virus; and b) absence of any virus-induced cytopathic effects at the light-microscopie level. Only 10%–12% or 20%–24% of cells in either neonatal skin or adolescent heart fibroblast cultures, respectively, could be infected by $CVB3_m$, as determined by immunofluorescence for capsid antigens [22]. Attachment and eclipse studies with purified $CVB3_m$ showed that for neonatal skin fibroblasts, only 41%–53% of a challenge inoculum of 100 PFU/cell bound to cells, and 12%–25% of the bound virus eclipsed in 3 h at 37 °, compared with binding of 85%–88% of virus to HeLa cells and approximately 70% undergoing eclipse within 1 h [20]. Thus, replication of $CVB3_m$ in the nonlytic fibroblast system is not as efficient as in the lytic HeLa cell system.

Virus Induction of Monocyte/Macrophage Chemoattractant in Primary Cultures of Murine Fibroblasts

Monocytes/macrophages are resident in heart tissues and are among the first inflammatory cells detected in the nascent myocardial lesion of $CVB3_m$-inoculated mice [11]. The heart also contains large numbers of fibroblasts, and we have pro-

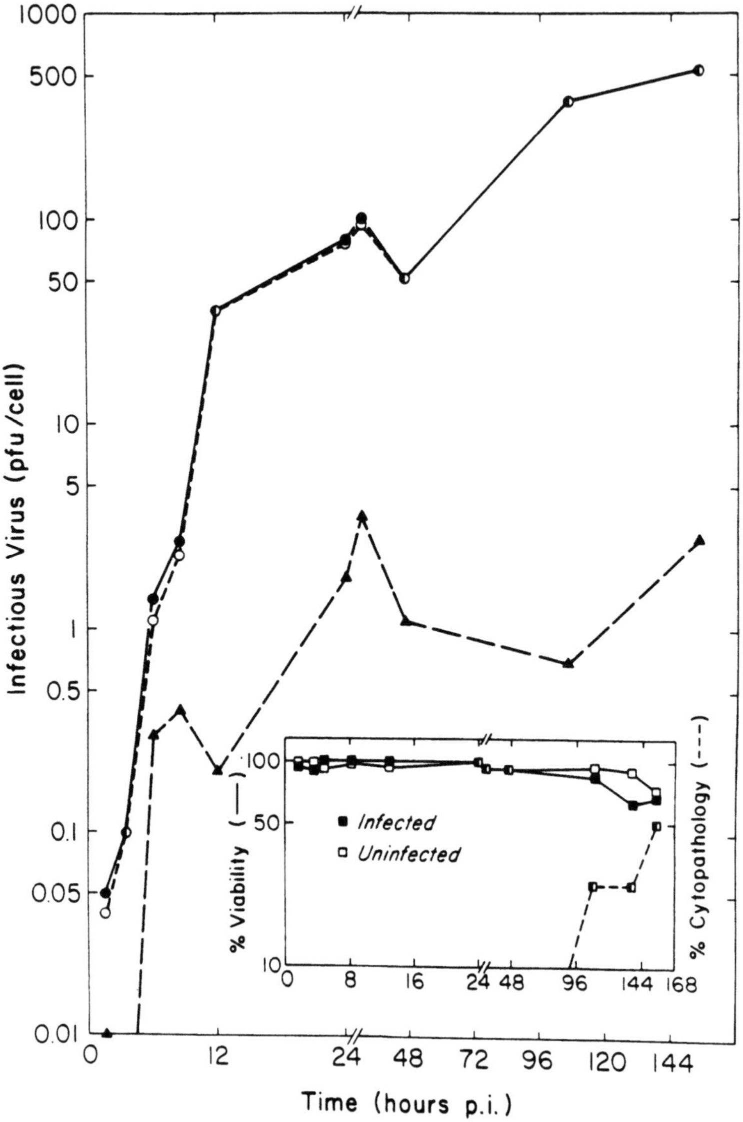

Fig. 1. Replication of CVB3 in primary cultures of neonatal skin fibroblasts. Cells were challenged with 150 PFU/cell CVB3 and after adsorption, washed three times and incubated in complete MEM. Duplicate cultures were assayed for titers of released (free) *(open circles)* or cell-associated *(triangles)* virus by plaque assay on HeLa cell cultures. Solid circles, total. *Inset:* percentage of viable cells in uninfected *(open squares)* or virus-challenged *(solid squares)* cultures assessed by a trypan blue dye exclusion assay for percentage of cells exhibiting cytopathology, as determined by microscopic observation (X40)

posed that these cells play a role in development of $CVB3_m$-induced focal myocardial lesions [8]. Fibroblasts and macrophages exchange numerous chemical signals, hence the following experiments. It was found that primary cultures of neonatal skin fibroblast cultures challenged with 100–150 PFU/cell $CVB3_m$ produce

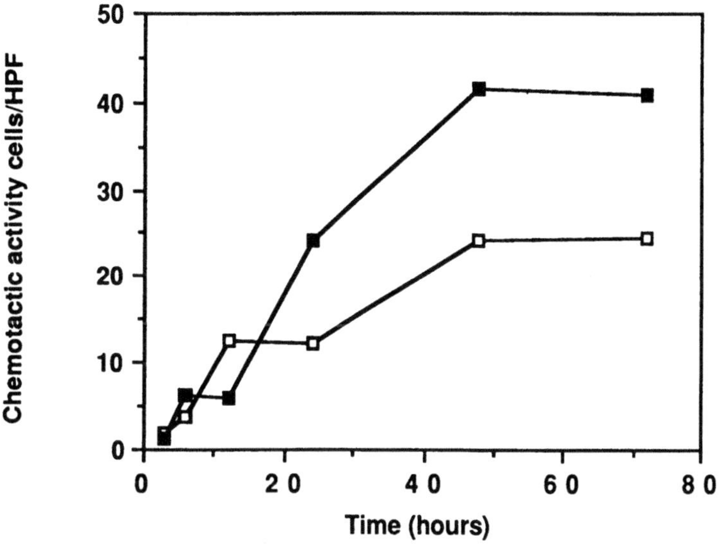

Fig. 2. Synthesis and release of a monocyte/macrophage chemotactic activity by CVB3-inoculated *(solid squares)* or normal *(open squares)* primary cultures of neonatal skin fibroblasts. Cell cultures (2–3x10^6 cells) were mock-infected or challenged with 100–150 PFU/cell. After adsorption, the cultures were washed and incubated in MEM-1% BSA. At intervals, supernatant fluids were removed and assayed for chemoattractant(s) for murine peritoneal macrophages using Boyden (blind well) chambers. Values plotted represent the mean of values from three separate cultures

and release increasing amounts of a chemotactic activity for macrophages over a 72-h period (Fig. 2). This activity is detectable by 6 h p.i., and production remains increased throughout the remaining 66 h of infection. In two additional experiments with neonatal skin fibroblasts and one with adolescent heart fibroblasts, infected cultures produced significantly greater levels (1.7- to 2.0-fold) of activity than uninfected cultures between 24 and 72 h p.i.. Cells challenged with ultraviolet light-inactivated virus did not produce increased levels of chemoattractant(s).

Virus Induction of Prostaglandin Synthesis in Murine Fibroblasts

Prostaglandins, leukotrienes, and other metabolites of arachidonic acid have been implicated as mediators of inflammation and regional blood flow/[19, 25, 32]. We have begun a systematic study of eicosanoids produced during nonlytic replication of $CVB3_m$ in murine fibroblasts because eicosanoids may be important inflammatory mediators in developing myocarditis. In initial experiments, neonatal skin fibroblast cultures challenged with 100–150 PFU/cell $CVB3_m$ were found to produce and release increased levels of PGE_2 compared to uninfected fibroblast cultures (Fig. 3). Virus-induced stimulation of PGE_2 levels in supernatant fluids was detected at 6 h p.i. Increased PGE_2 synthesis continued through 72 h

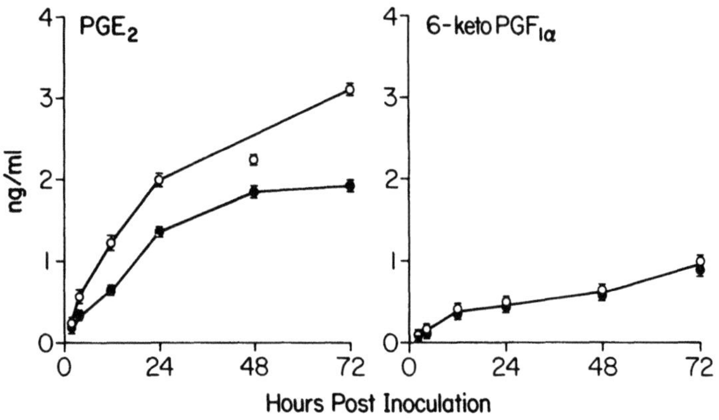

Fig. 3. Prostaglandin synthesis in CVB3-inoculated *(open circles)* or normal *(solid circles)* primary cultures of neonatal skin fibroblasts. Cell cultures (1–2x10^5 cells) were challenged with 100–150 PFU/cell and, after adsorption and washing three times, incubated in MEM-BSA. At various times, supernatant fluids were assayed directly for PGE_2 or 6-keto $PGF_{1\alpha}$ by RIA. Each value plotted represents the mean ± SEM of six separate cultures

p.i. where infected culture levels were nearly twice those of uninfected culture levels. On the other hand, these same cultures did not produce increased amounts of 6-keto $PGF_{1\alpha}$ in response to $CVB3_m$ infection. Similar results have been obtained using adolescent heart fibroblasts with respect to prostaglandin synthesis following $CVB3_m$ infection.

Virus-Induced Alterations in Synthesis of Arachidonic Acid Metabolites in Heart Tissues In Vivo

To examine the effects of $CVB3_m$ infection on cardiac arachidonic acid metabolism, initial experiments focused on isolating and quantifying specific eicosanoids produced by heart tissues obtained from both normal and CVB3-infected mice at 3 days p.i. (maximal viral titers in hearts) or 8 days p.i. (maximal myocardial lesion number). HPLC was used to identify the major products of cardiac arachidonic acid metabolism in normal and $CVB3_m$-infected hearts [31]. For these experiments, minced heart tissues were incubated with 1 μM ^{14}C-arachidonic acid for 30 min at 37 °C. Radiolabeled metabolites were separated by HPLC. Using this method, heart tissues obtained from normal or $CVB3_m$-infected mice at 8 days p.i. were shown to produce ^{14}C-labeled 6-keto $PGF_{1\alpha}$, PGE_2, 12-HETE, and 15-HETE [31].

RIA and TLC were then used to quantify the production of these eicosanoids in heart tissues obtained from mice at various times p.i. of $CVB3_m$. As shown in Table 1, $CVB3_m$ infection induces a 1.5-fold increase in 6-keto $PGF_{1\alpha}$ synthesis in heart tissues obtained 3 days p.i. By 8 days p.i., however, the production of this metabolite had returned to preinfection levels. On the other hand, $CVB3_m$ infection did not alter PGE_2 synthesis in heart tissues obtained from mice at 3 or 8 days

Table 1. Prostaglandin synthesis by normal and $CVB3_m$-infected mouse heart tissues[a]

Source of heart tissues	Prostaglandin synthesized 6-keto $PGF_{1\alpha}$ (pg/mg wet weight)	PGE_2 (pg/mg wet weight)
Normal	76.82 ± 12.75	89.00 ± 13.49
3 Days p. i.	115.10 ± 6.73[b]	80.95 ± 9.52
8 Days p. i.	72.50 ± 14.40	84.72 ± 12.49

[a] Groups of five adolescent male CD-1 mice were inoculated with MEM (normal) or CVB3 (1 x 10^5 PFU/ml) on day 0. On days 3 or 8 p.i., mice were sacrificed. Heart tissues were removed, rinsed in Dulbecco's phosphate-buffered saline and incubated in fresh buffer for 30 min at 37 °C. Prostaglandins released from heart tissues into the buffer were assayed by RIA. Data presented are mean values ±SEM

[b] $P < 0.05$, 3-day p. i. value vs. control value

p. i. Since viral titers in heart tissues are maximal at 3 days p. i. $CVB3_m$ interaction with cardiac endothelial cells may stimulate these cells to produce increased amounts of prostacyclin (detected as its stable metabolite 6-keto $PGF_{1\alpha}$).

HETE production by heart tissues obtained from mice following $CVB3_m$ infection was quantitated by TLC. Heart tissues obtained from mice at 3 days p. i. produced similar amounts of HETEs as those obtained from uninfected mice. As shown in Table 2, at 8 days p. i. heart tissues obtained from $CVB3_m$-infected mice produced 1.7-fold more HETEs than heart tissues from normal mice. Since inflammatory myocarditic lesion size and number is maximal at 8 days p. i., mononuclear cells with abundant lipoxygenase activity may be responsible for the increased production of HETEs by heart tissues from $CVB3_m$-infected mice at 8 days p. i. compared to normal heart tissues.

Table 2. Synthesis of 12- and 15-HETE by normal and $CVB3_m$-infected mouse heart tissues[a]

Source of heart tissues	6-keto-$PGF_{1\alpha}$ (pmol/mg wet weight)	PGE_2 (pmol/mg wet weight)	HETEs (pmol/mg wet weight)
Normal	8 ± 1	12 ± 2	38 ± 1
8 Days p. i.	6 ± 2	11 ± 4	65 ± 2[b]

[a] Groups of six adolescent male CD-1 mice were inoculated with MEM (normal) or CVB3 (1 x 10^5 PFU/ml) on day 0. On days 8 p. i., all mice were sacrificed. Heart tissues were removed, rinsed in Dulbecco's phosphate-buffered saline and incubated with 1 μM ^{14}C-arachidonic acid (1 μCi/mg tissue) for 30 min at 37 °C. Incubation buffer was removed, and the radiolabeled metabolites were extracted with organic solvents [30]. Radiolabeled prostaglandins and HETEs were separated by TLC. Data presented are means values ± SEM

[b] 12-HETE and 15-HETE; $P < 0.05$, 8-day p. i. value vs. normal value

Discussion

This study has demonstrated that $CVB3_m$ replication in primary murine fibroblast cultures or murine heart tissues can alter arachidonic acid metabolism. Identification and quantitation of virus-induced metabolites form the initial phase of studies designed to identify soluble mediators produced by $CVB3_m$-infected tissues which may influence the development of focal myocarditic lesions. A hypothetical scheme for $CVB3_m$ induction of focal lesions in the myocardium is suggested in Table 3. After a brief viremia of less than 1 day, we hypothesize that virus infects endothelial cells lining the vessels of the heart. These cells are most likely among the first to produce and release several types of pro-inflammatory mediators including prostacyclin. Increased prostacyclin production on day 3 p. i. may represent a physiologic "protective" response by the endothelium, induced by virus attachment and/or replication. Nascent virus then breaches the endothelial cell lining and gains access to the subadjacent fibroblasts where a small proportion (10%–20%) become infected and release virus to infect myocytes. During this prelesion phase, additional soluble pro-inflammatory mediators are produced and released by infected cells, including both β and γ interferons [6]. Previous studies have shown that administration of β interferon [21] or γ interferon [6] to mice 24 h p. i. of $CVB3_m$ did not reduce virus titers in heart tissues by 3 days p. i. but

Table 3. Hypothetical scheme outlining sequence of events suggested to occur at different times p. i. of a mouse which result in formation of focal lesions in the mouse myocardium

Time	Event	Pathology
	CVB3 inoculation	
Hours p. i.	Viremia	None
< 1 Day p. i.	Infection of endothelial cells (?) in heart vessels	Ultrastructural alterations?
1–3 Days p. i.	Infection of subadjacent cardiac fibroblasts (?) and myocytes (?) resulting in cell surface alterations and synthesis of pro-inflammatory soluble mediators locally	Myocyte necrosis
4–5 Days p. i.	Focal infiltration of macrophages, lymphocytes, and natural killer cells to sites of infected cells; synthesis and release of additional pro-inflammatory mediators	Nascent focal myocarditic lesions
6–9 Days p. i.	Continued leukocyte infiltration into focal sites of lesions in absence of virus; synthesis and release of additional soluble mediators	Development of myocarditic lesions

subsequently significantly reduced the number of myocardial lesions formed by 8 days p.i. These data suggested that interferon was acting through an anti-inflammatory rather than an antiviral mechanism (or mechanisms). Once macrophages, lymphocytes, and natural killer cells respond to the released chemical mediators and altered cell surfaces, focal lesion development begins. At or near the lesion site, the infiltrating cells respond to their microenvironment by releasing additional chemical mediators which foster lesion enlargement and further immunopathology.

There is no published evidence that endothelial cells in adolescent murine heart tissues can replicate CVB3. In studies to be reported elsewhere (G. E. Revtyak and C. J. Gauntt, unpublished data), $CVB3_m$ attached to but did not replicate in bovine coronary artery endothelial cells. However, $CVB3_m$ attachment stimulated the production and release of 6-keto $PGF_{1\alpha}$ that was detected by 4 h p.i. By 14 h p.i., levels were over 20-fold higher than in uninfected cell cultures (44ng/ml vs. ~2 ng/ml, respectively).

Fibroblasts compose 35%–75% of the total cell number of mammalian hearts and are a major component of connective tissue [4, 27, 45]. Production of macrophage chemoattractant(s) by $CVB3_m$-infected fibroblasts could be an early signal for infiltration of monocytes/macrophages resident in heart tissue. Recently, the picornavirus rhinovirus type 14 was found to stimulate synthesis and release of leukocyte chemotactic activity during infection of embryonic lung fibroblasts in culture [37]. Several types of cells infected with different viruses have been reported to produce and release chemotactic factors for peripheral blood leukocytes [16]. Fibroblasts infected with $CVB3_m$ also display α-l-fucosyl residues on their surface [22]. This surface alteration has also been detected on unidentified cells in sections of infected heart tissues taken from mice 3 days p.i. [11]. Activated macrophages are known to possess a receptor for mannosyl/fucosyl residues [34, 35], thus the virus-induced α-l-fucosyl residues may act as an additional signal to macrophages. Once at the site of an infected fibroblast, macrophages may contribute to lesion development through the release of chemotactic, collagen synthesis-stimulating, and growth-promoting factors for fibroblasts [14, 15, 26, 38, 40]. Of interest is the finding that human heart tissue biopsy specimens with myocardic lesions contain 29%–35% of total cells as fibroblasts, whereas only 15%–16% of total cells from infarcted heart tissues are fibroblasts [44].

Data showing that $CVB3_m$ stimulates synthesis and release of PGE_2 from murine fibroblasts have been reported in other virus-fibroblast systems. Human lung fibroblast cultures synthesized two- to three-fold more PGE_2 as well as three- to five-fold more thromboxane B_2 in response to infection by Newcastle disease virus [5]. Infection of mouse L929 fibroblasts with vaccinia virus required the presence of interferon to stimulate synthesis of PGE_2 as neither alone increased levels above uninfected cell cultures [33].

Some arachidonic acid metabolites are potent pro-inflammatory mediators with diverse properties. These mediators are thought to be synthesizd and act locally [25, 36]. Considerable work will be required to demonstrate their roles, if any, in $CVB3_m$-induced murine and human myocarditis.

Summary

Inoculation of adolescent male CD-1 mice with a myocarditic variant of $CVB3_m$ results in acute myocarditis. We hypothesize that induction of the focal inflammatory response results in part from virus-induced alterations in cellular production of soluble pro-inflammatory mediators. In support of this hypothesis, in vitro and in vivo studies detect virus induction of several mediators. Murine adolescent heart or neonatal skin fibroblasts challenged with $CVB3_m$ replicate virus in absence of cytopathology for 7–10 days. These cultures produce increased levels of a chemotactic activity for macrophages and increased levels of PGE_2 but not 6-keto $PGF_{1\alpha}$. Increased mediator levels were apparent by 6 h p.i. and remained through 72 h p.i. Mice were inoculated with $CVB3_m$ and heart tissues taken on days 3 and 8 p.i., minced and incubated with ^{14}C-arachidonic acid. HPLC identified radiolabeled PGE_2, 6-keto $PGF_{1\alpha}$, 15-HETE, and 12-HETE in all tissue samples. Quantitation of these metabolites by RIA and TLC showed: a) a 1.5-fold increased level of 6-keto $PGE_{1\alpha}$ was produced by tissues at 3 days p.i. compared to normal heart tissues, but tissues at 8 day p.i. produced similar low levels as normal tissues; b) PGE_2 levels were similar in all groups; and c) 12- and 15-HETE levels were similar in 3-day p.i. and control heart tissues, but 8-day p.i. tissues produced 1.7-fold higher levels of both HETEs than control tissues. Thus, infection of cells or tissues with $CVB3_m$ alters cellular production of some soluble mediators, including cardiac arachidonic acid metabolites. We hypothesize that these alterations most likely contribute to the virus-induced immunopathology.

Acknowledgements. We thank Mary Devadoss for excellent assistance in preparing this manuscript. This work was supported by Public Health Service Grant HL-21047 from the National Heart, Lung, and Blood Institute; the ERACE Foundation, Los Angeles, CA; American Heart Association, Texas Affiliate; and the Veterans Administration Research Advisory Groups.

References

1. Bendinelli M, Matteucci D, Toniolo A, Patane AM, Pistilli MP (1982) Impairment of immunocompetent mouse spleen cell functions by infection with coxsackievirus B3, J Infect Dis 146: 797–805
2. Campbell WB, Gomez-Sanchez CE Adams BV (1980) Role of prostaglandins in angiotensin-induced steroidogenesis: absence of an effect by prostaglandin E_2. Hypertension 2: 472–476
3. Campbell WB, Holland OB, Adams BV, Gomez-Sanchez CE (1982) Urinary excretion of prostaglandin E_2, prostaglandin $F_{2\alpha}$, and thromboxane B_2 in normotensive and hypertensive subject on varying sodium intakes. Hypertension 4: 735–741
4. David H (1977) Quantitative ultrastructural data of animal and human cells. Fischer, New York, p 395
5. Fitzpatrick FA, Stringfellow DA (1980) Virus and interferon effects on cellular prostaglandin biosynthesis. J Immunol 125: 431–437
6. Gauntt CJ, Godeny EK, Lutton CW (1988) Host factors regulating viral clearance. In: Huber SA (ed) Myocarditis: etiology, pathogenesis and treatment. Pathol and Immunopathol Res. (in press)

7. Gauntt CJ, Gomez PT, Duffey PS, Grant JA, Trent DW, Witherspoon SM, Paque R (1984) Characterization and myocarditic capabilities of coxsackievirus B3 variants in selected murine strains. J Virol 52: 598–605
8. Gauntt CJ, Lutton CW, Godeny EK, Witherspoon SM, Arizpe HM, Lanford RE (1987) Murine coxsackievirus B3 myocarditis. Eur Heart J 8 [Suppl J]: 393–397
9. Gauntt JC, Trousdale MD, La Badie DRL, Paque RE, Nealon T (1979) Properties of coxsackievirus B3 variants which are amyocarditic or myocarditic for mice. J Med Virol 3: 207–220
10. Godeny EK, Gauntt CJ (1986) Involvement of natural killer cells in coxsackievirus B3-induced murine myocarditis. J Immunol 137: 1695–1702
11. Godeny EK, Gauntt CJ (1987) In situ immune autoradiographic identification of cells in heart tissues of mice with coxsackievirus B3-induced myocarditis. Am J Pathol 129: 267–276
12. Godeny EK, Gauntt CJ (1987) Murine natural killer cells limit coxsackievirus B3 replication. J Immunol 139: 913–918
13. Gomez MP, Reyes MP, Smith F, Ho LK, Lerner AM (1980) Coxsackievirus B3-positive mononuclear leukocytes in peripheral blood of Swiss and athymic mice during infection. Proc Soc Exp Biol Med 165: 107–113
14. Gordon S (1986) Biology of the macrophage. J Cell Sci [Suppl] 4: 267–286
15. Gordon S, Crocker P, Morris L, Lee S-H, Perry VH, Hume D (1986) Localization and function of tissue macrophages. In: Biochemistry of macrophages. Ciba Found Sym 118: 54
16. Griffin DE (1984) The inflammatory response to acute viral infections In: Notkins AL, Oldstone MBA (eds) Concepts in viral pathogenesis Springer, Berlin Heidelberg New York, p 46
17. Herskowitz A, Traystman MD, Beisel KW (1986) Murine viral myocarditis — new insights into mechanism of disease. Heart Failure 2: 86–91
18. Lerner AM, Reyes MP (1985) Coxsackievirus myocarditis — with special reference to acute and chronic effects. Prog Cardiovasc Dis 27: 373–394
19. Lewis RA, Austen KF (1984) The biologically active leukotrienes. Biosynthesis, metabolism, receptors, functions and pharmacology. J Clin Invest 73: 889–897
20. Lutton CW (1986) Studies of pathogenic mechanisms of coxsackievirus B3 in a murine model of virus-induced myocarditis. Dissertation, University of Texas Health Science Center, San Antonio, Texas
21. Lutton CW, Gauntt CJ (1985) Ameliorating effect of IFN-β and anti-IFN-β on coxsackievirus B3-induced myocarditis in mice. J Interferon Res 5: 137–146
22. Lutton CW, Gauntt CJ (1986) Coxsackievirus B3-infection alters plasma membrane of neonatal skin fibroblasts. J Virol 60: 294–296
23. Lutton CW, Gudvangen RJ, Nealon TJ, Paque RE, Gauntt CJ (1985) Cellular immune responses in mice challenged with an amycarditic variant of coxsackievirus B3. J Med Virol 17: 345–357
24. Lyden DC, Olszewski J, Feran M, Job LP, Huber SA (1987) Coxsackievirus B3-induced myocarditis. Effect of sex steroids on viremia and infectivity of cardiocytes. Am J Pathol 126: 432–438
25. Marcus AJ (1984) The eicosanoids in biology and medicine. J Lipid Res 25: 1511–1516
26. Mensing H, Czarnetzki BM (1984) Leukotriene B4 induces in vitro fibroblast chemotaxis. J Invest Dermatol 82: 9–12
27. Morkin E, Ashford TP (1968) Myocardial DNA synthesis in experimental cardiac hypertrophy. Am J Physiol 215: 1409–1413
28. Paque RE, Gauntt CJ, Nealon TJ, Trousdale MD (1978) Assessment of cell-mediated hypersensitivity against coxsackievirus B3 viral-induced myocarditis utilizing hypertonic salt extracts of cardiac tissue. J Immunol 120: 1672–1678
29. Paque RE, Straus DC, Nealon TJ, Gauntt CJ (1979) Fractionation and immunologic assessment of KC1-extracted cardiac antigens in coxsackievirus B3 viral-induced myocarditis. J Immunol 124: 358–364
30. Revtyak GE, Johnson AR, Campbell WB (1988) Cultured coronary artery endothelial cells synthesize HETEs and prostacyclin. Am J Physiol 245: C8–C19

31. Revtyak GE, Fernandes AN, Fernandes GA, Arizpe HM, Gauntt CJ (1988) Alterations in cardiac arachidonic acid metabolism influence coxsackievirus B3-induced myocarditis in mice. Clin Res 36: 310A
32. Samuelsson B (1983) Leukotrienes: mediators of immediate hypersensitivity reactions and inflammation. Science 220: 568–575
33. Santoro MG, Jaffe BM, Paez E, Esteban M (1983) The relationship between the antiviral action of interferon and prostaglandins in virus-infected murine cells. Biochem Biophys Res Commun 116: 442–448
34. Stahl PD, Gordon S (1982) Expression of mannosyl-fucosyl receptor for endocytosis on cultured primary macrophages and their hybrids. J Cell Biol: 93: 49–56
35. Stahl PD, Wileman TE, Shepherd VL (1985) The mannose receptor of macrophages: a current perspective. In: Van Furth (ed) Mononuclear phagocytes. Characteristics, physiology and function, Martinus Nijoff, Dordrecht, p 59
36. Stjernschantz J (1984) The leukotrienes. Med Biol 62: 215–230
37. Turner RB (1988) Rhinovirus infection of human embryonic lung fibroblasts induces the production of a chemoattractant for polymorphonuclear leukocytes. J Infect Dis 157: 346–350
38. Tsukamoto Y, Helsel WE, Wahl SM (1981) Macrophage production of fibronectin, a chemoattractant for fibroblasts. J Immunol 127: 673–678
39. Valente AJ, Fowler SR, Sprague EA, Kelley JL, Suenram CA, Schwartz CJ (1984) Initial characterization of a peripheral blood mononuclear cell chemoattractant derived from cultured arterial smooth muscle cells. Am J Pathol 117: 409–417
40. Wahl SM, Wahl LM (1981) Modulation of fibroblast growth and function by monokines and lymphokines. Lymphokines 2: 179–201
41. Wilson FM, Miranda QR, Chason JL, Lerner AM (1969) Residual pathologic changes following murine coxsackievirus A and B myocarditis. Am J Pathol 55: 253–265
42. Woodruff JF (1980) Viral myocarditis: a review. Am J Pathol 101: 425–483
43. Woodruff JF, Woodruff JJ (1974) Involvement of T lymphocytes in the pathogenesis of coxsackievirus B3 heart disease. J Immunol 113: 1726–1734
44. Yu Z-X, Sekiguchi M, Hroe M, Take M, Hrosawa K (1984) Histopathological findings of acute and convalescent myocarditis obtained by serial endomyocardial biopsy. Jpn Circ J 48: 1368–1374
45. Zak R (1973) Cell proliferation during cardiac growth. Am J Cardiol 31: 211–219

Persistence of Echovirus Type 6 in Cultured Cells

V. F. RIGHTHAND

Wayne State University, School of Medicine, Department of Immunology and Microbiology, 540 E. Canfield Avenue, Detroit, MI 48201, USA

Introduction

Human enteroviruses cause a wide spectrum of systemic infections including myocarditis and encephalitis. Although these viruses are usually cytocidal and cause self-limiting infections, the enteroviruses can also establish persistent infections which result in chronic disease syndromes [3, 7, 9, 11, 14, 17–19]. Recent experiments by Kandolf [7] suggest that chronic dilated cardiomyopathies are associated with persistence of a variety of different enteroviruses, including echoviruses. In my laboratory, we have utilized in vitro models to study the cellular and molecular aspects of acute and chronic infections by the human enterovirus, echovirus type 6. The purpose of these studies is to investigate the mechanisms responsible for the stability and duration of viruses during either acute or chronic infections. The ultimate goal of these studies is to develop an approach for the prevention and/or treatment of acute and chronic enterovirus infections including viral heart disease.

In order to accomplish these studies, it was necessary to develop reproducible systems for lytic (acute) and persistent, nonlytic (chronic) infections. We selected the susceptible human amnion cell line (WISH) for the lytic system in which yields of $\geq 10^5$ virus particles are produced by each infected cell within 6–8 h after inoculation. Cellular protein synthesis is inhibited between 3 and 4 h after virus inoculation, and the cells do not usually survive the infection. Occasionally a few of the infected cells escape destruction, and we used these cells to establish a cell line persistently infected by echovirus 6.

Establishment of Persistent Echovirus 6 Infection in Cultured Human Cells

WISH cells, surviving infection by a high dose of plaque-purified echovirus 6 [10–30 plaque forming units (PFU)/cell], were regrown and rechallenged with a similar dose of virus. The survivors were cloned by the limiting dilution method [13]. Most (14/24) of the cell clones were resistant to lysis after a third exposure to virus. A line of one of these resistant clones was established [4] and has been maintained as monolayer cultures in my laboratory for more than 5 years in the absence of any extracellular factors (e. g., antiviral antibodies) that might limit virus infection. The persistently infected (PI) cells could not be distinguished from the

H.-P. Schultheiß (Ed.)
New Concepts in Viral Heart Disease

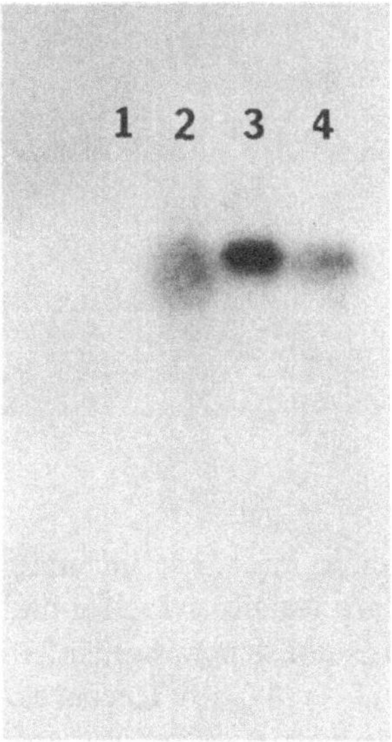

Fig. 1. Hybridization of echovirus-6 cDNA with Northern blots of RNA recovered from uninfected cells *(lane 1)*, acutely infected (10 PFU/cell; 4 h) cells *(lane 2)*, purified echovirus-6 virions *(lane 3)*, and PI cells *(lane 4)*. The RNAs were electrophoresed through methyl mercuric hydroxide-containing agarose gels and then transferred to diazotized paper. The Northern blots were treated with prehybridization buffer (4 h, 42 °C) before addition of the ^{32}P-labeled c DNA probe and incubation for 16 h at 42 °C. Nonspecifically bound probe was removed by five washes at 62 °C with decreasing concentrations of 2 x SSC (0.3 *M* NACl, 0.03 *M* trisodium citrate) to a final concentration of 0.1 x SSC. All washes except the last wash contained 0.1% sodium dodecyl sulfate. The washed blot was air dried and exposed to Fuji XR film at −70 °C for 2 days. Additional bands were not detected after exposure for 4 days

uninfected parental cells with respect to either morphology or growth rate. However, we demonstrated by hybridization and immunofluorescence that these cells contained viral RNA and synthesized viral proteins.

Radiolabeled echovirus 6 cDNA was used to probe for viral RNA from cytoplasmic extracts of PI cells and acutely infected WISH cells. RNA from uninfected WISH cells and echovirus 6 RNA were used as negative and positive controls, respectively. The viral cDNA hybridized with Northern blots of RNA from both acutely infected and PI cells, but not with RNA from uninfected cells (Fig. 1). The viral RNAs were the same size as genomic viral RNA; subgenomic species of viral RNA were not detected. Indirect immunofluorescent studies with monkey anti-echovirus-6 serum as the primary antibody and fluorescein-conjugated goat anti-monkey immunoglobulin G as the secondary antibody demonstrated that essentially all of the PI cells (86%) expressed viral antigens (Fig. 2). These results, along with the observations that persistence of echovirus-6 RNA and proteins occurred in the absence of antiviral antibodies and interferon and was not eliminated by these antiviral agents (4), indicated that we had established and maintained a steady state infection of echovirus-6 in cultured human cells.

Most of the previously described persistent infections by enteroviruses fit into the category of carrier state infections [1, 2, 8, 10, 12, 15, 16]. Carrier state infections involve a population of cells in which the majority of the cells are not infected. The small fraction of infected cells (≤1%) does not survive, and the newly produced virus is released into the extracellular fluid. Low concentrations of antiviral factors are necessary for maintenance of viral persistence because they limit the spread of virus to other cells. High concentrations of antiviral agents in the extracellular fluids can eventually eliminate these infections. Thus, in carrier state infections, viruses are transmitted horizontally from extracellular fluids to cells, whereas in steady state infections, viral persistence is maintained by transmission of viral genetic material during cellular division.

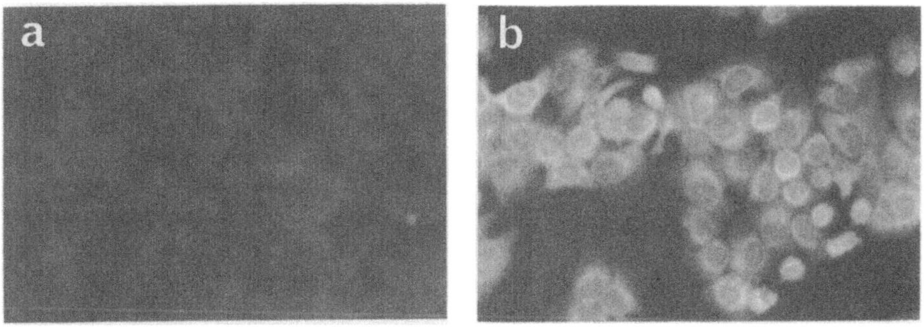

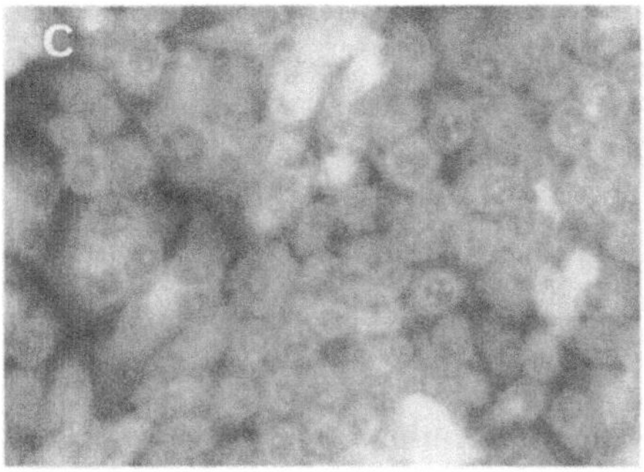

Fig. 2 a–c. Indirect immunofluorescent staining of uninfected **(a)** acutely infected **(b)** WISH cells, and PI **(c)** cells. Cells grown on coverslips were inoculated with either culture medium or echovirus-6 (16 PFU/cell) and incubated for 8 h at 37 °C before fixing with acetone at −4 °C (From [4])

Production of Nonlytic Virus Particles by PI Cells

Initially enough lytic echovirus-6 particles were produced by the PI cell line so that they could be identified by neutralization of infectivity by specific antiviral serum. However, infectious center assays suggested that only a small fraction (< 0.1%) of the viral population was lytic [4]. This possibility was investigated by comparison of physical properties of viruses recovered from lytic and nonlytic infections. Purified virus preparations from PI cells contained the same number of particles (6 x 10^{12}/ml) as preparations from parental WISH cells as estimated by ultraviolet light absorption at 260 nm and electron microscopy [4]. Yet only one lytic particle was detected in a population of 2 million virus particles recovered from the PI cells, instead of the several thousand virions detected in a comparable population of particles produced during a lytic infection. During serial passages, the virus population changed. Although all of the PI cells continued to synthesize viral RNA and proteins, only nonlytic virus particles were produced after 38 passages. Demonstration of these particles was dependent upon their recovery from linear cesium chloride gradients with the appropriate density (1.33 g/ml), ultraviolet light absorption (260 nm and 280 nm), and morphology as detected by electron microscopy. Both lytic and nonlytic virus particles were unenveloped icosahedrons with approximate diameters of 26 nm (Fig. 3). Identity of the particles as echovirus-6 was obtained from their specific interaction with antiviral serum in enzyme-linked immunoabsorbent assays (ELISA) and their immunoprecipitation by anti-echovirus-6 serum. Since the cultured cells did not undergo any visi-

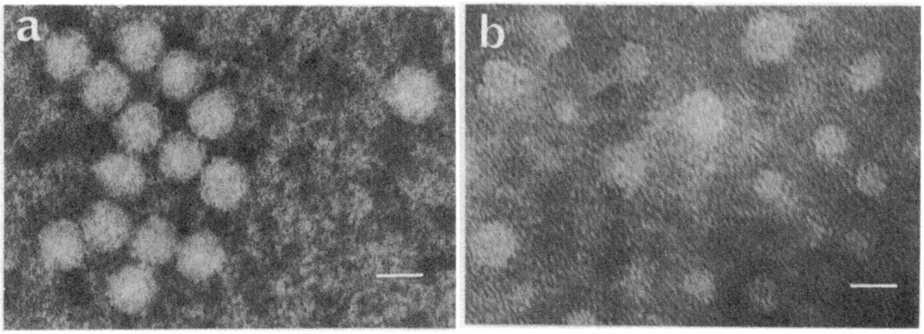

Fig. 3 a, b. Electron micrographs of standard, lytic echovirus-6 **(a)** and nonlytic echovirus-6 recovered from PI cells **(b)**. Purified virus preparations recovered from linear cesium chloride density gradients were dialyzed against phosphate-buffered saline (pH 7.2) and negatively stained with 1% phosphotungstic acid. The *bar* represents 25 nm

ble changes either upon establishment or after maintenance of a persistent infection, the existence of a viral infection would have been missed under ordinary circumstances, as might also be the case during chronic infections.

Characterization of Viruses Produced by PI Cells

The possibility that the nonlytic virus particles were defective was investigated. RNA was isolated from purified particles and analyzed in methyl mercury agarose gels. As shown in Fig. 4, viral RNA from purified particles produced by PI cells

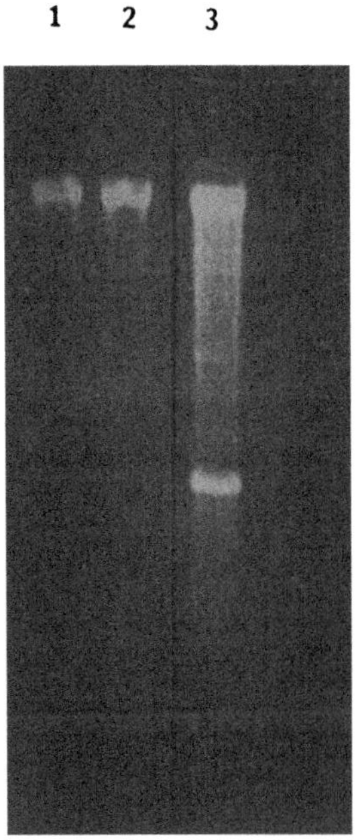

Fig. 4. Genome-sized viral RNA in particles recovered from PI cells. RNAs from purified lytic echovirus-6 produced by WISH cells *(lane 1)*, purified nonlytic echovirus-6 particles produced by PI cells *(lane 3)*, and RNA of the cow pea strain of tobacco mosaic virus *(lane 2)* were electrophoresed through methyl mercury agarose gels and stained with ethidium bromide. RNAs were extracted with phenol:chloroform:isoamyl alcohol (25:25:1)

Table 1. Infectivity of viral RNA from lytic and nonlytic echovirus 6 particles

	Virus Particles	
	Lytic	Nonlytic
RNA (ng/culture)[a]	40	300
PFU/culture	1.6×10^4	$< 1.0 \times 10^2$
PFU/ng RNA	400	< 1

[a] RNA transfections were carried out on monolayers in multiwells (4/sample) in the presence of 1 mg/ml DEAE-dextran
RNA concentrations were estimated by comparison to different concentrations of marker RNAs after electrophoresis in methyl mercury agarose gels

migrated to the same position as the RNA isolated from wild-type lytic particles. Also, approximately the same amount of RNA was recovered from the two different virus preparations containing comparable numbers of particles. Thus, the genome-sized viral RNA in PI cells was incorporated efficiently into virus particles. However, the RNA was defective since it did not initiate replication of lytic particles after transfection of parental WISH cells at 100-fold higher concentrations than that needed to demonstrate the infectivity of RNA from wild-type lytic virus (Table 1).

The protein profiles of the nonlytic virus particles were also examined and compared to those of lytic virus particles by sodium dodecyl sulfate-polyacrylamide gel electrophoresis of disrupted particles [4]. Immunoprecipitation and immunoblots with specific anti-echovirus-6 serum confirmed the viral origin of the polypeptides. The capsid polypeptides of lytic echovirus-6 had apparent molecular weights of 27, 25.8, 21, and 9.5 kilodaltons (kd) and were designated as VP1, VP3, VP2, and VP4, respectively (Fig. 5). Purified virus preparations also contained the 31.5-kd precursor polypeptide (VP0). Reduction in concentration of VP0 was accompanied by increased concentrations of VP2 and VP4. Monospecific antibodies, prepared against VP0, interacted in Western blots with VP2 and VP4 as well as with VP0. These results indicated a precursor relationship between VP0 and the two smaller polypeptides VP2 and VP4. Processing of VP0 is required for infectivity of enteroviruses and occurs during the final stage of maturation of the virus particle [6]. As demonstrated previously [4], the nonlytic particles contained VP0 but lacked the two smaller polypeptides even though they had the same protein concentration and density as lytic virus particles. Thus, the VP0 in particles from PI cells did not undergo cleavage.

Adsorption studies with radiolabeled virus preparations (^{35}S) demonstrated that the immature particles from PI cells could not attach to uninfected susceptible cells (WISH) which expressed receptors for lytic virus [4]. Thus, the RNA of the viruses from PI cells could neither gain entry to, nor be expressed in uninfected cells. These results demonstrated the defectiveness of the particles produced by PI cells and indicated that production of virus particles was not a prerequisite

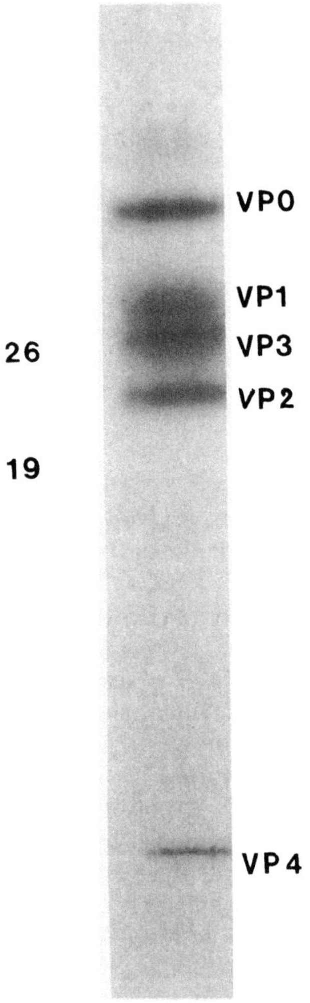

Fig. 5. Polypeptides of echovirus-6. ^{35}S-labeled virus was recovered from linear cesium chloride gradients, and the capsid polypeptides were resolved by electrophoresis in 12.5% sodium dodecyl sulfate polyacrylamide gels. The viral structural proteins are indicated on the *right*. The position of prestained molecular weight marker proteins, that were resolved in the same gel slab, are shown on the *left*

for persistence of the viral genome in cells. Recognition of this circumstance is important not only for diagnosis of persistent virus infections, but also for an understanding of the mechanisms of pathogenesis of chronic viral diseases.

Susceptibility of PI Cells to Virus Infection

The echovirus-6 preparation used to establish the steady state infection in PI cells did not destroy these cells after several weeks, even at concentrations 30-fold higher than necessary to disrupt the parental WISH cells in 24 h. The possibility that these cells were also resistant to destruction by other normally lytic viruses was investigated. Cell monolayers were inoculated with a related enterovirus (po-

Table 2. Comparison of virus titers in WISH and PI cells

	PFU/ml x 10^{-7}	
Virus	WISH	PI
Lytic echovirus 6	4.0	< 0.001
Defective echovirus 6[a]	0	0
Poliovirus I (Mahoney)	17.5	11.0
Vesicular stomatitis	1.2	0.7

[a] Approximately 10^9 virus particles produced by PI cells were used in the inoculum. Sets of four monolayers per cell line were used for assay of each virus preparation.

liovirus I) and the unrelated vesicular stomatitis virus (VSV) as well as with echovirus-6. The inoculated cultures were examined for development of cytopathology and plaque production during incubation at 37 °C (Table 2). Infected WISH cells and uninoculated PI cells served as positive and negative controls, respectively. Poliovirus I (Mahoney) and VSV (Indiana) were cytolytic for both PI and WISH cells, and similar yields of lytic virus were recovered from the two different cell lines. These results indicated that PI cells were capable of replicating and releasing cytolytic viruses. Thus, the resistance of PI cells to lysis by echovirus 6 was specific.

Effect of Interferon on Persistence of Echovirus 6 Genome

The PI cells were maintained and passaged serially 20 times in the presence of 5000 units of human leukocyte interferon. Untreated cells were handled in parallel. The supernates of the cells were tested routinely for lytic viral activity on uninfected WISH cells. Also, the cells, radiolabeled at 4 h after virus inoculation, were examined for the presence of ^{35}S-labeled viral proteins by sodium dodecyl sulfate-polyacrylamide gel electrophoresis of cytoplasmic extracts. Initially, the supernates of the PI cells contained lytic virus. Although untreated cells continued to produce lytic virus for an additional 15 passages, treatment with interferon inhibited the production of lytic virus by the fifth passage of PI cells. However, the PI cells continued to synthesize viral proteins and remained resistant to lysis after challenge with standard echovirus-6 (40 PFU/cell) for at least 26 additional serial passages after removal of interferon from the extracellular medium. These results indicated that interferon prevented production of lytic virus, but did not eliminate the persistent infection.

Resistance of PI Cells to Superinfection by Echovirus-6

Previously we reported that the survival of PI cells to challenge by high concentrations of lytic echovirus-6 was not due to interference by the defective nonlytic particles produced by PI cells, production of antiviral substances by PI cells, or lack of cellular receptors for attachment of virus particles [4]. The possibility existed that entry or uncoating of lytic viral RNA was prevented after attachment of echovirus-6 to PI cells. In order to examine this possibility, PI cells were inoculated with RNA extracted from lytic echovirus-6 and examined for cytopathology and virus production. Transfected WISH cell monolayers were included as positive controls for infectivity of the viral RNA. Uninoculated PI and WISH cells served as negative controls. Cytolysis occurred in WISH cells at 3 days after transfection of viral RNA, and all of the cultured cells were destroyed on day 4. The virus yield was 5.4 x 10^4 PFU/ml. Approximately 50% of the PI cells lysed in 4 days after transfection, and the virus yield was 4 x 10^4 PFU/ml. The surviving PI cells replicated in the presence of lytic virus and formed full monolayers within 7 days after transfection. These results indicated that the resistance of PI cells to superinfection by lytic echovirus-6 virions was due to a defect in uptake and/or uncoating of RNA after virus attachment. The viral specificity of this event suggests that the cellular change is dependent upon the presence and expression of defective echovirus-6 RNA. The significance of this oberservation is that persistence of a defective viral genome can alter cellular functions and may in certain instances preserve the integrity of infected cells in the presence of circulating homologous, competent viruses. However, enterovirus genomes undergo high rates of mutation in nature [5], and complementation between defective RNAs could result in production of lytic viruses with destruction of infected cells. Preliminary results in my laboratory support this possibility. Plaque morphology mutants of lytic virus were recently recovered from PI cells after prolonged maintenance of monolayer cultures at 37 °C (≥30 days). Currently, we are in the process of sequencing the genomes of lytic and nonlytic virus particles in order to identify regions of the viral RNA that are responsible for cytolysis. These studies will contribute to our understanding of viral pathogenesis.

Although some human viral infections may persist due to defects in the immune system and may represent carrier state infections, certain diseases caused by chronic viral infections, such as hepatitis, lymphomas, and degenerative diseases of the central nervous system, appear to be the result of steady state infections. Our studies indicate that human enteroviruses can also cause steady state infections which may play a role in chronic disease syndromes including certain heart diseases.

Summary

Echovirus-6 is one of many human enteroviruses which have been implicated in the etiology of heart disease. Persistence of the enteroviral RNA in the human myocardium may be related to the pathogenesis of chronic dilated cardiomyopathy. We have developed an in vitro model to study the persistence of the normally cytolytic echovirus-6 in cloned human cells and have maintained this system for more than 5 years. All of the cells contain viral RNA which is synthesized continuously. Some of the newly synthesized RNA remains cell associated, while the remainder of the viral RNA is incorporated into virus particles (≥ 1500 particles/cell). The persistence of the viral RNA confers resistance of infected cells to superinfection by homologous viruses and prevents the normal shutdown of protein synthesis during lytic virus infections. The steady state infection is maintained by the transfer of viral RNA to progeny cells during cellular division and is not dependent upon either the presence of antiviral factors in the extracellular media or the production of viral particles. Most of the virus particles produced by the PI cells are defective and cannot be detected by ordinary diagnostic techniques. These particles contain nonlytic RNA and an unprocessed capsid protein. Viral mutants appear during prolonged growth and division of the infected cells. Interferon treatment eliminates the production of lytic virus, but does not cure the cells of the infection. Thus, a system is available for the investigation of molecular events involved in pathogenesis of enteroviruses, which may provide an approach to treatment of enteroviral heart disease.

References

1. Carp RI (1981) Persistent infections of human lymphoid cells with poliovirus and development of temperature-sensitive mutants. Intervirology 15: 49–56
2. Crowell RL, Syverton JT (1961) The mammalian cell-virus relationship. VI. Sustained infection of HeLa cells by coxsackie B3 virus and effect on superinfection. J Exp Med 113: 419–435
3. Davis LE, Bodian D, Price D, Butler IJ, Vickers JH (1977) Chronic progressive poliomyelitis secondary to vaccination of an immunodeficient child. N Engl J Med 297: 241–245
4. Gibson JP, Righthand VF (1985) Persistence of echovirus 6 in cloned human cells. J Virol 54: 219–223
5. Holland JJ, Spindler K, Horodyski F, Grabau E, Nichol S, Vande Pol S (1982) Rapid evolution of RNA genomes. Science 215: 1577–1585
6. Jacobson M, Baltimore D (1958) Morphogenesis of poliovirus 1. Association of the viral RNA with the coat protein. J Mol Biol 33: 369–378
7. Kandolf R (1988) The impact of recombinant DNA technology on the study of enteroviral heart disease. In: Friedman H, Bendinelli M (eds) Coxsackieviruses — a general update. Plenum, New York
8. Kandolf R, Canu A, Hofschneider PH (1985) Coxsackie B3 virus can replicate in cultured human foetal heart cells and is inhibited by interferon. J Mol Cell Cardiol 17: 167–181
9. Kew OM, Nottay BK, Hatch MH, Nakano JH, Obijeski JF (1981) Multiple genetic changes can occur in the oral poliovaccines upon replication in humans. J Gen Virol 56: 337–347
10. Matteucci D, Paglianti M, Giangregori AM, Capobianchi MR, Dianzani F, Bendinelli M (1985) Group B coxsackieviruses readily establish persistent infections in human lymphoid cell lines. J Virol 56: 651–654
11. Miller JR (1981) Prolonged intracerebral infection with poliovirus in asymptomatic mice. Ann Neurol 9: 590–596
12. Pacsa S (1961) Poliovirus-carrying lines of HeLa cells: their establishment and sensitivity to virus. Acta Microbiol Acad Sci Hung 8: 329–332
13. Puck TT, Marcus PI (1955) A rapid method for viable cell titration and clone production with HeLa cells in tissue culture: the use of X-irradiated cells to supply conditioning factors. Proc Natl Acad Sci USA 41: 432–437
14. Rozhon EJ, Wilson AK, Jubelt B (1984) Characterization of genetic changes occurring in attenuated poliovirus 2 during persistent infection in mouse central nervous systems. J Virol 50: 137–144
15. Takemoto KK, Habel K (1959) Virus-cell relationship in a carrier culture of HeLa cells and coxsackie A9 virus. Virology 7: 28–44
16. Vogt M, Dulbecco R (1958) Properties of a HeLa cell culture with increased resistance to poliomyelitis. Virology 5: 425–434

17. Wilfert CM, Buckley, RH Mohanakamur T, Griffin JF, Katz SL, Whisnant JK, Eggleston PA, Moore M, Treadwell E, Oxman MN, Rosen FS (1977) Persistent and fatal nervous system ECHO virus infection in patients with agammaglobulinemia. N Engl J Med 296: 1485–1489
18. Wyatt HV (1973) Poliomyelitis in hypogammaglobulinemics. J Infect Dis 128: 802–806
19. Yoneyama T, Hagiawara A, Hara M, Shimojo H (1982) Alteration in oligonucleotide fingerprint patterns of the viral genome in poliovirus type 2 isolated from paralytic patients. Infect Immun 37: 46–53

Detection and Typing of Enteroviruses by Nucleic Acid Probes and Monoclonal Antibodies

T. HYYPIÄ[1], M. KALLAJOKI[2], and P. AUVINEN[1]

Departments of Virology[1] and Pathology[2], University of Turku, SF-20520 Turku, Finland

Introduction

Enteroviruses are currently detected using virus isolation followed by a neutralization test with a panel of antisera for specific typing of the virus strain. One of the problems in this method is the long time period needed for final diagnosis. In addition some coxsackievirus (CV) type A (CVA) strains can only be isolated in newborn mice. Gene technology and hybridoma methods offer new possibilities for development of reagents which can be used in enterovirus detection. The increasing knowledge on the genetic and antigenic structure, and mutual relatedness of enteroviruses further facilitates the applicability of these techniques.

Spot Hybridization

Enterovirus RNA was first detected by liquid hybridization [8] and more recently by solid support techniques [2, 4, 11]. In all these reports radioactively labeled probes were used. As a general rule, a passage in cell culture is needed prior to the test because the sensitivity of the methods presently used does not allow direct detection of enterovirus RNA in all isolation-positive clinical samples. However, the use of spot hybridization with broadly reacting nucleic acid probes makes possible rapid confirmation of enterovirus growth after virus isolation. New detection systems and the possibility to apply the in vitro amplification methods [12], which can multiply the number of target sequences up to 100000 times in a few hours, may permit direct demonstration of enteroviruses in clinical material in the future.

Enteroviruses consist of a protein capsid and an RNA genome which is approximately 7500 nucleotides long. One-tenth of the genome at the 5′ end is noncoding followed by regions coding for capsid and nonstructural proteins, and an approximately 100-nucleotide long 3′ end noncoding sequence. The presently available sequence data suggests that most of the 5′ end noncoding part and many regions of the nonstructural proteins are conserved among enteroviruses while the capsid region is more unique for a serotype. This was demonstrated for CV type B (CVB) [16] in a comparative analysis of the reactivity of cloned CVB3 cDNA fragments with the serotypes CVB 1–6. It was possible to localize the conserved and unique parts in the genome on the basis of reactivity in hybridization.

H.-P. Schultheiß (Ed.)
New Concepts in Viral Heart Disease

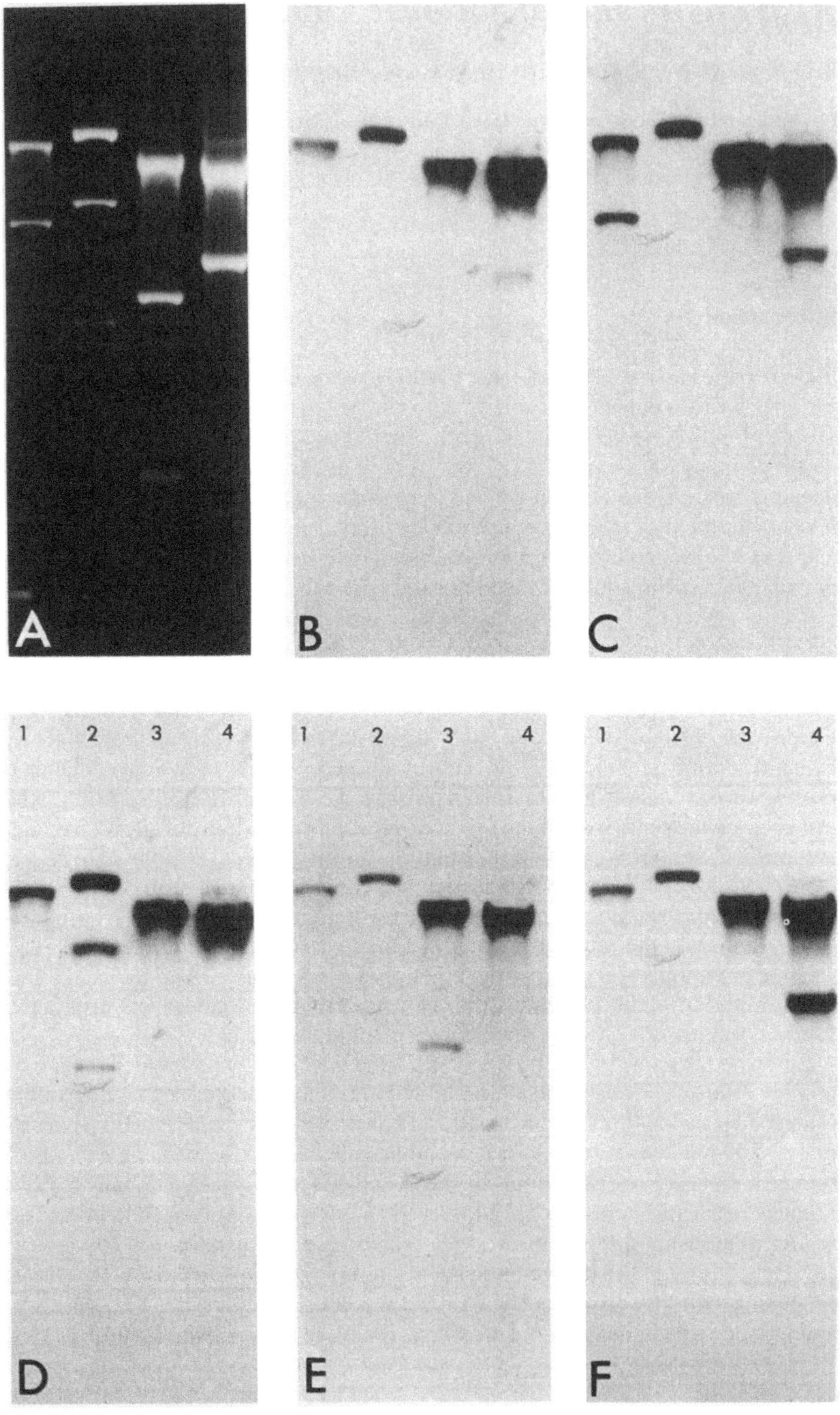
A
B
C
1 2 3 4
D
1 2 3 4
E
1 2 3 4
F

Hyypiä et al. [5] used cDNA probes containing 3′ end sequences of CVA21, CVB3, poliovirus 3, and enterovirus 70 genome for analysis of 48 enterovirus serotypes. The mutual cross-reactivity of these probes is shown in Fig. 1. More than 90% of the strains could be detected with this collection of reagents. The CVB3 probe covering 4.3 kb from the 3′ end of the genome detected the members of the CVB subgroup, virtually all the ECHO viruses and reacted also with some other serotypes. Poliovirus 3 and CVA21 probes had equal reactivity recognizing all polioviruses and also certain CVA serotypes, whereas enterovirus 70 probe was serotype specific.

Rotbart et al. [11] have used a cloned cDNA copy derived from the 5′ end of poliovirus 1 genome to detect members of each major enterovirus subgroup. Another probe could be used for more specific detection of poliovirus. Cova et al. [2] have reported the use of RNA probes for enterovirus detection. These probes have been produced by in vitro transcription using RNA polymerase reaction and bacterial promoters preceding the cloned poliovirus 1 fragments. In spite of the higher sensitivity of the RNA probes when compared to labeled cDNA, it was possible to detect enteroviruses directly only in a minority of clinical specimens.

We have recently extended the genetic analysis of enterovirus strains by using altogether 20 probes (Auvinen et al. submitted for publication). On the basis of these studies we have roughly divided enteroviruses into larger polio and CVB/ECHO subgroups, both containing members of CVA viruses. However, some serotypes, including at least ECHO virus 22 and enterovirus 70 are distantly related to other members of the group, which has to be taken into consideration when hybridization methods are used in diagnosis.

Specific regions of the genome can be detected either by using cloned cDNA probes or by synthetic oligonucleotides for the identification of serotypes. Often the first question after demonstration of an enterovirus in a clinical sample is whether it is a poliovirus or not. The answer can, for instance, be obtained by employing a short cloned sequence from the nonstructural region of polioviruses with the exception that some closely related CVA viruses may also react in the test (Auvinen et al. submitted for publication). Synthetic oligonucleotides can be used for a very specific analysis of the genome. Fig. 2 shows an example with a sequence selected from the poliovirus 5′ end noncoding region. In addition to polioviruses, CVA18 and, to a lesser extent, also enterovirus 70 give a signal. Oligonucleotides can be used for example for typing of CVB (U. Pettersson, personal communication). However, the extreme specificity of synthetic oligonucleotides may cause problems since point mutations which can affect the binding of the probe are common in enteroviruses. The development of optimal reagents therefore still needs more detailed knowledge on the biology of these important pathogens.

Fig. 1 A–F. Cross reactivity of enterovirus 3'end probes in Southern blotting. *Lane 1*, CVA21; *lane 2*, CVB3; *lane 3*, enterovirus 70; *lane 4*, poliovirus 3 probe DNA digested with appropriate restriction enzymes. **A** Gel stained with ethidium bromide to visualize all the DNA fragments; **B** pBR322 vector probe; **C** CVA21; **D** CVB3; **E** enterovirus 70; **F** poliovirus 3 probe

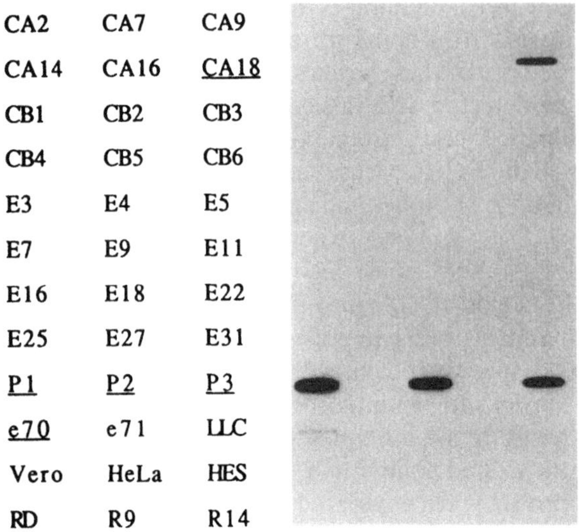

Fig. 2. Specificity of an oligonucleotide derived from the 5' noncoding region of polioviruses (5'-GTAGTCGGTTCCGCCACGG-3') in detection of enteroviruses. Cells were infected with virus serotypes as indicated, nucleic acids were prepared, applied onto a filter, and tested by hybridization using the 5' end-labeled oligonucleotide as a probe. The results were revealed by autoradiography on an X-ray film. *CA*, coxsackievirus A; *CB*, coxsackievirus B; *E*, ECHO virus; *P*, poliovirus; *e*, enterovirus; *R*, rhinovirus. *LLC, Vero, HeLa, HES,* and *RD* are uninfected cell controls

In Situ Hybridization

In situ hybridization methods are of special interest because they allow virus detection directly from tissue specimens. With this technique it is possible to identify the infected cell and to localize the hybridization signal in the cell. We have applied in situ hybridization for demonstration of CVB3-specific sequences in mouse tissues (BALB/c/a) after intraperitoneal inoculation of the virus (Vuorinen et al., submitted for publication). When the tissue samples were evaluated histopathologically, the animals were observed to develop acute necrotizing pancreatitis and mild focal myocarditis.

In situ hybridization was performed using paraffin sections of formaline-fixed tissues. The binding of sections onto microscope slides appeared to be a problem and often the sections did not remain attached during the entire hybridization procedure. Many investigators have tried to resolve this problem by using a variety of proteinaceous glues, polypeptides, or classical histochemical adhesion techniques. Our attempts with 1% Elmer's glue and polylysine- or Denhardt-coated slides were not satisfactory. However, using the organosilane methodology described by Maples [9] for use in immunocytochemistry, complete tissue adhesion was obtained.

Prehybridization treatment of the sections followed an earlier described protocol [13]. Proteinase K (1 μg/ml) digestion took place for 15 min at 37 °C. The cDNA probe used covered 4.3 kb from the 3′ end of the CVB3 virus genome [14] and it was labeled with ^{32}P in a nick translation reaction. Biotinylation was also performed using the nick translation reaction and biotin-11-deoxyuridine triphosphate (dUTP) as a deoxythymidine triphosphate (dTTP) analogue. The hybridization mixture [13] containing 0.1 μg/ml labeled probe was denaturated, cooled, and applied onto each slide, covered with plastic squares, and sealed with rubber cement. The slides were hybridized at 42°C for 18 h. After the hybridization procedure the tissue sections were washed thoroughly using standard methods.

For the detection of ^{32}P-labeled probe (Fig. 3a) autoradiography was performed using Kodak NTB-3 nuclear track emulsion, 1-3 days' exposure time, Kodak D-19 developer, and hematoxylin and eosin counterstaining. To detect the biotinylated probes, three different methods were used. We tested the applicability of streptavidin-Texas red conjugate (Fig. 3b) and two different methods based on the use of an antibiotin antibody. In the latter cases the goat anti-biotin was followed by biotinylated rabbit anti-goat antibody. Finally, biotin was detected using either avidin-peroxidase (Fig. 3c) or streptavidin-gold (5 nm particles; Fig. 3d) complexes. Peroxidase enzyme reaction was developed by using diaminobenzidine as chromogen and the colloidal gold was made visible in light microscopy by using a silver enhancement kit (Intence II, Janssen, Olen, Belgium). The slides were counterstained with hematoxylin. Controls of in situ hybridization included uninfected tissue, labeled pBR322 plasmid control probe, and detection system controls. Radioactive and biotin-based detection systems gave similar results in our mouse model. A positive hybridization signal was detected on days 1–5 post infection in the pancreatic tissue. The signal was localized to the acinar cells of the exocrine pancreas (Fig. 4), whereas islets of Langerhans and pancreatic ducts remained negative. The first lesions in the myocardium appeared 5 days after infection, but in situ hybridization results were negative in heart tissue.

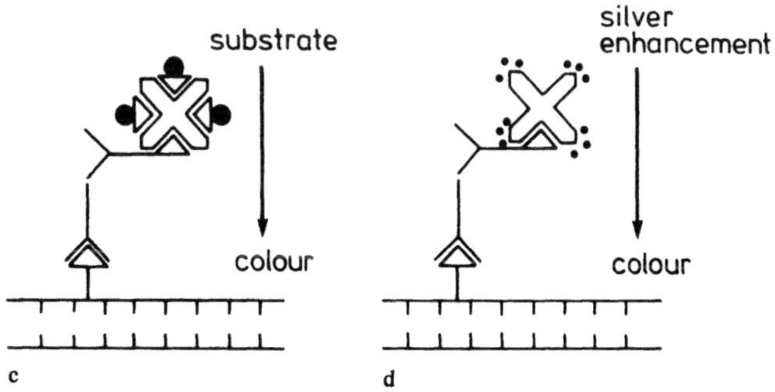

Fig. 3 a–d. Methods used in detection of the in situ hybridization signal. **a)** Radioactive label; **b)** biotinylated probe/streptavidin-Texas red conjugate; **c)** biotinylated probe/anti-biotin antibody/biotinylated secondary antibody/avidin peroxidase complex; **d)** biotinylated probe/antibiotin antibody/biotinylated secondary antibody/streptavidin-gold complex and silver enhancement.
Key: ∗, radioactive label; Δ, biotin; ☒, avidin or streptavidin; F, Texas red; ⅄, antibiotin primary antibody; ⅄, biotinylated secondary antibody; •, peroxidase; ∴, gold particles

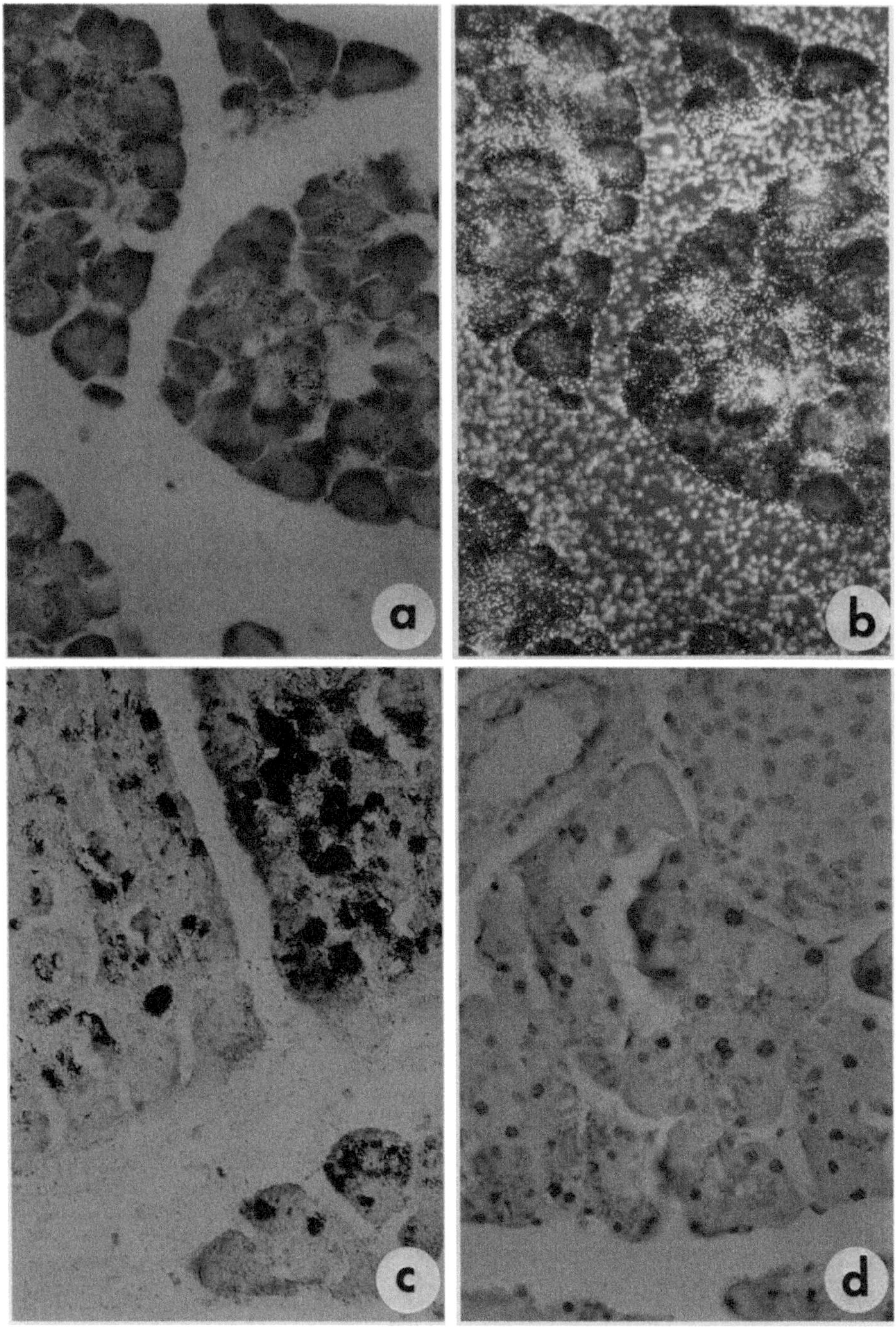

Fig. 4 a–d. Detection of CVB3 RNA by in situ hybridization in mouse pancreatic tissue 1 day after inoculation of the virus. **a)** and **b)** ^{32} P-Labeled probe detected by autoradiography in bright and dark field microscopy, respectively. Biotinylated probe was detected by anti-biotin antibody followed by biotinylated secondary antibody and either streptavidin/gold complex and silver enhancement **c)** or by avidin/peroxidase complex **d)** x 265

The sensitivity of all the detection systems was similar, but the localization of the signal was more precise with the nonradioactive methods. The ^{32}P label used makes short autoradiographic exposure times (1–3 days) possible compared to several weeks or even months with ^{3}H label. However, the resolution of the signal is much better with ^{3}H and biotin labels. Biotinylated probes have several advantages when compared to the radioactive ones: radioisotopes are rather expensive, inconvenient to handle, and the short half-life results in a restricted self-life of nucleic acid probes. Finally, the entire procedure with autoradiographic signal detection takes several days. These reasons make radioisotopes relatively impractical reagents for clinical laboratories. The use of biotinylated probes permits the whole in situ hybridization procedure to be completed in 1–2 days.

In our CVB3 mouse pancreatitis model, biotinylated probes proved to be as sensitive as the ^{32}P-labeled ones (Fig. 4). Of the biotin detection methods Texas red-conjugated streptavidin permits single step detection but demands the use of fluorescence microscope and does not produce permanent microscope preparates. An antibiotin-based system with streptavidin-gold/silver enhancement as the final step gives a black, clearly visible reaction product but also some background. In the avidin-peroxidase-complex procedure, the final reaction step gives a brown reaction product with exact localization and minimal background.

Other research groups have also applied the in situ hybridization technique in the detection of CVB3. Kandolf et al. [6] were able to detect CVB3 RNA in the affected myocardial cells of athymic mice (NMRI nu/nu) after intraperitoneal inoculation of the virus. However, we were not able to obtain a positive signal in the heart muscle of our mouse model in spite of the positive virus isolation and histologically detectable myocardial lesions. This may be due to the different models used because it is well known that the severity of CVB3-induced myocarditis is essentially dependent on the mouse strain used (for a review see [7]). Another difference is the use of cryocut sections by Kandolf et al. [6]. It is possible that probe penetration to the tissue is better in cryocuts than in the paraffin sections used by us.

Bowles et al. [1] have been able to detect enterovirus specific sequences by a slot blot hybridization assay in nine of 17 samples from patients with active healing myocarditis or dilatated cardiomyopathy with inflammatory changes. The in situ hybridization technique can, however, give more detailed information about the pathogenesis of infection by revealing the type and distribution of the infected cells. Recently Easton and Eglin [3] have detected CV RNA by in situ hybridization in six of 13 cardiac muscle specimens obtained from patients with a CV infection. The in situ hybridization methods described here are presently also used in the analysis of human biopsy samples.

Monoclonal Antibodies

The use of monoclonal antibodies together with synthetic peptides and X-ray crystallography has revealed the three-dimensional localization of poliovirus-neutralizing epitopes. These epitopes are, by definition, serotype specific, and antibodies detecting these structures are not practical in diagnostic use because a very large palette of reagents would be needed. Monoclonals detecting antigens common within the whole group or subgroups of enteroviruses would be more useful for routine purposes. The neutralizing monoclonal antibodies can, however, be used in analysis of antigenic variation of enterovirus isolates as has been shown for CVB4, for example [10].

We have produced monoclonal antibodies in mice against purified CVA9, CVB3, and ECHO 11 virions. The specificity of these reagents was tested by neutralization (NT), immunofluorescence (IF) and enzyme immunoassay (EIA). Neutralization titers up to 10^{-6} were observed in the ascites fluids and those with higher titers tended to have lower reactivity in IF and in EIA, especially in the case of CVA9. All except the five negative ones of the altogether 25 monoclonal antibodies were specifically positive against the parental strain and negative against 11 other enterovirus serotypes when tested by IF. Principally similar results were obtained by EIA using infected cells as antigen although a slight cross-reactivity was observed in some ascites fluids. Examples of the reactivity of the monoclonal antibodies is presented in Table 1. The high strain specificity can be explained by the use of purified nondenatured virions as antigens in immunization.

Recently Yousef et al. [18] have described production of a monoclonal antibody which reacts with all tested enteroviruses except hepatitis A virus. Heat-inactivated CVB5 was used as an antigen in immunization. The reactivity has been shown to be directed against the VP1 polypeptide. This finding opens new perspectives for the diagnostic procedures and shows that the form of the antigen used in immunization is an important factor for the reactivity of the antibodies.

The work on monoclonal antibodies has, without doubt, significantly increased our knowledge on the antigenic properties of enteroviruses, especially polioviruses. However, their applications in clinical diagnosis of enterovirus diseases still need further investigation. In the case of antigen detection assays the sensitivity may be another problem in addition to the high serotype specificity of most of the reagents. It has been reported that only 10% of poliovirus 3 isolation-positive stool specimens can be directly identified using a solid phase EIA with polyclonal antisera [17]. Similar results have also been obtained with other enteroviruses

Table 1. Examples of the reactivity of CVA9, CVB3, and ECHO virus 11 monoclonal antibodies in neutralization test, immunofluorescence, and enzyme immunoassay

Monoclonal antibody	NT	IF	EIA[1]
CVA9/1/4	10^{-3}	–	<20/<20
CVA9/1/7	10^{-1}	+	20 480/<20
CVB3/1/2	10^{-6}	+	20 480/ 20
CVB3/1/18	10^{-5}	–	5 210/ 320
E11/2/12	10^{-1}	+	20 480/ 20
E11/2/15	10^{-4}	+	1 280/<20

[1] EIA titer; infected/uninfected cells
E11, ECHO virus 11

(P. Halonen, personal communication). Therefore a passage in cell culture will most obviously be needed prior to the assay. However, EIA and IF would be rapid and convenient methods for confirmation of enterovirus growth in virus isolation if a suitable pattern of reagents were available.

Conclusions

The spot hybridization assay for enteroviruses in its present form is relatively rapid in confirmation of virus growth after cell culture isolation. The method has, however, to be sensitized before direct detection of enteroviruses in clinical samples can be routinely obtained. Development of new sensitive nonradioactive labels as well as in vitro amplification techniques will be needed before this approach can be used in clinical laboratories.

The use of in situ hybridization methods, especially with the rapid nonradioactive detection systems, offers a new possibility to detect viruses in tissue sections. When only a small proportion of cells in a tissue is infected, this method is more sensitive than the other hybridization techniques.

Monoclonal antibodies may become a useful tool in enterovirus diagnosis provided that standardized broadly reacting reagents become available. At the sensitivity level of assays presently used for antigen detection, a passage in cell culture may be needed prior to the identification of enteroviruses.

The expression of enterovirus antigens in bacteria has also to be considered seriously in enterovirus diagnostics. This approach has the special advantage that nonstructural proteins can be relatively easily produced [15]. These polypeptides would be valuable as broadly reacting antigens in the production of antiserum and for use in serological assays.

References

1. Bowles NE, Richardson PJ, Olsen EGJ, Archard LC (1986) Detection of Coxsackie-B-virus-specific RNA sequences in myocardial biopsy samples from patients with myocarditis and dilated cardiomyopathy. Lancet I: 1120–1123
2. Cova L, Kopecka H, Aymard M, Girard M (1988) Use of cRNA probes for the detection of enteroviruses by molecular hybridization. J Med Virol 24: 11–18
3. Easton AJ, Eglin RP (1988) The detection of coxsackievirus RNA in cardiac tissue by in situ hybridization. J Gen Virol 69: 285–291
4. Hyypiä T, Stålhandske P, Vainionpää R, Pettersson U (1984) Detection of enteroviruses by spot hybridization. J Clin Microbiol 19: 436–438
5. Hyypiä T, Maaronen M, Auvinen P, Stålhandske P, Pettersson U, Stanway G, Hughes P, Ryan M, Almond J, Stenvik M, Hovi T (1987) Nucleic acid sequence relationships between enterovirus serotypes. Mol Cell Probes 1: 169–176
6. Kandolf R, Ameis D, Kirschner P, Canu A, Hofschneider PH (1987) In situ detection of enteroviral genomes in myocardial cells by nucleic acid hybridization: an approach to the diagnosis of viral heart disease. Proc Natl Acad Sci USA 84: 6272–6276
7. Khatib R, Probert A, Reyes MP, Khatib G, Chason JL (1987) Mouse strain variation as a factor in the pathogenesis of coxsackievirus B3 murine myocarditis. J Gen Virol 68: 2981–2988

8. Kohne DE, Gibbs CJ, White L, Tracy SM, Meinke W, Smith RA (1981) Virus detection by nucleic acid hybridization: examination of normal and ALS tissues for the presence of poliovirus. J Gen Virol 56: 223–233
9. Maples JA (1985) A method for the covalent attachment of cells to glass slides for use in immunohistochemical assays. Am J Clin Pathol 83: 356–363
10. Prabhakar BS, Haspel MV, McClintock PR, Notkins AL (1982) High frequency of antigenic variants among naturally occurring human coxsackie B4 virus isolates identified by monoclonal antibodies. Nature 300: 374–376
11. Rotbart HA, Levin MJ, Villarreal LP (1984) Use of subgenomic poliovirus DNA hybridization probes to detect the major subgroups of enteroviruses. J Clin Microbiol 20: 1105–1108
12. Saiki RK, Scharf S, Faloona F, Mullis KB, Horn GT, Erlich HA, Arnheim N (1985) Enzymatic amplification of β-globin genomic sequences and restriction site analysis for diagnosis of sickle cell anemia. Science 230: 1350–1354
13. Sandberg M, Vuorio E (1987) Localization of types I, II, and III collagen mRNAs in developing human skeletal tissues by in situ hybridization. J Cell Biol 104: 1077–1084
14. Stålhandske POK, Lindberg M, Pettersson U (1984) Replicase gene of coxsackievirus B3. J Virol 51: 742–746
15. Strebel K, Beck E, Strohmaier K, Schaller H (1986) Characterization of foot-and-mouth disease virus gene products with antisera against bacterially synthesized fusion proteins. J Virol 57: 983–991
16. Tracy S (1984) A comparison of genomic homologies among the coxsackievirus B group: use of fragments of the cloned coxsackievirus B3 genome as probes. J Gen Virol 65: 2167–2172
17. Ukkonen P, Huovilainen A, Hovi T (1986) Detection of poliovirus antigen by enzyme immunoassay. J Clin Microbiol 24: 954–958
18. Yousef GE, Brown IN, Mowbray JF (1987) Derivation and biochemical characterisation of an enterovirus group-specific monoclonal antibody. Intervirology 28: 163–170

Generation of Enterovirus Group-Specific Antisera Using Bacterially Synthesized Coxsackievirus B3 Proteins

S. WERNER, H. SCHÖNKE, W. KLUMP, P. H. HOFSCHNEIDER, and R. KANDOLF

Department of Virus Research, Max-Planck-Institut für Biochemie, D-8033 Martinsried, FRG

Introduction

In North America and Europe, acute myocarditis is most commonly associated with infections by coxsackieviruses B (CVB) (types 1 to 5) [1, 10, 21, 26]. Other members of the human enteroviruses, e. g., various coxsackie A viruses and echoviruses have also been associated with human viral heart disease. These agents appear to be capable of producing dilated cardiomyopathy of acute onset or lead to life-threatening arrhythmias. Some of the acute cases of enterovirus heart disease may also evolve into a chronic form of dilated cardiomyopathy.

Although the clinical relevance of coxsackieviruses is increasingly recognized, there have been few studies concerning the molecular biology of these viruses or their interaction with host cells (2, 5). The genomic single-stranded RNA of coxsackieviruses consists of about 7500 nucleotides of positive polarity, covalently linked at the 5' end to a small virus-encoded protein VPg [4, 7]. The viral RNA, which is polyadenylated at its 3' end [3], functions as a mRNA in the cytoplasm of infected cells. It is translated into a polyprotein which is subsequently cleaved by viral proteases to the mature viral proteins [2, 3]. The four structural proteins VP1–VP4 (1A–1D, according to the systematic nomenclature of picornavirus proteins [22]) are generated from the amino terminal part of the polyprotein whereas the nonstructural proteins like the RNA-dependent polymerase (3D), VPg (3B), probably two proteases (2A and 3C), and several proteins of as yet unknown function are released from the carboxy terminal part.

We have recently developed an *in situ* hybridization assay using molecularly cloned CVB3 cDNA as a diagnostic probe [11, 12]. Because of the high degree of nucleic acid sequence homology among the numerous enterovirus serotypes, including the group A and B coxsackieviruses and the echoviruses, detection of these various agents commonly implicated in viral heart disease is possible in a single hybridization assay. Here we report on the generation of enterovirus group-specific antisera raised against bacterially synthesized CVB3 proteins which offer the unique possibility for an improved broad-spectrum diagnosis of enterovirus infection by antigen detection.

For the production of viral antigens, different parts of the CVB3 cDNA were subcloned into the vector pPLc24 and expressed in *Escherichia coli*. The expression products of three constructs were isolated and used for the immunization of rabbits. The specificity of the obtained antisera for native viral proteins was ana-

H.-P. Schultheiß (Ed.)
New Concepts in Viral Heart Disease

lyzed. In addition, the degree of cross-reactivity with viral capsid proteins of other enteroviruses was determined.

The *E. Coli* Expression System

The procaryotic expression vector pPlc24 [19] was used for the expression of distinct CVB3 antigens. It contains the strong left promotor P_L of bacteriophage λ, the ribosomal binding site, the start codon and the following 98 codons of the replicase gene of phage MS2. Transcription is controlled by the temperature-sensitive repressor cI857, which is active at 28 °C, but inactive at 42 °C [20]. This vector system is suitable for the expression of genes lacking expression signals or producing poisonous products.

By fusion of the reading frames of the vector-encoded MS2 replicase gene with the CVB3 coding sequences, fusion proteins can be expressed that contain MS2-specific amino acids at their amino terminus. Fusion of reading frames was achieved by two different methods which are shown in Fig. 1:

a) by direct blunt end ligation of end-repaired fragments of vector and insert (Fig. 1B); and
b) by introduction of an oligonucleotide adapter between the unrepaired fragments (Fig. 1C).

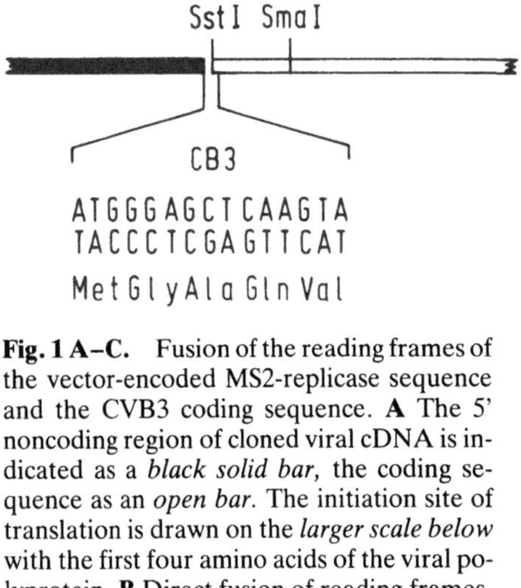

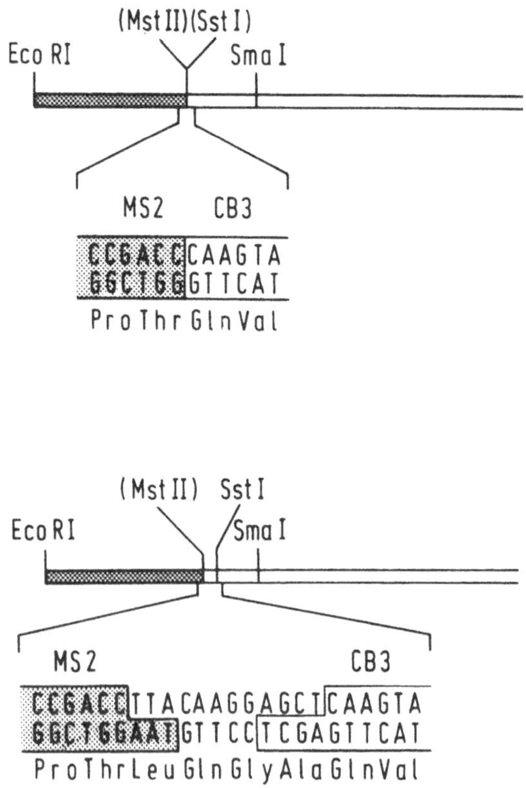

Fig. 1 A–C. Fusion of the reading frames of the vector-encoded MS2-replicase sequence and the CVB3 coding sequence. **A** The 5' noncoding region of cloned viral cDNA is indicated as a *black solid bar*, the coding sequence as an *open bar*. The initiation site of translation is drawn on the *larger scale below* with the first four amino acids of the viral polyprotein. **B** Direct fusion of reading frames. The vector-encoded MS2-replicase sequence is designated as a *screened bar*, the CVB3 coding sequence as an *open bar*. The fusion site is drawn on the *larger scale below*. Restriction sites which are lost by the fusion are put in *parenthesis*. **C** Fusion of the reading frames with the help of an oligonucleotide adapter. Vector-encoded sequence and CVB3-encoded sequence are designated as described in **B**. (From [25])

Expression of Distinct CVB3 Proteins

The expression of CVB3-specific proteins was performed as recently described [25]. Proteins of cell lysates were analyzed by $NaDodSO_4$-polyacrylamide gel electrophoresis [15] and immunoblotting [12, 22]. Expressed fusion proteins were detected by immunoblotting using an antiserum directed against the MS2-specific sequences which all expression products have in common at their amino terminal part [13].

To express the viral capsid proteins VP4 and VP2, the 1.35-kb *Sst*I-*Hind*III fragment of the infectious CVB3 cDNA (plasmid pCB3-M1) coding for VP4, VP2, and 115 amino acids of VP3, was inserted into pPLc24 (Fig. 2), whereby plasmid p1A–1C was obtained. A 58-kilodalton (kd) MS2-specific expression product was expected and found to be expressed, as is shown by immunoblotting (Fig. 3, lane 6). As the protein was expressed in large amounts, it could easily be detected in gels stained with Coomassie blue. Numerous MS2-specific proteins of

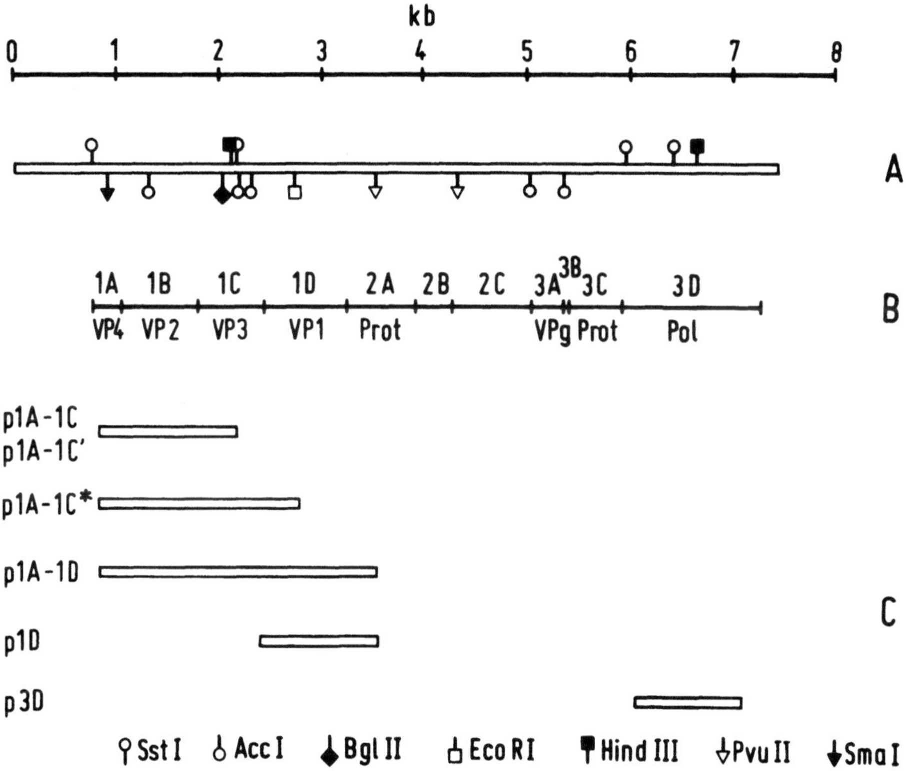

Fig. 2 A–C. Schematic representation of the construction of CVB3 subclones. **A** Restriction map of cloned infectious CVB3 cDNA. Restriction sites are indicated by *symbols:* ♀ *Sst* I; *Acc* I; *Bgl* II; *Eco* RI; *Hind* III; *Pvu* II; *Sma* I. **B** Viral polyprotein encoded by an open reading frame of about 6.6 kb. The location of post-translationally generated distinct viral proteins are indicated according to the systematic nomenclature of picornavirus proteins as proposed by Rueckert and Wimmer [22]. **C** CVB3 cDNA fragments which were inserted into the *E. coli* expression vector pPLc24. The expression plasmids obtained were designated as shown on the *left.* All cDNA fragments used for subcloning are drawn as *open bars*

a lower molecular weight were also detected and considered to be the result of a limited degradation of RNA or protein.

For the expression of sequences specific for VP3 and VP1, plasmids p1A–1C* and p1A–1D were constructed that encode either the capsid proteins VP4, VP2, VP3, and the amino terminal part of VP1 or the complete capsid protein precursor P1 (Fig. 2). The 81-kd expression product of p1A–1C* was obtained (Fig. 3, lane 4), but in significantly smaller amounts compared with the p1A–1C protein. In the case of p1A–1D the expected 110-kd primary translation product was not obtained, only several degradation products were detected (Fig. 3, lane 1). This finding indicates the presence of domains downstream from VP3 that confer instability in *E. coli*.

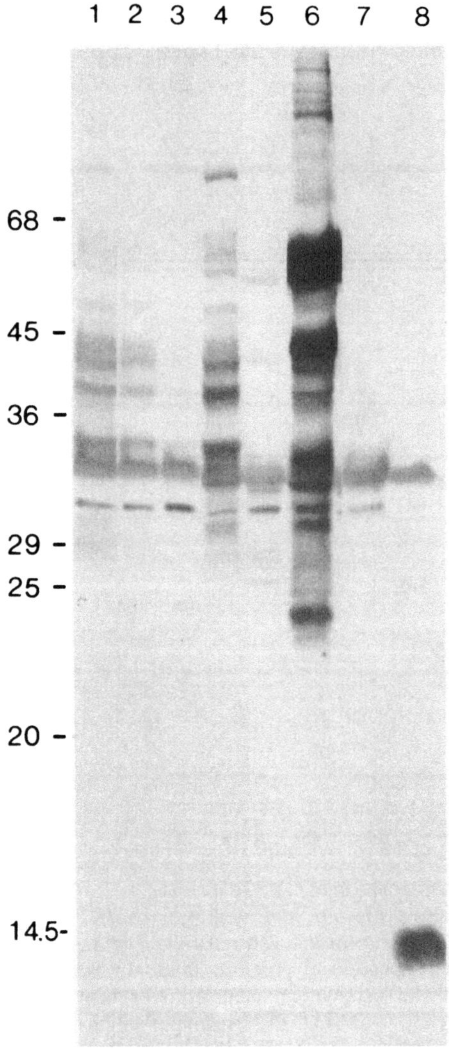

Fig. 3. Immunoblot analysis of expressed CVB3 fusion proteins. Bacterial cells were lysed 1 h after induction of expression. Cell lysates (25 μl) were applied to $NaDodSO_4$-polyacrylamide gel electrophoreses, blotted on nitrocellulose filter and incubated with aMS2, an antiserum which specifically recognizes the amino terminal MS2-replicase sequences of the fusion proteins. *Lane 1,* p1A–3D; *lane 2,* p1A–3C'; *lane 3,* p1A–1D; *lane 4,* p1A–1C*; *lane 5,* plD; *lane 6,* p1A–1C; *lane 7,* bacterial cells before induction of expression of p1A–1C*; *lane 8,* expression vector pPLc24 without CVB3 cDNA. Molecular weight markers (kd) are indicated on the *left.* (From [25])

Because VP1 could not be expressed stably as part of the capsid protein precursor, we subcloned the sequence coding for VP1 separately. A 1.2-kb *Acc*I-*Pvu*II fragment encoding 53 amino acids of VP3, the complete VP1, and 88 amino acids of 2A was inserted into pPLc24 (Fig. 2). After induction of expression from the newly constructed plasmid p1D, a 53-kd MS2-specific expression product was obtained in only low amounts (Fig. 3, lane 5). Therefore we conclude that the domain which is responsible for the instability of the fusion proteins must be present within VP1. The complete degradation of the p1A–1D primary translation product demonstrates that these domains are even more accessible when VP1 is expressed as part of the viral capsid precursor. The instability of VP1-specific proteins in *E. coli* seems to be a common phenomenon, as it was also observed when VP1 of poliovirus type 1 was expressed in the same system [6].

For the expression of a polymerase-specific antigen, the 1.08-kb *Bam*HI fragment from pCB3-M1 encoding 360 amino acids of the CVB3 polymerase 3D was inserted into pPLc24. The 51-kd expression product was obtained in high amounts (to be published).

Purification of Bacterially Synthesized CVB3 Proteins and Generation of Polyclonal Antisera

The expression products of p1A–1C, p1D, and p3D, which contain nonoverlapping fragments of the CVB3 polyprotein were chosen for the generation of antisera. These fusion proteins were purified from *E. coli* extracts as recently described [25]. First an urea extraction was performed, whereby the fusion proteins were enriched in the insoluble fraction. These proteins were separated by preparative polyacrylamide gel electrophoresis and the expression product was electroeluted [8]. In the case of the p1D fusion protein, where the enrichment achieved by urea extraction was not sufficient for further purification by preparative electrophoresis, most of the by-products were removed by a gel filtration step. A total of 150 μg of the purified expression product in 500 μl Freund's complete adjuvant was injected subcutaneously into rabbits. Rabbits were boostered with 50–75 μg of purified protein using the same procedure 12 and 35 days after the primary immunization.

The antisera obtained were titrated by an enzyme-linked immunosorbent assay, using the purified expression product as well as unpurified expression products from total cell extracts as antigens. Bacterial extracts without expression products served as a negative control. The antisera obtained were designated aVP4/2/3 (derived from the p1A–1C expression product), aVP1 (derived from the p1D expression product) and aPol (derived from the p3D expression product).

Specificity of Antisera for the Detection of Distinct Viral Proteins

The specificity of the antisera aVP4/2/3 and aVP1 was analyzed by immunoblotting using unpurified expression products as antigens. The antiserum aVP4/2/3

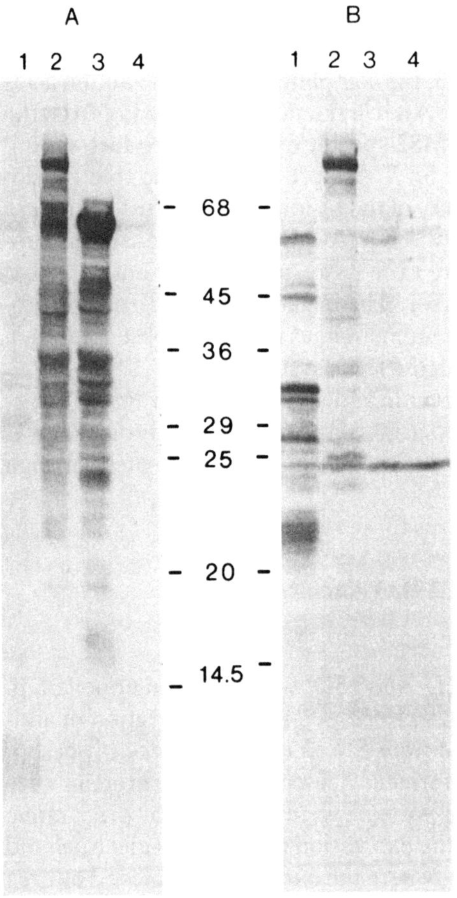

Fig. 4 A, B. Immunoblot analysis of expressed fusion proteins. $NaDodSO_4$-polyacrylamide gel electrophoresis and immunoblotting were performed exactly as in Fig. 3. Blots were treated with aVP4/2/3 **A** and aVP1 **B.** *Lane 1,* p1D; *lane 2,* p1A–1C*; *lane 3,* p1A–1C; *lane 4,* bacterial cells containing plasmid p1A–1C before induction of expression. Molecular weight markers (kd) are indicated

recognizes the homologous 58–kd antigen very well (Fig. 4, lane 3). Several proteins of a lower molecular weight — probably degradation products — must also be virus-specific, because bacterial extracts before induction of expression were not stained (Fig. 4, lane 4). The 81–kd p1A–1C* primary translation product which is homologous with the p1A–1C protein at its amino terminus, but contains additional carboxy terminal sequences, is also recognized. In contrast, only a very weak signal is obtained with the p1D expression product which does not contain overlapping CVB3 sequences. The weak cross-reactivity is due to the recognition of the MS2-specific part which all fusion proteins have in common.

aVP1 recognizes the unstable homologous p1D expression product, as well as the p1A–1C* fusion protein which also contains VP1-specific sequences. The cross-reactivity of this antiserum with the MS2-specific part is weak, as can be concluded from the very faint band obtained with the nonoverlapping p1A–1C expression product.

To analyze whether the antisera are able to detect native viral proteins, purified CVB3 was submitted to the immunoblot procedure. After treatment of the nitrocellulose filter with the antiserum aVP4/2/3, three bands were visible (Fig. 5, lane 1). By comparison of the molecular weight with those calculated from the nucleo-

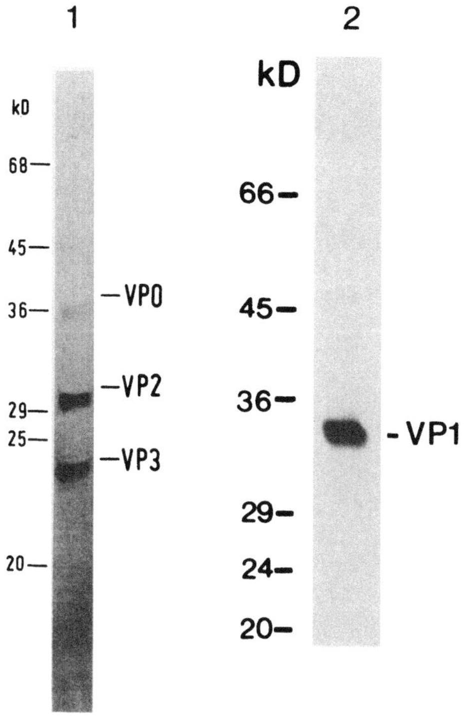

Fig. 5. Detection of CVB3 proteins using anti-VP1 and anti-VP4/2/3 antisera. Immunoblotting of capsid proteins obtained from CsCl gradient-purified CVB3. Viral proteins were separated by $NaDodSO_4$-polyacrylamide gel electrophoresis and transferred to a nitrocellulose filter. After treatment of the nitrocellulose with aVP4/2/3 and anti-rabbit IgG conjugated to peroxidase, the structural viral proteins VP3 (24 kd), VP2 (31 kd), and VP0 (precursor of VP4 and VP2, 37 kd) are detected *(lane 1)*. aVP1 specifically recognizes VP1 (33.5 kd) *(lane 2)*

tide sequence of the infectious CVB3 cDNA, the proteins were correlated to VP2 (31 kd), VP3 (24 kd), and VPO (37 kd). aVP1 recognizes a 33.5-kd protein (Fig. 5, lane 2). The molecular weight correlates with VP1, as calculated from the nucleotide sequence.

These results show that the antisera are able to detect distinct viral proteins. However, for an unequivocal diagnosis, the specific detection of viral polypeptides among eucaryotic proteins is necessary. We have therefore infected cell cultures with CVB3 and analyzed them by indirect immunofluorescence staining. Whereas noninfected cells were completely negative, CVB3-infected cells revealed a bright cytoplasmic fluorescence after treatment with all three antisera, as is shown for aPol in Fig. 6. These results demonstrate that the antisera which had been raised against bacterially synthesized fusion proteins are suitable for diagnostic purposes.

Polyclonal Antisera Raised Against Bacterially Expressed CVB3 Proteins are Enterovirus Group Specific

To estimate the degree of cross-reactivity of aVP4/2/3 and aVP1 with the capsid proteins of different enterovirus serotypes, Vero cells were infected with different enterovirus strains, including CVA9, the six CVB serotypes, echoviruses 11 and 12, and poliovirus type 1. Vesicular stomatitis virus (VSV) was used as a negative

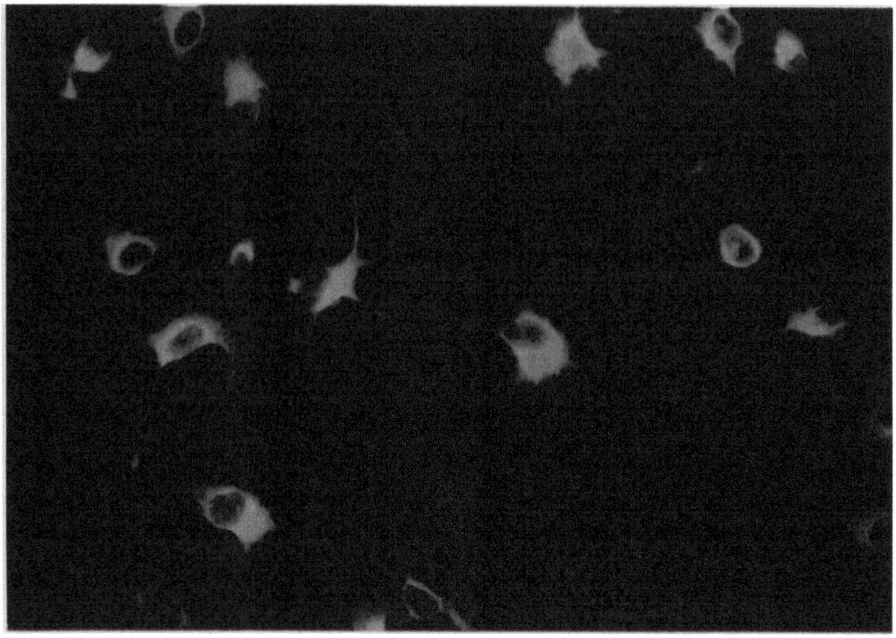

Fig. 6. Detection of CVB3-infected Vero cells by indirect immunofluorescence. Vero cells were seeded onto microscope slides and grown overnight. Cells were infected with CVB3 and incubated for 6 h. They were fixed with acetone, treated with aPol, followed by treatment with goat anti-rabbit/fluorescein isothiocyanate. Note bright cytoplasmic fluorescence of the infected cells. x250

control. Cells were infected with 10 plaque-forming units (PFU) per cell and harvested when about 100% of the cell population exhibited cytopathologic effects. Postmitochondrial supernatants were prepared and used as antigens in immunoblotting analysis. When 15 μg electrophoretically resolved total cellular protein of infected cells were blotted onto nitrocellulose, aVP4/2/3 revealed a remarkable cross-reactivity with the capsid proteins VP3, VP2 as well as VP0 (precursor of VP2 and VP4) of various enterovirus serotypes (Fig. 7, Table 1). VP0 and VP2 of all echoviruses and coxsackieviruses tested were recognized by this antiserum. In most cases, except CVB4 and CVB5, VP3 was also detected and even with poliovirus type 1 a faint band was visible. In contrast, cell cultures infected with the genetically unrelated VSV do not cross-react. In addition to aVP4/2/3, aVP1 also proved to be a useful group-specific reagent. As summarized in Table 1, strong signals were obtained with CVB1, CVB3, CVB5, echovirus 11, echovirus 12, whereas VP1 of CVB2, CVB6, and CVA9 revealed weak cross-reactivity. However, no significant reaction was observed with CVB4 and poliovirus type 1 using aVP1. These results demonstrate that the antisera are able to detect a broad spectrum of enteroviruses which might be implicated in viral heart disease.

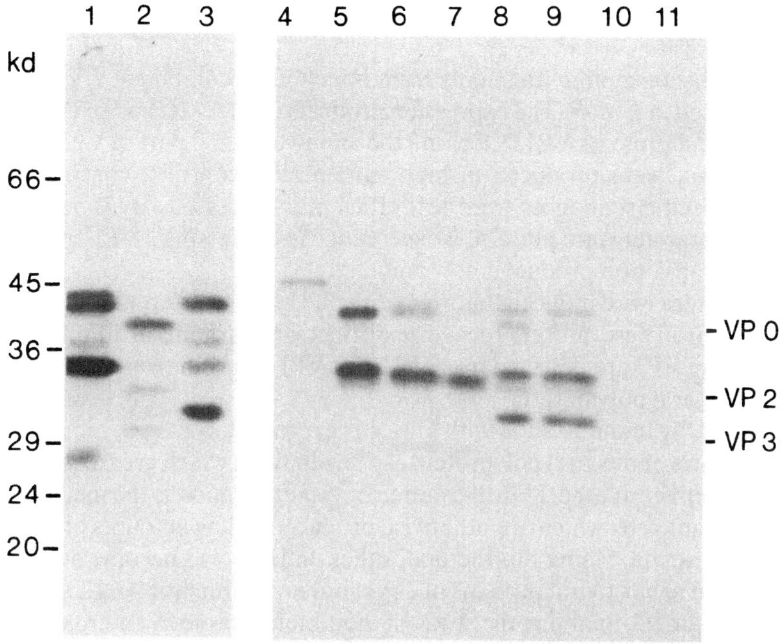

Fig. 7. Cross-reactivity of anti-VP4/2/3 antiserum with the capsid proteins of various enteroviruses. Electrophoretically resolved total cellular protein of infected cells (15 μg) were blotted on nitrocellulose. Treatment of the filter with aVP4/2/3 revealed a remarkable cross-reactivity of this antiserum with the appropriate capsid proteins (VP0, VP2, and VP3) of different coxsackieviruses and echoviruses. Note the weak reaction with CVB4 and especially PV1. *Lane 1,* CVB1; *lane 2,* CVB2; *lane 3,* CVB3; *lane 4,* CVB4; *lane 5,* CVB5; *lane 6,* CVB6; *lane 7,* CVA9; *lane 8,* echo 11; *lane 9,* echo 12; *lane 10,* PV1; *lane 11,* VSV

Table 1. Degree of cross-reactivity of anti-VP4/2/3 (a VP4/2/3) and anti-VP1 (aVP1) with the capsid proteins of various enteroviruses

		Coxsackievirus							Echovirus		Poliovirus
Antiserum	Antigen	B1	B2	B3	B4	B5	B6	A9	11	12	1
	VP0	+++	++	+++	+	++	++	(+)	+	+	0
aVP4/2/3	VP2	+++	+	++	(+)	+++	+++++		++	++	0
	VP3	++	+	+++	0	0	+	(+)	++	++	(+)
aVP1	VP1	+++	++	+++	(+)	+++	++	+	+++	+++	0

The reactions of the indicated antisera with the structural proteins of various enteroviruses are summarized. The reactions are graded: 0, no reaction; (+), very weak; +, weak; ++, strong; +++, very strong

Summary

Several subgenomic fragments from infectious recombinant CVB3 cDNA were expressed in *E. coli*. The expression products of p1A–1C and p3D, containing the structural proteins VP4, VP2, and the amino terminal part of VP3 or the viral polymerase, were produced in high amounts. In contrast, constructs containing VP1-specific sequences were less efficiently expressed. By subcloning the VP1 gene separately into pPLc24, we succeeded in expressing a VP1-specific 53-kd primary translation product in low amounts.

Antisera were induced that were directed against different parts of the viral polyprotein. These antisera recognize distinct viral proteins in different cell systems. VP2 and VP3 are detected by aVP4/2/3, aVP1 is specific for VP1 (Fig. 5) and aPol for the viral polymerase.

Thus, by immunization with antigens expressed in *E. coli*, antisera against distinct parts of the viral polyprotein can be obtained which are free from cross-reactivity with each other. Furthermore, *E. coli* expression is the method of choice to obtain antigens which are otherwise produced in low amounts or which are difficult to isolate. Using this method, other antisera will become available that are directed against viral proteins of as yet unknown function, e. g., against polypeptide 2B or 2C, enabling the study of their biologic role in enteroviral replication.

The enterovirus genus of viruses comprises at present 71 different serotypes, many of which are important human pathogens. Although a number of cross-relationships exist between different enteroviruses [17], the applicability of group-specific antigen detection of cardiotropic enteroviruses has met only modest success, because of the antigenic heterogeneity among and within the different serotypes. This fact has made immunohistochemistry in patients with a clinical suspicion of myocarditis a nearly impossible task. However, with our group-reactive antisera, the application of this technique might become a practical reality. Compared with hyperimmune antisera obtained by immunization of rabbits with purified enteroviral virions [18], the antisera described reveal an obviously broader spectrum of cross-reactivity. As shown by immunoblotting, aVP4/2/3 and aVP1 permit identification of various most commonly implicated agents of human viral heart disease, including the CVA and CVB as well as the echoviruses (Fig. 7, Table 1). This broad detection spectrum will greatly facilitate rapid identification of enteroviral infections by antigen detection. From the clinical point of view, the time-consuming antigenic typing of an etiologically implicated enterovirus strain appears to be of secondary importance and can be carried out in a second step.

The combination of the antisera with *in situ* hybridization is currently being pursued, making possible the simultaneous *in situ* detection of enteroviral RNA and the demonstration of distinct viral proteins at the single cell level.

Acknowledgements. We thank Helga Riesemann and Brigitte Schwaiger for excellent technical assistance. This work as supported in part by Grant Ka 593/2–1 from the Deutsche Forschungsgemeinschaft and by Grant 321–7291–BCT–0370 from the German Ministry for Research and Technology.

References

1. Abelmann WH (1973) Viral myocarditis and its sequelae. Ann Rev Med 24: 145–152
2. Chatterjee N, Tuchowski C (1981a) Comparison of capsid polypeptides of group B coxsackieviruses and polypeptide synthesis in infected cells. Arch Virol 70: 255–269
3. Chatterjee N, Tuchowski C (1981b) Translation of coxsackievirus B RNAs in a rabbit reticulocyte lysate: Characterization of the genome RNA, reaction conditions for translation, and analysis of the products. Arch Virol 70: 271–283
4. Chatterjee N, Nejman C (1985) Membrane-bound virions of coxsackievirus B4: Cellular localization, analysis of genomic RNA, genome-linked protein, and effect on host macromolecular synthesis. Arch Virol 84: 105–118
5. Crowell RL, Philipson L (1971) Specific alterations of coxsackievirus B3 eluted from HeLa cells. J Virol 8: 509–515
6. Enger-Valk BE, Jore J, Pouwels PH, Van der Marel P, Van Wezel TL (1984) Expression in *Escherichia coli* of capsid protein VP1 of poliovirus type 1. In: Modern approaches to vaccines. (R. M. Chanock, R. A. Lerner, eds.) Cold Spring Harbor: 173–175
7. Hewlett MJ, Florkiewicz RZ (1980) Sequence of picornavirus RNAs containing radioiodinated 5′-linked peptide reveals a conserved 5′-sequence. Proc Natl Acad Sci USA 77: 303–307
8. Hunkapillar MW, Lujan E, Ostrander F, Hood LE (1983) Isolation of microgram quantities of proteins from polyacrylamide gels for amino acid sequence analysis. In: Meth Enzym 91: 227–236
9. Hyppiä T, Stålhandske P, Vainionpää R, Petterson U (1984) Detection of enteroviruses by spot hybridization. J Clin Microbiol 19: 436–438
10. Kandolf R (1988) The impact of recombinant DNA technology on the study of enterovirus heart disease. In: Coxsackieviruses – A General Update. (M. Bendinelli and H. Friedmann, eds.) Plenum Press, New York, pp. 293–318
11. Kandolf R, Hofschneider PH (1985) Molecular cloning of the genome of a cardiotropic coxsackie B3 virus: Full-length reverse-transcribed cDNA generates infectious virus in mammalian cells. Proc Natl Acad Sci USA 82: 4818–4822
12. Kandolf R, Ameis D, Kirschner P, Canu A, Hofschneider PH (1987) In situ detection of enteroviral genomes in myocardial cells by nucleic acid hybridization: An approach to the diagnosis of viral heart disease. Proc Natl Acad Aci USA 84: 6272–6276
13. Klump W (1985) Untersuchungen zur proteolytischen Wirkung der Protease des Maul- und Klausenseuchevirus: Etablierung eines Modellsystems in *E. Coli*. Dissertation, Fakultät für Chemie und Pharmazie der Ludwig-Maximilians-Universität, München
14. Klump W, Marquardt O, Hofschneider PH (1984) Biologically active protease of foot-and-mouth disease virus is expressed from cloned viral cDNA in *Escherichia coli*. Proc Natl Acad Sci USA 81: 3351–3355
15. Maizel J (1971) Polyacrylamide gel electrophoreses of viral proteins. In: Meth Virol 5:179–246
16. Melnick JL (1985) Enteroviruses: Polioviruses, coxsackieviruses, echoviruses, and newer enteroviruses. In: Virology (B. N. Fields, ed.), Raven Press, New York, 739–794
17. Melnick JL, Wenner HA, Philips CA (1979) Enteroviruses in Diagnostic Procedures for Viral, Rickettsial and Chlamydial Infections, eds. Lennette EH, and Schmidt NJ (Am Public Health Assoc., Washington, DC), 471–534
18. Mertens T, Pika U, Eggers HJ (1983) Cross antigenicity among enteroviruses as revealed by immunoblot technique. Virology 129: 431–442
19. Remaut E, Stanssens P, Fiers W (1981) Plasmid vectors for high-efficiency expression controlled by the P_L promoter of coliphage lambda. Gene 15: 81–93
20. Remaut E, Tsao H, Fiers W (1983) Improved plasmid vectors with a thermoinducible expression and temperature-regulated runaway replication. Gene 22: 103–113
21. Reyes MP, Lerner AM (1985) Coxsackievirus myocarditis – with special reference to acute and chronic effects. Prog Cardiovasc Dis 27: 373–394
22. Rueckert RR, Wimmer E (1984) Systematic nomenclature of picornavirus proteins. J Virol 50: 957–959

23. Towbin H, Staehelin T, Gordon J (1979) Electrophoretic transfer of proteins from polyacrylamide gels to nitrocellulose sheets: Procedure and some applications. Proc Natl Acad Sci USA 76: 4350–4354
24. Tracy S (1984) A comparison of genomic homologies among the coxsackievirus B group: Use of fragments of the cloned coxsackievirus B3 genome as probes. J Gen Virol 65: 2167–2172
25. Werner S, Klump W, Schönke H, Hofschneider PH, Kandolf R (1988) Expression of coxsackievirus B3 capsid proteins in *Escherichia coli* and generation of virus-specific antisera. DNA7, Vol. 5: 307–316
26. Woodruff JF (1980) Viral myocarditis. Amer J Pathol 101: 427–478

III. Virus and Immunity

Myocarditis: A Postinfectious Autoimmune Disease

N. R. ROSE, N. NEU[1], D. A. NEUMANN, and A. HERSKOWITZ

Department of Immunology and Infectious Diseases, The Johns Hopkins University, School of Hygiene and Public Health, 615 North Wolfe Street, Baltimore, Maryland 21205, USA
[1] Institute of General and Experimental Pathology, University of Innsbruck, Fritz-Pregl-Str. 3, 6020 Innsbruck, Austria

Introduction

An association of autoimmune disease with infection, including viral infection, is strongly supported by clinical observation and epidemiologic study. A cause-and-effect relationship, however, between a particular infectious agent and a specific autoimmune response has been difficult to establish. A number of plausible mechanisms have been proposed, including molecular mimicry, changes induced in endogenous antigens, and disturbance in the host's immune response. Coxsackievirus (CV) infections frequently initiate autoimmune response in humans and provide a tool for understanding postinfection autoimmune disease. The first step is to delineate the specificity of the autoimmune response and identify the corresponding antigens. In the case of postinfectious autoimmunity, this task may be complicated by the fact that only a minor proportion of patients go on to develop significant autoimmune disease. Analysis of the pathogenetic potential of the autoimmune response requires establishing a malleable model.

Characterization of Postviral Myocarditis

CV group B (CVB) is the most frequent cause of human viral myocarditis. Since susceptibility of a host to viral infection is thought to be under genetic control, we postulated that murine susceptibility to myocarditis induced by CVB type 3 (CVB3) would similarly be under genetic control. In initial experiments, inbred strains of mice which differed in either the major histocompatibility complex (MHC) or in the background (i.e., non-MHC) genome were inoculated with 10^5 CVB3 at 4 weeks of age. Our initial studies indicated that both MHC and non-MHC genes play a role in determining the extent of myocardial injury after CVB3 infection [6, 7].

From these preliminary data, a larger study was undertaken to examine the inherent predisposition of different inbred strains of mice to develop CVB3-induced myocarditis [8]. The A.BY/SnJ (H-2^b), A.SW/SnJ (H-2^s), A.CA/SnJ (H-2^f), B10.S/SgSf (H-2^s), B10.PL/SgSf (H-2^u), and C3H.NB/SnJ (H-2^p) strains were

H.-P. Schultheiß (Ed.)
New Concepts in Viral Heart Disease

found to vary widely in the extent and duration of viremia, in the temporal appearance and titer of neutralizing antibody, and in the prevalence, severity, and duration of active myocarditis. Four strains, A.BY, A.SW, A.CA, and C3H.NB, were most severely affected. Their immunologic profile was characterized by high levels of viremia, delayed production of neutralizing antibody, and delayed clearance of virus from the heart. In contrast, the B10.S and B10.PL strains exhibited lower levels of viremia, consistently low neutralizing antibody levels, and rapid clearance of virus from the heart. A detailed study of the histopathology of CVB3 myocarditis [2] in these six strains for the first 45 days after virus inoculation was then undertaken. Although there were dramatic differences in the two strains, the histologic pattern of the early phases of myocarditis (days 1–9) was essentially the same. Focal zones of myocyte necrosis involving polymorphonuclear leukocytes as well as foci of myocytes with contraction band injury were the predominant early manifestations of direct viral injury. At this stage, the inflammatory process is not associated with evidence of global myocardial injury such as myocyte hypertrophy or ventricular dilation. The same strains continued to show inflammation after day 10. This "late-phase-myocarditis" is histologically different from the early-phase disease and inflammation in these late-phase hearts peaked at a time when *no* virus could be recovered from them. The pathology of late-phase myocarditis was characterized by a diffuse interstitial mononuclear cell myocarditis which was similar to the pattern of inflammation seen in human myocarditis. This diffuse interstitial myocarditis was seen mainly in mice with the A background genome and was associated with both tissue-bound IgG immunoglobulin and circulating heart IgG autoantibodies. The influence of the major histocompatibility complex on autoimmune myocarditis was further demonstrated by examining a larger panel of congenic mouse strains at 21 days after infection. Autoimmune myocarditis was present in all the A/J and BALB/c H-2 congenics along with the DBA/2J strain, but was absent in C57BL/6J and the B10 MHC congenics. These data substantiate the previous conclusion of Wolfgram et al. [8] that a gene (genes) other than the MHC controls the predisposition to autoimmune myocarditis. However, the appearance of myocarditis in three of five C3H MHC congenics suggests that H-2 does influence the magnitude of the immune response.

The presence of heart autoantibodies provided a means of identifying heart antigen(s) that were associated with the late-phase autoimmune myocarditis. The identification and characterization of these autoantibodies as well as their relationship to the development of late-phase myocarditis became the next priority of our investigations.

Identification of Circulating Heart Antibodies

Initial studies using indirect immunofluorescence in our laboratory [7] first revealed circulating heart autoantibodies in mouse strains genetically predisposed to late-phase myocarditis. These circulating antibodies were IgG immunoglobulins, could only be completely absorbed with heart tissue, and produced both antisarcolemmal and anticontractile protein staining patterns in indirect immunofluorescence.

Characterization of Circulating Heart Antibodies

Sera from mouse strains which were susceptible (e.g., mainly those with the A background genome) as well as those which were resistant (e.g., mainly those with the B10 background) to late-phase myocarditis were used to identify target myocardial autoantigens [1]. This panel of serum pools was then tested against solubilized extracts from whole heart and skeletal muscles. Results of western immunoblotting analyses demonstrated that antibodies to *cardiac myosin heavy chain* were a prominent feature in the sera of strains which developed late-phase myocarditis. Myocardial autoantibodies were not identified in strains which did not develop late-phase myocarditis. Subsequently, the late-phase sera were assayed against low-salt, high-salt, and a number of detergent extracts of heart and skeletal muscle (these *sequential* buffers enrich for cytostolic, contractile, and membrane-associated antigens). Myosin antibodies still showed the most prominent reaction, although other autoantigens were detected. Within the limits of the immunofluorescent and immunochemical assays, absorption with purified cardiac myosin removed the vast majority of heart reactivity. Both sarcolemmal and contractile protein staining patterns by indirect immunofluorescence were abolished by absorption with cardiac myosin. These studies suggested that myosin was the major autoantigen in CVB3-induced autoimmune myocarditis.

Specificity of Heart Autoantibodies to Cardiac Myosin Isoform

We subsequently developed an enzyme-linked immunoabsorbent assay (ELISA) to quantitate the autoantibodies and to test for myosin isoform specificity. Initially, pooled serum from A.CA (susceptible) and B10.PL (resistant) mice obtained at 15–21 days after CVB3 infection was tested for antibodies to cardiac myosin [3]. This interval was chosen because mice susceptible to late-phase myocarditis show a peak of heart autoantibody at this time. As expected from previous work, sera from CVB3-infected mice as well as from normal mice were positive for cardiac myosin autoantibodies. Sera from infected A.CA mice, though, showed a significantly higher level of IgG antibody to cardiac myosin than infected B10.PL strain and normal, noninfected mice. These latter two groups had mainly IgM antibodies to cardiac myosin. The level of IgM antibodies to cardiac myosin was identical for all groups. The three serum pools were additionally analyzed for their IgG antibodies to heart, skeletal, and brain myosin isoforms. The results indicated that sera from the susceptible A.CA mice exhibited more IgG antibody against heart myosin than against skeletal muscle or brain myosins. Sera from infected B10.PL and noninfected mice had low levels of myosin IgG antibodies. Competitive inhibition assays using all three myosins as inhibitors suggested that sera from late-phase myocarditis contained at least two populations of myosin antibodies. One reacted exclusively with heart myosin and was an IgG, whereas the second was an IgM autoantibody population which cross-reacted with all three myosins. These results suggested that the predominance of IgG antibodies to cardiac myosin may be a unique feature of late-phase myocarditis and could be considered a serologic marker for autoimmune heart disease.

Autoimmune Myocarditis Induced with Immunization with Purified Cardiac Myosin Heavy Chain

By immunizing a panel of mouse strains with defined susceptibility to CVB3-induced myocarditis, we found that cardiac myosin was capable of inducing myocarditis and that this model of myocarditis parallels CVB3 myocarditis in many respects [5]. We selected four mouse strains susceptible to late-phase myocarditis, A.SW (H-2^s), A/J (H-2^a), A.BY (H-2^b), and A.CA (H-2^f), and three resistant strains, C57BL/6J (H-2^b), C57BL/10J (H-2^b) and B10.A/SgSnJ (H-2^a). Six-week-old mice were injected twice with varying doses (20, 50, and 100 μg) of cardiac or skeletal myosin at 7-day intervals. The myosins were dissolved in sodium pyrophosphate and emulsified in complete Freund's adjuvant. Serum samples were taken on days 7, 14, and 21 after immunization for IgG antibodies to heart and skeletal muscle. Immunization with cardiac myosin elicited high-titered IgG antibodies to both cardiac and skeletal muscle myosin in A.SW, A.CA, and A/J mice. In these three strains, the myosin antibody titers ranged between 5120 and 10240 on day 21, whereas the titers in C57BL/10J and C57BL/6J mice averaged about 600 which is no different than controls immunized with adjuvant alone. Intermediate titers of myosin IgG antibodies were found in A.BY and B10.A mice (approximately 1280). Paralleling autoantibody production, immunization with 100 μg cardiac myosin also induced a severe autoimmune myocarditis in the majority of A.SW, A.CA, and A/J mice. The prevalence of myocarditis in the low responders (C57BL/10J and C57BL/6J) was 0%, while the prevalence in the intermediate responders (A.BY and B10.A) was 29% and 47%, respectively. A.BY mice, despite their susceptible A background genome, were relatively poor responders to cardiac myosin by virtue of the H-2^b alleles. The interplay between H-2 and non-H-2 genes is thus evident. Immunization with skeletal muscle myosin did not induce myocarditis in any of the mouse strains tested, and no histologic signs of myositis were detected. The myosin antibody titers induced with skeletal muscle were low in all strains, averaging about 300. Absorption of pooled sera with skeletal muscle myosin only removed 50% of the antiheart myosin reactivity, and competitive inhibition assays confirmed specificity of the residual autoantibody for heart myosin.

Histologic and Immunopathologic Similarities Between Late-Phase Post-CVB3 Myocarditis and Myosin-Induced Autoimmune Myocarditis

Hearts from both late-phase myocarditis and myosin-induced myocarditis appear grossly dilated. There are active small foci of myocyte necrosis, although the characteristic histologic feature is a prominent interstitial mononuclear cell infiltrate throughout the myocardium. Morphologic evidence of diffuse myocardial cell hypertrophy, interstitial fibrosis and myocyte degenerative changes of myocytolysis are equally evident in susceptible strains by light microscopy. Electron-microscopic evaluation confirms a similar development of diffuse myopathic changes even in areas away from foci of inflammation. By electron microscopy, morphologic features of diffuse myocyte injury consistent with cardiomyopathic change are:

(a) nuclear pleomorphism characteristic of myocyte hypertrophy; (b) loss of myofibrillar density; and (c) evidence of mitochondrial degeneration, such as significant loss of mitochondrial cristae density and the presence of dense bodies within scattered mitochondria. These myopathic changes are similar to those seen in human cardiomyopathy and indicate that in both models of autoimmune myocarditis there is evidence of general myocardial cell involvement. To study further the relationship between autoimmunity and the development of cardiomyopathy, we examined the hearts from the large panel of congenic strains for the presence of morphologic evidence of diffuse myopathic changes. Results showed that during the acute phase of myocarditis (days 5–7) no cardiomyopathic changes were observed. By day 21 after CVB3 infection, all of the A congenics (i.e., A/J, A.BY/SnJ, A.CA/SnJ, and A.SW/SnJ) and the BALB/cByJ, BALB.B/Kh, and BALB.K/Kh strains exhibited cardiomyopathic changes. Fibrosis, myocyte hypertrophy, and myocytolysis were also observed in the DBA/2J mice. No such pathologic findings were found in the C57BL/6J inbred line, in nine of the ten B10 H-2 congenics or in the C3H MHC congenics. The B10.WB/SgSf ($H\text{-}2^j$) displayed evidence of diffuse myopathy. With the exception of the B10.WB/SgSf strain, there was a correlation between the autoimmune myocarditis and the development of diffuse myopathic changes. Of the ten lines which developed autoimmune myocarditis, seven exhibited cardiomyopathic changes. Thus, we believe that the late-phase myocarditis associated with autoimmunity is also associated with the development of diffuse cardiomyopathic histologic changes.

Immunohistochemical staining of the myocardium in both late-phase myocarditis (day 14) and in myosin-induced myocarditis (days 14, 21) reveals sarcolemmal staining of IgG antibody. Hearts from myosin-immunized mice also show evidence of cytoplasmic staining suggestive of reactivity with contractile proteins. This may reflect the relatively higher titers of myosin antibodies in the myosin-immunized mice, compared to their CVB3-inoculated counterparts. Preliminary data characterizing the interstitial infiltrate in post-CVB3 late-phase myocarditis revealed the lesions to contain scattered $L3T4^+$ and $Thy1^+$ cells, but the majority of the interstitial cells are $Thy1^-$. Scattered $Lyt2^+$ cells are seen, and 40% of the interstitial cells stain with macrophage markers Mac1 and Mac3. Preliminary data from mice with myosin-induced autoimmune myocarditis suggest that there are increased numbers of T cells ($Thy1^+$) as compared to the virus-induced late-phase disease. Similar numbers of Mac1 and Mac3 positive cells are noted in both models. Complement (C3) is deposited within the myocardium and interleukin 2 (IL-2)-positive cells are scattered throughout the interstitium. A number of the interstitial mononuclear cells do not seem to stain with established markers, suggesting that some interstitial cells may be natural killer (NK) cells.

Another striking change in late-phase myocarditis is increased expression of class I antigens of the major histocompatibility complex by the myocytes. Class II MHC antigens, which are not normally demonstrable in the myocardium, are visible by immunohistochemical staining of the cardiac endothelium, some myocytes, and some interstitial dendritic cells.

Cross-Reactivity of Cardiac Myosin Autoantibodies and CVB3 Virus

The identification of a major autoantigen in autoimmune myocarditis provided the opportunity to address the question of whether there were antigens shared by cardiac myosin and the CVB3 virus that might initiate autoimmune myocarditis. A crude heart extract and a CVB3 concentrate were first separated by sodium dodecyl sulfate-polyacrylamide gel electrophoresis (SDS-PAGE) and electroblotted onto nitrocellulose paper [4]. The representative protein stain showed numerous bands in the crude heart extract, the most prominent of which is a 200-kilo dalton (kd) band which had been previously described by Alvarez et al. [1]. The bands found in the CVB3 preparation displayed the typical pattern described by others for the CVB3 capsid proteins. The pair of bands with a relative molecular weight of 30 kd represents the capsid proteins designated as VP1 and VP2, and the band of about 20 kd is VP3. These Western-blotted heart and virus preparations were then tested for their reactivity with serum pools obtained from A.CA mice with CVB3-induced myocarditis. The major reaction was with the 200-kd band in the heart extract. In the CVB3 preparation, the serum recognized the viral capsid protein VP2. To determine whether myosin antibodies in post-CVB3 sera cross-reacted with the CVB3 proteins, the serum pools were absorbed on cardiac myosin-coated Sepharose beads and compared to the two antigen preparations by western immunoblotting. Cardiac myosin removed all of the heart reactivity, confirming our previous observations; however, the reactivity with CVB3 was unaffected. Conversely, affinity-purified myosin antibodies reacted with the 200-kd protein in the heart extract, but did not react with the CVB3 preparation. These findings suggest that cardiac myosin heavy chain autoantibodies found in mice with CVB3-induced autoimmune myocarditis do not cross-react with CVB3 antigens detectable by western immunoblotting.

Further studies have begun to examine sera of post-CVB3 infected mice for reactivity to cardiac antigens other than myosin. For these studies, a three-step solubilization to whole heart extract was used: a low-ionic-strength-buffer, high-ionic-strength buffer (enriches for cytoplasmic contractile proteins), and SDS-PAGE sample buffer (enriches for membrane-bound constituents). After absorption of the postinfectious sera with skeletal and cardiac myosins, several other myocardial autoantigens were detected. Two predominant bands were found in the high-ionic-strength buffer and had the relative molecular weights of 44 kd and 26 kd. These bands were detected in both absorbed and unabsorbed sera. Two additional bands, 55 kd and 48 kd, were detected in myosin-absorbed sera in the SDS detergent extract. Postinfectious sera from 13 MHC congenic strains which produce myocardial autoantibodies were analyzed, and their reactive patterns were identified.

Relative Roles of T Cell Immunity and Humoral Immunity in Autoimmune Myocarditis

With data supporting the hypothesis that cardiac myosin heavy chain is released or exposed during viral injury and was responsible for at least part of the pathogenesis of autoimmune myocarditis, we turned to investigations designed to determine the role of T-cell immunity in autoimmune myocarditis. Preliminary adoptive transfer experiments were conducted in which spleen cells from cardiac myosin-immunized A strain mice were transferred into irradiated and nonirradiated syngeneic mice. There was no evidence of myocardial lesions among the recipients 11 days after cell transfer. Animals irradiated prior to cell transfer had higher titers of antimyosin antibody than did nonirradiated recipients. Donor cells stimulated in vitro with ConA prior to transfer had little influence on antibody titers.

To determine the sites of autoantibody production, mononuclear cells (MNC) were isolated from the spleen, lymph nodes, and heart of A/J mice immunized with cardiac myosin. They were tested for secretion of myosin antibody using the dot ELISA. In a typical experiment, $470/10^6$ MNC were found in the spleen, $353/10^6$ MNC in a lymph node pool, and $160/10^6$ MNC in the heart. The cardiac infiltrate itself, therefore, contributes significantly to antibody production.

Circulating IgG Heart Antibodies in Myocarditis: Functional Effects on Isolated Myocytes

A puzzling aspect of myocarditis is that it produces only small foci of myocardial inflammation and necrosis, yet it is frequently associated with global myocardial dysfunction. We had previously described the reactivity of heart autoantibodies to both myocyte surface antigens (sarcolemmal pattern of reactivity), as well as reactivity with cytoplasmic antigens such as myosin heavy chain (contractile protein pattern of reactivity). It was our hypothesis that these autoantibodies bind to myocytes throughout the heart and produce global effects on myocardial function. Our experience with serum from both murine and human myocarditis revealed that circulating heart antibodies cross-react with antigens on rat myocytes and stain with the characteristic sarcolemmal and/or contractile protein patterns. We used enzymatically dispersed, calcium-tolerant cardiac myocytes from adult rats. To quantify cell contraction, the image of a single cell was projected on a video monitor through a camera, and a video dimension analyzer monitored a selected raster line for differences in light intensity between the cell and the surrounding field. The addition of mouse IgG from cardiac myosin-induced myocarditis was evaluated on the extent of shortening during twitches induced by field depolarization. There is an immediate, marked negative inotropic effect. After steady state is reached in 3 min, the negative effect is stable and persists throughout the 30-min period of observation.

Myosin Isoenzymes in Myocarditis

Since myosin antibodies develop late after viral infection, we investigated whether a shift in myosin isoenzymes takes place. The distribution of cardiac myosin isoenzymes in mice appears unaffected by CVB3 infection. Preliminary experiments have attempted to cleave the V1 isoform of myosin with trypsin to determine if autoantibodies are directed toward certain cleavage fragments of myosin. Further experiments are currently in progress.

Parallel Human Studies

Sera from 120 Johns Hopkins Hospital patients with suspected myocarditis were screened for heart antibodies using immunofluorescence; 34% had circulating IgG heart-reactive antibodies in titered amounts not seen with control sera. Preliminary experiments using Western immunoblots have revealed a number of unique heart-reactive antibodies. Myosin antibodies have been commonly seen.

Summary

Following CV-induced myocarditis, certain strains of mice were found to develop progressive mononuclear myocarditis associated with the production of heart-specific autoantibodies. Both H-2 and non-H-2 genes control susceptibility to the autoimmune sequela of the inital viral infection. One prominent antigenic target of the autoimmune response is cardiac myosin heavy chain. Purified murine myosin induced autoimmune myocarditis in genetically susceptible strains of mice. The immunopathology of the myosin-induced disease differs in certain respects from CV-induced myocarditis in that the myosin-induced infiltrate is largely T cells and macrophages, whereas the postinfectious infiltrate contains many null cells, possibly representing an NK population. Both models, however, are characterized by extensive IgG and complement deposition in the heart. Expression of class II major histocompatibility antigen is also evident in the myocardium. The evidence implicates this well-defined endogenous antigen in the pathogenesis of postviral myocarditis in experimental animals and opens the door to defining the mechanism of post-infectious autoimmune disease in human populations.

Acknowledgement. This study was supported by PHS grant from the National Institutes of Health, No. HL-33878.

References

1. Alvarez FL, Neu N, Rose NR, Craig SW, Beisel KW (1987) Heart specific autoantibodies induced by Coxsackievirus B_3: identification of heart autoantigens. Clin Immunol Immunopathol 43: 129–139
2. Herskowitz A, Wolfgram LJ, Rose NR, Beisel KW (1987) Coxsackievirus B_3 murine myocarditis — histopathologic spectrum in genetically defined inbred strains. J Am Coll Cardiol 9: 1311–1319
3. Neu N, Beisel KW, Traystman MD, Rose NR (1987) Autoantibodies specific for the cardiac myosin isoform are found in mice susceptible to Coxsackievirus B_3 induced myocarditis. J Immunol 138: 2488–2492
4. Neu N, Craig SW, Beisel KW, Rose NR (1987) Coxsackievirus induced autoimmune myocarditis in mice: cardiac myosin autoantibodies do not cross-react with the virus. Clin Exp Immunol 69: 566–574

5. Neu N, Rose NR, Beisel KW, Herskowitz A, Gurri-Glass G, Craig SW (1987) Cardiac myosin induces myocarditis in genetically predisposed mice. J Immunol 139: 3630–3636
6. Rose NR, Wolfgram LJ, Herskowitz A (1986) Two distinct phases of Coxsackievirus B_3-induced myocarditis. Ann NY Acad Sci 475: 146–156
7. Wolfgram LJ, Beisel KW, Rose NR (1985) Heart-specific autoantibodies following murine Coxsackievirus B_3 myocarditis. J Exp Med 161: 1112–1121
8. Wolfgram LJ, Beisel KW, Herskowitz A, Rose NR (1986) Variations in the susceptibility to Coxsackievirus B_3-induced myocarditis among different strains of mice. J Immunol 136: 1846–1852

Viral Myocarditis: Immunogenetic and Autoimmune Aspects

K. W. BEISEL[1], and M. D. TRAYSTMAN[2]

[1] Department of Pathology and Microbiology, University of Nebraska Medical Center, Omaha, NE 68105-1065,USA
[2] Department of Pediatrics, The Johns Hopkins University School of Medicine, Department, Baltimore, MD 21205, USA

Introduction

Chronic inflammatory heart disease has been suggested as a sequela of infection by a variety of microbial agents in man [31]. Several lines of evidence suggest that myocardial autoimmunity is the major pathogenic mechanism in chronic inflammation of the myocardium. Maisch et al. [15] have demonstrated that patients with viral myocarditis develop autoantibodies to cardiac contractile and sarcolemmal proteins. Cardiac-specific autoantibodies to the mitochondrial adenosine di-/triphosphate (ADP/ATP) translocator protein have also been identified in patients with dilated cardiomyopathy [24]. This relationship between viral heart disease and myocardial autoimmunity has been documented using the mouse model. The presence of heart-specific autoantibodies and chronic, late-phase myocarditis suggest that a sequela of coxsackievirus type B3 (CVB3)-induced myocarditis in mice is an autoimmune-based disease [30]. Other findings similarly support the involvement of humoral and cellular autoimmune mechanisms in the pathogenesis [10, 14].

The genetic composition of the host in relationship to virus-induced myocarditis has received limited attention. Before 1980, five mouse strains were predominately used in the investigation of viral heart disease. Only a few laboratories have initiated investigations into the genetic control of the host's susceptibility to viral myocarditis [1, 5, 30]. A genetic component in susceptibility to inflammatory heart disease has been suggested from clinical and experimental work. Autoimmune myocarditis has been shown to be in linkage disequilibrium with the HLA-DR4 allele [1]. Huber and colleagues [10, 14] suggest that host genetics play a critical role in not only the development of myocardial autoimmunity but in the predominant immunopathogenic mechanism utilized. This laboratory has documented that CVB3-induced autoimmune myocarditis is under multigenic control by major histocompatibility complex *(MHC)* and non-*MHC* genes [5, 30].

The hypothesis for the studies presented herein is that the genetic control of chronic myocarditis is determined by genes associated with immune responsiveness, and that this form of myocardial disease is an autoimmune process. Murine genes which play an integral role in immune responses are the genes for immunoglobulins (i.e., *Igh, Igκ* and *Igλ*), the T cell receptor (i.e., *Tcrα, Tcrβ, Tcrγ*, and *Tcrδ*), and the MHC class I and II glycoproteins (i.e., *K, D, L*, $A_\alpha A_\beta$, and $E_\alpha E_\beta$).

H.-P. Schultheiß (Ed.)
New Concepts in Viral Heart Disease

Thus, these genes should be responsible for determining susceptibility to chronic inflammatory heart disease of an autoimmune origin.

Methods and Materials

Animals

A large panel of *MHC* congenic and recombinant inbred (RI) mouse strains were used for experimentation at 2 weeks of age. Breeding stocks of the 23 *MHC* congenic strains were obtained from either Jackson Laboratory, Bar Harbor, ME or Dr. C. S. David, Department of Immunology, Mayo Medical School, Rochester, Minn. Twenty-five lines of the AXB/BXA RI strain were generously supplied by Dr. M. N. Nesbitt, Department of Biology, University of California, San Diego, La Jolla, Calif. Both the breeding stock and experimental animals were maintained in microisolators in our animal facility with food and water being provided *ad libitum*.

Virus

CVB3 (Nancy) which was obtained from Dr. A. M. Lerner was maintained in the laboratory by passage on Vero cells. Virus titers were determined by both the plaque forming assay [29] and $TCID_{50}$ analysis [30].

Infection and Necropsy

All mice received intraperitoneally 10^5 plaque-forming units (PFU) in a volume of 0.1 ml of CVB3 at 2 weeks of age. Pups and mothers were maintained in a separate room from the breeding stock. Experimental mice were bled and then killed at 21 days after infection. Individual serum samples were collected and frozen at − 70 °C until used. Hearts were removed and immediately immersed in cold phosphate-buffered saline (PBS), pH 7.2, with 30 m*M* KCl to arrest the heart in diastole. A section of the heart was fixed for 24 h in 10% formalin-PBS, pH 7.4 and then histologically prepared for further evaluation.

Histology

Two transverse slices of fixed heart tissue were sectioned, stained with hematoxylin and eosin, and then examined for pattern and quantity of lesions. Areas of focal myocyte necrosis were semiquantified using a grading system based on a 0–4 scale [9]. The focal lesions were then assessed to be either active or resolving in nature. Each heart was also examined for the presence or absence of a diffuse interstitial mononuclear cell infiltrate.

Immunofluorescence

Indirect immunofluorescence was done according to Wolfgram et al. [30] to characterize the presence, titer, and reactivity patterns of heart-reactive autoantibodies. Serum was initially diluted 1/10 and then tested against normal A/J heart, skeletal muscle, kidney, and liver. Representative serum pools were made from each mouse strain.

Antigen Preparations and Immunochemistry

Cardiac sequential differential extracts were prepared using a modified procedure of Alvarez et al. [2]. Three fractionation steps were performed and at each step protease inhibitors were added. Each extraction procedure was done at 4 °C for 15 min with constant mixing followed by centrifugation at 100000 x *G* (33000 rpm) in a Beckman Ti-50 rotor at 4 °C for 1 h. The resulting pellet was used for the next extraction. Hearts were pulverized under liquid N_2 and then solubilized using 5 v/w of a low-ionic-strength buffer (20 m*M* Tris-HCl, 1 m*M* ethylenediaminetetra-acetic acid (EDTA), pH 9.0. The resulting pellet was then used for extraction with a high-ionic-strength buffer (20 m*M* Tris-HCl, 0.6 *M* KCl, 2 m*M* ATP, 2 m*M* $MgCl_2$, pH 7.4). Supernatant from this extraction step was dialyzed against dH_2O for 2 h in the presence of the protease inhibitors. The insoluble material left from high-ionic-strength extraction was washed once with dH_2O and then resuspended in 2X Sodium dodecyl sulfate-polyacrylamide gel electrophoresis (SDS-PAGE) sample buffer. All extracts were aliquoted and stored in liquid N_2 until used.

Murine skeletal and cardiac myosin was prepared according to the procedure of Alvarez et al. [2]. Purified myosin was bound to CNBr-sepharose 4B using the protocol outlined by Pharmacia, Inc., Piscataway, NJ. Adsorption of myosin antibodies was determined using an enzyme-linked immunosorbent assay (ELISA) procedure [20].

Western Immunoblotting

A discontinuous 7.5% SDS-PAGE system [29] was used to separate proteins from the three sequential differential extractions. Prestained, high molecular weight standards (BRL, Inc. Bethesda, MD) were used. The myocardial proteins were electrophoretically transferred to 0.45 μm nitrocellulose paper for 2.5 h at 80 mV using a cooling water bath. Following transfer, the nitrocellulose was incubated in 50 m*M* Tris, 150 m*M* NaCl, pH 7.4 containing 1% gelatin and was used immediately or stored at 4 °C for no more than 1 month. Immunoblotting was done using a blocking solution of 50 m*M* Tris, 150 m*M* NaCl, 0.5 m EDTA, 0.25% gelatin, pH 7.4 which contained 5% bovine and normal goat sera. The nitrocellulose was then treated with a 1/40 dilution of the pooled postinfection sera, followed by an appropriate dilution of an affinity-purified goat anti-mouse IgG specific immunoglobulin (Southern Biotechnology Associates, Inc., Birmingham,

AL) which was conjugated to horseradish peroxidase. A diamine benzidine substrate solution was used for staining.

Results

Genetic Control of Chronic Myocarditis

The observation from this laboratory [5, 30] were extended to document that both *MHC* and non-*MHC* genes played a role in the late phase of virus-induced myocarditis. Comparisons of four *H-2* congenic series of mouse strains demonstrated a multigenic influence by *MHC* and non-*MHC* genes upon late disease. Each *MHC* congenic series consisted of mouse strains having the same genetic composition except at a segment of chromosome 17 on which the *MHC* is located. A strong *MHC* association could be observed in the C3H *H-2* series, where disease was observed in C3H.SW *(H-2^b)*, C3H.JK *(H-2^j)*, C3H.Q *(H-2^q)*, and C3H.NB *(H-2^p)* but not in the C3H/HeSnJ *(H-2^k)* strain. All the A/J [i.e., A/J *(H-2^a)*, A.BY *(H-2^b)*, A.CA *(H-2^f)* and A.SW *(H-2^s)]* and BALB/c congenics [i.e., BALB.B *(H-2^b)*, BALB/cByJ *(H-2^d)* and BALB.K *(H-2^k)]* developed a chronic myocarditis and produced heart-specific autoantibodies, while the B10 congenic strains [i.e. B10 *(H-2^b)*, B10.D2/n *(H-2^d)*, B10.M *(H-2^f)*, B10.WB *(H-2^j)*, B10.BR *(H-2^k)*, B10.Q *(H-2^q)*, B10.RIII *(H-2^r)*, B10.S *(H-2^s)*, B10.PL *(H-2^u)*, and B10.SM *(H-2^v)]* did not. These results suggested that a non-*MHC* gene(s) was the primary genetic factor in determining the outcome of disease and was also able to mask the influence of the *H-2* complex upon myocardial autoimmunity.

A non-*MHC* effect was originally documented when two strains which both carry the *H-2^s* haplotype were compared [30]. The resistant B10.S/SgSf animals had healing lesions and did not develop myocardial autoantibodies, while the susceptible A.SW/SnJ strain had inflamed, mineralized lesion, a diffuse interstitial mononuclear cell infiltrate in the myocardium and heart-specific autoantibodies in the serum. Two groups of *MHC* congenic strains were examined which shared the same *H-2* haplotype but derived the remainder of their genome from different genetic sources. As shown in Table 1, three of the five strains which carry the *b* haplotype developed chronic disease. This disparity in late phase myocarditis was also observed in the *H-2^d* congenics. The BALB/cByJ and DBA/2J strains had the characteristic features of late-phase disease, whereas the B10.D2/n animals were recovering from the early virus-induced phase of myocarditis.

A second approach to investigating genetic influences on disease is the use of RI strains. RI strains were developed and have been successfully used to determine the chromosomal association and location of a wide variety of genetic traits [19]. The number and chromosomal location of the non-*MHC* gene(s) which control or influence autoantibody production and development of chronic myocarditis were determined by examination of 25 randomly chosen AXB/BXA RI lines. The AXB/BXA RI lines were derived from the A/J (susceptible) and the C57BL/6J (resistant) progenitor lines. A line is established by inbreeding for at least 20 generations of randomly chosen F_2 mice. Each line is distinguished by the letters of the progenitor strain and a number (e.g., AXB10), where the A indicates that

Table 1. Comparison of non-MHC influences on CVB3-induced autoimmune myocarditis

Strain	*H-2* Haplotype	Heart ASA[a] Prevalence[b] *(n)*	(%)	Interstitial Infiltrate
A.BY/SnJ	b	50/98	51	Present
BALB.B/Kh	b	4/32	17	Present
C3H.SW/SnJ	b	7/22	32	Present
C57BL/6J	b	0/51	0	Absent
C57BL/10J	b	0/118	0	Absent
BALB/cByJ	d	5/39	13	Present
B10.D2/nSnJ	d	0/21	0	Absent
DBA/2J	d	12/26	46	Present

[a] ASA, antisarcolemmal activity
[b] Expressed as the number of positive over the total number examined

a female from A/J (A) was mated with a C57BL/6J male (B). In these studies the strain distribution pattern (SDP), based on the phenotypic trait assayed, was compared to the progenitor strain to determine the allelic inheritance. Traits with similar SDP are considered to be linked. Examination of the RI lines for autoimmune myocarditis 21 days after infection showed that 10 out of 24 lines exhibited disease. Lines AXB 2, 6, 10, 21, 23 and BXA 6, 11, 23, 24, 25 had late disease, whereas AXB 1, 5, 7, 8, 15, 17, 18, 19, 20, 25 and BXA 8, 10, 12, 19, 22 did not. Thereby, this approximate 1:1 segregation ratio indicated that at least one gene (*amp* — autoimmune myocardial pathology) controls the expression of autoimmune pathology for this virus-induced disease. The *amp* gene has been designated as a recessive gene, since this trait is not expressed in the (B6 x A/J) F_1 mice [29]. The SDP for the *amp* trait suggested a linkage with the nucleoside phosphorylase-2 *(Np-2)* and the *Tcrα* genes on chromosome 14. Map distances of 7 (95% confidence interval CI ± 28) and 2–3 (95% CI ± 18), respectively, were found. These observations are important since the *amp* gene(s) is associated with the *Tcr* gene complex which is an essential component in the immune system. Additional studies must be done to establish that *amp* is a specific *Tcrα* gene.

Postinfection, autoimmune sera were then examined for their reactivity with sarcolemma and cardiac myosin. From the SDP of these autoantibodies, the genetic control of the anticardiac responses could be ascertained. The segregation pattern of the heart-specific autoantibody response, as determined by indirect immunofluorescence, suggested that this trait is controlled by at least two genes. One gene appeared to control the production of cardiac sarcolemmal autoantibodies, and the second determines a cardiac myosin-specific autoantibody response. The gene(s) controlling the sarcolemmal autoantibodies has been designated as *ama* (autoimmune myocardial autoantibodies). In the F_1 studies this trait was demonstrated to be recessively inherited. The AXB 2, 6, 15, 23 and BXA 6, 8, 19, 22, 23, 24, 25 produced these antibodies, while the remainder of the lines had no detectable antisarcolemmal activity. From this SDP this trait was localized to chromosome 14 and showed a linkage association with the *Np-2, Tcrα*, and *amp* loci. It will be important to determine the sarcolemmal antigen targeted by these antibodies since it will provide new insights into the pathogenic mechanisms of this disease.

Next, the ability of these RI lines to elicit a cardiac myosin-specific response was tested. Approximately 50% (8 out of 17) of the lines produced cardiac myosin-specific autoantibodies, i.e. AXB 2, 17, 18, 23, 25 and BXA 23, 24, 25. Thus, there appears to be at least one gene (i.e. *Cma;* cardiac myosin autoantibodies) controlling this autoantibody response. The SDP did not correlate with any of the known genetic markers (i.e., >150) so far identified for the AXB/BXA lines, including the *ama*- and *amp*-encoded traits. However, there was a weak association with the *H-2* complex.

Immunochemical Identification of Heart Autoantigens

We have recently examined a variety of postinfection, autoimmune sera from mouse strains which develop the CVB3-induced autoimmune myocarditis for their reactivity to cardiac myosin and other myocardial antigens. In these experiments, both the method of the sequential differential solubilization of mouse hearts and the staining procedure for immunodetection were modified to increase efficiency and sensitivity from our previously published procedures [2]. As shown in Fig. 1, the majority of reactivity of these autoimmune sera was with the intact myosin heavy chain and various myosin heavy chain subfragments. Myosin heavy chain can be broken down into heavy meromyosin, head (S1), rod, light meromyosin and S2 by light proteolytic digestion [3]. These various breakdown components were found in the high-ionic-strength extract and to a minor extent in the SDS extract. The antimyosin activity could be detected toward all these fragments. This indicates that there are numerous serologic determinants on myosin being recognized by the autoimmune response.

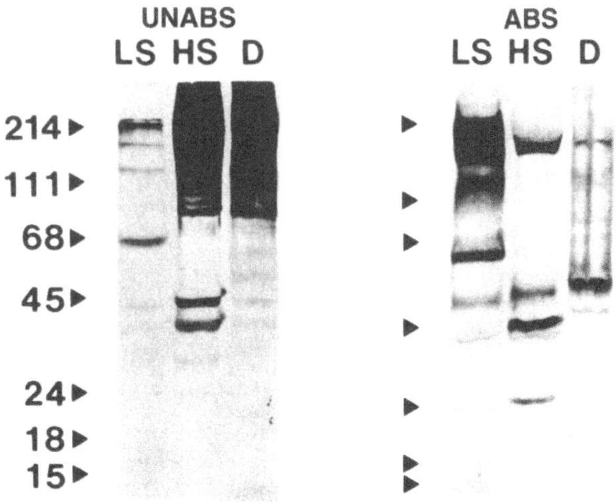

Fig. 1. *Immunochemical analysis of C3H.JK autoantibody reactivity to heart.* Low-ionic-strength *(LS)*, high-ionic-strength *(HS)* with pyrophosphate and SDS detergent *(D)* extracts of mouse hearts were run on 7.5% polyacrylamide gels. Western immunoblotting was performed using a pooled serum obtained 21 days post infection. The pooled serum was absorbed with cardiac myosin *(ABS)* and compared with the unabsorbed autoimmune serum *(UNABS)*. The developing antibody was a horseradish peroxidase-labeled goat anti-mouse IgG-specific reagent. Molecular weight standards are expressed in kd

After absorption of the postinfection, autoimmune sera with both skeletal and cardiac myosins, several other myocardial autoantigens were detected, whereas all the antimyosin activity had been removed. Two predominant bands were found in the high-ionic-strength buffer and had the relative molecular weights (M_r) ± SEM of 44263 ± 2527 and 26538 ± 2527. Preliminary studies have indicated that these proteins are not α-actin or one of the cardiac myosin light chains (unpublished data). Antibodies to these myocardial antigens were detected in both the absorbed and unabsorbed sera. A panel of postinfection sera from 13 *MHC* congenic strains which produce myocardial autoantibodies were analyzed and their reactive patterns were semiquantified. All these CVB3-infected strains produced antibodies to myosin and the 44 and 26 kd proteins. Unique bands with M_r of approximately 18 & 20, 47, 38 and 23 kd were also identified in the autoimmune sera from A/J, BALB.K/Kh, C3H. JK, and DBA/2J mice, respectively. The unique band of 23 kd M_r recognized by DBA/2J postinfection sera was present in the detergent-solubilized extract, while the remaining unique cardiac autoantigens were found in the high-ionic-strength extract. Absorption with cardiac myosin showed that these bands were not associated with myosin.

Two additional bands, 48183 ± 1103 and 55859 ± 1163, were detected with myosin-absorbed sera in the SDS detergent extract. These two bands are similar to those described by Paul [22] and Beyer et al. [6] for cardiac gap junction proteins. Since the protease inhibitor phenylmethyl-sulfonylfluoride (PMSF) was added, the degradation product of 26 kd band was not present as expected [16]. We have prepared enriched membrane fractions containing gap junction proteins following the procedure of Manjunath and Page [16]. A very small quantity of mouse liver and cardiac muscle was obtained. Coomassie staining of 7.5% PAGE gels demonstrated that we had enriched for two proteins of ~48 and 56 kd. Western immunodetection of the blotted liver and cardiac muscle jap junction preparations was done using postinfection sera from A/J, A.BY/SnJ, C57BL/6J mice and normal sera from A/J mice. Only the cardiac 48-kd protein was stained and reactivity against this protein was observed with the A/J and A.BY/SnJ autoantisera. No reactivity was apparent with the C57BL/6J and normal mouse sera. These data indicate that cardiac gap junction proteins are recognized as a sequela to infection with CVB3 in mouse strains which develop autoimmune myocarditis.

Histologic Evidence for Cardiomyopathy

As a potential sequela of inflammatory heart disease, morphologic features indicative of cardiomyopathy were evaluated in the hearts from this panel of congenic strains. Histologic evidence of cardiomyopathic lesions in hearts was assessed by fibrosis, myocyte hypertrophy, and myocytolysis. These histologic features were usually associated with a mineralized focal lesion. Results showed that during the acute phase of myocarditis (days 5 and 7) cardiomyopathic lesions were not observed (A. Herskowitz, M. D. Traystman, and K. W. Beisel, unpublished data). By 21 days after CVB3 infection, only the A/J, A.BY/SnJ, BALB/cByJ, BALB.K, and DBA/2J strains exhibited these changes in the myocardium. Studies by Matsumori and Kawai [17] have demonstrated the development of dilated cardiomyopathy in encephalomyocarditis virus-infected DBA/2J mice. No pathologic changes were found in the C57BL/6J inbred line, in nine of the ten B10 *H-2* congenics, and in the five C3H *MHC* congenics. The B10.WB/SgSf *(H-2^j^)* displayed evidence of myopathic disease. With the exception of the B10.WB/SgSf, there was a correlation between the autoimmune myocarditis and the presence of cardiomyopathic changes. Of the 11 lines which develop autoimmune myocarditis, six exhibited cardiomyopathic changes. Only the pathology in B10.WB/SgSf cannot be explained as being caused by an autoimmune process, since this strain develops neither the heart-specific autoantibodies nor the diffuse, chronic myocardial pathology. We project from these observations that the development of cardiomyopathy may emerge as a sequela to chronic autoimmune myocarditis.

Discussion

The murine model of viral myocarditis has been developed to examine the mechanisms of viral and immunologic pathogenesis and the subsequent testing of various therapeutic modalities. Yet, over the years, viral myocarditis has begun to be appreciated as a complex disease involving a number of pathogenic pathways. The usefulness of the present genetic studies indicates the complexities and dynamics behind this disease process. As depicted in Fig. 2, there are at least six distinct endpoints in viral myocarditis, and almost all of these endpoints can be represented by a number of mouse strains. In the murine model of viral myocarditis, all the animals can be infected with coxsackievirus and cardiac tissue injury does occur in all mouse strains, but the degree of myocardium affected will vary [5, 29, 30]. After an ensuing generalized inflammatory response to virus and the dama-

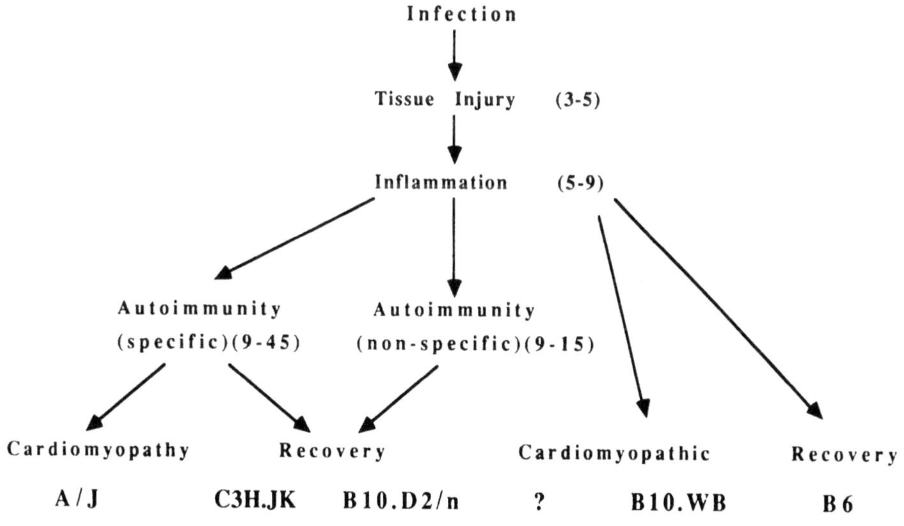

Fig. 2. *Sequela to CVB3-induced myocarditis — representative strains.* Six distinct pathologic patterns are presented which represent different sequelae to CVB3 infection. The *numbers presented in parentheses* are the days after infection when the corresponding parameters were observed

ged tissue, a number of pathways can be predicted. Chronic inflammation of the heart can result from cardiac-specific autoimmunity and subsequently followed by either cardiomyopathic changes in the heart or recovery. These are represented by the A/J and C3H.JK mouse strains, respectively. A transient, generalized muscle autoimmune response has been observed in B10.D2/n and occurs between the 1st and 2nd weeks after infection as evidenced by cross-reactive myosin autoantibodies and a widespread inflammation of the myocardium. By 3 weeks postinfection, the diffuse inflammation in the myocardium of B10.D2/n has disappeared and the focal lesions have begun to resolve. So far, we have not identified a mouse strain that has the transient disease with concurrent cardiomyopathic lesions. Recovery from the early virus-induced lesions without continuing myocarditis is represented by the C57BL/6J and C57BL/10J strains. The B10.WB strain is similar to the C57BL/6 strain except it develops cardiomyopathic changes in the heart tissue. Examination of these mouse strains should facilitate investigations into the relationship of cardiac autoimmunity with myocarditis and cardiomyopathy.

A genetic association of chronic inflammatory heart disease with the *Tcrα* and the *MHC* has been found. Both of these gene complexes are basic components of the immune response system. Thus, we feel that chronic inflammatory heart disease is, for the most part, an immune response-mediated disease process. Figure 3 is a representation of the appropriate interaction of the Tcr, MHC glycoprotein and autoantigen to elicit a response. The interaction of the Tcr heterodimer, the MHC product, and antigen is the basic component of an immune response [25]. For the Tcr the variable regions of the Tcr chains constitute the combining site for antigen and MHC [13]. Studies of peptide antigen binding to class II molecules

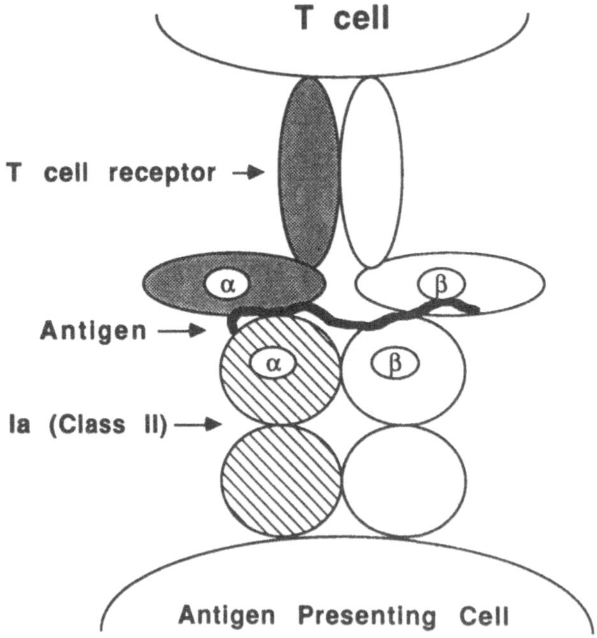

Fig. 3. *Interaction of Tcr, MHC, and autoantigen.* The Tcr is located on the T cell and is dimer comprised of an alpha and beta chain. The MHC glycoprotein which is representated by an Ia or class II molecule is present on an antigen-presenting cell. An alpha and beta chain constitutes the functional Ia molecule. A peptide fragment, presumably from a myocardial autoantigen, is shown to be bound by both the Tcr and MHC molecules. In addition to this interaction, the Tcr also recognize sites on the MHC molecule

have established that peptide-Ia interactions occur in the absence of the Tcr and varies with the allele of Ia [7]. Residues in the immunogenic peptide independently control the association with MHC and with specific Tcr binding [7], while regions of MHC determine binding with antigen and interaction with the Tcr [23]. These are the basic principles behind our observations and the association of *MHC* and *Tcr* genes with autoimmune myocarditis. The absence of myocardial autoimmunity can be explained in the B10 congenic series by the lack of an appropriate Tcr or the inability to produce T cell clones. An alternative possibility is that the MHC product(s) is unable, or has a low affinity, to bind with the autoantigen. In the C3H/HeSnJ mouse strain the Tcr repertoire appears to be ineffective in recognizing the autoantigen(s) in the context of the $H\text{-}2^k$ gene products. Development of autoimmune myocarditis in the A/J and BALB/c *MHC* congenics, DBA/2J, and the majority of the C3H *H-2* congenic strains suggests that all three components can interact appropriately to elicit these antimyocardial responses. Thus, the MHC products can bind with autoantigen and there are available T cell clones with the appropriate Tcr. The complexity of this dynamic molecular interaction is reflected in the presence of autoantibodies to unique myocardial autoantigens in the A/J, BALB.K, C3H.JK, and DBA/2 strains.

Much has been learned concerning the genetic, biochemical, functional, evolutionary, and developmental characteristics of the Tcr. In addition, its association with accessory molecules and interactions with the MHC class I and II molecules and antigens are under intense investigation. The genetic association of the *Tcr* and *MHC* genes with autoimmune disease has come from several primary lines of evidence. In insulin-dependent diabetes mellitus [28], pemphigus vulgaris [26] and myasthenia gravis [27] restriction fragment length polymorphism has been used to identify specific class II alleles which have a strong association with disease. Sequence analyses of HLA-DQ_β alleles indicate that susceptibility or resistance can be predicted by the amino acid present at residue 57 of the beta chain [28]. Susceptibility in NOD mice also correlates with a serine residue at position 57 in the A_β chain [28]. In murine experimentally induced thyroiditis, the class II gene(s) appears to control the initiation of disease [4] whereas the class I genes influence the severity of lesions [12]. Recently, the *D* region-encoded tumor necrosis factor-alpha gene has been shown to influence the severity of murine autoimmune "lupus" nephritis [11]. This indicates that genes associated with the MHC, other than class I and II, can influence susceptibility to autoimmune disease. The other approach used autoreactive T cell clones and/or hybridomas to examine the usage of the variable region genes and Tcr chains. For the most part, autoreactive lymphocytes obtained from patients with autoimmune disease appear to be oligoclonal in nature. Two studies have been reported that show a genomic association with *Ir*-gene related phenomena. These are the genetic control of the immune response toward H-Y antigen [8] and of immune-mediated demyelination induced by Theiler's murine encephalomyelitis virus (TMEV) [18]. These *Ir* genes have a linkage with the Tcr_α and Tcr_β complexes, respectively.

Several areas of essential investigation can be suggested from these studies in the association of autoimmunity with chronic heart disease. The central problem, which still needs to be resolved, is the identification of myocardial autoantigen(s).

Our data have indicated that there are a limited number of antigens being recognized by the humoral autoimmune response. However, each of these antigens must be examined for their pertinence and role in heart autoimmunity at both the humoral and cellular levels. The work of Neu et al. [20, 21] in identifying cardiac myosin as a potential autoantigen has already established the guidelines for demonstrating other myocardial antigens in disease. The establishment and usage of genetic parameters which determine the development of disease will lead to a better understanding of the pathogenic mechanisms and provide useful prognostic markers of susceptibility. Finally, the relationship between virus and induction of autoimmunity of the heart can be examined to determine the precise nature of this association.

Acknowledgements. We wish to thank our colleagues A. Herskowitz and M. N. Nesbitt for their participitation and efforts in various studies presented herein. Also, we appreciate the helpful discussions of the data with L. Chow, B. M. McManus and N. R. Rose. We would like to acknowledge the laboratory assistance of S. Schonlaw. This work was supported in part by the American Heart Association, Maryland Affiliate, and by USPHS grants HL-30144 and HL-38276.

References

1. Anderson JL, Carlquist JF, Lutz JR, DeWitt CW, Hammond EH (1982) HLA A, B and DR typing in idiopathic dilated cardiomyopathy: a search for immune response factors. Am J Cardiol 53: 1320–1330
2. Alvarez FL, Neu N, Rose NR, Craig SW, Beisel KW (1987) Heart-specific autoantibodies induced by coxsackievirus B3: idenfication of heart autoantigen. Clin Immunol Immunopathol 43: 129–139
3. Beisel KW, Traystman MN (1988) Autoimmune myocarditis: a murine model. In: Bigazzi PE, Wick G. Wicher K (eds) Progress in autoimmunity. I. Organ-specific autoimmunity. Dekker, New York (in press)
4. Beisel KW, David CS, Giraldo AA, Kong YM, Rose NR (1982) Regulation of experimental autoimmune thyroiditis: mapping of susceptibility to the *I-A* subregion of the mouse *H-2*. Immunogenetics 15: 427–430
5. Beisel KW, Wolfgram, Herskowitz A, Rose NR (1985) Differences in severity of coxsackievirus B3-induced myocarditis among *H-2* congenic mouse strains. In: Skamene E (ed) Genetic control of host resistance of infection and malignancy, vol 3. Liss, New York, pp 195–205
6. Beyer EC, Paul DL, Goodenough DA (1987) Connexin 43: a protein from rat heart homologous to a gap junction protein from liver. J Cell Biol 105: 2621–2629
7. Buus S, Sette A, Grey HM (1987) The interaction between protein-derived immunogenic peptides and Ia. Immunol Rev 98: 115–141
8. Epstein R, Sham G, Womach J, Yague J, Palmer E, Cohen M (1986) The cytotoxic T cell response to the male-specific histocompatibility antigen (H-Y) is controlled by two dominant immune response genes, one in the MHC, the other in the *Tar*-locus. J Exp Med 163: 759–773
9. Herskowitz A, Wolfgram LJ, Rose NR, Beisel KW (1987) Coxsackievirus B3 murine myocarditis — pathologic spectrum of myocarditis in genetically defined inbred strains. J AM Coll Cardiol 9: 1311–1319
10. Huber SA, Lodge PA (1986) Coxsackievirus B-3 myocarditis. Identification of different pathogenic mechanisms in DBA/2 and BALB/c mice. Am J Pathol 122: 284–291
11. Jacob CO, McDevitt HO (1988) Tumour necrosis factor-α in murine autoimmune "lupus" nephritis. Nature 331: 356–358

12. Kong YM, David CS, Giraldo AA, ElRehewy M, Rose NR (1979) Regulation of autoimmune response to mouse thyroglobulin: influence of *H-2D*-end genes. J Immunol 123: 15–18
13. Kronenberg M, Siu G, Hood LE, Shastri N (1986) The molecular genetics of the T-cell antigen receptor and T-cell antigen recognition. Ann Rev Immunol 4: 529–591
14. Lodge PA, Herzum M, Olszweski J, Huber SA (1987) Coxsackievirus B-3 myocarditis — acute and chronic forms of the disease caused by different immunopathogenic mechanisms. Am J Pathol 128: 455–463
15. Maisch B, Deeg P, Liebau G, Kochsiek K (1983) Diagnostic relevance of humoral and cell-mediated immune reactions in patients with viral myocarditis. Clin Exp Immunol 48: 533–545
16. Majunath C, Page E (1985) Cell biology and protein composition of cardiac gap junctions. Am J Physiol 248: H783–H791
17. Matsumori A, Kawai C (1982) An animal model of congestive (dilated) cardiomyopathy: dilatation and hypertrophy of the heart in the chronic stage in DBA/2 mice with myocarditis caused by encephalomyocarditis virus. Circulation 66: 355–360
18. Melvold RW, Jokinen DM, Knobler RL, Lipton HL (1987) Variations in genetic control of susceptibility to Theiler's murine encephalomyelitis virus (TMEV) induced demyelinating disease. I. Differences between susceptible SJL/J and resistant BALB/c strains map near the T cell β-chain constant gene on chromosome 6. J Immunol 138: 1429–1433
19. Nesbitt MN, Skamene E (1984) Recombinant inbred mouse strains derived from A/J and C57BL/6J: a tool for the study of genetic mechanisms in host resistance to infection and malignancy. J Leuk Biol 36: 357–364
20. Neu N, Beisel KW, Traystman MD, Rose NR, Craig SW (1987) Autoantibodies specific for the cardiac myosin isoform are found in mice susceptible to coxsackievirus B3 induced myocarditis. J Immunol 138: 2482–2492
21. Neu N, Rose NR, Beisel KW, Herskowitz A, Gurri-Glass G, Craig SW (1987) Cardiac myosin induces myocarditis in genetically predisposed mice. J Immunol 139: 3030–3036
22. Paul DL (1985) Antibodies against liver gap junction 27-kD protein is tissue specific and cross-reacts with a 54-kD protein. In: Bennett MVL, Spray DC (eds) Gap junctions. Cold Spring Harbor Laboratories, Cold Spring Harbor, New York, pp 107–122
23. Ronchese F, Schwartz RH, Germain RN (1987) Functionally distinct subsites on a class II major histocompatibility complex molecule. Nature 329: 254–256
24. Schultheiss HP, Bolte HD (1985) Immunological analysis of autoantibodies against the adenine nucleotide translocator in dilated cardiomyopathy. J Mol Cell Cardiol 17: 603–617
25. Schwartz RH (1985) T-lymphocyte recognition of antigen in association with gene products of the major histocompatibility complex. Ann Rev Immunol 3: 237–261
26. Sinha AA, Brautbar C, Szafer F, Friedmann A, Tzfoni E, Todd JA, Steinman L, McDevitt HO (1987) A newly characterized HLA DQ_β allele associated with pemphigus vulgaris. Science 329: 1026–1029
27. Smith CIE, Borgonovo L, Carlsson B, Hammarstrom L, Rabbitts TH (1987) Molecular probing of disease susceptibility in myasthenia gravis patients: an analysis of T-cell receptor and HLA class II genes using restriction fragment length polymorphism. Ann NY Acad Sci 505: 388–397
28. Todd JA, Bell JI, McDevitt HO (1987) HLA-DQ_β gene contributes to susceptibility and resistance to insulin-dependent diabetes mellitus. Nature 329: 599–604
29. Traystman MD (1987) Genetic regulation of CVB3-induced autoimmune myocarditis. Thesis, The Johns Hopkins University, Baltimore Md
30. Wolfgram LJ, Beisel KW, Herskowitz A, Rose NR (1986) Variations in the susceptibility to coxsackievirus B_3-induced myocarditis among different strains of mice. J Immunol 136: 1846–1852
31. Woodruff JF (1980) Viral myocarditis. Am J Pathol 101: 428–479

Is Coxsackievirus B3-Induced Myocarditis in A/J Mice Mediated by an Autoimmune Response to Cardiac Myosin?

N. Neu[1], G. Krömer[1], B. Ploier[2], and N. R. Rose[2]

[1] Institut für Allgemeine und Experimentelle Pathologie der Universität Innsbruck, Fritz-Pregl-Str. 3, A-6020 Innsbruck, Austria
[2] Department of Immunology and Infectious Diseases, The Johns Hopkins University, Baltimore, MD 21205, USA

Introduction

The possible postinfectious nature of certain autoimmune diseases has gained considerable interest during recent years. Several diseases, inculding diabetes mellitus and multiple sclerosis, have been attributed to virus-induced autoimmunity [1, 2], although a cause-and-effect relationship has not been directly demonstrated.

Similarly, an autoimmune response is thought to play a role in chronic myocarditis, where group B coxsackieviruses (CVB) are suspected of constituting the most frequent etiological agents [3]. However, the wide spectrum of pathology observed in humans makes an unequivocal diagnosis very difficult [4]. The diversity of postinfectious myocarditis is also seen experimentally. There are a few well-established animal models of CVB3-induced myocarditis. One developed more than 15 years ago by Lerner and Wilson [5] showed the importance of CVB3 infection in producing myocardial lesions in the murine system. Other pioneering work conducted by Woodruff [6] and later extended by Huber's group [7, 8] suggested that myocardial lesions in CVB3-infected BALB/c mice are in part due to a cellular autoimmune response.

In an extensive investigation of the genetics of susceptibility to CVB3-induced myocarditis, Wolfgram et al. [9, 10, 11] infected a large number of A/J and C57BL/10 *H-2* congenic mouse strains. They found that, a few days after infection, the virus produced focal myocardial lesions in all strains tested. While most of the C57BL/10 *H-2* congenics recovered rapidly, mice on the A/J background developed persistent myocarditis although no infectious virus could be isolated at that timepoint. This so-called late disease was characterized by a diffuse interstitial cell infiltrate and by circulating heart autoantibodies. Subsequent biochemical analyses revealed that these antibodies were predominantly directed against the heavy chain of cardiac myosin [12, 13].

These findings raised the question of how CVB3 can induce an autoimmune response to cardiac myosin. A simple model would be molecular mimicry, but so far we have not been able to demonstrate a cross-reactivity of the myosin antibodies with the virus [14]. An alternative mechanism, not excluding the former, would be an autoimmune response to cardiac myosin after this antigen has been released due to virus-mediated cell damage. Persistent extracellular exposure of myosin

H.-P. Schultheiß (Ed.)
New Concepts in Viral Heart Disease

following muscle injury, for instance after myocardial infarction, is a well-described phenomenon and facilitates myocardial imaging using radiolabeled myosin antibodies [15, 16]. If the antimyosin response in CVB3-infected A/J *H-2* congenic mice is indeed the pathogenic factor acting in "late phase" myocarditis, it should be possible to induce the disease with cardiac myosin alone, i. e., in the absence of virus. On the other hand, mice which are resistant to CVB3-induced autoimmune myocarditis should be unresponsive to cardiac myosin.

Materials and Methods

Preparation of Cardiac Myosin and Cardiac Heary Meromyosin The purification of cardiac myosin from mice has been described earlier [14]. Heavy meromyosin (HMM) was prepared from purified cardiac myosin by limited tryptic digestion as described by Schwartz et al. [17]. Following digestion, the proteins were dialyzed against phosphate-buffered saline (PBS, pH 7.2), and the insoluble material was removed by ultracentrifugation. For immunizations the crude HMM preparation was further purified to remove any undigested myosin as described [17].

Immunizations, Serology, Histology. Six-week-old mice were immunized twice with 100 μg of cardiac myosin (or HMM) in complete Freund's adjuvant (CFA) as described [18]. Sera were obtained 3 weeks after the first immunization and were tested for IgG autoantibodies to cardiac myosin in the enzyme-linked immunosorbent assay (ELISA) [18]. The hearts were embedded in paraffin and examined histologically.

Western Immunoblotting. Crude, water-soluble HMM preparations were size-separated by sodium dodecylsulfate polyacrylamide gel electrophoresis (SDS-PAGE) and electroblotted onto nitrocellulose paper [12]. The immunoblotting procedure has been described in detail [14].

Cell Suspensions for Spot-ELISA. Spleens and lymph nodes from animals immunized with 100 μg cardiac myosin were gently pressed through a 200-mesh sieve into RPMI 1640 followed by two washes in RPMI. Inflammatory cells infiltrating the hearts were released enzymatically. Briefly, hearts were cut into small pieces and digested in 0.2% collagenase for 30 min at 37 °C . The suspension was then washed in phosphate buffered saline (PBS) and further digested with 0.1% trypsin. Released leukocytes were separated from debris and myocardial cells by Ficoll-Isopaque density centrifugation.

Spot-ELISA. Autoantibody-producing cells were enumerated by using the spot-ELISA technique as described by Sedgwick and Holt [19]. Briefly, cells in RPMI 1640 containing 5% fetal calf serum (FCS) were plated in 24-well tissue culture plates (Nunc, Denmark) and incubated overnight at 37 °C . The plates were then washed with PBS-Tween followed by addition of an alkaline phosphatase rabbit-antimouse IgG conjugate (Sigma Chemical Co.). After 1 h at room temperature, the plates were washed and developed with 0.6% gelling agarose containing

5-bromo-4-chloro-3-indolyl phosphate (BCIP, Sigma). After 3 h, blue spots corresponding to antibody-producing cells were counted.

Statistics. Statistical differences between the autoantibody titers were evaluated by using a three-way nested analysis of variance. Differences in the prevalence of heart disease were determined by $\varkappa^2$ analysis.

Results

In agreement with our previous results [18] we demonstrate here that purified cardiac myosin does indeed induce severe myocarditis, and preferentially in those strains of mice which are highly susceptible to virus-induced autoimmunity (Table 1, Fig. 1). In addition, there are close parallels between the myosin autoantibody response after CVB3 infection and that observed after myosin immunization (Table 2). A.SW and A/J mice develop high myosin autoantibody titers after CVB3 infection [9, 10] and after injection with purified cardiac myosin. A.BY mice are intermediate responders in both cases, whereas C57BL/10 mice are nonresponders as compared to animals injected with adjuvant alone. The increased autoantibody titer in CFA-injected controls was most probably due to polyclonal stimulation mediated by the adjuvant [20]. In high-responder mice, one myosin autoantibody population was specific for the cardiac isoform. No myocarditis and no increased myosin autoantibody titers were observed after immunization with skeletal muscle myosin (data not shown).

In a further set of experiments, spleens, lymph nodes, and inflammatory infiltrates from the hearts were examined for myosin antibody-producing cells. This analysis was performed by plating single cell suspensions from these organs in plastic dishes coated with cardiac myosin. Zones of bound antibody, each of which corresponded to an antibody-producing cell, were visualized as spots by means of an ELISA technique. Cells producing autoantibodies to cardiac myosin were not only found in the spleen and lymph nodes, but also in the heart.

Table 1. Myocarditis after immunization with cardiac myosin in mice susceptible and resistant to CVB3-induced myocarditis

Strain	Susceptibility to CVB3-induced autoimmune myocarditis[a]	Prevalence of myocarditis after immunization with cardiac myosin[b] *(n)*	(%)
A.SW/SnJ	High	15/16	94
A/J	High	12/14	86
A.BY/SnJ	High/intermediate	4/17[c]	24
C57BL/10J	Low	0/18	0

[a] According to Wolfgram [10] and Wolfgram et al. [9]
[b] Twenty-one days after the first immunization
[c] Significantly different from C57BL/10 ($P < 0.01$) and from A.SW and A/J ($P < 0.01$)

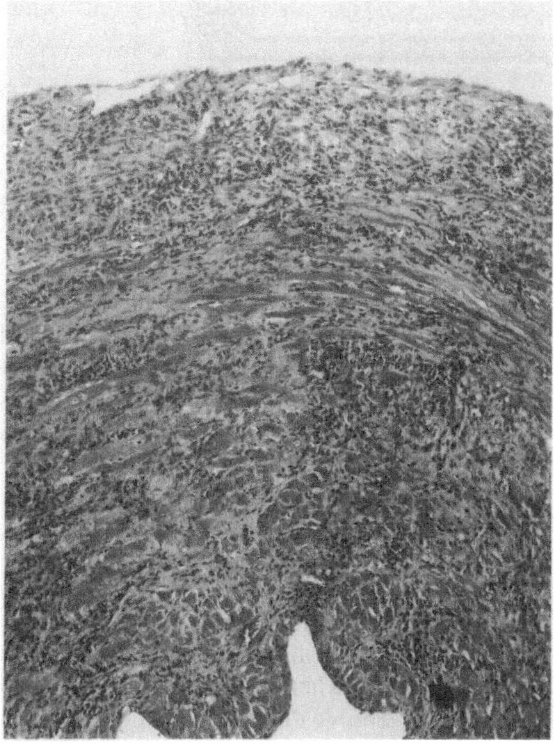

Fig. 1. Histological section of a heart from an A/J mouse immunized with 100 μ cardiac myosin. Note the diffuse interstitial infiltrate. H & E, x60

Table 2. Myosin autoantibody response after immunization with cardiac myosin in mice susceptible and resistant to CVB3-induced myocarditis

Strain	Heart-autoantibody response after CVB2 infection[a]	log2 anti-myosin IgG titer after immunization with cardiac myosin[b] ± SD[b]
A.SW/SnJ	High	13.9 ± 1.5[c]
A/J	High	14.4 ± 1.5[c]
A.BY/SnJ	Intermediate/low	11.2 ± 1.2[d]
C57BL/10J	Low	9.2 ± 1.0[e]

[a] According to Wolfgram [10] Wolfgram et al. [9, 11]
[b] Serum was obtained from the same animals as shown in Table 1. Titers were determined by ELISA using cardiac myosin as test antigen
[c] Significantly different from A.BY and C57BL/10 ($P < 0.001$)
[d] Significantly different from controls injected with adjuvant alone ($P < 0.01$)
[e] Not significantly different from controls injected with adjuvant

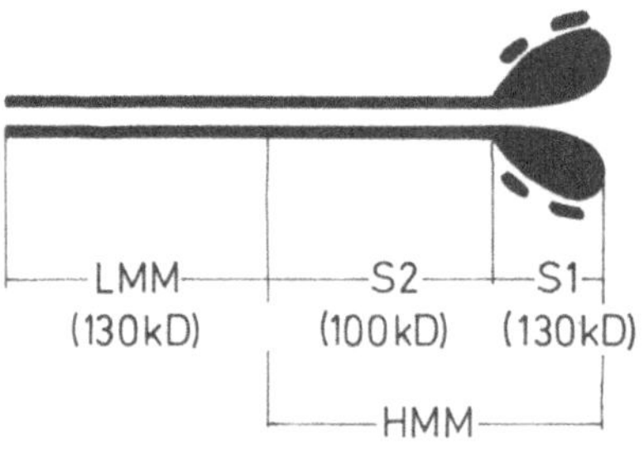

Fig. 2. Schematic drawing of the myosin molecule. The positions of the proteolytic digestion fragments heavy meromyosin *(HMM)*, light meromyosin *(LMM)*, and of the HMM subfragments S1 and S2 are indicated. The light chains are attached to the globular headpiece

Cardiac myosin is a relatively large protein and is composed of a pair of heavy chains and two pairs of light chains. So far, three "classical" proteolytic myosin fragments have been described [21]. These are HMM, consisting of the globular head (S1) and part of the rod (S2), and light meromyosin (LMM, Fig. 2). HMM is the water-soluble part of the molecule, whereas LMM is, like intact myosin, soluble at higher ionic strength only. To circumvent experimental difficulties resulting from the poor solubility of myosin at physiological salt concentrations, we examined whether HMM has similar antigenic properties to the whole molecule.

Using Western immunoblotting we demonstrated that sera from A/J mice immunized with undigested cardiac myosin are reactive not only with cardiac HMM (prepared by tryptic digestion), but also with smaller HMM fragments. Subsequently, we tested whether HMM was also capable of inducing autoimmune myocarditis in susceptible mice. A/J and C57BL/10 mice were immunized with HMM following the same schedule which was previously used for immunization with cardiac myosin [18]. Histological examination revealed that HMM is, like the whole myosin molecule, capable of inducing autoimmune myocarditis in A/J, but not in C57BL/10 mice.

Discussion

In the present study we have shown that mouse strains which are susceptible to CVB3-induced myocarditis are also susceptible to autoimmune myocarditis induced by cardiac myosin. Thus, the genetic predisposition to elicit an autoimmune response to cardiac myosin might be at least one crucial factor in determining the susceptibility to the late phase of the viral disease.

Beside the parallels between myocarditis induced by CVB3 and that induced by cardiac myosin, a major difference was observed in one of the strains tested earlier, namely B10.A. B10.A mice do not develop chronic disease after CVB3 infection, although they are intermediate responders when immunized with cardiac myosin [18]. An explanation for this apparent discrepancy may be that the early virus-mediated lesions tend to be mild in these mice [10] and may therefore be insufficient to induce an autoimmune response. The severity of the early lesions, which differs from strain to strain [9, 10], might well affect the amount of released autoantigen, thereby influencing the late autoimmune disease process. Moreover, we cannot entirely exclude the presence of virus or virus-related products during the late phase of the disease. This possibility is currently being investigated using the very sensitive in situ hybridization technique.

As reported for other autoimmune diseases, e.g., Hashimoto's thyroiditis [22], autoantibodies appear to be produced at the site of the autoaggressive attack, i.e., within the inflammatory heart infiltrate. Autoantibodies produced by infiltrating plasma cells may thus directly exert their pathogenic effect via antibody-mediated cytotoxicity or complement fixation. There are some indications that heart autoantibodies in patients with viral myocarditis [23] and in CVB3-infected DBA/2 mice [24] are pathogenic, but the myosin autoantibodies in our murine model have not yet been analyzed for possible pathogenicity.

The capability of HMM to induce autoimmune myocarditis indicates that this fragment contains immunogenic epitopes. Although we cannot exclude that LMM has similar antigenic properties to HMM when injected into susceptible mice, it is evident that the immunogenic epitopes are unique to the cardiac myosin isoform, because immunization with skeletal muscle myosin proved to be ineffective [18].

The finding that HMM is immunogenic for genetically predisposed mice is not only of biochemical, but also of practical interest. Due to its poor solubility at physiological salt concentrations, myosin is not suitable for certain analyses, for instance in vitro lymphocyte proliferation. These problems can now be circumvented by using HMM because it is water soluble and seems to have the same degree of antigenicity as the whole myosin molecule. We have recently been able to demonstrate specific in vitro lymphocyte responses to HMM (unpublished data). Such experiments may be extremely useful for analyzing the autoimmune phase of CVB3-induced myocarditis at the cellular level.

The ability to induce autoimmune myocarditis with an identified self-antigen provides us with a simple murine model for analysis of autoimmune heart disease. The value of this new model will be determined by the extent of similarity of myocarditis induced by CVB3 and that induced by immunization with cardiac myosin. The question of whether immunization with this antigen does mimic the late phase of the viral disease forms the basis for our current investigations.

Summary

Several *H-2* congenic mouse strains on the A/J background are susceptible to autoimmune myocarditis following CVB3-infection, whereas most C57BL/10 *H-2* congenics are resistant. In susceptible mice, the disease is associated with autoantibodies specific for cardiac myosin. Moreover, cardiac myosin itself is capable of inducing autoimmune myocarditis, preferentially in mouse strains which are highly susceptible to myocarditis following CVB3 infection. Spot-ELISA assays demonstrated the presence of myosin antibody-producing cells at the site of the autoaggressive attack, i.e., within the inflammatory heart infiltrate. Because immunization with skeletal muscle myosin was ineffective, we concluded that the immunogenic epitopes are unique to the cardiac myosin isoform. More detailed biochemical analyses revealed that heart autoantibodies in susceptible mice are reactive with HMM, the water-soluble fragment of the myosin molecule. In addition, cardiac HMM can induce myocarditis when injected into A/J mice in a similar manner to the whole myosin molecule. Because of the parallels between myocarditis induced by CVB3 and that induced by cardiac myosin or HMM, this animal model might be a valuable tool for analyzing autoimmune mechanisms in certain forms of viral heart disease.

Acknowledgements. This work was supported by the Kamillo-Eisner Foundation, by the Austrian Research Council (N. N.), and by a PHS grant from the National Institute of Health, HL-33878 (N. R. R.).

References

1. Yoon JW (1983) Viruses in the pathogenesis of type I diabetes mellitus. Curr Probl Clin Biochem 12: 11–44
2. Waksman BH, Reynolds WE (1984) Multiple sclerosis as a disease of immune regulation. Proc Soc Exp Biol Med 175: 282–294
3. Hirchman SZ, Hammer GS (1972) Coxsackievirus myopericarditis: a microbiological and clinical review. Am J Cardiol 34: 224–232
4. James TN (1983) Myocarditis and cardiomyopathy. N Engl J Med 308: 39–41
5. Lerner AM, Wilson FM (1973) Virus myocardiopathy. Prog Med Virol 15: 63–91
6. Woodruff JF (1980) Viral myocarditis: a review. Am J Pathol 101: 426–478
7. Huber SA, Lodge PA (1984) Coxsackievirus B3 myocarditis in BALB/c mice: evidence for autoimmunity to myocyte antigens. Am J Pathol 116: 21–29
8. Huber SA, Job LP (1983) Cellular immune mechanisms in coxsackievirus group B, type 3 induced myocarditis in BALB/c mice. Adv Exp Med Biol 161: 491–508
9. Wolfgram LJ, Beisel KW, Herskowitz A, Rose NR (1986) Variations in the susceptibility to coxsackievirus-induced myocarditis among different strains of mice. J Immunol 136: 1846–1852
10. Wolfgram LJ (1986) Variations in the susceptibility to coxsackievirus B3-induced myocarditis among different strains of mice. Thesis, The Johns Hopkins University, Baltimore
11. Wolfgram LJ, Beisel KW, Rose NR (1985) Heart-specific autoantibodies following murine coxsackievirus B3 myocarditis. J Exp Med 161: 1112–1121
12. Alvarez FL, Neu N, Rose NR, Craig SW, Beisel KW (1987) Heart-specific autoantibodies induced by coxsackievirus B3: identification of heart autoantigens. Clin Immunol Immunopathol 43: 129–139
13. Neu N, Beisel KW, Traystman MD, Rose NR, Craig SW (1987) Autoantibodies specific for the cardiac myosin isoform are found in mice susceptible to coxsackievirus B3-induced myocarditis. J Immunol 138: 2488–2492
14. Neu N, Craig SW, Rose NR, Alvarez FL, Beisel KW (1987) Coxsackievirus induced myocarditis in mice: cardiac myosin autoantibodies do not cross-react with the virus. Clin Exp Immunol 69: 566–574
15. Haber E (1982) Immunological probes in cardiovascular disease. Br Heart J 47: 1–10
16. Khaw BA, Matis JA, Melincoff G, Strauss HW, Gold HK, Haber E (1984) Monoclonal antibody to cardiac myosin: imaging of myocardial infarction. Hybridoma 3: 11–23
17. Schwartz K, Bouveret P, Sebag C, Leger J, Swynghedauw B (1977) Immunochemial evidence for the species-specificity of mammalian cardiac myosin and heavy meromyosin. Biochim Biophys Acta 425: 24–36
18. Neu N, Rose NR, Beisel KW, Herskowitz A, Gurri-Glass G, Craig SW (1987) Cardiac myosin induces myocarditis in genetically predisposed mice. J Immunol 139: 3630–3636
19. Sedgwick JD, Holt PG (1983) A solid phase immunoenzymatic technique for the enumeration of specific antibody-secreting cells. J Immunol Methods 57: 301–309

20. Rosenberg YJ (1981) The ability of nonspecific T-cell stimulators to induce helper-cell-dependent increases in either polyclonal or isotype-restricted Ig-production in vivo. Cell Immunol 61: 416–424
21. Leger JJ, Berson G, Delcayre C, Klotz C, Schwartz K, Leger J, Stephens M, Swynghedauw B (1975) Heart contractile proteins. Biochimie 57: 1249–1273
22. Krömer G, Sundick RS, Schauenstein K, Hala K, Wick G (1985) Analysis of lymphocytes infiltrating the thyroid gland of obese strain chickens. J Immunol 135: 2452–2457
23. Maisch B, Trostel-Soeder R, Stechemesser E, Berg PA, Kochsiek K (1982) Diagnostic relevance of humoral and cell-mediated immune reactions in patients with acute viral myocarditis. Clin Exp Immunol 48: 533–545
24. Huber SA, Lodge PA (1986) Coxsackievirus B3 myocarditis. Identification of different pathogenic mechanisms in DBA/2 and BALB/c mice. Am J Pathol 122: 284–291

Virus-Induced Autoimmunity: Cross-Reactivity of Antiviral Antibodies with Self-Components

B. S. PRABHAKAR[1], J. SRINIVASAPPA[1], K. W. BEISEL[2], and A. L. NOTKINS[1]

[1] Laboratory of Oral Medicine, National Institute of Dental Research, National Institutes of Health, Bethesda, MD 20892, USA
[2] Department of Pathology and Microbiology University of Nebraska Medical Center Omaha, NE 68105, USA

Introduction

There are a number of important human diseases that have an autoimmune component. In some of these diseases, the autoimmune response is very broad, with autoantibodies present against a number of tissues and organs. For example, in systemic lupus erythematosus, autoantibodies against DNA, RNA, cytoplasmic proteins, lymphocytes, and erythrocytes are often found. Similarly, patients with polyendocrinopathy have autoantibodies to pancreas, thyroid, pituitary, and gastric mucosa. In contrast, patients with myasthenia gravis show a far more restricted autoimmune component with the predominant response directed against the acetylcholine receptor.

Several hypotheses, ranging from genetic susceptibility to environmental factors, have been proposed to explain the triggering of autoimmune responses. We know from several studies that viruses can sometimes be involved. Autoantibodies have been found in the sera of experimental animals following infection with viruses, e.g., reovirus, vaccinia, Sendai, rinderpest, Aleutian mink disease virus, and retroviruses [1–7].

In humans, autoantibodies against normal lymphocytes have been observed after infection with rubella, measles, and Epstein-Barr virus (EBV) [8]. Autoantibodies to nuclei, smooth muscle, and renal tubules are often found in patients with influenza or recurrent herpes simplex virus infection [9–11]. Other examples include antibodies to myelin basic protein in patients with subacute sclerosing panencephalitis [12], antibody to platelets in patients with rubella [13], and rheumatoid factor in patients with cytomegalovirus and hepatitis virus infections [14, 15]. Similarly, autoantibodies to the myocardium have been observed subsequent to coxsackievirus infections. As in experimental animals, these antibodies appear transiently after viral infection and are of low titer [16]. There are a number of other reports where autoantibody production is associated with viral infection [17]. There are several possible mechanisms by which a virus might induce autoimmunity [17]. Four of these are discussed below.

H.-P. Schultheiß (Ed.)
New Concepts in Viral Heart Disease

Mechanisms of Virus-Induced Autoimmunity

Viruses Acting on Target Cells

In this case the virus makes the infected cell appear foreign to the host [7, 18]. If the host antigens are incorporated into the virus, then the antiviral response could cross-react with the host component. In addition, viruses can insert, expose, or modify host antigens on the cell surface. The viral infection can also lead to the release of sequestered antigens which are not usually encountered by the host's immune system.

Viruses Acting on Effector Cells

A second possibility is that the virus triggers autoimmunity by acting on the immune system. The virus may stimulate or destroy certain subsets of lymphocytes or macrophages which in turn may lead to the production of autoantibodies. Immunological abnormalities exist following viral infections [19]. One of the best examples is EBV infection. Only the B lymphocytes have receptors for this virus. In vitro EBV can infect and transform B lymphocytes into immortalized lymphoblastoid cell lines. These cells are capable of secreting immunoglobulins, some of which are autoantibodies. Autoantibodies that react with cells in the thyroid, pituitary, stomach, and pancreas have been prepared by EBV transformation of B cells. In vivo EBV can induce B cell proliferation leading to autoantibody production. However, the EBV infected cells are promptly eliminated by the host's cytotoxic T cell response against the virus. It is known that lymphocytes infected with EBV in vivo can spontaneously proliferate when placed in culture. Recent studies from our laboratory have clearly demonstrated autoantibody production by these spontaneously proliferating cells [20, 21]. This may explain the autoantibodies seen in patients with defective immunoregulation of EBV infection (e.g., those with infectious mononucleosis, AIDS, and malaria).

Anti-idiotypic Antibodies Against Antiviral Antibodies

A third mechanism by which virus could trigger an autoimmune response is by eliciting an anti-idiotypic antibody response. For example, Nepom et al. [22] have shown that anti-idiotypic antibodies made against a monoclonal antibody to reovirus type 3 hemagglutinin reacted with receptors for reovirus on the surface of lymphocytes and nerve cells. Similar observations have been made with anti-idiotypic antibodies to antihormonal monoclonal antibodies [23]. If such a response could occur in vivo because of the immune network, then it could result in autoimmunity [24].

Molecular Mimicry

A fourth mechanism by which a virus could trigger an autoimmune response is molecular mimicry [17, 25]. Different molecules share similar structures either in the amino acid sequence or in their tertiary conformation. These molecules could be as distinct as viruses and host antigens. Structural similarities have frequently been noted between a number of both RNA and DNA viruses and "self" proteins. In addition, antiviral antibodies which cross-react with host tissues have been found in several studies and may play a role in the pathogenesis of autoimmune diseases [17, 25]. This particular mechanism of virus-induced autoimmunity is discussed in detail in the next section.

Molecular Mimicry: Reactivity of Antiviral Antibodies with Normal Tissues

Although the idea that antibodies raised against viruses may cross-react with self antigens is not a new one, definitive proof has been difficult to obtain. However, the availability of large numbers of monoclonal antibodies has allowed this hypothesis to be tested. For example, a monoclonal antibody to the phosphoprotein of measles virus and another to a herpesvirus protein react with intermediate filaments of normal cells [26]. These antibodies are thought to bind to different determinants on the vimentin molecule. Similarly, monoclonal antibodies to SV40 T antigen were found to react with proteins in normal host cells [27]. Monoclonal antibodies to a nonstructural protein of Japanese encephalitis virus bound to nuclear antigen in uninfected cells [28] and another monoclonal antibody to HTLV-1 reacted with normal thymic epithelium [29].

In our laboratory we have screened 635 antiviral monoclonal antibodies against a panel of 14 normal mouse tissues [30]. These monoclonal antibodies were made against 11 different viruses (Table 1) including both DNA and RNA viruses. Some examples of reactivity are shown in Fig. 1. Of a total of 34 antibodies against Japanese encephalitis virus, six gave positive reaction with normal tissues. All six showed some reactivity with the anterior pituitary. Three of these also reacted with ductal cells in the salivary gland, and two others reacted with interstitial cells of the testis and chief cells in the stomach respectively. Three out of 74 anti-LCM virus monoclonal antibodies also reacted with anterior pituitary. In addition, one of these also reacted with testis and ovaries, and two others reacted with salivary gland. Five out of 39 antimeasles virus monoclonal antibodies reacted with normal cells. Three reacted with cells in the anterior pituitary, one with the cells in gastric mucosa, and one with the nuclei of neurons in the hippocampus and granular layer of the cerebellum. Eighty antirabies antibodies were tested, and one of the two positive antibodies reacted with cells in the anterior pituitary, islet of langerhans and stomach. The other positive antibody reacted with cells in the salivary gland and the anterior pituitary. Thirty-seven antivesicular stomatitis virus antibodies were tested, and two of them were found to react with cells in the stomach and salivary glands, respectively. One of the 20 anti-HSV monoclonal antibodies reacted with pancreatic islets, salivary gland, and nuclei of spermatogonia. Sixteen antivaccinia monoclonal antibodies were tested, and only one of them reac-

Table 1. Reactivity of monoclonal antiviral antibodies with normal tissues

Monoclonal antiviral antibodies to:	Monoclonal antibodies screened (*n*)	Monoclonal antibodies that reacted with normal tissue (*n*)
CVB4	66	1
JEV	34	6
LCM	174	3
Measles	39	5
Rabies	80	2
Vesicular stomatitis	37	2
HSV (type 1)	20	1
Vaccinia	16	1
Dengue	132	0
Cytomegalovirus		
-human	23	0
-murine	14	0
	635	21

Each of the monoclonal antibodies was tested against a panel of 14 different normal mouse tissues. For details see [30].

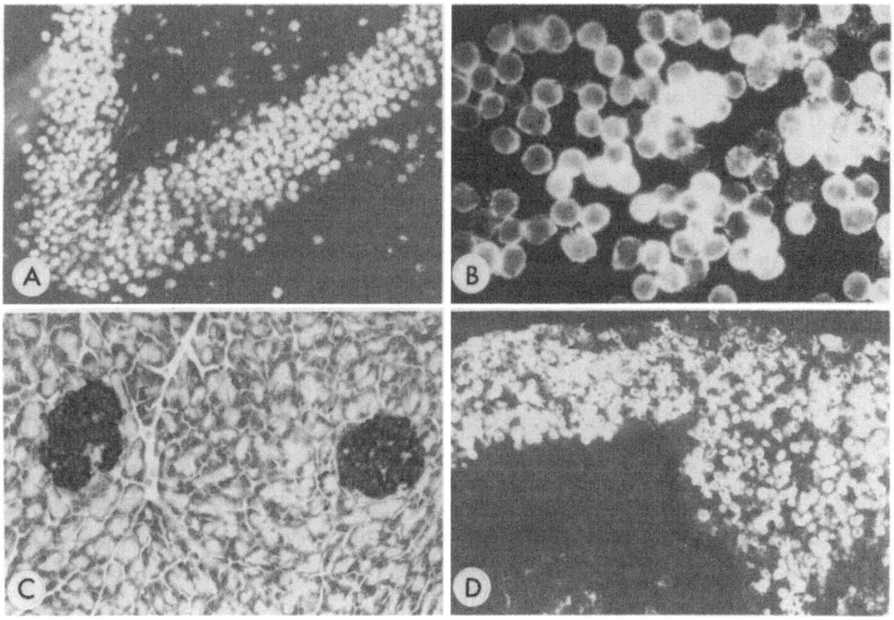

Fig. 1 A–D. Reactivity of antiviral antibodies with normal host antigens. **A** Reactivity of an antimeasles virus monoclonal antibody with nuclei of cells in the hippocampus. **B** Reactivity of another antimeasles antibody with T lymphocytes. **C** An anti-HSV monoclonal antibody reacts with cells in the islets of Langerhans. **D** An anti-JEV monoclonal antibody reacts with cells in the anterior pituitary. For more details see [30]

ted with normal tissues. This antibody reacted with cells in the salivary glands. None of the antibodies to dengue virus, and cytomegaloviruses reacted with normal tissues. Approximately 3.5% of the antiviral monoclonal antibodies reacted with self antigens and several of these antibodies reacted with more than one organ. These studies clearly showed that cross-reactivity of antiviral antibodies with normal host antigens is relatively common.

Cross-reactivity of Neutralizing Anticoxsackievirus B4 Monoclonal Antibody with the Myocardium

Coxsackieviruses (CV) are known to produce myocarditis in both humans and experimental animals [31]. Although the cytolytic effects of the virus are the main cause of damage to the heart, it is thought that an autoimmune component may also contribute to the tissue injury [32, 33]. Autoantibodies to the heart have been reported in a number of heart conditions including rheumatic carditis [34], postpericardiotomy [35], and postinfarction syndromes [34]. Similar antibodies have been observed in both animals and humans after CV infections [36, 37]. However, it is not known whether these antibodies arise as a result of myocardial damage or represent antiviral antibodies that cross-react with the heart. To test this, we screened, by indirect immunofluorescence, 66 neutralizing mouse monoclonal antibodies to CV type B4 (CVB4) [38, 39] for their ability to react with normal mouse tissues. Only one of the antibodies, designated 356-1, reacted with the heart but not with any of the other tissues tested (Fig. 2A).

This antibody belongs to the IgG2a subclass and had a titer of 1:128 against CVB4 in a microneutralization assay. Further characterization of this antibody showed that it binds to a moderately conserved epitope on CVB4 which is expressed in approximately 70% of all the clinical isolates tested [40]. Radiolabeled virus-infected cell extracts were immunoprecipitated using 356-1. The precipitated proteins were subjected to sodium dodecyl sulfate-polyacrylamide gel dectropharesis (SDS-PAGE), then electropheretically separated proteins were visualized by exposing the gel to X-ray film. These studies showed that the 356-1 was capable of immunoprecipitating all the viral polypeptides [41] perhaps due to aggregation of virus proteins. However, when purified virus polypeptides, separated by SDS-PAGE and then electropheretically transferred to nitrocellulose membrane, were reacted with 356-1, it bound specifically to VP1 and not to any of the other viral polypeptides.

By immunoperoxidase staining, the antibody 356-1 reacted with transverse striations in longitudinal sections of mouse heart. Polarized light microscopy revealed that the stained striae were A bands of the myofibrils (Fig. 2E and F). Fab fragments of antibody 356-1 yielded identical results indicating that it was the antibody-combining site that reacted with the myocardium.

This antibody was further tested on frozen sections of heart and skeletal muscle from other species (Table 2). It reacted strongly with the atrium of the mouse, rat, hamster, rabbit (Fig. 2B), guinea pig (Fig. 2C), and chicken, but not with that of monkey or human (Fig. 2D). The antibody also reacted strongly with the ventricles of mouse, rat, and hamster, weakly with the ventricles of rabbit, and not at all

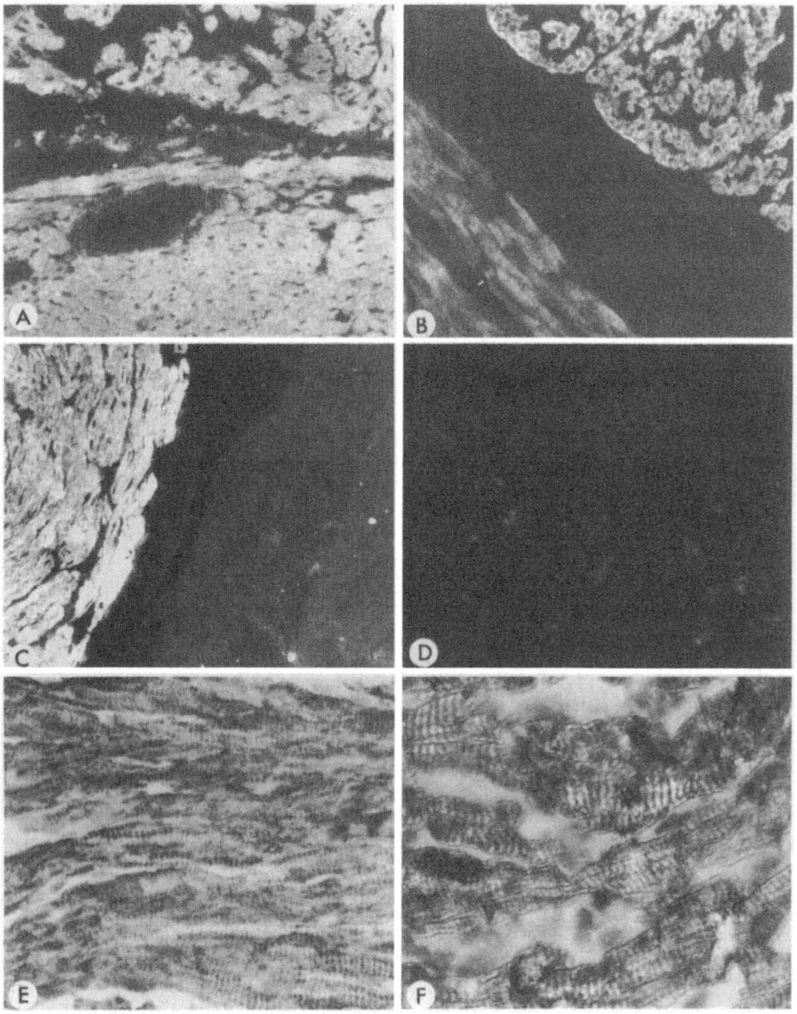

Fig. 2 A–F. Reactivity of monoclonal antibody 356–1 with heart by indirect immunofluorescence **(A-D)** and immunoperoxidase **(E,F)** staining. **A** Mouse heart: both atrial and ventricular myofibers show brilliant fluorescence. **B** Rabbit heart: atrial myofibers stain strongly, but reactivity with ventricle is weak. **C** guinea pig heart: atrial myofibers stain strongly, but ventricle shows no staining. **D** negative reaction with human ventricle. **E** staining of transverse striations in longitudinal section of chicken atrium. **F** Polarized light shows that the A bands of myofibrils are stained.

with the ventricles of guinea pig, chicken, monkey, or human. This antibody did not react with skeletal or smooth muscle from any of these species [39]. These observations are consistent with earlier reports of structural differences between the atrial and ventricular myosins [42, 43] and between the cardiac, skeletal, and smooth muscle myosins [44] of these species.

Table 2. Reactivity pattern of the anti-CVB4 monoclonal antibody 356-1

Tissue	Animal species							
	Mouse	Rat	Hamster	Rabbit	Guineapig	Chicken	Monkey	Human
Myocardium								
Atrium	++	++	++	++	++	++	--	--
Ventricle	++	++	++	+	--	--	--	--
Skeletal muscle	--	--	--	--	--	--	--	--
Smooth muscle	--	--	--	--	--	--	--	--

To determine the nature of the antigen with which 356-1 was reacting, we used, initially, sequential solubilization to enrich for murine cardiac and skeletal muscle myosins. These enriched proteins were subjected to SDS-PAGE and blotted onto nitrocellulose. The Western blots of the separated proteins were reacted with 356-1, and a control antimyosin antibody was designated 6C1. The antibody 356-1 reacted very strongly with cardiac muscle extract and weakly with skeletal muscle extract. These observations were further confirmed by testing 356-1 on plates coated with purified cardiac and skeletal muscle myosins. These experiments clearly showed that 356-1 has a much higher affinity for cardiac myosin than skeletal myosin (manuscript in preparation). The lack of reactivity of the monoclonal antibody 356-1 with skeletal muscle by immunofluorescence staining could be explained by this low affinity for skeletal muscle myosin.

Our observation that molecular mimicry exists between myocardium and CVB4 is similar to results of other groups. Maisch et al. [37, 45] have indicated that CVB, influenza, and mumps viral antigens can absorb the antisarcolemmic cytotoxic activity from the sera of patients with viral myocarditis. In Chagas' disease, cross-reactivity between *Trypanosoma cruzi* and a myocardial sarcoplasmic reticulum antigen, the Ca^{2+}-dependent ATPase, has been identified [46]. This phenomenon has now also been seen in streptococcally induced rheumatic heart disease [47, 48]. Using monoclonal antibodies to streptococcal M protein, cross-reactivity with the myosin heavy chain has been demonstrated [49]. Our study adds to the growing body of evidence that molecular mimicry may be a mechanism in viral myocarditis.

Perspective

The idea that an antibody to an infectious agent might react with normal host tissues is not new, but definitive proof has been difficult to obtain. By exhaustively absorbing sera with appropriate antigens, evidence was obtained that, in patients with rheumatic heart disease, antibody to streptococcal M protein reacted with normal heart [50], and, in patients with Chagas' disease, antibody to *T. cruzi* reacted with normal heart and nerves [51]. Monoclonal antibodies against streptococcus and *T. cruzi* have now provided definitive proof of cross-reactivity with normal heart. Recently, several monoclonal antiviral antibodies that react with normal tissues have also been described. We have systematically screened more than 600 monoclonal antiviral antibodies against as many as 14 different organs. Because each organ contains a variety of different cell types, and each cell type contains thousands of different antigens, the screening procedure covered a large segment of the antigenic repertoire of the host. These experiments showed that a high percentage (3.5%) of monoclonal antiviral antibodies reacted with normal tissues. This high yield is not surprising when one considers that an antibody is capable of recognizing an antigenic determinant as short as four to six amino acids [27, 52]. Thus, considering the tens of thousands of proteins in an organism, the probability of a similar sequence appearing by chance in two unrelated molecules is not insignificant. Search for overlapping sequences in 2511 protein sequences, consisting of approximately 470000 residues, in Dayhoff protein data base Wilson et al. [53] have found 2469 hexamers, 186 septamers, and 17 octamers with sequence homologies. This frequency is much higher than the frequency one would predict based upon random distribution of all the 20 amino acids. Moreover, since an antibody reacts with the three-dimensional configuration of a molecule, the antibody could bind, but with a different affinity, to similar but not necessarily identical sequences [54, 55]. From our studies we conclude that molecular mimicry is a common phenomenon. As the search continues, more examples will certainly be found. The role cross-reacting antibodies actually play in pathogenesis is still not known. This will undoubtedly depend on the nature and location of the autoantigens (e.g., cell surface versus cytoplasm) and on the properties of the antibodies. Cross-reacting antibodies, however, need not always be deleterious and conceivably could be one of the explanations for long-lasting antiviral immunity after certain viral infections [56, 57]. B lymphocytes that make antiviral antibodies which cross-react with normal host cell antigens may continue to be stimulated by those host antigens long after the viral infection has disappeared. If this turns out to be the case, molecular mimicry might in part account for the enigma of persistent immunity in the absence of continued exposure to virus.

Acknowledgement. We would like to thank Mrs. Karen Hardman for her dedication and skill in typing this manuscript.

References

1. Onodera T, Toniolo A, Ray UR, Jenson AB, Knazek RA, Notkins AL (1981) Virus-induced diabetes mellitus. XX. Polyendocrinopathy and autoimmunity. J Exp Med 153: 1457–1473
2. Dales S, Fujinami RS, Oldstone MBA (1983) Infection with vaccinia tavors the selection of hybridomas synthesizing autoantibodies against intermediate filaments, one of them cross-reacting with virus hemmagglutinin. J Immunol 131: 1546–1553

3. Kay MMB (1979) Parainfluenza infection of aged mice results in autoimmune disease. Clin Immunol Immunopathol 12: 301–315
4. Fukuda A, Yamanouchi K (1981) Comparison of autoimmunity induction with virulent and attenuated rinderpest virus in rabbits. Jpn J Med Sci Biol 34: 149–159
5. Hahn EC, Kenyon AJ (1980) Anti-deoxyribonucleic acid antibody associated with persistent infection of mink with Aleutian disease virus. Infect Immun 29: 452–458
6. Varet B, Cannat A, Gisselbrecht S (1977) Genetic control of antinuclear antibodies in mice infected with Rauscher Leukemia virus. Cancer Res 37: 1115–1118
7. Hirsch MS, Proffitt MR (1975) Autoimmunity in viral infection. In: Notkins AL (ed) Viral immunology and immunopathology. Academic, New York, p 419
8. Linder E, Kurki P, Anderson LC (1979) Autoantibody to "intermediate filaments" in infectious mononucleosis. Clin Immunol Immunopathol 14: 411–417
9. Mottironi VD, Terasaki PI (1970) Lymphocytotoxins in disease. I. Infectious mononucleosis, rubella and measles. In: Terasaki PI (ed) Histocompatibility testing. Munksgaard, Copenhagen, p 301
10. Loza-Tulinmowska M, Semkow R, Michalak T, Nowoslawski A (1976) Autoantibodies in sera of influenza patients. Acta Virol 20: 202–207
11. Wutzker P, Farber I, Ulbricht A (1978) Autoantibodies in herpes virus infections. Acta Virol 22: 342
12. Panitch HS, Hooper CJ, Johnson KP (1980) CSF antibody to myelin basic protein. Arch Neurol 37: 206–209
13. Kahane S, Dvilanksy A, Estok L, Nathan I, Zolotov Z, Sarov I (1981) Detection of antiplatelet antibodies in patients with idiopathic thrombocytopenic purpura (ITP) and in patients with rubella and herpes group viral infections. Clin Exp Immunol 44: 49–56
14. Stagno S, Pass RF, Reynolds DW, Moore MA, Nahmias AJ, Alford CA (1980) Comparative study of diagnostic procedures for congenital cytomegalovirus infection. Pediatrics 65: 251–257
15. Markensen JA, Daniels CA, Notkins AL, Hoofnagle JH, Gerety J, Barker LF (1975) The interaction of rheumatoid factor with hepatitis B surface antigen-antibody complex. Clin Exp Immunol 19: 209–217
16. Toh BH, Yildiz A, Sotelo J, Osung O, Holborow EJ, Kanakoudi F, Small JV (1979) Viral infections and IgM autoantibodies to cytoplasmic intermediate filaments. Clin Exp Immunol 37: 76–82
17. Notkins AL, Onodera T, Prabhakar B (1984) Virus-induced autoimmunity. In: Notkins AL, Oldstone MBA (eds) Concepts in viral pathogenesis. Springer, Berlin Heidelberg New York, pp 210–215
18. Eaten MD (1980) Autoimmunity induced by syngeneic membranes carrying irreversibly adsorbed paramyxovirus. Infect Immun 27: 855–861
19. Woodruff JF, Woodruff JJ (1975) The effect of viral infections on the function of the immune system. In: Notkins AL (ed) Viral immunology and immunopathology. Academic, New York, p 393
20. Allaway GP, Srinivasappa J, Miller FN, Prabhakar BS, Notkins AL (1987) Autoantibody production by human B lymphocytes which spontaneously proliferate. In: Levine PH, Nonoyama H, Pearson G, Glaser R (eds) Epstein-Barr virus and human disease. Humana, Clifton, pp 287–291
21. Allaway GP, Srinivasappa J, Miller FN, Prabhakar BS, Notkins AL (1985) Spontaneously proliferating human B lymphocytes make autoantibodies. J Infect Dis 157: 968–972
22. Nepom JT, Weiner HL, Dichter MA, Tardieu M, Spriggs DR, Gramm CF, Powers ML, Fields BN, Greene MI (1982) Identification of a hemagglutinin-specific idiotype associated with reovirus recognition shared by lymphoid and neural cells. J Exp Med 155: 155–167
23. Sege K, Peterson PA (1978) Use of anti-idiotypic antibodies as cell surface receptor probes. Proc Natl Acad Sci USA 75: 2443–2447
24. Plotz PH (1983) Autoantibodies are anti-idiotype antibodies to antiviral antibodies. Lancet II: 824–826
25. Oldstone MBA, Notkins AL (1986) Molecular mimicry. In: Notkins AL, Oldstone MBA (eds) Concepts in viral pathogenesis, vol II. Springer, Berlin Heidelberg New York, pp 372–379

26. Fujinami RS, Oldstone MBA, Wroblewska Z, Frankel ME, Koprowski H (1983) Molecular mimicry in virus infection: crossreaction of measles virus phosphoprotein or of herpes simplex virus protein with human intermediate filaments. Proc Natl Acad Sci USA 80: 2346–2350
27. Crawford L, Leppard K, Lane D, Harlow E (1982) Cellular proteins reactive with monoclonal antibodies directed against simian virus 40 T-antigen. J Virol 42: 612–620
28. Gould EA, Chanas AC, Buckley A, Clegg CS (1983) Monoclonal immunoglobulin M antibody to Japanese encephalitis virus that can react with a nuclear antigen in mammalian cells. Infect Immun 41: 774–779
29. Haynes BF, Robert-Guroff M, Metzger RS, Franchini G, Kalyanaraman VS, Palker TJ, Gallo RC (1983) Monoclonal antibody against human T cell leukemia virus p19 defines a human thymic epithelial antigen acquired during ontogeny. J Exp Med 157: 907–920
30. Srinivasappa J, Saegusa J, Prabhakar BS, Gentry MK, Buchmeier J, Wiktor TJ, Koprowski H, Oldstone MBA, Notkins AL (1986) Molecular mimicry: frequency of reactivity of monoclonal antiviral antibodies with normal tissues. J Virol 57: 397–401
31. Woodruff JF (1980) Viral myocarditis: a review. Am J Pathol 101: 426–484
32. Wilson FM, Miranda QR, Chason JL, Lerner AM (1969) Residual pathologic changes following murine coxsackie A and B myocarditis. Am J Pathol 55: 253–265
33. Huber SA, Lodge PA (1984) Coxsackievirus B3 myocarditis in BALB/c mice: evidence for autoimmunity to myocyte antigens. Am J Pathol 116: 21–29
34. Kaplan MH, Frengley JD (1969) Autoimmunity to the heart in cardiac disease. Current concepts of the relation of autoimmunity to rheumatic fever, postcardiotomy and postinfarction syndromes and cardiomyopathies. Am J Cardiol 24: 459–473
35. Maisch B, Berg PA, Kochsiek K (1979) Clinical significance of immunopathological findings in patients with post-pericardiotomy syndrome I. Relevance of antibody pattern. Clin Exp Immunol 38: 189–197
36. Wolfgram LJ, Beisel KW, Rose NR (1985) Heart-specific autoantibodies following murine coxsackievirus B3 myocarditis. J Exp Med 161: 1112–1121
37. Maisch B, Trostel-Soeder R, Stechemesser E, Berg PA, Kochsiek K (1982) Diagnostic relevance of humoral and cell-mediated immune reactions in patients with acute viral myocarditis. Clin Exp Immunol 48: 533–545
38. Prabhakar BS, Haspel MV, McClintock PR, Notkins AL (1982) High frequency of antigenic variants among naturally occurring human coxsackie B4 virus isolates identified by monoclonal antibodies. Nature 300: 374–376
39. Saegusa J, Prabhakar BS, Essani K, McClintock PR, Fukuda Y, Ferrans VJ, Notkins AL (1986) Monoclonal antibody to coxsackievirus B4 reacts with myocardium. J Infect Dis 153: 372–373
40. Prabhakar BS, Menegus MA, Notkins AL (1985) Detection of conserved and non-conserved epitopes on coxsackievirus B4: frequency of antigenic changes. Virology 146: 302–306
41. Prabhakar BS, Srinivasappa J, Ray UR (1987) Selection of coxsackievirus B4 variants with monoclonal antibodies results in attenuation. J Gen Virol 68: 865–869
42. Sartore S, Pierobon-Bormioli S, Schiaffino S (1978) Immunohistochemical evidence for myosin polymorphism in the chicken heart. Nature 274: 82–83
43. Flink IL, Rader JH, Banerjee SK, Morkin E (1978) Atrial and ventricular cardiac myosins contain different heavy chain species. FEBS Lett 94: 125–130
44. Whalen RG, Sell SM, Eriksson A, Thornell LE (1982) Myosin subunit types in skeletal and cardiac issues and their developmental distribution. Dev Biol 91: 478–484
45. Maisch B, Berg PA, Kochsiek K (1980) Autoantibodies and serum inhibition factors (SIF) in patients with myocarditis. Klin Wochenschr 58: 219
46. Sadiqursky M, Acosta AM, Santos-Buch CA (1982) Muscle sarcoplasmic reticulum antigen shared by Trypanosoma cruzi. Am J Trop Med Hyg 31: 934
47. Kaplan MH, Suchy ML (1964) Immunologic relation of streptococcal and tissue antigens. II. Cross reaction of antisera to mammalian heart tissue with a cell wall constituent of certain strains of group A streptococci. J Exp Med 119: 643

48. Zabriskie JB, Freimer EH (1966) An immunological relationship between the group A streptococcus and mammalian muscle. J Exp Med 124: 661
49. Cunningham MW, Hall NK, Krishner KK, Spanier AM (1986) A study of anti-group A streptococcal monoclonal antibodies cross-reactive with myosin. J Immunol 136: 293
50. van de Rijn I, Zabriskie JB, McCarty M (1977) Group A streptococcal antigens cross-reactive with myocardium: purification of heart-reactive antibody and isolation and characterization of the streptococcal antigen. J Exp Med 146: 579–599
51. Khoury EL, Ritacco V, Cossio PM, Laguens RP, Szarfman A, Diez C, Avana RM (1979) Circulating antibodies to peripheral nerve in American trypanosomiasis (Chagas' disease). Clin Exp Immunol 36: 8–15
52. Atassi MZ (1975) Antigenic structure of myoglobin: the complete immunochemical anatomy of protein and conclusions relating to antigenic structures of proteins. Immunochemistry 12: 423–438
53. Wilson I, Niman H, Houghten R, Chetenson A, Connolly M, Lerner RA (1984) The structure on an antigenic determinant in protein. Cell 37: 767–778
54. Lane D. Koprowski H. (1982), Molecular recognition and the future of monoclonal antibodies. Nature 296: 200–202
55. Tainer JA, Getzoff ED, Alexander H, Houghten RA, Olson AJ, Lerner RA, Hendrickson WA, (1984), The reactivity of antipeptide antibodies is a function of the atomic mobility of sites in a protein. Nature 312: 127–134.
56. Ahmed R, Oldstone MBA (1984), Mechanisms and biological implication of virus-induced polyclonal B-cell activation. In: Notkins AL, Oldstone MBA (eds) Concepts in viral pathogenesis. Springer, Berlin Heidelberg New York, pp 231–238.
57. Mandel B, (1984), Mechanisms of virus neutralization. In: Notkins AL, Oldstone MBA (eds) Concepts in viral pathogenesis. Springer, Berlin Heidelberg New York, pp 32–38.

Immunopathogenic Mechanisms in Experimental Myocarditis: Evidence for Autoimmunity to the Virus Receptor and Antigenic Mimicry between the Virus and Cardiocyte

S. A. HUBER[1], K. SIMPSON[1], A. WELLER[1], and M. HERZUM[2]

[1] Department of Pathology, University of Vermont, Burlington, VT 05405, USA
[2] Medizinische Klinik der Universität Würzburg, 8700 Würzburg, FRG

Introduction

The past 2 decades have witnessed remarkable advances in the understanding of the pathogenesis of myocarditis and dilated cardiomyopathy. Animal models of picornavirus-induced heart disease have helped to elucidate many of the factors involved in disease susceptibility and cardiac injury although much remains to be learned. The disease process can basically be divided into three distinct phases: viral infection, inflammation, and fibrosis in dilated cardiomyopathy.

Viral Infection

Virus entering the body must interact with specific receptors on cardiocytes leading to infection of the cells [2, 3, 25]. Active infection appears to be an absolute requirement for myocarditis induction since inoculation of animals with inactivated virus fails to initiate inflammation. Lack of receptors on selected tissues or species usually prevents viral infection and replication. Studies indicate that those mouse strains showing reduced capability to support cardiac viral infection also fail to develop myocarditis [5, 11, 19, 30]. In this case, resistant animals presumably lack specific genes coding for the receptor. However, under other circumstances, animals having the appropriate genes may not express them. Male and female mice show dramatic differences in disease susceptibility with males developing severe cardiac injury and females showing minimal heart disease [8, 20, 21, 33]. Susceptibility correlates to localization of the virus in the myocardium within hours of in oculation. Accelerated localization in males results from enhanced virus receptor expression in cardiocytes. In vitro studies further show that sex-associated hormones modulate receptor expression with androgens increasing and estrogens decreasing cell surface receptors as much as four-fold.

The viral phase of the disease is transient. Infectious virus concentrations in the heart are maximal 3–4 days after in oculation and decrease rapidly thereafter reaching undetectable levels within 2–3 weeks [9, 17, 26, 33]. Whether this observation truly means virus is eliminated from the heart or remains as incomplete and noninfectious genomic material is not clear. Persistent coxsackievirus (CV) infections have been reported as lasting many months in fibroblast cultures [22]. Such low-grade infections might be easily missed in vivo. Additionally, various investi-

H.-P. Schultheiß (Ed.)
New Concepts in Viral Heart Disease

gators using in situ hybridization techniques apparently identify CV genomes in myocardial biopsies when infectious virus should be absent [1, 13]. The importance of determining whether virus persists in the heart relates to elucidating the mechanisms of cardiac injury. One might assume that infection of cardiocytes alone could produce significant disruption of myocardial function, either by directly killing heart cells, or through interference with cellular metabolism. However, cardiac injury in experimental models is minimal during periods of greatest virus titers in the heart and increases only in direct association with inflammatory cell infiltration of the myocardium [9, 32]. Such a pattern suggests that immune rather than viral mechanisms participate predominately in disease pathogenesis, and emphasis in therapy should be directed toward suppressing immunity rather than inhibiting virus replication. Yet, if virus remains even at low levels in the heart, its effect on cardiocytes could explain myocardial dysfunction. In this case, therapy should include antiviral reagents.

Inflammation

Inflammatory cell infiltration of the heart begins approximately 5 days after infection and consists of lymphocytes, macrophage, and fibroblasts [6, 14]. Of the lymphocytic component, most cells in early lesions belong to the immature T cell subset expressing pan T cell markers but lacking the T helper (L3T4) and cytolytic/suppressor (Lyt-2) differentiation antigens [15, 16]. As the disease progresses, these effector cells become more prominent. Woodruff and Woodruff [31] first demonstrated the importance of immune mechanisms in CV myocarditis. Animals made T cell deficient show remarkable resistance to myocarditis induction, yet virus replication and clearance from the hearts of immunologically intact and compromised animals occur normally. These studies prove three points which still remain valid:

a) viral infection of myocytes may produce subtle aberrations in cardiac function but are not responsible for the overt myocardial necrosis normally observed in the disease;
b) T cell-dependent functions are not involved in controlling CV infections; and
c) inflammation and myocardial necrosis result either directly or indirectly from T cell-dependent immunity.

Fibrosis and Dilated Cardiomyopathy

Although in certain genetically predisposed individuals, myocarditis can become chronic and last for extended periods, others develop only acute disease with active inflammation persisting only several weeks [7, 17, 18, 26]. More importantly, whether inflammation occurs either in acute or chronic form, not all individuals with myocarditis ultimately develop apparent dilated cardiomyopathy. Of the three myocarditic CV type B3 (CVB3) variants presently empolyed in murine models, the virus strain used by Woodruff [31, 32, 10, 11] and his colleagues induces severe cardiac injury but no evidence of dilated cardiomyopathy. This contrasts remarkably with the variant used by Reyes et al. and Lerner and Reyes [17, 26] which has been extensively documented as causing severely dysfunctional and hypertrophied hearts. Why certain viruses should preferentially induce dilated cardiomyopathy is a mystery. Perhaps the variants induce distinct forms of immunopathogenic mechanisms with factors produced by some effectors disrupting cardiac architecture and function more than others. Alternatively, the viruses themselves may affect cardiocytes differently. Certainly this area will become an important area for future study.

Immunopathogenic Mechanisms

Initial work concentrated on cellular immune mechanisms of cardiac injury. Cytolytic T lymphocytes (CTL) arise in peripheral lymphoid organs of BALB/c mice shortly after infection reaching maximal activity at times corresponding to peak myocarditis [9]. Three distinct CTL populations occur in CV-infected mice recognizing normally expressed cardiac antigens (autoreactive CTL; ACTL), virus epitopes (virus-specific CTL; VCTL), and cellular neoantigens induced by the infection process (metabolic CTL; MCTL) [4, 10, 12]. Table 1 compares these effectors. How each CTL population functions in disease pathogenesis is not certain. In a limited study, myocarditic potential of viruses correlated with their ability to interfere with cellular metabolism and induce MCTL. However, while in vivo depletion of either CD4 or CD8 cells provides partial protection from cardiac damage, clearly the latter cell containing the ACTL population is primarily pathogenic in this model [4, 10]. Furthermore, only the ACTL population induces myocarditis when transferred into uninfected animals [12], an important observation since much of both the clinical and experimental disease occurs when infectious virus and therefore, presumably, virus-infected and metabolically altered target cells are scarce. Some evidence suggests that CD4 cells function in myocarditis primarily as lymphokine producers which either recruit antigen-presenting cells into the heart or enhance clonal expansion of autoreactive T cells. In this

Table 1. Characteristics of CTL populations

	ACTL	VCTL	MCTL
1. T cell subset	CD8	CD4	CD4
2. MHC restriction	Class I	Class II	Class II
3. Tissue specificity	Uninfected cardiocytes (heart specific)	CVB3-infected tissues (not heart specific)	Metabolically inhibited cells
4. Putative antigen specificity	Virus receptor	Virus polypeptides	Stress proteins
5. Pathogenic potential	High	Low	Unknown

manner, minimal CD4 responses may be required for ACTL sensitization, but only the autoimmune T cell mediates direct myocardial injury (M. Guthrie, unpublished observations).

Although CD8 cells predominantly cause myocarditis in BALB/c mice, CD4-dependent immunopathogenic mechanisms apparently function in genetically dissimilar animals [11, 18]. One might assume that humoral autoimmunity plays a major role in these strains since various investigators have demonstrated correlation of disease with specific heart-reactive autoantibodies (HRA) and shown that complement depletion in animals with CD4-dependent disease abrogates

Table 2. Characterization of MAb Reactivity to Tissue Membranes and Virus

	Reactivity by immunoblot					
MAb	Heart	Skeletal muscle	Liver	HeLa cell	L cell	CVB3
7A2	+++	+++	+++	±	++	+
7A3	+++	++	++	±	++	0
8A6	0	0	0	0	+	
10A1	++++	0	++	+++	0	0

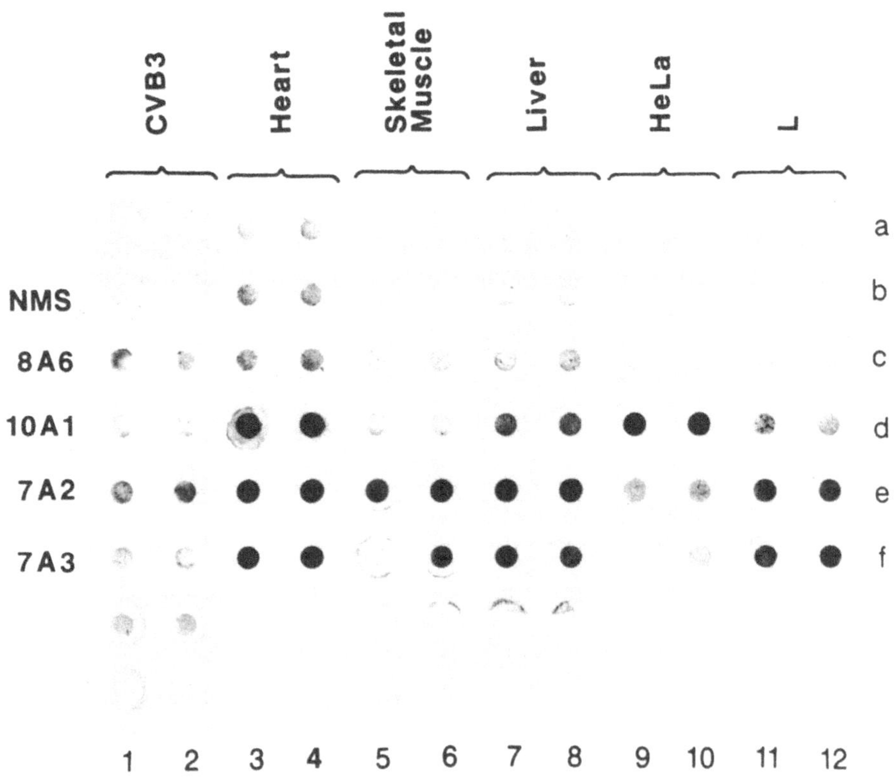

myocarditis [11, 23, 27, 30]. However, no one has yet conclusively proven antibody-mediated myocarditis by adoptively transferring disease with immune serum, monoclonal HRA, or B lymphocytes.

We have produced four monoclonal antibodies (MAbs) from CVB3-infected mice and have investigated their effect on viral infection and myocyte viability in vivo and in vitro. All four MAbs belong to the IgM subclass. One (8A6), which was selected for its ability to prevent cytopathic effect of CVB3 to HeLa cells, binds to virus but shows no reactivity by immunoblot to membrane preparations derived from a number of different murine tissues (Fig. 1 and Table 2). In contrast, 10A1, a MAb selected for reactivity to cultured cardiocytes, additionally binds to liver and HeLa cell membranes, which also share CVB3 receptors, but not to preparations from virus-resistant cells or CVB3. The MAbs 7A2 and 7A3 cross-reactively bind to both tissue and virion, undoubtedly indicating antigenic mimicry as described earlier for CVB4 by Notkins et al. [24]. When tested for ability to interfere with infection of susceptible cells, 8A6 and 10A1 both showed remarkable activity, but 8A6 only inhibited in a virus neutralization assay where MAb binds directly to virus, and 10A1 only inhibited when preincubated on cells (Table 3). Neither 7A2 nor 7A3 had any substantial effect on viral infection. Interestingly, none of the MAbs were directly injurious to cardiocytes alone or in the presence of complement, but 10A1 completely abrogated ACTL-mediated lysis of these targets while 7A2 showed definite synergism between cellular and humoral autoimmunity since ACTL-mediated cytotoxicity tripled in the presence of this antibody (Table 4). The effects noted with 10A1 and 7A2 cannot be attribu-

Table 3. Characterization of MAbs

MAB	Inhibition	
	Virus neutralization[a] (%)	Inhibition of myocyte infection (%)
7A2	−6 ± 1	12 ± 3
7A3	−10 ± 3	7 ± 2
8A6	100 ± 0[b]	9 ± 2
10A1	10 ± 1	66 ± 7[b]

[a] CVB3 (100 plaque-forming units; PFU) was mixed and incubated with 10 μg MAb for 30 min then added to HeLa cell monolayers. Plaques were determined 2 days later. Results represent mean decrease in PFU compared to virus incubated with 5% normal mouse serum (NMS) ± SEM

[b] Significantly different from NMS control at $P \leq 0.05$

[c] Five hundred myocytes were incubated for 90 min with 10 μg MAb, washed, then incubated 2 h at 37°C with 5 x 10^4 PFU CVB3, washed again, treated with hyperimmune anti-CVB3 to remove unincorporated virus, trypsinized and dispensed (100 cells) onto HeLa cell monolayers. Plaques, observed 2 days later, represented number of myocytes infected. Control cultures containing NMS showed that 75% of myocytes became infected. Results represent mean decrease of myocytes infected ± SEM

Table 4. Effect of MAb on myocyte cytotoxicity

MAb	Lysis with: MAb + complement (%)	MAb + normal lymphocytes (%)	MAb + ACTL (%)
NMS	0 ± 0	13 ± 2	40 ± 3
7A2	0 ± 0	4 ± 2	60 ± 1*
7A3	0 ± 0	0 ± 0	47 ± 1
8A6	0 ± 0	4 ± 1	45 ± 3
10A1	0 ± 0	0 ± 0	0 ± 0*

BALB/c myocytes (3 x 10^3) were ^{51}Cr-labeled and incubated for 18 h with either 5% NMS or 10 μg MAb and either 20% rabbit complement or 3 x 10^5 lymphocytes. Normal lymphocytes were derived from uninfected syngeneic mice. ATCL were derived as reported by Huber and Lodge (1984). Results represent mean ± SEM
* Significantly different from NMS at $P \leq 0.05$

ted exclusively to their ability to react to myocytes since 7A3 also binds to heart antigens yet shows no modulation of either infection or cell destruction. The effects of the various MAbs in vitro might not indicate their role in the disease process in vivo, however. Therefore, the MAbs were injected into either uninfected or CVB3-infected recipients (Table 5). None of the MAbs induced any detectable myocardial injury in uninfected mice. Type 8A6, as expected of a virus-neutralizing antibody, completely abrogated both infection and inflammation. Type 10A1 also prevented myocarditis, but while virus concentrations were reduced in the heart, they were not eliminated. Thus, the antibody to the virus receptor was significantly more effective against immunity than infection. Type 7A2 greatly exacerbated disease in vivo since mortality in treated animals was nearly three times that in infected controls. These results indicate several conclusions:

a) both ACTL and 10A1 presumably react with the virus receptor or a closely related molecule;
b) a rather heterogeneous HRA response occurs after CVB3 infection; and
c) HRA may greatly alter disease pathogenesis, either suppressing or aggravating the disease.

Proposed Mechanisms of Autoimmunity

Autoimmunity may arise in different ways. Antigenic mimicry between an infectious agent and tissues have been well documented and is clearly one mechanism [24, 28, 29]. This method probably explains how two of the MAbs arise. However, it cannot explain the putative antibody to the virus receptor. Autoimmunity directed to the virus receptor may be coincidental, but it is more likely that this is not

Table 5. Effect of MAb in vivo in CVB3-infected and uninfected BALB/c mice

MAb	Animal mortality		Cardiac virus titer (log 10)		Myocarditis	
	Uninfected (%)	CVB3-infected (%)	Uninfected	CVB3-infected	Uninfected	CVB3-infected
None	0 ± 0	47[b]	0 ± 0	7.8 ± 0.1	0 ± 0	1.4 ± 0.4
7A2	0 ± 0	82	0 ± 0	8.3 ± 0.2	0 ± 0	1.7 ± 0.3
7A3	0 ± 0	50	0 ± 0	0 ± 0[b]	0 ± 0	1.8 ± 0.3
8A6	0 ± 0[b]	0 ± 0	0 ± 0	0 ± 0	0.1 ± 0.1[b]	
10A1	0 ± 0	8[b]	0 ± 0	3.5 ± 0.2[b]	0 ± 0	0.2 ± 0.1[b]

BALB/c mice received either 1.8 x 10^5 PFU CVB3 in 0.5 ml PBS or PBS alone, then 2 mg MAb on days, 1, 3, and 5 after infection and were sacrificed on day 7. Results represent mean ± SEM

[a] Inflammation is based on a 0–4 scale with 0 representing no myocarditis and 4 representing approximately 30% of the myocardium inflamed

[b] Significantly different from control (no MAb) at $P \leq 0.05$

so. Immunity to virus epitopes may ultimately lead to anti-idiotypic responses. Where the initial reaction recognizes important epitopes on the virus for binding to the cellular receptor, an anti-idiotypic response should react to the receptor itself. In humoral autoimmunity, this results in 10A1, an antibody preventing viral infection by attaching to the cellular receptor. In cellular immunity, this may result in the ACTL, an effector which might be induced to the antigen receptor on virus-specific T cells. Only future studies will determine the validity of this concept.

Acknowledgement. This work was supported in part by Public Health Service Grants-in-Aid HL 28833 and HL 33256 from the National Institutes of Health and American Heart Association Grant-in-Aid 87-0834.

References

1. Bowels NE, Dubowitz V, Sewry CA, Archard LC (1987) Dermatomyositis, polymyositis and coxsackie B virus infection. Lancet I: 1004–1007
2. Crowell RL, Landau BJ (1970) Picornaviridae: enteroviruses-coxsackieviruses. In: Hsuing GD, Green RH (eds) CRC handbook series in clinical laboratory science, virology and rickettsiology, vol 1, part 1. CRC, Ft Lauderdale, pp 135–155
3. Crowell RL, Landau BJ (1983) Receptors in the initiation of picornavirus infections. In: H. Fraenkel-Conrat, Wagner RR (eds) Comprehensive virology, vol 18. Plenum, New York, pp 1–42
4. Estrin M, Smith C, Huber S (1987) Coxsackievirus B3 myocarditis: autoimmunity is L3T4+ T helper cell and IL-2 independent in BALB/c mice. Am J Pathol 127: 335–341

5. Gauntt CJ, Gomez PT, Duffey PS, Grant JA, Trent DW, Witherspoon SM, Paque RE (1984) Characterization and myocarditic capabilities of coxsackievirus B3 variants in selected murine strains. J Virol 52: 598–605
6. Godeny EK, Gauntt CJ (1987) In situ immune autoradiographic identification of cells in heart tissues of mice with coxsackievirus B3-induced myocarditis. Am J Pathol 129: 267–276
7. Herskowitz A, Wolfgram LJ, Rose NR, Beisel KW (1987) Coxsackievirus B3 murine myocarditis-marked strain differences in histopathologic features of early and late phase myocarditis. J Am Coll Cardiol 9: 1311–1319
8. Huber SA, Job LP, Auld KP (1982) Influence of sex hormones on coxsackie B3 virus infection in BALB/c mice. Cell Immunol 67: 173–189
9. Huber SA, Job LP (1983) Cellular immune mechanisms in coxsackievirus group B, type 3 induced myocarditis in BALB/c mice. Adv Exp Med Biol 161: 491–507
10. Huber SA, Lodge PA (1984) Coxsackievirus B3 myocarditis in BALB/c mice: evidence for autoimmunity to myocyte antigens. Am J Pathol 116: 21–29
11. Huber SA, Lodge PA (1986) Coxsackievirus B3 myocarditis. Identification of different pathogenic mechanisms in DBA/2 and BALB/c mice. Am J Pathol 122: 284–291
12. Huber SA, Lodge PA, Herzum M, Estrin M, Olszewski J (1987) The role of T lymphocytes in the pathogenesis of coxsackievirus B3 myocarditis. In: Kawai C, Abelmann WH, Matsumori A (eds) Pathogenesis of myocarditis and cardiomyopathy. Cardiomyopathy update 1. University of Tokyo Press, Tokyo, pp 9–21
13. Kandolf R, Ameis D, Kirschner P, Canu A, Hofschneider PH (1987) In situ detection of enteroviral genomes in myocardial cells by nucleic acid hybridization: an approach to the diagnosis of viral heart disease. Proc Natl Acad Sci USA 84: 6272–6276
14. Kishimoto C, Kuribayashi K, Masuda T, Tomioka N, Kawai C (1985) Experimental and immunological studies in viral myocarditis and dilated cardiomyopathy: significance of T lymphocytes in the severity of myocarditis. In: Sekiguchi M, Olsen EGJ, Goodwin JF (eds) Myocarditis and related disorders. Springer, Berlin Heidelberg New York, pp 218–220
15. Kishimoto C, Kuribayashi F, Fuhuma K, Masuda T, Tomiska N, Abelmann WH, Kawai C (1987) Immunologic identification of lymphocyte subsets in experimental murine myocarditis with encephalomyocarditic virus. Circ Res 61: 715–725
16. Kishimoto C, Misaki T, Crumpacker CS, Abelmann WH (1988) Serial immunologic identification of lymphocyte subsets in murine coxsackievirus B3 myocarditis: different kinetics and significance of lymphocyte subsets in the heart and peripheral blood. Circulation 77: 645–653
17. Lerner AM, Reyes MP (1985) Coxsackievirus myocarditis — with special reference to acute and chronic effects. Prog Cardiovasc Dis 27: 374–394
18. Lodge PA, Herzum M, Huber SA (1987) Coxsackievirus B3 induced myocarditis: acute and chronic forms of the disease. Am J Pathol 128: 455–463
19. Lutton CW, Gudvangen RJ, Nealon TJ, Paque RE, Gauntt CJ (1985) Cellular immune responses in mice challenged with an a myocarditic variant of coxsackievirus B3. J Med Virol 17: 345–357
20. Lyden DC, Huber SA (1984) Aggravation of coxsackievirus, group B, type 3-induced myocarditis and increase in cellular immunity to myocarditis and increase incellular immunity to myocyte antigens inpregnant BALB/c mice and animals treated with progesterone. Cell Immunol 87: 462–472
21. Lyden DC, Feran M, Olszewski J, Job LP, Huber SA (1987) Coxsackievirus B3 induced myocarditis: effect of sex steroids on viremia and infectivity of cardiocytes. Am J Pathol 126: 432–438
22. Matteucci D, Paglianti M, Giangregori AM, Capobianchi MR, Dianzani F, Bendinelli M (1985) Group B coxsackieviruses readily establish persistent infections in human lymphoid cell lines. J Virol 56: 651–654
23. Neu N, Beisel KW, Traystman MD, Rose NR, Craig SW (1987) Autoantibodies specific for the cardiac myosin isoform are found in mice susceptible to coxsackievirus B3 induced myocarditis. J Immunol 138: 2488–2492

24. Notkins AL, Onodera T, Prabhakar B (1984) Virus-induced autoimmunity. In: Notkins AL, Oldstone MBA (eds) Concepts in viral pathogenesis. Springer, Berlin Heidelberg New York, pp 210–215
25. Reagan KJ, Goldberg B, Crowell RL (1984) Altered receptor specificity of coxsackievirus B3 after growth in rhabdomyosarcoma cells. J Virol 49: 635–640
26. Reyes MP, Ho KL, Smith F, Lerner AM (1981) A mouse model of dilated-type cardiomyopathy due to coxsackievirus B3. J Infect Dis 144: 232–236
27. Rose NR, Beisel KW, Herskowitz A, Neu N, Wolfgram LJ, Alvare FL, Traystman MD, Craig SW (1987) Cardiac myosin and autoimmune myocarditis. In: Evered D, Whelan S (eds) Autoimmunity and autoimmune disease. Ciba Foundation Symposium 129. Wiley, Chichester 129: 3–24
28. Saegusa J, Prabhaker BS, Essani K, McClintock PR, Fukuda Y, Ferrans VJ, Notkins AL (1986) Monoclonal antibody to coxsackievirus B_4 reacts with myocardium. J Infect Dis 153: 372–373
29. Van de Rijn I, Zabriskie JB, McCarty M (1977) Group A streptoccal antigens cross-reactive with myocardium: purification of heart-reactive antibody and isolation and characterization of the streptococcal antigen. J Exp Med 146: 579–599
30. Wolfgram LJ, Beisel KW, Herskowitz A, Rose NR (1986) Variations in the susceptibility to coxsackievirus B3-induced myocarditis among different strains of mice. J Immunol 136: 1846–1852
31. Woodruff JF, Woodruff JJ (1974) Involvement of T lymphocytes in the pathogenesis of coxsackievirus B3 heart disease. J Immunol 133: 1726
32. Woodruff JF, Wong CY, Woodruff JJ (1977) Cytotoxic T cells in coxsackieviral disease. In: Weksler ME, Litwin SO, Riggio RR, Siskind GW (eds) Immune effector mechanisms in disease. Grune and Stratton, New York, pp 207–237
33. Woodruff JF (1980) Viral myocarditis: a review. Am J Pathol 101: 425–484

Coxsackie B3 Myocarditis: Genetic Aspects of Different Immunopathogenic Mechanisms in BALB/c and DBA/2 Mice. Antigenic Specificity of Heart-Reactive Antibodies in DBA/2 Mice

M. HERZUM[1], S. A. HUBER[2], and B. MAISCH[1]

[1] Department of Internal Medicine, University of Würzburg, Josef-Schneider-Str. 2, D-8700 Würzburg, FRG
[2] Department of Pathology, University of Vermont, Medical Alumni Building, Burlington Vermont, 05405-0068,USA

Introduction

To date myocarditis has been diagnosed by demonstrating cellular infiltrations and myocytolysis in the myocardium [1]. Serology and in situ hybridization have generated substantial evidence that viruses, especially enteroviruses, are major causitive agents of the disease [3, 8].

Animal models [9] and studies in patients with myocarditis [10, 12] have demonstrated the importance of cellular and humoral immune reactions to cardiac tissue in this disease. Differences in susceptibility to myocarditis and the pathogenic mechanisms, partially related to the major histocompatibility complex, have been shown in Coxsackie B3 and in encephalomyocarditis virus-induced murine myocarditis [2]. Whereas Coxsackie B3 myocarditis in BALB/c mice depends almost exclusively on CD8-positive cytolytic T cells [6], CD4-positive cells and heart-reactive antibodies, especially of the IgG isotype, seem to produce myocarditic lesions in DBA/2 mice infected with Coxsackie B3 virus [7].

The following article describes studies on the genetic inheritance of different immunopathogenic mechanisms in DBA/2 and BALB/c mice and antigens of cardiac tissue that are recognized by circulating antibodies in DBA/2 mice.

Material and Methods

Animals

BALB/c mice originally purchased from Cumberland Farms, Clinton, Tennessee were bred at the University of Vermont. DBA/2 mice were bought from Jackson Laboratories, Bar Harbour, Maine. F^1 hybrids between BALB/c and DBA/2 mice and backcrosses were established at the University of Vermont.

H.-P. Schultheiß (Ed.)
New Concepts in Viral Heart Disease

Virus

Coxsackie B3 virus (Nancy strain) was passaged through HeLa cell cultures. Male animals were infected with 6×10^4 plaque-forming units (PFU) of the virus intraperitoneally and sacrificed on day 7 after inoculation.

Monoclonal Antibodies

The hybridomas GK1.5, producing a monoclonal antibody to L3T4-bearing helper T cells, and 2.43, producing a monoclonal antibody to Lyt2.2-positive cytolytic T cells in BALB/c mice, were obtained through Dr. S. Sriram, University of Vermont. The hybridoma HB 129, producing a monoclonal antibody to Lyt2.1-positive cytolytic T cells in DBA/2 mice, was purchased from American Type Culture Collection (ATCC). The animals were injected with 2 mg monoclonal antibody on days -2, -1, and 1 relative to day 0 of virus inoculation.

Histology

The number of myocarditic lesions was scored on hemotoxylin and eosin stained cross-sections of the hearts on a scale of 0–4.

Myocarditis Virus Titers

The titer of replicating Coxackie B3 virus in the myocardium was determined as described in detail elsewhere [4].

Immunoblot

Hearts of DBA/2 mice, infected 7 days before with 6×10^4 PFU Coxsackie B3, and hearts of uninfected animals were homogenized and lysed with 0.5% deoxycholate Na and 1% Triton X-100 [11]. The lysate was centrifuged at 1000 *g*, the supernatant electrophoresed on a 12.5% sodium dodacyl sulfate-polyacrylamide gel and transblotted to nitrocellulose (NC) paper. Single strips of the NC paper were incubated with control or experimental sera of the animals at a dilution of 1:75. Bound IgM and IgG were detected using the biotin-avidin system (vectastain).

Results and Discussion

Table 1 summarizes findings with regard to immunopathogenic mechanisms in BALB/c and DBA/2 mice: in DBA/2 mice Coxsackie B3-induced myocarditis is dependent on CD4-positive T lymphocytes. Depletion of the complement system by cobra venom factor largely protects the animals from the disease. The demon-

Table 1. Coxsackie B3-induced myocarditis in DBA/2 and BALB/c mice depends upon different immunopathogenic mechanisms

	Protection by depletion of:		
Strain	T lymphocyte subset	Complement	IgG deposited in the heart
DBA/2	T helper cells (CD4 positive)	Yes	Yes
BALB/c	T cytolytic cells (CD8 positive)	No	No

stration of heart-reactive antibodies, especially of the IgG isotype, favors the idea that a humoral response mediates the myocarditic process [7]. In contrast, BALB/c mice developed extensive myocarditis which can be abolished by eliminating the CD8-positive cytolytic T cells. Those cytolytic T lymphocytes can also be demonstrated by in vitro cytolysis of cardiac myocytes [6]. Cobra venom factor has no beneficial effect for the animals.

Table 2. Preliminary studies on the genetic inheritance of immunopathogenic mechanisms

Strain	Monoclonal antibody treatment	Protected total (*n*)	Protected (%)
BALB/c	–	3/13	23
	α-CD4	0/10	0
	α-CD8	7/8	88
	α-CD4 + α-CD8	8/8	100
DBA/2	–	0/10	0
	α-CD4	8/8	100
	α-CD8	0/16	0
	α-CD4 + α-CD8	11/12	92
(BALB/c x DBA/2) F1	-	1/12	8
	α-CD4	0/14	0
	α-CD8	1/12	8
	α-CD4 + α-CD8	11/13	85
F1 x BALB/c	–	0/12	0
	α-CD4	1/15	7
	α-CD8	11/28	39
F1 x DBA/2	–	1/21	5
	α-CD4	10/24	42
	α-CD8	0/10	0

At least three explanations can be proposed for these different immunopathogenic mechanisms:

1. Differences in antigen(s) for T lymphocytes and antibodies between those two strains.
2. The antigen processing and capability of presenting the antigens onto class II and class I molecules of the major histocompatibility complex.
3. The antigen's specific T cell receptors of T lymphocytes of the two strains favor one or the other pathomechanism.

The following experiments addressed all three of these hypotheses. Table 2 shows that BALB/c mice develop myocarditis depending on CD8-positive cells. In DBA/2 mice, only monoclonal antibody to CD4-positive cells protects the animals from myocarditis, whereas monoclonal antibody to CD8 has no visible effect. In the F^1 hybrid between the two strains, complete protection can only be achieved by eliminating both the T lymphcyte subsets, the CD4-positive and CD8-positive cells, pointing to an equal contribution of both subsets to the immunopathogenesis of the disease. Crossing the F^1 back to either parent shows that one single gene must determine the different pathogenic mechanisms in both strains. As would be expected from a single gene inheritance, 50% of the animals of either backcross hybrid gain protection from eliminating one T cell subpopulation. The difference in pathogenic T cell subpopulation does not seem to lie in the expression of antigens, which the lymphocytes recognize, nor in the capability of antigen presentation in connection with class II or class I of the major histocompatiblity complex. In T cell-deficient BALB/c mice, the adoptive transfer of T lymphocytes of DBA/2 animals exhibit the same way of inducing myocarditis by CD4-positive lymphocytes as in their syngeineic strain. Transferring T cells from BALB/c mice back into the syngeineic environment yields the expected, repeatedly shown, immunopathogenesis through CD8-positive cells (Table 3). Thus

Table 3. Adoptive transfer of myocarditis into BALB/c T cell-deficient mice with T cells from BALB/c and DBA/2 donors

Source fo T cells transferred	Monoclonal antibody treatment	Protected total (*n*)	Protected (%)
–	–	3/3	100
BALB/c	–	0/4	0
	α-CD4	0/4	0
	α-CD8	4/4	100
	α-CD4 + α-CD8	4/4	100
DBA/2	–	0/4	0
	α-CD4	4/4	100
	α-CD8	1/4	25
	α-CD4 + α-CD8	4/4	100

Thus antigens evoked by a Coxsackie B3 infection in BALB/c hearts are processed, presented, and recognized exclusively by CD4-positive cells in DBA/2 animals. The antigen processing and presentation in BALB/c mice is not limited, favoring a class I-restricted CD8-positive cytolytic T cell. Class II-restricted CD4-positive cells of DBA/2 mice readily find their antigens in this strain of mice. The T cell receptor of lymphocytes in either strain — perhaps through deletions in the gene — most likely restricts the strains to either immune response, not permitting the interaction between T cell receptor and class I in DBA/2 mice or class II in BALB/c mice, respectively.

Preliminary immunoblot analysis of antigens recognized by IgM and IgG antibodies in DBA/2 mice after Coxsackie B3 infection revealed multiple proteins of interest (Fig. 1 and 2). Various bands detected by the antibodies in the serum of infected animals lighten up, ranging particularly from a size of 20–43 kilodalton (kd) using infected and uninfected hearts as antigenic targets. There is one protein band with a size of about 60 kd which is constantly detected by autoantibodies in the immune sera (n=4), but not in normal sera (n=8). The reaction is stronger for the IgM isotype as would be expected for a recent onset of the infection. The weak detection by IgG antibodies shows that the switch of immunoglobulin is already present on day 7. More sera have to be tested, and the proteins, recognized by immune sera, characterized to understand their significance in normal cell physiology and their role in the immunopathogenic process wehn they are targets for the immune system. The differences of the antigens recognized by antibodies, even in syngeineic animals of the same age infected with the same virus at the same time, may be due to differences in the amount of myocardial necrosis, releasing proteins which give rise to antibodies, and by different virus titers in the heart (Table 4). Those differences may also be due to a polyclonal nonspecific stiumlation of plasma cells during a generalized infection, as is the acute Coxackie B3 infection.

Summary

In Coxsackie B3-virus induced myocariditis, different immunopathogenic mechanisms mediate information in BALB/c and DBA/2 mice. In DBA/2 mice heart-reactive antibodies seem to be of major importance in developing myocardial lesions, whereas in BALB/c mice cellular immune reactions lead to the disease. Genetically this difference seems to be inherited in one single gene encoding for

Table 4. Myocardial virus titers and myocarditic lesions on day 7 after inoculating DBA/2 mice with 6 x 10^4 PFU Coxsackie B3 virus

Group	n	Virus titers (mean ± SD)	Myocarditis lesions (mean ± SD)
A	8	0 ± 0	0 ± 0
B	4	4.82 ± 0.41	2.15 ± 0.71

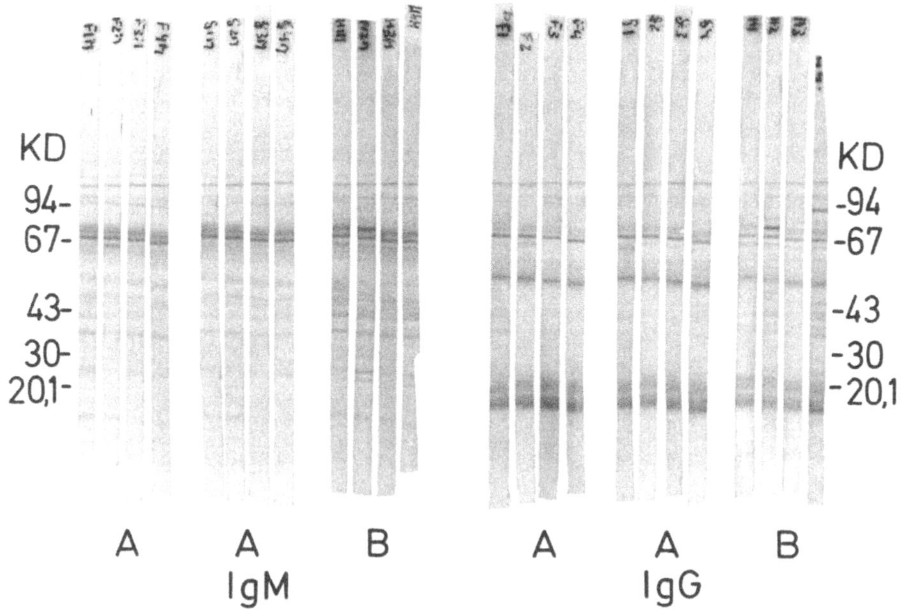

Fig. 1. Antibodies of Coxsackie B3 virus-infected animals directed to virally infected cardiac tissue. *A*, uninfected control; *B*, infected group

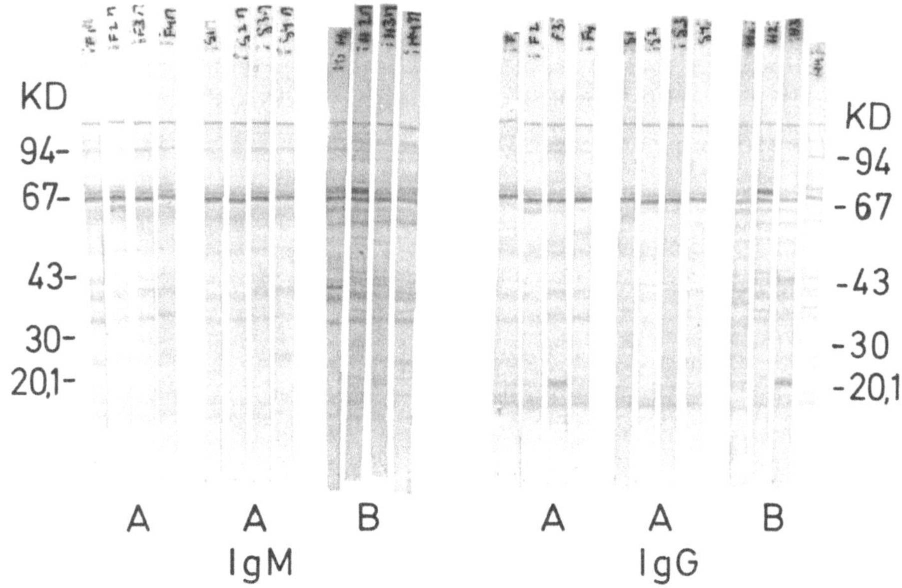

Fig. 2. Autoantibodies of Coxsackie B3 virus-infected animals to cardiac tissue. *A*, uninfected control; *B*, infected group

parts of the T cell receptor. Preliminary studies on the antigens, that heart-reactive antibodies in DBA/2 mice recognize, reveal different proteins detected by circulating antibodies after Coxsackie B3 infection.

References

1. Billingham ME (1984) The diagnostic criteria of myocarditis by endomyocardial biopsy. In: Sekiguchi M, Olsen EGJ, Goodwin JF (eds), Myocarditis and related disorders. Proceedings of the International Symposium on Cardiomyopathy and Myocarditis. Springer, Berlin Heidelberg New York, pp 133–137
2. Buie C, Lodge PA, Herzum M, Huber SA (1987) Genetics of Coxsackie virus B3 and encephalomyocarditis virus-induced myocarditis in mice. Eur Heart J [Suppl J] 8: 389–401
3. Cambridge G, McArthur CGC, Waterson AP, Goodwin JF, Oakley C (1979) Antibodies to Coxsackie B viruses in congestive cardiomyopathy. Br Heart J 16: 692–696
4. Crowell RL, Syverton JT (1961) The mammalian cell-virus relationship. IV Sustained infection of HeLa cells by Coxsackie B3 virus and effector superinfection. J Exp Med 113: 419–435
5. Guthrie M, Lodge PA, Huber SA (1984) Cardiac injury in myocarditis induced by Coxsackie virus group B type 3 in BALB/c mice is mediated by Lyt2+ cytolytic lymphocytes. Cell Immunol 82: 558–567
6. Huber SA, Lodge PA (1984) Coxsackievirus B-3 myocarditis in BALB/c mice: evidence for autoimmunity to myocyte antigens Am J Pathol 116: 21–29
7. Huber SA, Lodge PA (1986) Coxsackievirus B-3 myocarditis: Identification of different pathogenic mechanisms in DBA/2 and BALB/c mice. Am J Pathol 122: 284–291
8. Kandolf R, Ameis D, Kirschner P, Canu A, Hofschneider PH (1987) In situ detection of enteroviral genomes in myocardial cells by nucleic acid hybridization: an approach to the diagnosis of viral heart disease. Proc Natl Acad Sci USA 84: 6272–6276
9. Lodge PA, Herzum M, Olszweski J, Huber SA (1987) Coxsackievirus B-3 myocarditis: acute and chronic forms of the disease caused by different immunopathogenic mechanisms Am J Pathol 128: 455–463
10. Maisch B (1986) Immunologic regulator and effector functions in perimyocarditis, postmyocarditis heart muscle disease and dilated cardiomyopathy. Basic Res Cardiol [Suppl.] 81: 217–242
11. Mapoles JE, Krah DL, Crowell RL (1985) Purification of a Hela cell recepter protein for group B Coxsackieviruses. J Virol 55: 560–566

Interactions of Group B Coxsackieviruses with Immunocytes

P. G. CONALDI, D. MATTEUCCI, E. SOLDAINI, M. GUIDI, and M. BENDINELLI

Department of Biomedicine, Virology Section, University of Pisa, Via S. Zeno 35, I-56100 Pisa, Italy

Introduction

The role of group B coxsackieviruses (CVB) in inducing human heart disease is well established: these viruses are probably the most frequent cause of infectious acute myopericarditis and have recently been implicated in the etiology of chronic cardiac diseases [8, 42]. The range of illnesses caused by CVB is, however, much wider and more diverse, and the clinical importance of these pathogens is increasingly more apparent [3].

CVB-induced histopathology is characterized by focal or gross invasion of several organs including myocardium, brain, pancreas, and liver. Virus-mediated lysis of the permissive cells leads to the development of necrotic lesions surrounded by an inflammatory infiltrate of mononuclear cells. However, as exhaustively discussed in other chapters of this book, experimental models do clearly prove that immunological factors not only influence the susceptibility to and recovery from CVB infection, but also participate in the generation of CVB-induced disease. In addition, CVB have been shown to alter the host's immune system functions and to cause a profound involution of both central and peripheral lymphoid organs [3]. Hence, elucidation of the interactions CVB establish with immunocytes is of paramount interest for a precise understanding of CVB pathophysiology. Here we briefly review existing evidence on the role of immune mechanisms in CVB infections and then discuss our own studies on the interplay between CVB and immunocompetent cells.

Role of Immune Responses in CVB Infections

Host Resistance

Both innate and adaptive immune mechanisms appear to participate in a host's resistance to CVB infection. Data obtained in murine models indicate that the cells of monocyte-macrophage lineage afford an efficient protection against CVB. Susceptibility was shown to be deeply influenced by the functional state of macrophages [33], and the impairment of the reticuloendothelial system resulted in more severe infections [22]. Seemingly, macrophages act at first by limiting CVB spread to target organs, but also contribute to terminate the infection and

H.-P. Schultheiß (Ed.)
New Concepts in Viral Heart Disease

clear invaded tissues from the virus [41]. Accordingly, the administration of drugs which inhibit the production of inflammatory mediators by macrophages has been shown to markedly exacerbate CVB3 myocardiopathy in mice [12].

Natural killer (NK) cell-mediated cytotoxicity is another mechanism of innate immunity which appears to contribute to host defense against CVB, as activated NK have recently been shown to limit CVB3 replication both in vivo and in vitro. Activated NK cells were also seen to lyse CVB3-infected cells preferentially. This was an unprecedented finding because nonenveloped viruses, such as CVB, do not usually incorporate viral proteins onto the cell membrane. CVB3-infected cells do, however, express subtle changes on their surface which might serve as recognition target for NK cells [15].

The role of interferon (IFN) in CVB infection has been extensively discussed elsewhere [14].

What role, if any, the mononuclear phagocytic system plays in natural CVB infection of humans is not known. To our knowledge the only relevant finding is the observation that human peripheral blood mononuclear cells (BMNC) effected a reduction of CVB3 infectivity in vitro, possibly due to monocytes [13]. However, CVB have been shown to interact with other human phagocytes, such as the neutrophils, causing plasma membrane changes which enhance granulocyte adhesion to endothelial cells and impair their chemotactic response [25]. These virus-induced alterations might account for the mild transient granulocytopenia observed during CVB infections of man and for the scarcity of polymorphonuclear cells recruited by the inflammatory reaction at sites of viral replication [43].

Resistance to CVB is largely mediated by humoral immune responsiveness. In mice, virus-specific antibodies were proven crucial in reducing the gravity of infection [9, 33]. As recently shown, the incidence of CVB-induced pathology is related to the host's haplotype of the major histocompatibility complex which also appears to control timing and extent of neutralizing antibody production [40]. In humans, the pattern of humoral immune response against CVB is quite varied and characterized by the synthesis of both type- and group-specific antibodies of the IgA, IgM and IgG classes. Differences in antibody production might determine type and severity of the clinical manifestations induced by CVB infection [35, 39]. The beneficial effects of humoral immunity are brought about in collaboration with macrophages [41] and are potentiated by the concomitant production of IFN [10].

Cell-mediated immune (CMI) responses do not seem to be crucial in host resistance against CVB, as mice congenitally or artificially depleted of T lymphocytes showed a normal capability of terminating CVB infections [20]. These results confirm clinical observations that CVB-induced diseases occur with increased frequency in individuals with antibody deficiency syndromes, while they are not particularly common among patients affected by T cell immunodeficiencies [30]. However, a role for T cells in anti-CVB defenses is suggested by experimental evidence of prolonged CVB detectability in the heart of T cell deficient mice [36]. The importance of CMI responses in CVB infections is emphasized by recent results showing that tissue invasion and lethality by CVB3 is highly enhanced in mice by prior infection with immunosuppressive murine retroviruses. CVB3-induced cardiac and pancreatic lesions in these mice were also quite different from those produced in normal immunocompetent animals. These differences were attributed to suppression of CMI reactions because CVB3 specific antibody responses appeared unimpaired in retrovirus-immunosuppressed mice [38].

In summary, host invasion by CVB appears to be controlled by a number of defense mechanisms acting in concert. Antibody responses are, however, of pivotal importance.

Tissue Damage

The pathways which lead to CVB-induced pathogenesis are ill understood. Experimental models of CVB3 myocarditis have consistently indicated that direct viral cytopathogenicity cannot be solely responsible for the induction of heart disease. Thus, recent studies have suggested that following CVB3 infection of mice, cardiac lesions develop in two waves. According to this view, the first pathologic changes would develop as a consequence of viral replication in situ and the ensuing inflammatory response, while the subsequent injury would result from immunologically mediated mechanisms.

From the 2nd week of infection onwards, heart-specific autoantibodies have been detected in the serum of selected strains of mice infected with CVB3. Since such antibodies showed specificity for cardiac myosin and did not cross-react with viral epitopes, it has been proposed that their production is triggered by this myosin, becoming accessible to the immune system as a result of virus-induced cardiomyocyte necrosis [31]. On the other hand, the possibility that the autoimmune processes are triggered by virus-induced neoantigens in the heart [32] has not been formally excluded. Molecular mimicry between the infecting virus and heart antigenic determinants could also contribute to CVB-induced autoimmunity since monoclonal antibodies specific to CVB4 have been seen to react with normal myocardiocytes of several animals species [34]. Antibodies reacting with viral antigens and heart myolemma and able to lyse viable cardiocytes in the presence of complement have been detected in the sera of patients with CVB3- and CVB4-associated myocarditis [26].

Murine models have demonstrated that also CMI responses may take part in CVB-induced histopathology [18]. Early studies had suggested that tissue damage could be produced by T lymphocyte-activated macrophages [42] as well as by antibody-dependent cytotoxic cells [19]. More recent data have emphasized the role of cytotoxic T lymphocytes (CTL). Infection with a myocarditic variant of CVB3 has been shown to induce the development of two different populations of these effector cells [18]. Autoreactive CTL which damaged uninfected heart cells were mainly detectable early during infection. Virus-specific CTL which preferentially lysed CVB3-infected myocytes were instead produced at later stages of infection [21]. The significance of these two cytotoxic T cell populations in pathogenesis is still under scrutiny. Autoreactive CTL produced marked myocarditis when transfused into T lymphocyte-deficient infected or uninfected animals [18], thus suggesting that cell-mediated autoimmunity might exert a significant role in CVB-induced heart diseases.

In summary, there is circumstantial evidence that autoimmunity and other immunopathologic mechanisms contribute to overall CVB-induced cardiac damage in mice. Yet, evidence is not so firm as to exclude other possible interpretations [24]. In addition, next to nothing is presently known about CMI responses of man to CVB infections [26]. Further investigations are clearly needed to speculate whether similar mechanisms are operative also in CVB infection of the natural human host.

Alterations of the Immune System Induced by CVB

CVB infections can profoundly perturb the functions of the immune system. This may result in depressed host immune responsiveness and in a progressive generalized involution of lymphoid organs [7].

Immunodepression in Mice

That immunodeficiency can be a major consequence of CVB infection was first noted in randomly bred adult Swiss mice [4]. CVB-infected animals exhibited reduced antibody responses to a variety of antigens and, in the case of CVB3 challenge, also an altered CMI reactivity. Closer scrutiny of CVB3-infected BALB/c mice showed that immunodepression is most likely sustained by an impairment of macrophage accessory functions. Spleen cells of infected animals also exhibited suppressor T cell activity, but this change could not be correlated with the extent of immunodepression. B lymphocyte functions appeared essentially normal [5].

Lymphoid Involution in Mice

After the stage of functional immune impairment, a progressive involution of central and peripheral lymphoid organs may also develop in CVB3-infected mice [28]. Lymphoid involution was virus dose- and clone-dependent and could be abolished by pretreating the animals with CVB-specific antiserum. Host susceptibility to this viral effect was also found to be genetically determined, but scarcely influenced by gender and age.

The thymus is the first organ to show pathologic changes which than rapidly extend to the spleen and lymph nodes. An involution of the cortex represents the early feature of thymus injury: the cortical tissue is progressively reduced and the organ architecture is so deeply altered as to render the localization of the corticomedullar junction difficult. One week after infection, epithelial elements of the thymus become more prominent and fibrosis develops. Histopathology of spleen and lymph nodes is also characterized by cellular depletion and fibrosis: decreased cellularity is nonselective, and the residual cells normally remain viable. In the spleen various classes and subclasses of lymphocytes were seen to decline in similar fashions and, likewise, B- and T-dependent areas of lymph nodes are equally affected.

Systematic examination of the lymphoid organs of infected mice evidenced neither necrotic changes which might be due to viral cytopathology, nor significant numbers of CVB3-producing cells. This is in accordance with in vitro evidence discussed below that murine lymphoreticular cells of adult mice are restrictive to CVB3 replication. Likewise, indirect mechanisms afforded by stress-mediated secretion of corticosteroid hormones could be excluded since adrenalectomy prior to infection failed to prevent the development of atrophy.

No satisfactory explanations have so far been found for CVB3-induced lymphoid involution. The effect is exacerbated by the administration of immunopotentiating agents such as thymosin, isoprinosine, bestatin [28], and lipopolysaccharide (unpublished results). This, coupled with the occasional observation of a complement-dependent lymphocytotoxic activity in the serum of infected mice, might indicate that the phenomenon is at least partially autoaggressive in nature.

Immunodepression in Humans

Although there is no clearcut evidence that CVB might be immunodepressive in man, clinical reports exist which show that CVB can worsen concomitant infections and facilitate disease production by opportunistic agents in humans [17]. In a follow-up of the animal studies discussed above we have started to examine the ability of CVB to suppress human peripheral BMNC functions in vitro. The results obtained so far show that peripheral BMNC from adult donors, when exposed to the six CVB serotypes, present a reduction of mitogen-driven proliferation whose extent varied from 20% to 50% depending not only on virus dose and serotype, but also on the BMNC donor. The effect of CVB6 was particularly pronounced and consistent.

Individual variations of BMNC susceptibility to CVB-induced suppression were unrelated to age and gender and also to the immune state against CVB (as assessed by neutralizing antibody assay). The presence of specific anti-alpha and gamma IFN antibody did not modify the degree of suppression, thus showing that the effect is not mediated by the antiproliferative properties of IFN, which might be produced by the BMNC in response to CVB challenge. The suppression was only partially restored by the addition of exogenous interleukin-2 [11].

CVB Replication in Immunocytes

There are several documented mechanisms whereby viruses can perturb the host's immune responsiveness [2, 29]. Many viruses infect immunocytes productively and this may result in a variable degree of cell dysfunction and even cell death. Alternatively, viral infections may immunodepress by indirect mechanisms such as, for example, the release by the infected nonlymphoid tissues of soluble factors that modulate immune responsiveness. A further, less well-known possibility is that viruses interact with immunocompetent cells without undergoing detectable productive replication and nevertheless disrupting their functional properties.

In an attempt to understand the mechanisms by which CVB can interfere with immunologic funtions and induce lymphoid involution, we have studied the permissiveness of the cells of the immune system to these viruses.

Murine Immunocytes

In our hands, stimulated and unstimulated thymus, spleen, and lymph node cells of adult mice have proved absolutely restrictive to CVB invasion and replication both in vivo and, after recent explant, in vitro [5, 28]. Thus, the high infectivity titers which are found in the lymphoid organs of CVB-infected mice are not attributable to local production but result either from viremia or from drainage of other tissues. Therefore, the report [16] of virus-positive circulating BMNC in mice infected with CVB3, which were tentatively identified as T cells, remains unexplained. In vitro cultured macrophages also failed to support CVB growth [14a]. The role exerted by endogenous IFN in maintaining the lymphoreticular cells in such a nonpermissive state for CVB has not been investigated despite indications from other picornaviral infections that IFN might be of importance [1].

Human Lymphoid Cell Lines

CVB challenge of human continuous cell lines of B and T origin very often resulted in the production of abundant virus [6, 27]. Viral yields varied considerably depending on the lymphoid cell line and the viral serotype under test. However, in most CVB cell combinations, virus-induced cell killing was moderate to minimal. In the light of the prominent inhibitory effects exerted by this virus on human peripheral BMNC, it is of interest that CVB6 was the only serotype which destroyed all the cell lines tested. On the whole, the results suggested that the lines of B cell origin were somewhat more susceptible to CVB than T cell lines. Subsequent results (unpublished) have shown that B lymphocytes from healthy donors infected soon after immortalization in vitro with Epstein-Barr virus are permissive to CVB.

On propagation, many infected lymphoid cell lines continued to grow and shed virus for prolonged periods of time, without cell proliferation appearing to be affected. In other words, CVB exhibited a marked tendency to establish persistent infections in these cells. Such a property was unexpected because CVB are known to be highly cytocidal viruses. The T cell line MOLT-4 has been particularly studied in this respect. In the MOLT-4 cultures persistently infected with CVB 1–5, viral yields were generally high and fluctuated with no evident pattern and without clear relationship to cell viability. The virus-cell equilibrium sustaining CVB persistence proved substantially stable as it was unaffected by a number of treatments which modified functional and structural features of the cells [6]. However, persistence was readily cured by the addition of exogenous IFN or virus-specific antibody to the culture media. Infectious center assay showed that viral replication was detectable only in a small proportion of the cell population, with a tendency to decrease with increasing time of propagation.

Collectively, the data indicate that CVB persistence in human lymphoid cell lines is maintained by a "carrier culture" mechanism characterized by the presence and continuous replenishment of a tiny minority of permissive cells which replicate and spread the virus. Cell-to-cell transmission of the virus does not seem to be important since extracellular virus is essential in maintaining the chronic infection.

Extensive investigations have so far failed to identify factors which may explain why, at any given time, only a small proportion of lymphoid cells is permissive to CVB, thus maintaining persistence. No evidence has been found for the involvement of soluble inhibitors, such as IFN, or defective interfering particles whose importance has clearly been demonstrated in persistence of other viruses in vitro. Likewise, viral and cellular genetic factors do not appear to be involved.

Fresh Human Blood Mononuclear Cells

We have just begun to investigate the susceptibility to CVB of fresh human BMNC. After phytohemagglutinin (PHA) stimulation, cordal BMNC could be readily infected. Reminiscent of what was seen in human lymphoid lines, cordal BMNC showed different degrees of permissiveness to the six CVB serotypes, and viral multiplication did not affect appreciably the viability of the cultures. As detected by infectious center assay, only a small population of the cordal BMNC were productively infected by the virus. The nature of these cells is presently under examination.

The results have differed considerably when peripheral BMNC of adult individuals have been challenged with CVB. These cells proved only sparadically susceptible to productive infection, even when preactivated with mitogens and cultured in the presence of anti-IFN antibody. Susceptibility to viral challenge did not correlate with the age of the donors, nor with their immune state against the infecting CVB serotype (unpublished results).

We are currently investigating the factors which determine the variations in susceptibility to CVB of peripheral BMNC from adult donors. For example, we still do not know whether they reflect differences in viral absorption or intracellular events.

Conclusions

As discussed in other chapters of this book, there are many aspects of CVB-induced pathogenesis which deserve further investigation. Among these, the interactions CVB establish with the immune system of the infected host is certainly a major one. Indeed, our present understanding of this matter is far from satisfactory, although sufficient to indicate that CVB-immunocyte interplay is crucial not only in terms of host resistance/susceptibility to the infection, but also from the standpoint of disease production mechanisms. Thus, for example, the evidence summarized here shows that CVB can substantially impair functional and anatomic integrity of the host's immune system without undergoing extensive replication in lymphoreticular cells, but the mechanisms involved remain essentially obscure.

Until recently CVB had been considered relatively unsophisticated pathogens. As a result of clinical and experimental observations it is now becoming increasingly apparent that these widespread human viruses are instead implicated in a number of acute and chronic disease which are produced by complex pathogenetic processes. Many such illnesses are immunologically mediated, at least in part. It is especially in this newly emerging landscape that the importance of CVB-immunocyte interplay should be evaluated.

Thus, for example, the ability of CVB to readily establish persistence in human lymphoid cells should be kept in mind when considering recent observations that CVB can persist in the tissues of infected patients much longer than previously believed [37]. Similarly, the ability of CVB to inhibit certain immune functions might explain the opportunistic infections which have been described in CVB-infected patients [7, 17]. Moreover, the higher susceptibility to CVB of cordal versus adult BMNC might contribute to explaining the severity of CVB infections in children infected during the neonatal period [23].

In conclusion, we believe that a better insight of the principles which regulate CVB-immunocyte interactions might open new dimensions to our understanding of CVB-induced pathogenesis and hopefully pave the way to the development of rational instruments for therapeutic immune intervention.

Acknowledgements. We are grateful to Lucia Barontini, Giulietta Cerretini, and Annalisa Montagnani for skillful assistance. This work was supported in part by grants from the Italian Research Council and Ministry of Education.

References

1. Belardelli F, Vignaux F, Proietti E, Gresser I (1984) Injection of mice with antibody to interferon renders peritoneal macrophages permissive for vesicular stomatitis virus and encephalomyocarditis virus. Proc Natl Acad Sci USA 81: 602–606
2. Bendinelli M (1984) Immunomodulation in viral infections: virus or infection-induced? In: Fudenber HH, Whitten HD, Ambrogi F (eds) Immunopotentiation. New frontiers and advances. Plenum, New York, pp 161–195
3. Bendinelli M, Friedman H (eds) (1988) Coxsackie viruses: a general update. Plenum, New York
4. Bendinelli M, Ruschi A, Campa M, Toniolo A (1975) Depression of humoral and cell-mediated immune responses by coxsackieviruses in mice. Experientia 31: 1227–1229
5. Bendinelli M, Matteucci D, Toniolo A, Patanè AM, Pistillo MP (1982) Impairment of immunocompetent mouse spleen cell functions by infection with coxsackievirus B3. J Infect Dis 146: 797–805
6. Bendinelli M, Matteucci D, Conaldi PG, Giangregorio AM, Capobianchi MR, Dianzani F (1987) Mechanisms of group B coxsackie virus persistence in human cells. Eur Heart J 8 [Suppl J]: 441–444
7. Bendinelli M, Conaldi PG, Matteucci D (1988) Interactions with the immune system. In: Bendinelli M, Friedman H (eds) Coxsackieviruses: a general update. Plenum, New York, pp 81–102
8. Bolte H-D (ed) (1984) Viral heart disease. Springer, Berlin Heidelberg New York

9. Chaturvedi UC, Tandon HO, Mathur A (1978) Control of in vitro and in vivo spread of coxsackievirus B4 infection by sensitized spleen cells and antibody. J Infect Dis 138: 181–190
10. Cho CT, Feng KK, McCarthy VP, Lenahan MF (1982) Role of antiviral antibodies in resistance against coxsackievirus B3 infection: interaction between preexisting antibodies and an interferon inducer. Infect Immun 37: 720–727
11. Conaldi PG, Matteucci D, Soldaini E, Bendinelli M (1988) Interactions of group B coxsakkieviruses with human lymphoid cells. Annual Meeting of American Society for Microbiology, p 326
12. Constanzo-Nordin MR, Reap EA, O'Connell JB, Robinson JA, Scanlon PJ (1985) A nonsteroid anti-inflammatory drug exacerbates coxsackie B-3 murine myocarditis. J Am Coll Cardiol 6: 1078–1082
13. Denman AM, Rager-Zisman B, Merigan TC, Tyrrell DAJ (1974) Replication or inactivation of different viruses by human lymphocyte preparation. Infect Immun 9: 373–376
14. Dianzani F, Capobianchi MR, Matteucci D, Bendinelli M (1988) The role of interferon in picornavirus infections. In: Bendinelli M, Friedman H (eds) Coxsackieviruses: a general update. Plenum, New York, pp 65–80
14a. Gauntt CJ, Trousdale MD, La Badie DRL, Paque RE, Nealon T (1979) Properties of coxsackievirus B3 variants which are amyocarditic or myocarditic for mice. J Med Virol 3: 207–220
15. Godeny EK, Gauntt CJ (1987) Murine natural killer cells limit coxsackievirus B3 replication. J Immunol 139: 913–918
16. Gomez MP, Reyes MP, Smith F, Ho LK, Lerner AM (1980) Coxsackievirus B3-positive mononuclear leukocytes in peripheral blood of Swiss and athymic mice during infections. Proc Soc Exp Biol Med 165: 107–113
17. Gururaj VJ, Marsh WW, Aiyar SR (1984) Fulminant pulmonary coccidiom cosis in association with coxsackie B4 infection. Clin Pediatr 24: 406–408
18. Huber SA (1988) The role of immune mechanisms in pathogenesis. In: Bendinelli M, Friedman H (eds) Coxsackieviruses: a general updat. Plenum, New York, pp 103–116
19. Huber SA, Job LP, Auld KR, Woodruff JF (1981) Sex-related differences in the rapid production of cytotoxic spleen cells active against uninfected myofibers during coxsackievirus B-3 infection. J Immunol 126: 1336–1340
20. Huber SA, Job LP (1983) Cellular immune mechanisms in coxsackievirus group B, type 3 induced myocarditis in Balb/ c mice. In: Spitzer JJ (ed) Myocardial injury. Plenum, New York, pp 491–508
21. Huber Sa, Lodge PA (1984) Coxsackievirus B-3 myocarditis in Balb/c mice. Evidence for autoimmunity to myocyte antigens. Am J Pathol 116: 21–29
22. Kabiri M, Basiri E, Kadivar D (1978) Potentiation of coxsackievirus B3 infection in adult mice pretreated with a gold salt. J Med Virol 3: 125–136
23. Kaplan MH (1988) Coxsackie infection in children under three months of age. In: Bendinelli M, Friedman H (eds) Coxsackieviruses: a general update. Plenum, New York, pp 241–251
24. Khatib R, Probert A, Reyes M, Khatib G, Chason JL (1987) Mouse strain-related variation as a factor in the pathogenesis of coxsackievirus B3 murine myocarditis. J Gen Virol 68: 2981–2988
25. Kirkpatrick CJ, Bultmann BD, Gruler H (1985) Interaction between enteroviruses and human endothelial cells in vitro. Alterations in the physical properties of endothelial cell plasma membrane and adhesion of human granulocytes. Am J Pathol 118: 15–25
26. Maisch B, Trostel-Soeder R, Stechemesser F, Berg PA, Kochsiek K (1982) Diagnostic relevance of humoral and cell-mediated immune reactions with acute viral myocarditis. Clin Exp Immunol 48: 533–545
27. Matteucci D, Paglianti M, Giangregorio AM, Capobianchi MR, Dianzani F, Bendinelli M (1985) Group B coxsackieviruses readily establish persistent infection in human lymphoid cell lines. J Virol 56: 651–654
28. Matteucci D, Toniolo A, Conaldi PG, Basolo F, Gori Z, Bendinelli M (1985) Systemic lymphoid atrophy in coxsackievirus B3-infected mice: effects of virus and immunopotentiating agents. J Infect Dis 151: 1100-1108
29. McChesney MB, Oldstone MBA (1987) Viruses perturb lymphocyte functions: selected principles characterizing virus-induced immunosuppression. Annu Rev Immunol 5: 279–304

30. Moore M, Morens DM (1984) Enteroviruses, including polioviruses. In: Belshe RB (ed) Textbook of human virology. PSG, Littleton, pp 407–483
31. Neu N, Craig SW, Rose NR, Alvarez F, Beisel KW (1987) Coxsackievirus induced myocarditis in mice: cardiac myosin autoantibodies do not cross-react with the virus. Clin Exp Immunol 69: 566–574
32. Paque RE, Gauntt CJ, Nealon TJ, Trousdale HD (1978) Assessment of cell-mediated hypersensitivity against coxsackie B3 viral-induced myocarditis utilizing hypertonic salt-extracts of cardiac tissue. J Immunol 120: 1672–1678
33. Rager-Zisman B, Allison AC (1973) The role of antibody and host cells in the resistance of mice against infection by coxsackie B-3 virus. J Gen Virol 19: 329–338
34. Saegusa J, Prabhakar BS, Essani K, McClintok PR, Fukuda Y, Ferrans VJ, Notkins AL (1986) Monoclonal antibody to coxsackievirus B4 reacts with miodardium. J Infect Dis 153: 372–373
35. Schernthaner G, Banatvala JE, Scherbaum W, Bryant J, Borkenstein M, Schober E, Mayr WR (1985) Coxsackie-B-virus-specific IgM responses, complement-fixing islet-cell antibodies, HLA DR antigens, and C-peptide secretion in insulin-dependent diabetes mellitus. Lancet II: 630–632
36. Schnurr DP, Cao Y, Schimdt NJ (1984) Coxsackievirus B3 persistence and myocarditis in N:NIH (S) II nu/nu and +/nu mice. J Gen Virol 65: 1197–1201
37. Schnurr DP, Schimdt (1988) Coxsackievirus persistent infections. In: Bendinelli M, Friedman H (eds) Coxsackieviruses: a general update. Plenum, New York, pp 181–201
38. Specter S, Basolo F, Bendinelli M (1987) Retroviruses as immunosuppressive agents. In: Good RA, Lindenlaub E (eds) The nature, cellular, and biochemical basis and management of immunodeficiences. Schattauer, Stuttgart, pp 127–145
39. Tilzey AJ, Signy M, Banatvala JE (1986) Persistent coxsackie B virus specific IgM response in patients with recurrent pericarditis. Lancet I: 1491–1492
40. Wolfgram LJ, Beisel KW, Herskowitz A, Rose NR (1986) Variations in the susceptibility to coxsackievirus B3-induced myocarditis among different strains of mice. J Immunol 136: 1846–1854
41. Woodruff JF (1979) Lack of correlation between neutralizing antibody production and suppression of coxsackievirus B-3 replication in target organs: evidence for involvement of monuclear inflammatory cells in host defences. J Immunol 123: 31–36
42. Woodruff JF (1980) Viral myocarditis. Am J Pathol 101: 426–479
43. Young NA (1979) Coxsackievirus and echovirus. In: Mandell GC, Douglas RG, Bennett JE (eds) Principles and practice of infectious disease, vol. II. Wiley, New York, pp 1104–1120

Early Cellular Infiltrates in Coxsackievirus B3 Murine Myocarditis

L. H. CHOW[1], C. J. GAUNTT[2], Y. L. YE[1], and B. M. MCMANUS[1]

[1] Department of Pathology and Microbiology, The University of Nebraska Medical Center, 42nd and Dewey Avenue, Omaha, NE 68105-1065, USA
[2] Department of Microbiology, University of Texas Health Science Center at San Antonio, 7703 Floyd Curl Drive, San Antonio, Texas, 78284-7758, USA

Introduction

The pathologic basis of murine myocarditis induced by coxsackievirus type B3 (CVB3) places necessary emphasis on the inflammatory infiltrate [11, 21]. In efforts to establish the phenotypic identity of infiltrating cells, investigators have used a variety of methods, including selective depletion or enrichment of defined cellular subsets [5, 10, 16], semiquantitative in situ immune autoradiography [8], immuno-electron microscopy [3], and immunohistochemistry [4, 19]. Such studies, however, have yet to provide serially comprehensive and quantitative data pertinent to the evolution of the inflammatory response in the earliest phase of the disease. Better understanding of the way in which phenotypically different subpopulations of immune cells invade the myocardium may help to explain the potential genesis of autoimmunity inducible in various strains of mice infected with CVB3 [1, 13]. The ever-expanding complement of monoclonal antibodies directed against cell surface antigens, with greater and greater specificity for cell type and state of activation, permits progressively more distinctive characterization.

Immunohistochemical methods for examining myocardial inflammatory cells have been standardized for routine use in our laboratory [2, 15] and by other groups [20]. We empolyed this technique with a panel of monoclonal antibodies (MAb) to lymphocyte and monocyte markers to characterize the early cellular events involved in the development of CVB3-induced myocarditis in CD1 mice, focusing on the period from the inoculation of virus to the peak in inflammatory injury 8 days post inoculation (p.i.) [8].

Materials and Methods

Tissues

Formaldehyde-fixed and freshly frozen heart tissues, obtained from normal and infected mice, were prepared by Dr. Gauntt as previously described [8]. Briefly, adolescent (4-6-week-old), male, semi-inbred, CD1 mice were intraperitoneally inoclutated with Woodruff's myocarditic variant of CVB3 (Nancy strain) at 10^5 plaque-forming units per animal. Seven to eight mice were sacrificed on each of

H.-P. Schultheiß (Ed.)
New Concepts in Viral Heart Disease

days 2, 4, 6, and 8 p.i., and hearts from each time point were distributed for fixation in 10% neutral buffered formaldehyde solution (four mice), or for rapid freezing (three to four mice) in Tissue-Tek OCT embedding medium (Ames Division, Miles Laboratories, Inc.).

Histology and Immunohistochemistry

Formaldehyde-fixed, paraffin-embedded, 1–2-mm thick, transverse ventricular slices (five to six per heart) were sectioned and stained with hematoxylin and eosin (H&E; six levels), Masson's trichrome stain (two levels), Dahl's calcium stain (two levels), and methyl green pyronine to reflect lymphoblastic activity by cytoplasmic RNA content (two levels).

Inflammatory infiltration was immunohistochemically characterized in frozen tissues using an avidin-biotin immunoperoxidase technique. From each of two transverse ventricular slices per heart, contiguous 5 μm cryostat sections were placed on gelatin-coated glass slides and incubated for 30 min at 24 °C with a panel of rat anti-mouse MAb provided by Dr. Gauntt, and by Dr. G. A. Perry of the Department of Anatomy, University of Nebraska Medical Center. Characteristics of these primary MAb were published elsewhere [6, 8] and are summarized in Table 1. The remainder of the procedure (Vectastain ABC kit, Vector Laboratories, Inc., Burlingame, California) was accomplished in three steps as follows, each separated by washing with 0.1 *M* phosphate-buffered saline: (a) biotinylated goat anti-rat immunoglobulin was applied for 30 min; (b) avidin DH: biotinylated horseradish peroxidase H complex was added for 30 min; and (c) diaminobenzidine was used as chromogen to indicate a positive reaction, counterstained by

Table 1. Murine immune cell markers recognized by monoclonal antisera

Marker	Cellular specificity	Structure recognized
Thy1.2	T cells	Differentiation antigen
L3T4	Helper T cells	Accessory molecule for target recognition
Lyt2	Cytolytic/suppressor T cells	Accessory molecule for target recognition
IL-2R	Activated T cells[b]	IL-2 receptor
Mac1	Macrophages	Differentiation antigen
B220	B cells	Differentiation antigen
Mel14[a]	T and B cells	Putative homing receptor

[a] Term designates the monoclonal antibody
[b] IL-2R may be found in low density on B cells

methyl green. Subjacent levels of frozen tissue from each ventricular slice were stained with H&E without MAb.

Cell Counts and Image Analysis

Cells defined immunohistochemically were examined by light microscopy and counted at x 330–500 magnification on the entire ventricular section for each phenotypic marker. Myocardial cross- sectional area was calculated by computer-assisted image analysis, using an AMS optomax V central processing unit, equipped with software for digitization. This allowed user-confirmed and user-defined fields of interest, based on visually displayed input from an Ikegami ITC-510 video camera, attached to an Olympus BH2 microscope and an SPlan Fl1 1x objective. The Optomax unit was further interfaced with an IBM-XT personal computer for data stroage and analysis.

Computation and Statistics

An ABstat statistical package (release 5.03; Anderson-Bell, Parker, Colorado) was used with an IBM personal computer to calculate infiltrate densities (cells per square millimeter), means and standard deviations, Pearson's correlation coefficient matrix, and Student's unpaired *T* test (two-tailed) for the comparison of means.

Results

Routine Histology (Days 0–8)

H&E staining of ventricular myocardium from control, uninfected mice demonstrated sparse, but consistently present, macrophages in perivascular and interstitial spaces, along with occasional mast cells and small lymphocytes. One of four animals at 2 days p.i. showed a trace accumulation of focal mononuclear cells (MNC) without any evidence of myocyte involvement.

By day 4, however, sections in all infected mice revealed mild, focal myocarditis (Fig. 1) of mixed cellularity, including polymorphonuclear leukocytes. Foci were usually clearly demarcated and small, sometimes involving single myofibers. Myocyte degeneration ranged from vacuolization to frank necrosis, but isolated hypereosinophilia of myofibers in the absence of cellular infiltration was not seen on day 4, nor at any other time in the study. Early calcification was observed in a single mouse at sites adjacent to, but not encompassed by, active myocarditis. Myocarditic lesions were often near blood vessels and produced a pronounced vasculopathy in an intramural artery of one animal.

Focal myocarditis was moderate in animals sacrificed 6 days p.i. The cellular infiltrate was almost entirely comprised of mixed MNC. Tracking of MNC along interstitial planes, in addition to focal accumulation, was noted in a single animal.

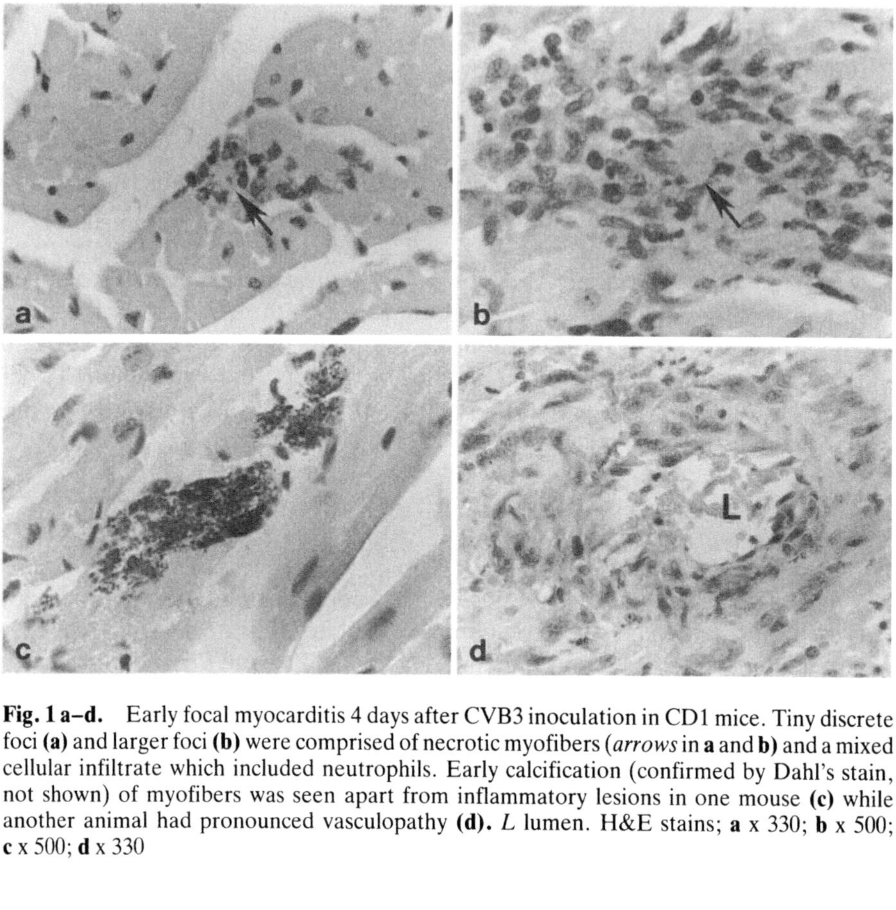

Fig. 1 a–d. Early focal myocarditis 4 days after CVB3 inoculation in CD1 mice. Tiny discrete foci **(a)** and larger foci **(b)** were comprised of necrotic myofibers (*arrows* in **a** and **b**) and a mixed cellular infiltrate which included neutrophils. Early calcification (confirmed by Dahl's stain, not shown) of myofibers was seen apart from inflammatory lesions in one mouse **(c)** while another animal had pronounced vasculopathy **(d)**. *L* lumen. H&E stains; **a** x 330; **b** x 500; **c** x 500; **d** x 330

Calcification, present in necrotic zones, varied in extent from animal to animal. Active myocarditis continued on day 8, with two mice showing severe disease (Fig. 2) and the other two affected only mildly. Infiltrate morphology and lesional characteristics were largely similar to day 6, and no significant fibrosis was yet apparent. Staining with methyl green pyronine was rarely positive at any time studied, and positive staining was invariably faint. Inflamed or necrotic myocardium included an average of 3.0% (range 2.0%–5.5% in all fives slices) of the ventricular cross-sectional area by digital image analysis in a mouse having severe myocarditis on day 8.

Immunohistochemical Staining

Uninfected Controls

The ventricular myocardium of uninfected mice contained low numbers of immunohistochemically marked cells, as listed below (cells per square millimeter; mean ± SD): 1.0 ± 0.8 (Thy1.2$^+$); 1.8 ± 1.5 (L3T4$^+$); 0.6 ± 0.5 (Lyt2$^+$); 0.1 ± 0.1

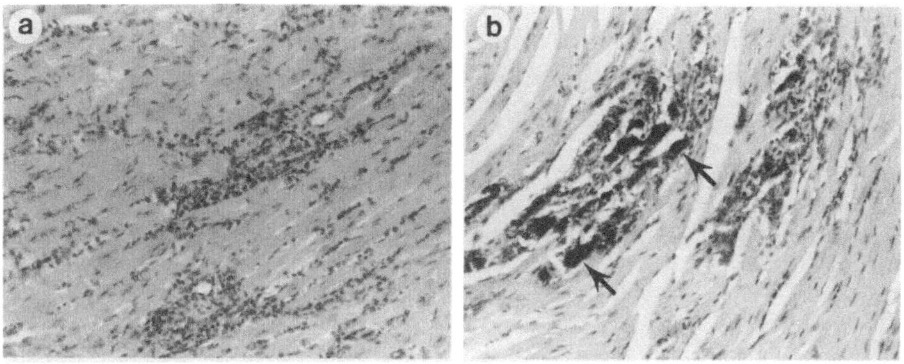

Fig. 2 a, b. Severe myocarditis 8 days after CVB3 inoculation in CD1 mice. Multifocal disease **(a)** was characterized by a mononuclear infiltrate and in certain mice by moderate calcification *(arrows)* within necrotic zones **(b)**. The affected myocardium in **a** represented 5.5% of the total ventricular cross-sectional area. H&E stains; **a** x 125; **b** x 125

(interleukin-2 receptor; IL-2R$^+$); 3.6 ± 1.4 (Mac1$^+$); 2.6 ± 1.0 (B220$^+$); and 0.2 ± 0.1 (Mel14$^+$). Thus, a slight predominance of macrophages was found. Lymphocyte counts compared well to the results of others [15], who showed a total of 3.6 lymphocytes per square millimeter in uninflamed human myocardium using similar methods.

Infiltrate Populations in Myocarditis

T cells (Thy1.2$^+$; Fig. 3) rose dramatically during the development of acute myocarditis, reaching a density of 81.4 ± 40.6 cells per square millimeter in the ventricular myocardium on day 8. T cell numbers surpassed those of B cells and macrophages after the onset of active disease on day 4 (Fig. 4). The modest elevations in B cells and macrophages peaked on day 6, at 6.5 ± 1.7 and 14.6 ± 7.6 cells per square millimeter, respectively. Despite variation between animals, general concordance in the pattern of cellular infiltration was observed; furthermore, many individual mice had consistently strong or weak results for different cellular phenotypes.

All T cell subsets were elevated during the course of disease, with changes discernible by day 4 (Fig. 5). Correlations among the different T subpopulations were strong: $r = 0.83$, 0.93, and 0.88 for Thy1.2 versus L3T4, Lyt2, and IL-2R, respectively. However, the phenotypic ratio of L3T4/Lyt2 gradually diminished and eventually reversed: 3.5 ± 0.8 (uninfected); 2.2 ± 1.4 (day 2); 2.3 ± 1.2 (day 4); 1.1 ± 0.3 (day 6); and 0.9 ± 0.4 (day 8; $P < 0.01$ vs. uninfected).

When mean inflammatory cell densities were compared to respective control values in each phenotype, the behavior of cells bearing Thy1.2, Lyt2, IL-2R, and the Mel14 homing antigen became easily distinguished (Fig. 6). Greater than tenfold increases in both IL-2R$^+$ and Mel14$^+$ cell types were recognized as early as day 4 p.i., and eventual elevations exceeded 80-fold. However, while IL-2R$^+$ cells accumulated in parallel with Thy1.2 from day 4 to day 8, the Mel14$^+$ subset peaked sharply on day 6 and quickly subsided (Fig. 7).

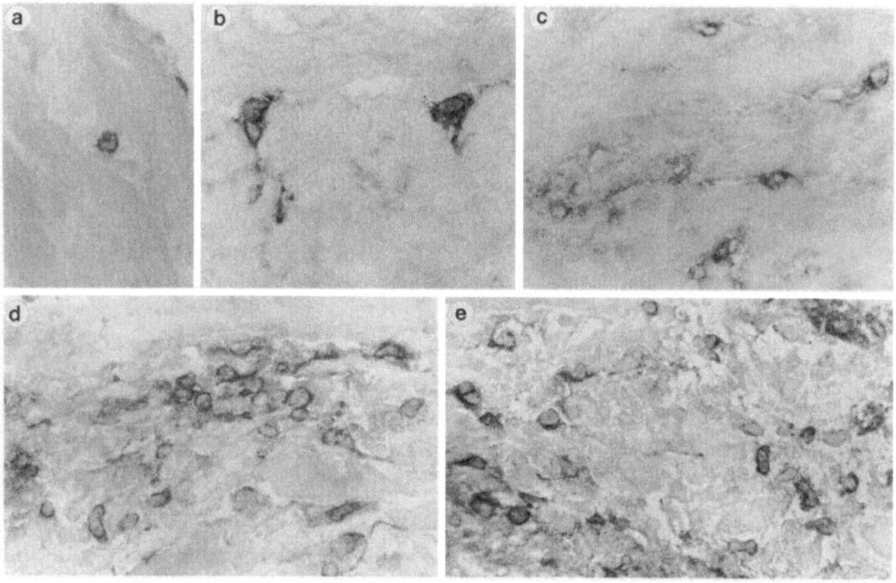

Fig. 3 a–e. Thy1.2$^+$ cells in the myocardium by immunohistochemistry. While Thy1.2$^+$ (pan T) cells were rare in the control, uninfected myocardium of CD1 mice **(a)**, progressive infiltration of these cells was noted on days 2 **(b)**, 4 **(c)**, 6 **(d)**, and 8 **(e)** after coxsackievirus B3 inoculation. Increasing areas of myocyte necrosis and disruption were present on days 4, 6, and 8. Immunoperoxidase stains; x500

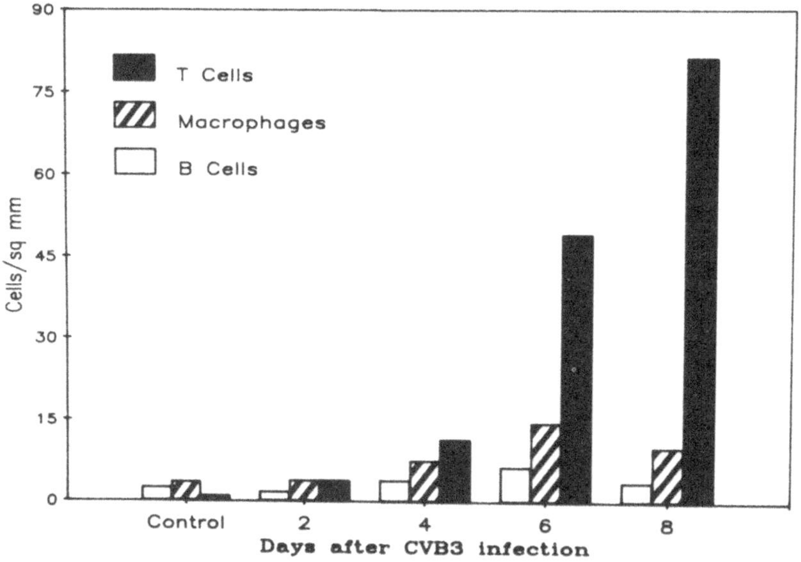

Fig. 4. Infiltrating T cells (Thy1.2$^+$; *solid bars)*, immunohistochemically stained and counted, outnumbered macrophages (Mac1$^+$; *hatched bars)* and B cells (B220$^+$; *open bars)* after the onset of acute myocarditis on day 4 p. i. of CVB3 in CD1 mice

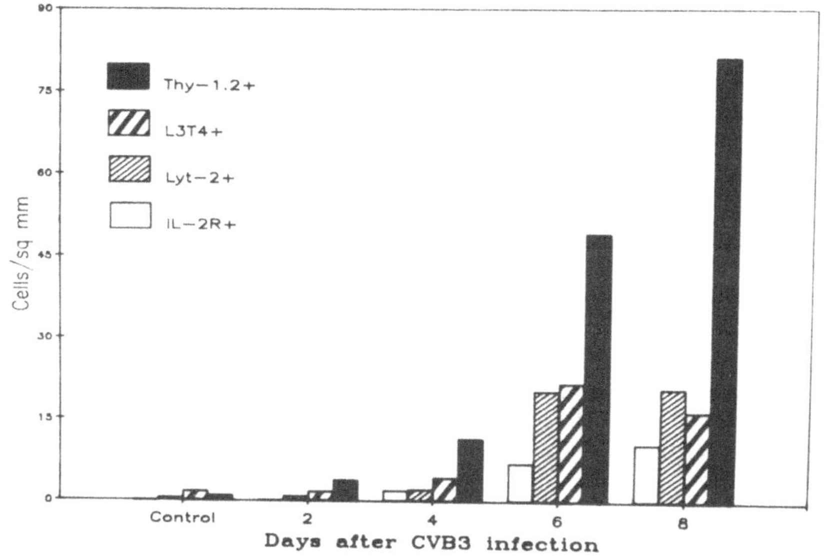

Fig. 5. Infiltrating T cell subsets, including those stained immunohistochemically for Thy1.2 (pan T; *solid bars*), L3T4 (helper T; *broadly hatched bars*), Lyt2 (cytolytic/suppressor T; *narrowl hatched bars*), and IL-2R *(open bars)* were elevated during acute myocarditis in CD1 mice. Changes were discernible as early as 4 days p. i. of CVB3

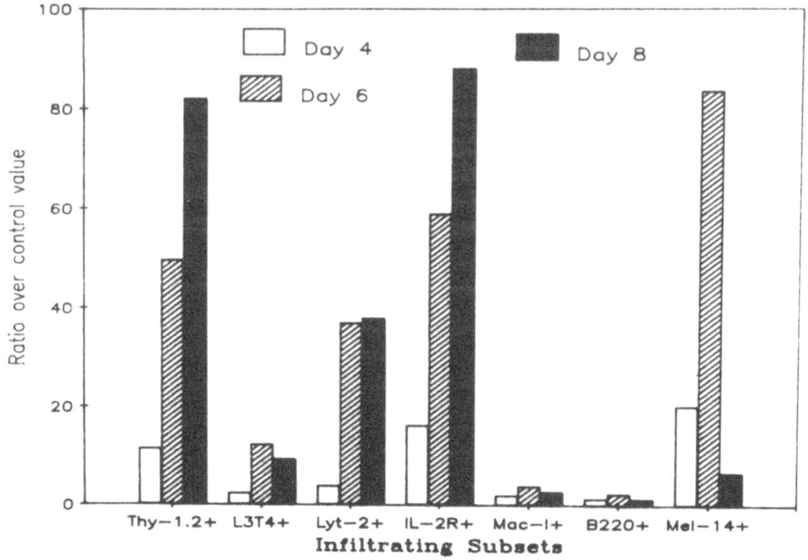

Fig. 6. The density of each immunohistochemically defined subpopulation in the inflamed myocardium was expressed as a ratio over the respective control value from uninfected CD1 mice. A rise was notable for cells bearing Thy1.2 (pan T), Lyt2 (cytolytic/suppressor T), IL-2R, and the putative homing receptor marked by the monoclonal antibody, Mel14. A distinctive and rapid decline in the latter marker followed its peak on day 6 p. i. of CVB3. L3T4 (helper T), Mac1 (macrophages), B220 (B cells). *Open bars*, day 4; *hatched bars*, day 6; *solid bars*, day 8

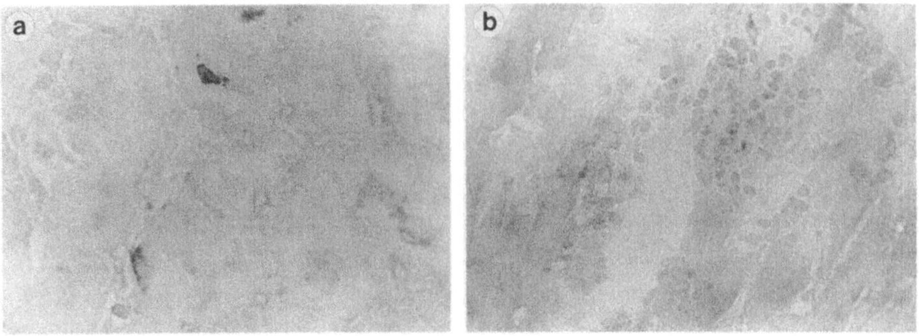

Fig. 7 a, b. Peak incidence of Mel14$^+$ cells 6 days after CVB3 inoculation. The monoclonal antibody Mel14 marks a putative homing receptor on lymphocytes. Unlike Thy1.2 (pan T), Mel14 positivity was greatest on day 6 **(a)** and subsided by day 8 **(b)**. Immunoperoxidase stains; x330

Discussion

Definition of the immunohistochemical profile of invading inflammatory cells, and their apparent states of immune activation, provides new and corroborative information regarding the peak time of target organ involvement in the CD1 mouse model of CVB3-induced myocarditis. The general similarity between animals suggests a common mechanistic cascade, from early viremia to cell death and myocardial inflammatory infiltration. Variability between animals may reflect the influence of several factors, including the adequacy of infection, viral clearance, and immune amplification.

Good agreement is noted between the present data and that recently documented semi-quantitatively by autoradiography [8]. Prominent elevations in Thy1.2 and Lyt2 occurred on days 7–8, and L3T4$^+$ cells, while comprising a smaller proportion of the infiltrate, were consistently increased after the 5th day. A singular elevation in Mac1 on day 6 in the previous study is consistent with current findings. Since CVB3 replicates in high titers in the myocardium of CD1 mice between day 3 and day 5 p.i. and persists beyond day 8 [7], substantial temporal overlap exists between cellular infiltration and the virus load in the myocardium.

Experience in BALB/c mice suggests an important functional role for Lyt2$^+$ cells, which represented cytolytic effectors in studies conducted by Guthrie et al. [10], Lodge et al. [16], and Estrin and Huber [5]. In other strains, however, such as DBA/2, pathogenesis seemed more dependent on L3T4$^+$ cells and IgG [16]. Discrepant observations by Rubski et al. [19] suggested an absence of both Lyt2$^+$ and L3T4$^+$ cellular responses in CVB3-infected BALB/c mice.

Deguchi et al. have noted on two occasions [3, 4] that Lyt1$^+$ cells peaked at approximately 14 days p.i. in the CVB3-infected C3H mouse. While this cell population may be less discrete (encompassing L3T4$^+$ and Lyt2^2 cells as well as some B cells), Lyt1$^+$ cells appeared to these investigators to be intimately related to macrophages at the ultrastructural level, supporting their involvement in CVB3-related immunopathogenesis.

Asialo GM_1 positive cells, presumptive natural killers (NK), were as numerous as T cells by autoradiography from days 6–9 in CD1 mice [8], with NK activity confirmed by YAC-1 lysis in vitro. Deguchi et al. [4] likewise demonstrated a peak in asialo GM_1 positivity on day 9 in the myocardium of C3H mice infected with CVB3. However, the implication that NK cells may play an injurious role in either of these mouse strains stands somewhat in contrast to earlier observations by Godeny and Gauntt [9] that NK function may be important for protection at an earlier time (day 3) following CVB3 infection. In myosin-induced experimental autoimmune myocarditis in A/J mice, Neumann et al. [17] concluded that the majority of the inflammatory cells, which stained immunohistochemically positive for leukocyte common antigen, but negative for Thy1.2 and Lyt2, was consistent with NK cells.

Prior to the current observations, no information was available on the Mel14 marker in murine myocarditis. The Mel14 MAb, which has been studied extensively by Gallatin et al. [6], blocks the binding of circulating lymphocytes to specialized endothelium in high endothelial venules (HEV) of lymph nodes. It marks a putative homing receptor, on the surface of mature lymphocytes, that can be traced to intrathymic development [18]. It is expressed on recent thymic emigrants, but is quickly lost upon in vivo antigenic, or in vitro mitogenic, stimulation (reviewed in [14]). The kinetics of Mel14 positivity in our data is compatible with such rapid down-regulation. In part, extravasation of $Mel14^+$ cells into the myocardium may have resulted from vascular changes at the capillary level after CVB3 infection. However, there is evidence for a larger family of closely related, organ-specific, HEV-reactive molecules that are responsible for selective lymphocyte migration into target organs [14]; the presence of Mel14 positivity in CVB3-infected hearts may reflect the co-expression of multiple homing molecules, or cross-antigenicity between closely related ones.

It is not possible in our data to distinguish between virus-specific and myocyte-specific (autoimmune) inflammation. Others have suggested a diffuse, interstitial, inflammatory pattern as being distinctive in the presumptively immunopathic, late phase (after day 9) of CVB3-induced myocarditis in recombinant inbred mice [11]; from six strains examined, the four that showed this late morphologic pattern were those in which heart-specific autoantibodies were found. On the other hand, myocyte-specific autoreactive T cells have been demonstrated in vitro as early as day 7 in BALB/c mice [12] associated primarily with focal cellular infiltrates in the myocardium.

In summary, available observations in a variety of inbred and semi-inbred strains of mice, on balance, and in conjunction with our own data, suggest a well-orchestrated intensification of a T cell predominant, Lyt2-rich, cellular response which is paralleled by increased expression of IL-2 receptors, and an increasing dominance of effector over helper cells by day 8 in CVB3-induced myocarditis.

Acknowledgement. We thank Jerrold Remmenga for his invaluable assistance in the preparation, processing, and photographic documentation of all tissues.

References

1. Beisel KW, Traystman MD (1988) Autoimmune myocarditis: a murine model. In: Bigazzi PE, Wick G, Wicher K (eds) Organ-specific autoimmunity. Dekker, New York (in press) (Progress in autoimmunity, vol 1)
2. Cassling RS, Linder J, Sears TD, Waller BF, Rogler WC, Wilson JE, Kugler JD, Kay DH, Dillon JC, Slack JD, McManus BM (1985) Quantitative evaluation of inflammation in biopsy specimens from idiopathically failing or irritable hearts: experience in 80 pediatric and adult patients. Am Heart J 110: 713–720
3. Deguchi H, Kitaura Y, Hayashi T, Kotaka M, Morita H, Kawamura K (1986) Immunohistochemical study of the myocardium in murine Coxsackie B3 virus myocarditis using monoclonal antibodies: significance of Lyt1 antigen-bearing lymphocytes in cell-mediated immunity. Jpn Circ J 50: 1268–1274
4. Deguchi H, Hayashi T, Kitaura Y, Kawamura K (1988) In situ analysis of T cell subsets and natural killer cells in the myocardium of mice with Coxsackie B3 virus myocarditis. J Am Coll Cardiol 11: 95A
5. Estrin M, Huber SA (1987) Coxsackievirus B3-induced myocarditis: autoimmunity is L3T4+ T helper cell and IL-2 independent in BALB/c mice. Am J Pathol 127: 335–341
6. Gallatin WM, Weissman IL, Butcher EC (1983) A cell-surface molecule involved in organ-specific homing of lymphocytes. Nature 304: 30–34
7. Gauntt CJ, Trousdale MD, LaBadie DRL, Paque RE, Nealon T (1979) Properties of coxsackievirus B3 variants which are amyocarditic or myocarditic for mice. J Med Virol 3: 207–220
8. Godeny EK, Gauntt CJ (1987) In situ immune autoradiographic identification of cells in heart tissues of mice with coxsackievirus B3-induced myocarditis. Am J Pathol 129: 267–276
9. Godeny EK, Gauntt CJ (1987) Murine natural killer cells limit Coxsackie virus B3 replication. J Immunol 139: 913–918
10. Guthrie M, Lodge PA, Huber SA (1984) Cardiac injury in myocarditis induced by coxsackievirus group B, type 3 in BALB/c mice is mediated by Lyt2+ cytolytic lymphocytes. Cell Immunol 88: 558–567
11. Herskowitz A, Wolfgram LJ, Rose NR, Beisel KW (1987) Coxsackievirus B3 murine myocarditis: a pathologic spectrum of myocarditis in genetically defined inbred strains. J Am Coll Cardiol 9: 1311–1319
12. Huber SA, Lodge PA (1984) Coxsackievirus B3 myocarditis in BALB/c mice: evidence for autoimmunity to myocyte antigens. Am J Pathol 116: 21–29
13. Huber SA, Lodge PA (1986) Coxsackievirus B3 myocarditis: identification of different pathogenic mechanisms in DBA/2 and BALB/c mice. Am J Pathol 122: 284–291
14. Jalkanen S, Reichert RA, Gallatin WM, Bargatze RF, Weissman IL, Butcher EC (1986) Homing receptors and the control of lymphocyte migration. In: Moller G (ed) Population dynamics of lymphocytes. Munksgaard, Copenhagen, pp 39–60 (Immunological reviews, vol 91)
15. Linder J, Cassling RS, Rogler WC, Wilson JE, Markin RS, Sears TD, McManus BM (1985) Immunohistochemical characterization of lymphocytes in uninflamed ventricular myocardium. Arch Pathol Lab Med 109: 917–920
16. Lodge PA, Herzum M, Olszewski J, Huber SA (1987) Coxsackievirus B3 myocarditis: acute and chronic forms of the disease caused by different immunopathogenic mechanisms. Am J Pathol 128: 455–463
17. Neumann DA, Herskowitz A, Rose NR (1988) Immunohistochemical characterization of cardiac myosin-induced autoimmune myocarditis in mice. FASEB J 2: A1080
18. Reichert RA, Jerabek L, Gallatin WM, Butcher EC, Weissman IL (1986) Ontogeny of lymphocyte homing receptor expression in the mouse thymus. J Immunol 136: 3535–3542
19. Rybski J, Grogan T, Barret JM (1988) Focal myocyte IA expression in early coxsackievirus B3 myocarditis, analogous to autoimmune thyroiditis. Lab Invest 58: 79A
20. Schnitt SJ, Ciano PS, Schoen FJ (1987) Quantitation of lymphocytes in endomyocardial biopsies: use and limitations of antibodies to leukocyte common antigen. Hum Pathol 18: 796–800
21. Woodruff JF (1980) Viral myocarditis. Am J Pathol 101: 426–484

IV. Characterization of Autoimmunological Mechanisms

Beta-Adrenoceptor Autoantibodies in Idiopathic Dilated Cardiomyopathy

C. J. LIMAS, and C. LIMAS

Department of Medicine (Cardiovascular Section) and Laboratory Medicine and Pathology, University of Minnesota School of Medicine, Minneapolis, MN 55455, USA

Several recent studies have indicated that the density of β-adrenoceptors in cardiac membranes is decreased in patients with clinical heart failure [1–4], and that this decline selectively affects the β_1-adrenoceptor subtype [2, 4]. Since β-adrenergic pathways are involved in the control of positive inotropism, loss of responsiveness to β-antagonists has important adverse effects on the performance of the failing myocardium. In addition, decline in β-receptor numbers restricts the effectiveness of β-agonists as therapeutic agents. The elucidation of the pathogenesis of β-receptor changes in heart failure, therefore, has therapeutic as well as conceptual implications.

The identification of an agonist-mediated endocytotic pathway of β-adrenoceptors [5] and reports of symphathetic nervous system activation in heart failure [6] have led to the suggestion that cardiac β-receptors are desensitized in heart failure by the high levels of norepinephrine released endogenously at the nerve endings [1, 2]. Little evidence, however, has been forthcoming in support of this hypothesis. Specifically, it has not been shown that the concentration of norepinephrine achieved locally is high enough to desensitize cardiac β-receptors. Arterial and coronary sinus norepinephrine concentrations are typically in the nanomolar range and, although it is assumed that local (myocardial) concentrations are much higher, there is considerable uncertainty about the precise levels achieved and how these would relate to an endocytotic mechanism. The correlation between plasma or tissue norepinephrine and the density of cardiac β-adrenoceptors has been modest at best [3]. Finally, the expected redistribution of receptors from a cell surface to an intracellular pool predicated by the endocytotic model is not borne out in direct measurements [3].

The major drawback of a unifying theory about the pathogenesis of β-receptor decline is that it ignores the diversity of pathogenetic mechanisms involved in mediating the clinical syndrome of heart failure and the role that the etiology may play in determining β-receptor function.

In view of these considerations, we decided to evaluate the possible role of anti-β-receptor antibodies in dilated cardiomyopathy. Selection of this clinical entity was predicated upon the proposed role of immunologic factors in its initiation and/or maintenance, as exemplified by the presence of autoantibodies against several cell constituents [7, 8] and evidence for impaired cellular immunity [9–11]. Previous studies on lymphocyte function in patients with cardiomyopathy have reported a number of defect [9–11] but have failed to identify a unique or specific

H.-P. Schultheiß (Ed.)
New Concepts in Viral Heart Disease

abnormality, probably because of the heterogeneity of the disease, the modifying influences of neurohumoral adaptations secondary to heart failure, and the nature of the functional parameters examined. Similarly, tests for nonspecific antigenic responses to myocardial cell components or cytotoxicity have not been discriminating enough to be of either diagnostic or conceptual usefulness. We chose to focus on the possible presence of antigenic response against a specific membrane protein, the β-adrenoceptor, which is known to be affected by the cardiomyopathic process, to gain further insight into the pathogenesis of this disease.

Autoantibodies were sought in three experiment groups: patients with idiopathic dilated cardiomyopathy (IDC) ($n = 48$), patients with ischemic or valvular heart disease ($n = 36$), and normal controls ($n = 25$). The clinical and hemodynamic severity of heart disease in the two patient groups was comparable (Table 1).

For autoantibody detection, sera were routinely diluted 100-fold and were used to detect inhibition of [^{3}H]dihydroalprenolol binding to rat cardiac membranes. In the standard assay, 0.1–0.2 mg cardiac membranes in 50 m*M* Tris-HCl (pH.5) – 5 m*M* $MgCl_2$ – 0.1 m*M* phenylmethylsulfonyl fluoride (PMSF) – 5 μg leupeptin were preincubated at 4 °C for 4 h with 100 μl of diluted serum. The (^{3}H]dihydroalprenolol-binding reaction was then carried out at 37 °C for 15 min in the presence or absence of 10 μm propranolol as previously described [12]. B_{max} and K_D values were calculated from Scatchard plots [13] of binding data.

Figure 1 compares the extent of binding inhibition in the three experimental groups. A striking difference is noted between the IDC (32%) and ischemic/valvular disease (7%) groups in the incidence of [^{3}H]dihydroalprenolol binding inhibition. It should be stressed, however, that this difference is largely quantitative since, at lower serum dilutions, a higher percentage of noncardiomyopathy patients were positive (although there was still a significant difference between the two groups). These findings suggest that patients with IDC have a higher titer of antireceptor antibodies. The presence of low-titer autoantibodies even in normal subjects has been reported for several putative autoimmune disease and may be part of the normal immune repertoire[14]. It is likely that, even normally, such autoantibodies participate in the regulation of β-adrenergic pathways. In patients with dilated cardiomyopathy, generation of autoantibodies is probably amplified and their modification of β-adrenoceptor function enhanced accordingly.

Table 1. Clinical and hemodynamit profile of patients with idiopathic dilated cardiomypathy (group I) and ischemic or valvular heart disease (group II)

	Group I	Group II
Age (years)	47 ± 5	52 ± 6
Sex (F/M)	12/36	10/26
Ejection fraction (%)	34 ± 5	40 ± 6
Pulmonary Capillary Wedge Pressure (mmHg)	25 ± 4	22 ± 3
Cardiac index (liters/min per m^2)	2.3 ± 0.3	2.4 ± 0.3
Plasma norepinephrine (pg/ml)	416 ± 93	402 ± 78

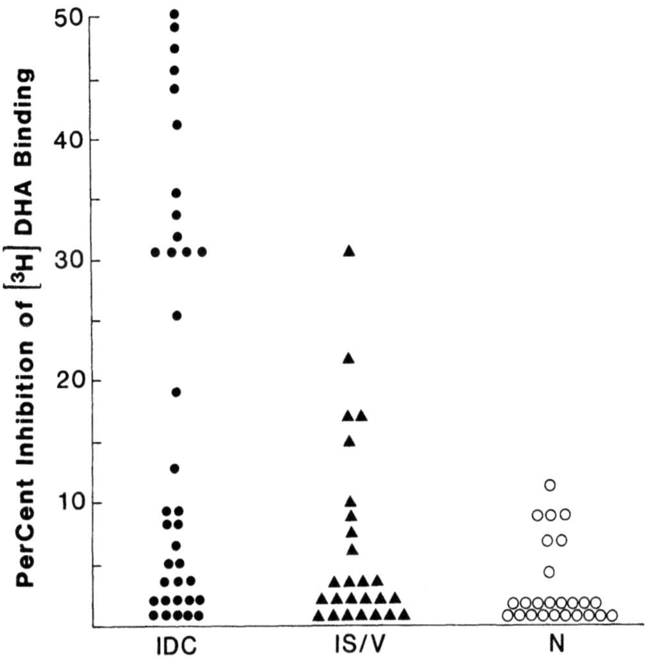

Fig. 1. Comparison of the extent of [^{3}H]dihydroalprenolol *(DHA)* binding inhibition in the experimental groups. *IDC,* idiopathic dilated cardiomyopathy patients; *IS/V,* patients with ischemic or valvular heart disease; *N,* normals

Some of the characteristics of the inhibitory serum activity were then explored. As shown in Table 2, only the density of the β-receptors and not their affinities were affected by patient sera. Furthermore, this inhibitory activity was prevented by preincubating the serum with human anti-IgG antibodies. Cardiac α_1-adrenoceptors, on the other hand, were not affected. Interestingly, the inhibitory effect

Table 2. Effect of serum from cardiomyopathic patients on cardiac and lung β-adrenoceptors and cardiac α_1-adrenoceptors

	β-adrenoceptors		α-adrenoceptors	
Experimental conditions	Bmax (fmol/mg)	K_D (n*M*)	Bmax (fmol/mg)	K_D (n*M*)
Control cardiac membranes	67 ± 5	2.0 ± 0.1	51 ± 4	0.4 ± 0.06
+ Serum (1:50)	40 ± 3*	1.9 ± 0.2	48 ± 3	0.4 ± 0.05
+ Serum preincubated	60 ± 4	2.1 ± 0.1	52 ± 3	0.3 ± 0.06
Control lung membranes	365 ± 16	3.2 ± 0.2		
+ Serum (1:50)	290 ± 13	2.1 ± 0.1		
+ Serum preincubated with anti-IgG	349 ± 12	1.9 ± 0.3		

Results represent mean ± SE for six experiments
* $P < 0.01$, compared to contols

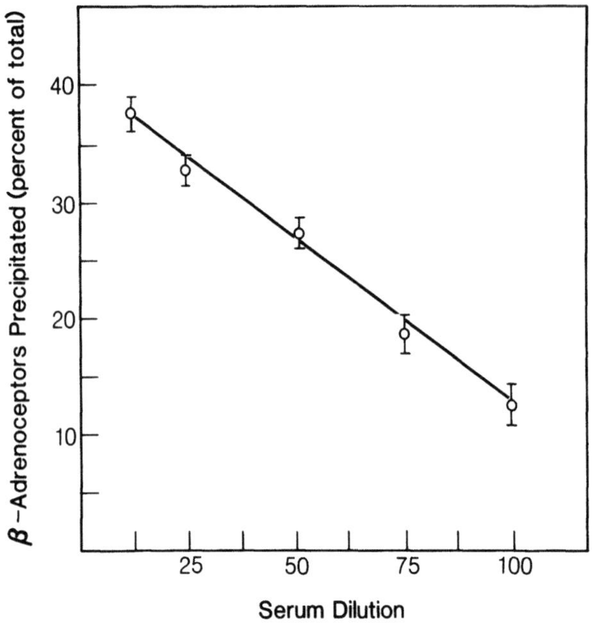

Fig. 2. Immunoprecipitation of β-adrenoreceptors from solubilized cardiac membranes by sera from cardiomyopathy patients. Membranes were first prelabeled with [^{3}]dihydroalprenolol and then solubilized with 1% digitonin in the presence of protease inhibitors. After centrifugation at 48000 *g* for 1 h, the supernatant was incubated at 4 °C for 6 h with diluted patient sera and, then, with *Staphylococcus aureus* cells (Pansorbin, Calbiochem) to precipitate the β-receptors

of any given serum was much less in lung membranes (which contain mostly β_2-adrenoceptors) than for cardiac membranes (mainly $\beta_1$adrenoceptors). This may indicate that the putative autoantibodies are directed preferentially against β_1-adrenoceptors.

In addition to inhibition of [^{3}H]dihydroalprenolol binding, sera from cardiomyopathy patients could precipitate receptors from solubilized cardiac membranes (Fig. 2). Following preincubation with [^{3}H]dihydroalprenolol, cardiac membranes were solubilized with 1% digitonin and then incubated with different serum dilutions from cardiomyopathy patients. It is evident that serum from IDC patients can quantitatively precipitate cardiac β-receptors. We cannot, as yet, tell whether this will be true of all patients with ligand inhibitory activity. It is likely that the titers for immunoprecipitation and ligand binding inhibition will differ significantly in individual patients.

These results strongly suggest that IDC may join the list of autoimmune diseases, such as Graves' disease, myasthenia gravis, insulin-dependent diabetes mellitus, and allergic rhinitis for which antireceptor antibodies have been identified and are thought to play a role in the clinical manifestations of the disease [15–18]. It is not possible, on the basis of current knowledge, to define the mechanisms by which autoantibodies interfere with β-receptor function. Preliminary results suggest that they alter the interaction with adenylate cyclase or agonist-induced cycling of the receptor.

Since IDC probably represents a heterogeneous disease group, it was of some interest to identify immune-associated markers for the presence of anti-receptor antibodies. In a group of 82 patients with dilated cardiomyopathy, the prevalence of HLA antigens was compared to normals. As shown in Table 3, the frequency

Table 3. Comparison of the frequencies of class I HLA antigens in 617 normal controls and 82 idipathic dilated cardiomyopathy patients

	HLA-A				HLA-B				HLA-C		
	IDC		C		IDC		C		IDC		C
	(n)	(%)	(%)		*(n)*	(%)	(%)		*(n)*	(%)	(%)
A1	29	35.4	27.4	B7	20	24.4	24.8	Cw1	1	1.2	7.5
A2	43	52.4	49.3	B8	16	19.5	19.6	Cw2	6	7.3	9.7
A3	24	29.3	25.0	B13	7	8.5	3.0	Cw3	13	15.8	20.1
A11	9	11.0	11.5	B14	6	7.3	7.9	Cw4	10	12.2	22.1
Aw24	11	13.4	19.4	B27	8	9.8	7.6	Cw5	6	7.3	11.7
A28	8	9.8	8.3	Bw35	3	3.7	16.4				
Aw30	8	9.8	4.0	Bw39	3	3.7	4.2				
Aw32	4	4.9	7.8	Bw44	29	35.4*	23.2				
				Bw51	7	8.5	9.9				
				Bw62	11	13.4	10.9				

* $P = 0.03$, compared to controls

Table 4. Comparison of the frequencies of HLA-DR antigens in normal controls and idiopathic dilated cardiomyopathy patients

	EDC ($n = 80$)		C ($n = 511$)
	(n)	(%)	(%)
DR1	13	16.3	20.0
DR2	21	26.3	31.5
DR3	20	25.0	23.1
Dr4	32	40.0*	24.0
DR5	13	16.3	18.4
DR6	17	21.3	20.0
Dr7	16	20.0	22.5
Dr8	3	3.8	3.4

* $P < 0.01$, compared to controls
C, controls

of class I HLA antigens was not significantly different between normals and IDC patients, with the sole exception of HLA-Bw44 which was significantly higher in cardiomyopathy patients. Similarly, the distribution of class II antigens was significantly different for HLA-DR4 (40% in cardiomyopathy vs. 24% in normals, $P < 0.001$). This corresponds to a relative risk of 2.25 and etiologic factor of 0.25 (Table 4). Furthermore, when the presence of autoantibodies against the β-receptor

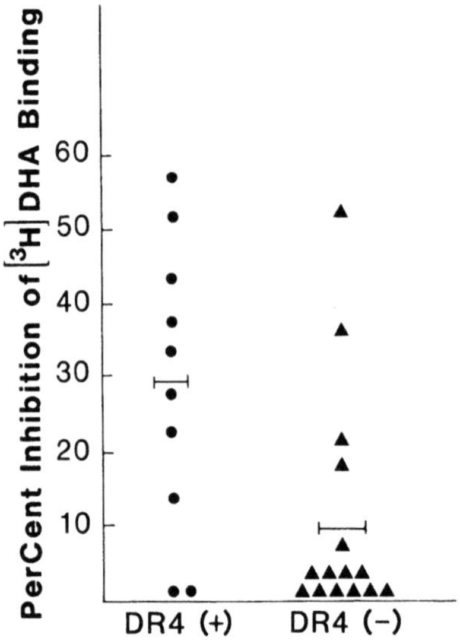

Fig. 3. Correlation between HLA-DR4 detectability and the presence of autoantibodies against the cardiac β_1-adrenoceptor in patients with dilated cardiomyopathy

was correlated with the distribution of HLA antigens in 30 consecutive cardiomyopathy patients, a striking difference was noted (Fig. 3). Only three of the 15 HLA-DR4-negative patients (20%) were positive for autoantibodies, compared to seven of ten HLA-DR4-positive patients (77%). Although this is a relatively small series, it does suggest that the presence of autoantibodies in cardiomyopathy may be under the influence of the major histocompatibility locus.

The role that autoantibodies play in the regulation of cardiac β-receptors in clinical heart failure has not been defined yet and is likely to be hetergeneous. For example, competition for ligand binding demonstrated in this study may not always result in decline of adenylate cyclase activity and does not elucidate the influence of autoantibodies on postreceptor events. In accordance with experience from other receptor systems, it is likely that either inhibitory or stimulatory effects can be expected within the same disease group, reflecting pathogenetic heterogeneity. Effects on agonist-induced receptor cycling and/or degradation are also likely and may determine the net influence on β-receptor-mediated physiologic effects.

Finally, it may be profitable to speculate on the therapeutic implications of the autoantibodies in IDC. Our initial findings suggest that immunologic factors are important in only a subset of patients with this disease. Furthermore, it is likely that other autoantibodies with pathogenetic significance (e.g., against calcium channels or other integral membrane proteins) are present in the same patient subset. Strategies to prevent or modify the immunologic component in these patients may prevent or arrest progression of the disease process and significantly reduce morbidity and mortality.

Summary

It is well known that the density of cardiac βadrenoceptors is decreased in IDC, but the mechanisms involved have not been elucidated. In view of the reported involvement of immune dysfunction in the pathogenesis of dilated cardiomyopathy, we examined the possible presence of autoantibodies directed against cardiac βj-adrenoceptors. Sera from 48 patients with dilated cardiomyopathy, 36 patients with ischemic or valvular heart disease, and 25 normal subject were compared. In 15 of the cardiomyopathy patients (32%), five of the patients with ischemic/valvulary disease (7.5%), and none of the normals, 1:100 dilution of the sera inhibited the binding of [^{3}H]dihydroalprenolol to cardiac membranes. Such inhibition was not seen when [^{3}H] prazosin was used to label cardiac α_1-receptors and was considerably less for lung, compared to heart, βadrenoceptors. Positive sera could also immunoprecipitate the β-receptor from solubilized cardiac membranes. Immune-associated genetic factors controlling the presence of β-receptor autoantibodies were then sought. In patients with IDC, ther frequency of HLA antigens Bw44 and DR4 were significantly higher than in normal controls or patients with ischemic cardiomyopathy (40% compared to 24%, $P < 0.01$; relative risk 2.25; etiologic factor 0.25). In addition, about two-thirds of the HLA-DR4 positive patients had antireceptor autoantibodies compared to only 25% of the HLA-DR4-negative cardiomyopathy patients. These results indicate that a substantial percentage of patients with dilated cardiomyopathy have antibodies against the cardiac β_1-adrenoceptor, and that these autoantibodies are under the control of the major histocompatibility locus.

References

1. Bristow MR, Ginsburg R, Minobe W, Cubicciotti RS, Sageman NS, Lurie K, Billingham ME, Harrison DC, Stinson EG (1982) Decreased catecholamine sensitivity and β-adrenergic receptors density in failing human hearts. N Engl J Med 307: 205–211
2. Bristow MR, Ginsburg R, Umans V, Fowler M, Minobe W, Rasmussen R, Zera P, Menlove R, Shah P, Jamieson S, Stinson EB (1986) β_1- and β_2-adrenergic receptor subpopulations in nonfailing human ventricular myocardium. Coupling of both receptor subtypes to muscle contraction and selective β_1-receptor down-regulation in heart failure. Circ Res 59: 297–309
3. Limas CJ, Goldenberg IF, Pierpont GL (1988) Intracellular β-adrenoceptors distribution in the failing myocardium. J Am Cardiol 11: 117A
4. Golf S, Andersen D, Hannson V (1986) Beta adrenoceptor density and adenylate cyclase response in right atrial and left ventricular myocardium of patients with mitral valve disease. Cardiovasc Res 20: 331–336
5. Limas CJ, Limas C (1984) Rapid recovery of cardiac β-adrenergic receptors following isoproterenol-induced "down" regulation. Circ Res 55: 524–531
6. Cohn JN, Levine TB, Francis GS, Goldsmith S (1981) Neurohumoral control mechanisms in congestive heart failure. Am Heart J 102: 509–516
7. Das SK, Cassidy JT, Petty RE (1972) Antibodies against heart muscle and nuclear constituents in cardiomyopathy. Am Heart J 83: 159–166
8. Schultheiss HP, Bolte HD (1985) Immunological analysis of autoantibodies against the adenine nucleotide translocator in dilated cardiomyopathy. J Mol Cell Cardiol 12: 603–611
9. Franceschini R, Petilo A, Corrazza M, Nizzo MC, Azzolini A, Gianrossi R (1983) Lymphocyte blastogenic response in dilated cardiomyopathy. IRCS Med Sci 11: 1019–1023

10. Jacobs B, Matsuda Y, Deodhar S, Shirey E (1979) Cell-mediated cytoxicity to cardiac cells of lymphocytes from patients with primary myocardial disease. Am J Clin Pathol 72: 1–4
11. Anderson JL, Carlquist JF, Hammond EH (1982) Deficient natural killer cell activity in patients with idiopathic dilated cardiomyopathy. Lancet II: 11124
12. Limas CJ, Limas C (1987) Altered intracellular adrenoceptor distribution in the myocardium of spontaneously hypertensive rats. Am J Physiol 253: H904–H908
13. Scatchard G (1949) The attraction of proteins for small molecules and ions. Ann NY Acad Sci 51: 660–672
14. Bottazzo GF, Todd I, Mirakian R, Belfiore A, Pujol-Borrell R (1986) Organ-specific autoimmunity: a 1986 overview. Immunol Rev 94: 137–156
15. Flier JS, Kahn CR, Jarrett DB, Roth J (1976) Characteristics of antibodies to the insulin receptor: a cause of insulin-resistant diabetes in man. J Clin Invest 58: 1142–1149
16. Lindstrom JM, Segbold ME, Lemmon VA (1976) Antibody to acetylcholine receptor in myasthenia gravis: prevalence, clinical correlation and diagnostic value. Neurology (Minneapolis) 26: 1054–1059
17. Rees-Smith B, Hall R (1974) Thyroid-stimulating immunoglobulins in Graves' disease. Lancet II: 427–430
18. Venter JG, Fraser CM, Harrison LC (1980) Autoantibodies to β_2-adrenergic receptors: a possible cause of adrenergic hyporesponsiveness in allergic rhinitis and asthma. Science 207: 1361–1363

Antibodies Against the Adenosine Di-/Triphosphate Carrier Cross-react with the Ca Channel – Functional and Biochemical Data

G. Ulrich, U. Kühl, B. Melzner, I. Janda, B. Schäfer, and H.-P. Schultheiss

Department of Internal Medicine, Klinikum Großhadern, University of Munich, Marchioninistraße 15, D-8000 München 70, FRG

Introduction

Heart-reactive antibodies against cell surface and intracellular components have been described in myocarditis [1, 6, 12]. During recent years studies by our group have characterized the adenosine di-/triphosphate (ADP/ATP) carrier of the inner mitochondrial membrane as an autoantigen in myocarditis and dilated cardiomyopathy [14, 15, 18]. The ADP/ATP carrier is a highly specific transport system for the transfer of ATP to the cytosol with its energy-consuming processes and the return of ADP to the inner mitochondrial space for rephosphorylation [7]. Autoantibodies to the ADP/ATP carrier were shown to be organ specific [15]. They inhibit the nucleotide transport not only in vitro [15, 18], but also in vivo [18, 20] (also Schultheiss et al., this volume).

In this study we investigated the effects of the antibodies primarily directed against the ADP/ATP carrier on isolated adult rat cardiac myocytes. The antibodies were shown to bind specifically to the cell surface and to impair the contractility of the myocytes. Moreover, exposure of the myocytes to a high concentration of the antibodies resulted in cell damage and death, which was prevented by the addition of Ca channel blockers [19, 25]. After isolation of the Ca channel, the antibodies against the ADP/ATP carrier were found to bind specifically to the subunits of the protein complex indicating a cross-reactivity between the ADP/ATP carrier ot the inner mitochondrial membrane and the Ca channel of the cell surface.

These studies provide evidence for an interaction of the anti-ADP/ATP carrier antibodies with the Ca channel and suggest a new mechanism for cytotoxicity which is not complement dependent or cell mediated.

Material and Methods

Antibodies against the isolated ADP/ATP carrier were raised in rabbits as described previously [16]. Adult rat cardiac myocytes were isolated using the collagenase and hyaluronidase digestion method [5]. Cell surface staining of isolated myocytes was carried out by the immunofluorescence technique. The cells were incubated with affinity-purified antibodies to the ADP/ATP carrier (antibody dilution 1:1000, 30 min). For control experiments, the antibodies were replaced by

H.-P. Schultheiß (Ed.)
New Concepts in Viral Heart Disease

rabbit preimmune IgG or by antibodies to the ADP/ATP carrier previously adsorbed to the purified antigen (50 μg antigen/1ml antibody dilution 1:1000, 3 h, 0°C). A secondary antibody (goat anti-rabbit IgG) conjugated to fluoresceine isothiocyanate was used (dilution 1:40) to detect bound antibodies by fluorescence microscopy (Schultheiss et al., this volume).

Antibody binding was also tested by a radioimmunobinding assay. Myocytes (2 x 10^5) were incubated with antibodies (affinity-purified IgG fraction, 1:1000) with different concentrations of antibodies and for different times. ^{125}I-protein A (specific activity: 9.8 μCi/μg, NEN) was used to label bound antibodies (20000 cpm, 2 x 10^5 cells, 1 h). The data were calculated for 1 x 10^6 cells for comparative quantification [19].

Contraction velocity of the isolated myocytes was measured in an electro-optic system that registers contractions of externally paced myocytes, expressed as amplitude in time [26]. Contraction velocity was registered after exposure of myocytes to the antibodies and to preimmune sera (IgG, antibody dilution 1:100). Cytotoxicity was measured in cell suspensions [19] (5 x 10^4 myocytes) after incubation

- With varying concentrations of antibody: 1:25–1:5000 (1 m*M* CA^{2+}, 1 h)
- With varying times up to 3 h (antibody concentration 1:100, 1 m*M* Ca^{2+})
- For 3 h after preadsorption of the antibody with its antigen (100 μg antigen/1ml antibody dilution 1:100, 3 h)
- With varying Ca^{2+} concentrations: 0, 0.1, 0.5, 1 m*M* (antibody concentration 1:100, 3 h)
- After addition of the Ca channel blockers nifedipine, nitrendipine and verapamil at a concentration of 10^{-6} *M* (antibody dilution 1:100, 1 m*M* Ca^{2+}, 3 h)
- After addition of the beta-receptor antagonist propranolol (10^{-6} *M*) (antibody dilution 1:100, 1 m*M* Ca^{2+}, 3 h)

The Ca channel protein complex was isolated according to the method of Leung et al. [9, 10] and Sharp et al. [21] from bovine skeletal muscle. Gel electrophoresis was carried out on 5%–18% polyacrylamide slab gels and stained with silver nitrate [8, 13]. Immunoblot analysis was performed on nitrocellulose [24]. Bound antibodies were detected by ^{125}I-protein A and visualized after exposure to an X-ray film plate. To detect autoantibodies which bind to the ADP/ATP carrier in sera of patients, a solid phase radioimmunoassay was carried out according to Schultheiss [16, 17].

Results and Discussion

Recently presented data suggest that autoimmunity to the ADP/ATP carrier may contribute to the pathophysiology of viral heart disease by causing an antibody-mediated imbalance between energy delivery and demand. A prerequisite for an interaction of this antibody — primarily directed against a mitochondrial protein — with the myocardial cell is to show a specific binding of the antibodies to the surface of cardiac myocytes. As demonstrated by indirect immunofluorescence (Fig. 1a) the antibody can be detected as a dot-like pattern on the surface of the living isolated cardiac myocyte. This sarcolemmal staining was abolished by

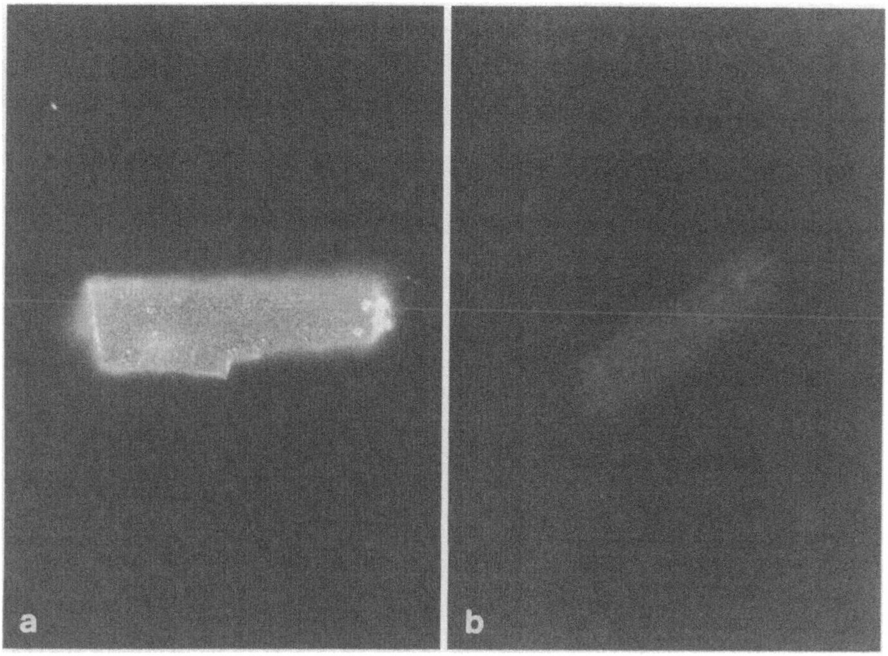

Fig. 1 a, b. Immunofluorescence on isolated adult rat cardiac myocytes. **a** Cell surface staining of isolated adult rat cardiac myocytes with anti-ADP/ATP carrier antibodies; **b** disappearance of cell surface reaction after neutralization of antibodies with the purified ADP/ATP carrier

preincubation of the antibodies with the purified antigen, the ADP/ATP carrier (Fig. 1b). Preimmune sera did not react with the cell surface [19].

A cross-reactivity between antibodies to the ADP/ATP carrier and a cell surface protein was also demonstrated by radioimmunobinding assay [19, 25]. Figure 2a shows a time-dependent binding of the antibodies to myocytes. After 60 min most of the antibodies were bound to the cell surface. Prolonged incubation for up 180 min did not lead to a significant further increase of antibody binding. Preimmune serum IgG (control) did not bind to the myocytes. This antibody binding was also concentration dependent. Antibody dilution from 1:500 to 1:6000 resulted in a concentration-dependent reduction of the antibody binding. At a serum dilution of 1:5000 no significant binding could be detected [19]. The specificity of the antigen-antibody reaction was demonstrated by neutralization of the antibody with its antigen (Fig. 2b). Preincubation of the antibody with increasing amounts of the isolated ADP/ATP carrier (1–1500 μg) resulted in a concentration-dependent decrease of the antibody binding.

The results so far suggest that antibodies to the ADP/ATP carrier specifically bind to a cross-reacting epitope on the cell surface. As it is known from in vivo experiments that these antibodies cause an imbalance between energy supply and demand [18] (also Schultheiss et al., this volume), an inhibitory effect on contraction velocity as an indirect parameter for an energy deficit was supposed. For that

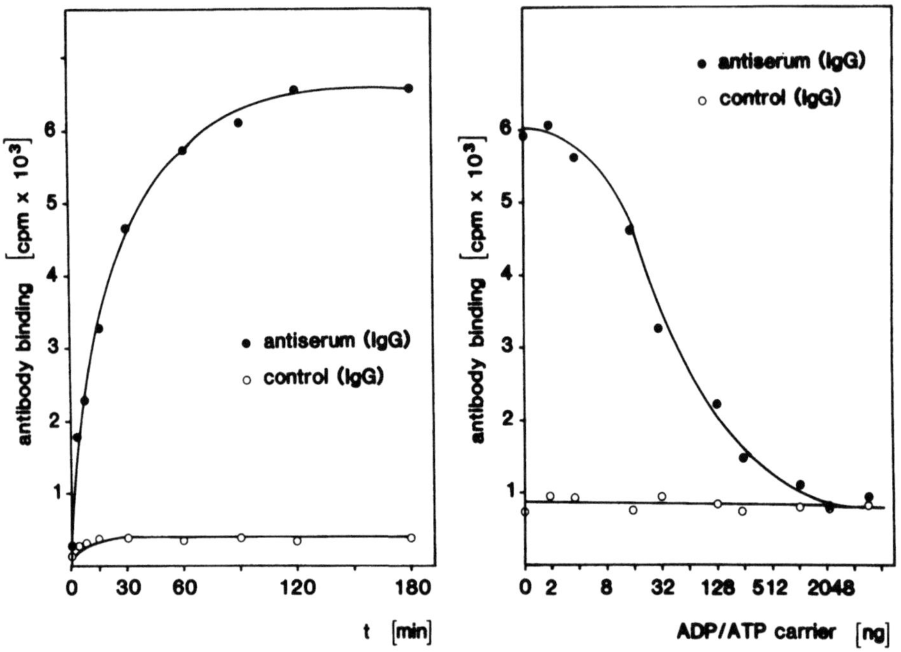

Fig. 2 a, b. Binding of antibodies against the ADP/ATP carrier to isolated cardiac myocytes in the radioimmunobinding assay: time-dependent binding **(a)** and inhibition of binding after neutralization of the antibodies with purified ADP/ATP carrier **(b)**

purpose cultered adult rat myocytes were examined for their contraction characteristics under the influence of antibodies. As shown in Fig. 3, externally paced myocytes beat regularly after addition of preimune sera (control). Five minutes after the addition of antibodies against the ADP/ATP carrier (affinity-purified IgG), contraction velocity increased, and the cells began to beat arrhythmically. This finding was clearly inconsistent with an antibody-mediated disturbance of energy supply. After 10 min of incubation time contraction velocity decreased markedly, finally ending up in a cessation of contraction. This acute cytotoxic effect takes place within 15–30 min.

This antibody-mediated cytotoxicity could be objected by a cytotoxicity assay. Incubation of cardiac myocytes with antibodies against the ADP/ATP carrier in a concentration higher than 1:1000 resulted in a concentration-dependent decrease of cell viability. As shown in Fig. 4a, intact isolated adult rat cardiac myocytes showed rod-shaped morphology. Cross-striation was clearly visible, and the cell surface appeared smooth. After addition of antibodies, deterioration of myocytes involved, in sequence: rapid spontaneous contractions. bleb formation, contracture to almost cuboid shape, disarrangement of myofibrils, cell rounding and finally complete loss of membrane integrity (Fig. 4b,c). At the end stage the cells took up trypan blue as a marker of cell death [19]. This cytotoxic effect was concentration dependent and was completely abolished by preincubation of the antibodies with the ADP/ATP carrier (Schultheiss et al., this volume).

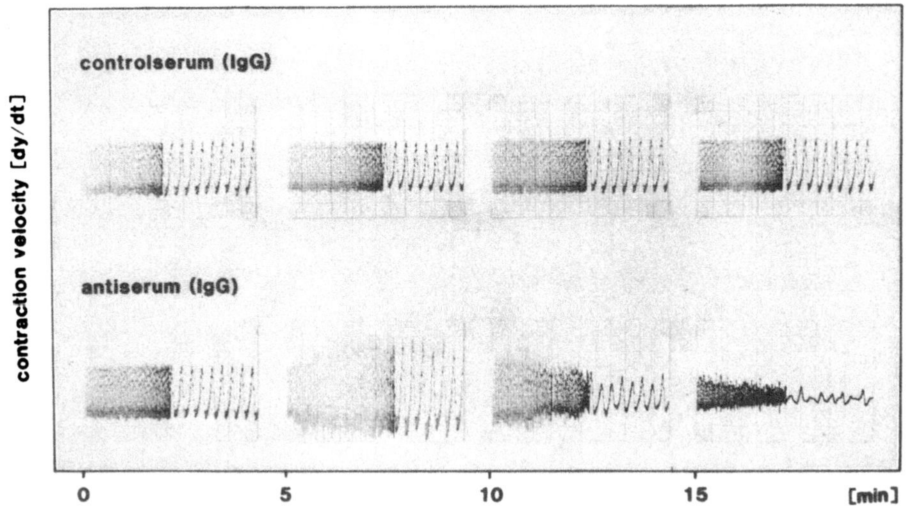

Fig. 3. Contraction characteristics of externally paced cardiac myocytes after addition of preimmune serum *(controlserum)* and antibodies to the ADP/ATP carrier *(antiserum)*

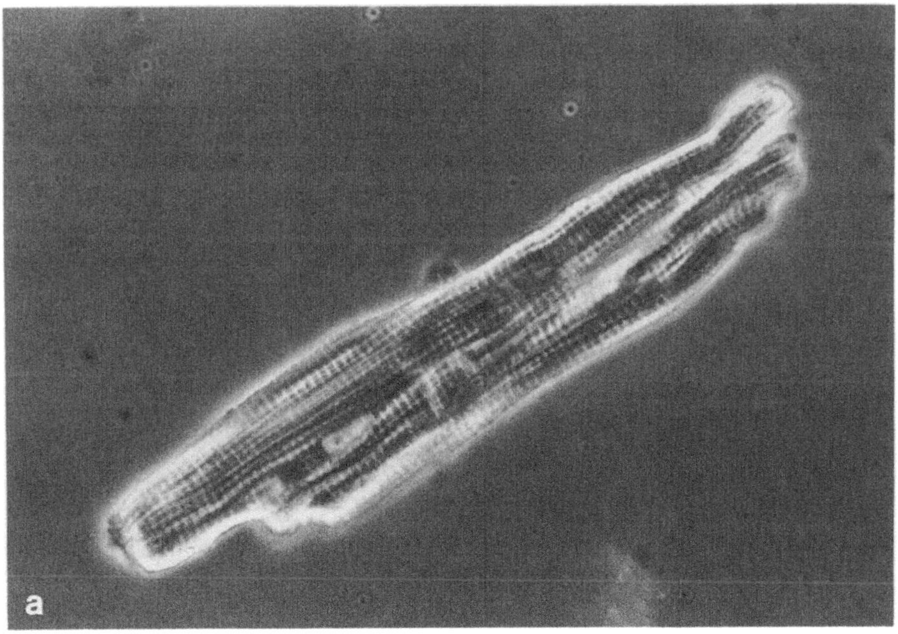

Fig. 4

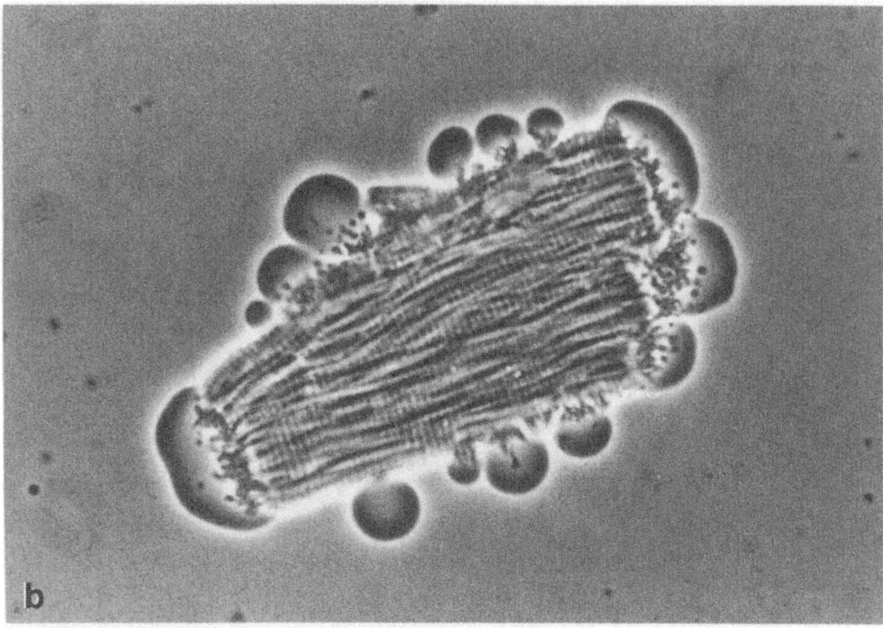

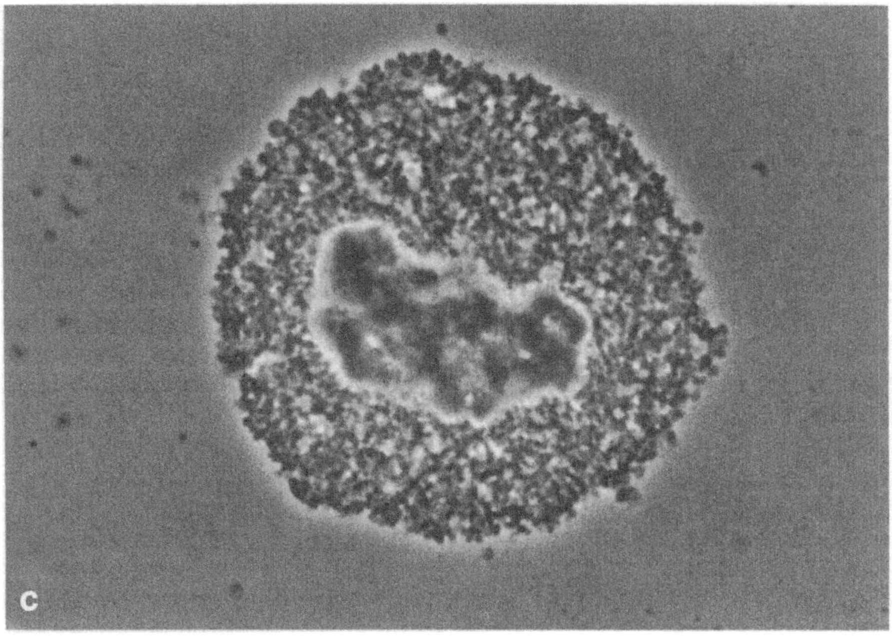

Fig. 4 a-c. Appearance of isolated cardiac myocytes during incubation with antibodies to the ADP/ATP carrier. **a** Normal, rod-shaped cells; **b** contracture, bleb formation, disarragement of myofibrils; **c** final cell death

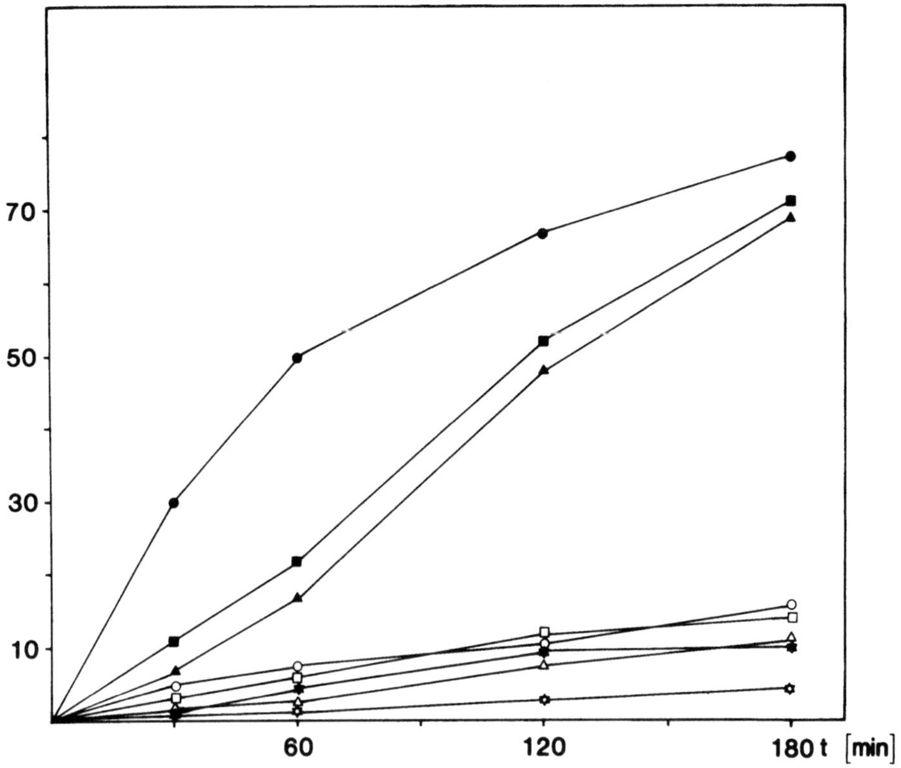

Fig. 5. Calcium concentration-dependent mortality of cardiac myocytes incubated with antibodies against the ADP/ATP carrier: Ca^{2+} concentration 1.0 m*M* + antiserum *(solid circles); open circles,* control. Ca^{2+} concentration 0.5 m*M* + antiserum *(solid squares); open squares,* control. Ca^{2+} concentration 0.1 m*M* + antiserum *(solid triangles); open triangles,* control. Ca^{2+} concentration 0 + antiserum *(circled asterisks); asterisks,* control. Antibody (IgG) dilution: 1:100

Experimental data of our group indicate that endocytosis of the antibodies and their deposition at the intracellular target requires much more time (data not shown). Therefore it has to be assumed that the effects described are not caused by an inhibition of the ADP/ATP carrier function but by a direct interaction of the antibodies with the cell surface.

Further characterization of the mechanism involved in the antibody-mediated cytotoxicity showed that this effect was strictly Ca dependent [19]. As demonstrated in Fig. 5, the mortality of the myocytes increased up to 80% in the presence of 1 m*M* Ca^{2+} within 3 h. In contrast, only 10% of the myocytes incubated with preimmune IgG (control) did not survive 3 h. Reduction of external Ca^{2+} reduced mortality in a concentration-dependent way. In the presence of antibodies and nominally no Ca, only 10% of the cells died within 3 h (Fig. 5).

To support the role of Ca as a possible mediator of the cytotoxic effect, we added Ca channel blockers together with the antibodies against the ADP/ATP carrier (1:100) to the myocyte suspension (Fig. 6). Addition of nifedipine and nitrendipine (both 10^{-6} M) protected the cells very efficiently, as did verapamil (10^{-6} M) to a lesser extent, although the Ca^{2+} concentration was 1 mM. Without antibodies none of the Ca channel blockers had a significant influence on cell viability. The use of the beta-receptor antagonist propranolol (10^{-6} M) had no protective effect.

The results so far suggest that antibodies to the ADP/ATP carrier bind specifically to the cell surface of cardiac myocytes and enhance Ca permeability leading to cell death. To gain further evidence for a direct interaction of the antibodies with the Ca channel, we isolated the Ca channel protein complex according to the method of Leung et al. [9, 10] and Sharp et al. [2]. After sodium dodecyl sulfate

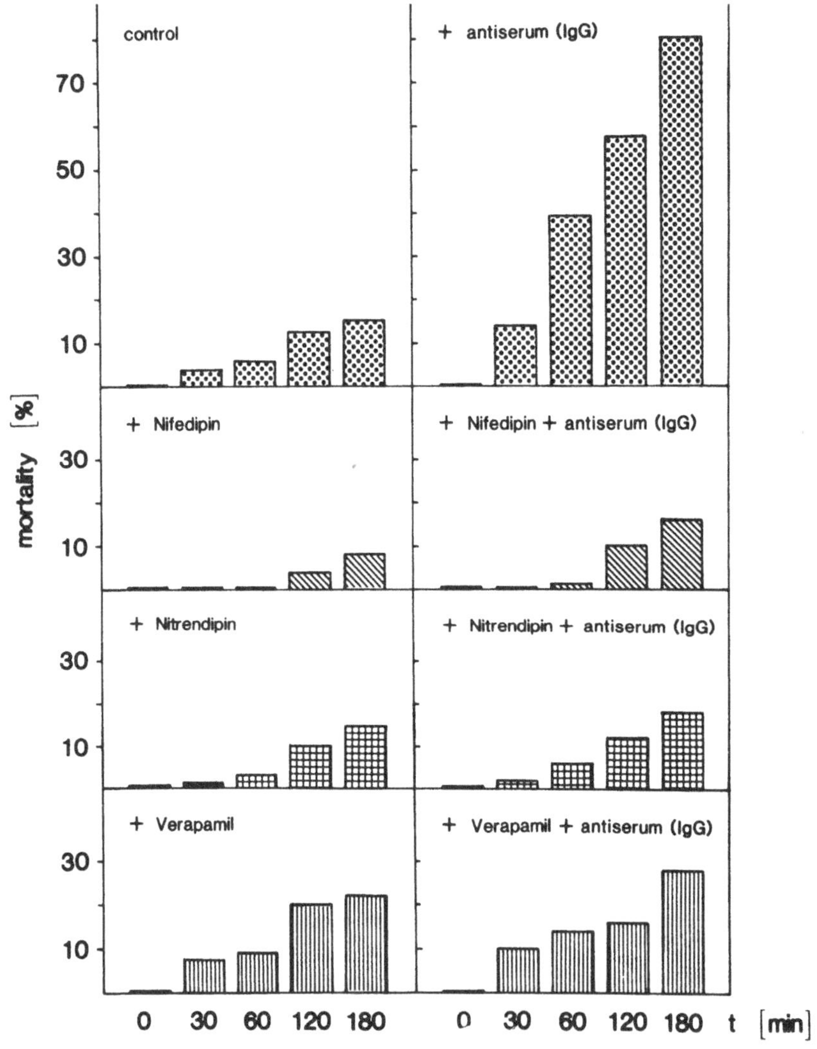

Fig. 6. Mortality of cardiac myocytes after addition of Ca channel blockers (10^{-6} M) in the presence of preimmune serum (control, *left*) and of antibodies to the ADP/ATP carrier (antiserum, *right*)

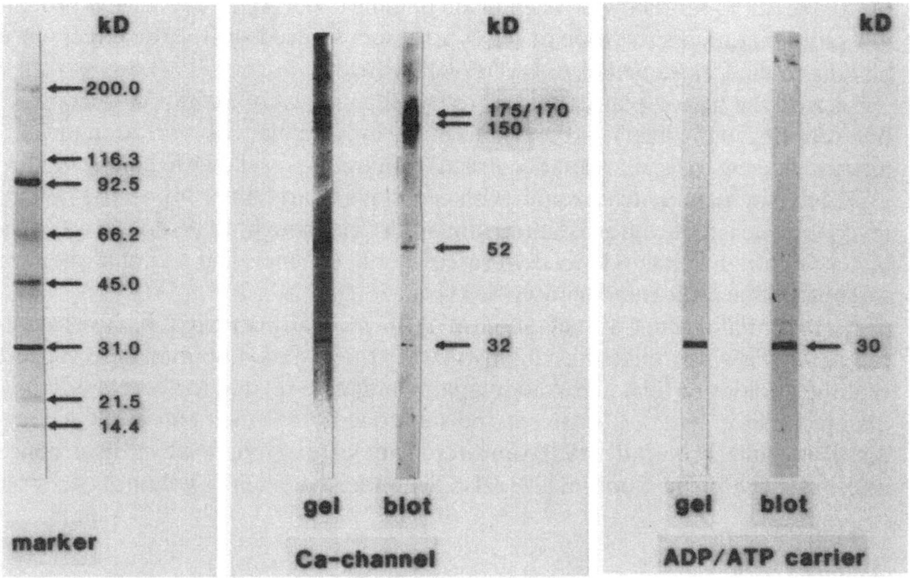

Fig. 7. Sodium dodecyl sulfate gel electrophoresis of the Ca channel and the ADP/ATP carrier and subsequent immunoblot analysis with affinity-purified antibodies to the ADP/ATP carrier

gel electrophoresis and transfer of the protein to nitrocellulose, the immunoblot analysis showed that the antibodies against the ADP/ATP carrier bind to the Ca channel subunits as well as to their purified antigen (Fig. 7). These data clearly indicate a cross-reactivity between the ADP/ATP carrier of the inner mitochondrial membrane and the Ca channel of the cell surface.

The 170-kilodalton α_1-subunit of the Ca channel was reported to include the dihydropyridinge binding site [21]. The 52-kilodalton β-subunit and the 32-kilodalton γ-subunit are thought to be involved in Ca channel gating [2]. From these data it can be speculated that the antibodies against the ADP/ATP carrier interfer with the gating mechanism of the Ca channel. This was confirmed by patch-clamp experiments which clearly showed an enhancement of Ca^{2+} influx through the voltage-dependent Ca channel as well as a prolongation of the open time of the channel [19] (also Morad et al., this volume).

The data so far reported are derived from experimental antibodies raised in rabbits. Studies by our group described autoantibodies to the ADP/ATP carrier in myocarditis [14, 16]. According to the clinical diagnosis and to the histology (carried out by Professor E. G. J. Olsen, London) we have so far selected 18 patients with myocarditis. Five patients with coronary heart disease and 20 heathly blood donors were taken as a control group. IgG fractions from sera of these patients were investigated for their cytotoxicity and their binding to the Ca channel.

Of 18 myocarditis sera, 13 showed a cytotoxic effect on cardiac myocytes which was markedly reduced by Ca channel blockers (nifedipine, nitrendipine, 10^{-6} *M*); these 13 sera bound at least to one subunit of the Ca channel protein complex.

Moreover, all 13 sera showed a significant binding to the ADP/ATP carrier protein in a radioimmunoassay. None of the control sera showed a cytotoxic effect nor a binding to the Ca channel or the ADP/ATP carrier.

Because the Ca ion plays a central role in the process of excitation — contraction coupling in the heart — alterations in the intracellular concentration of Ca^{2+} may provide the basis for cardiac contractile failure, as was shown in animal models [3, 22] and in human myocardium with end-stage heart failure [4]. In the human myopathic heart, elevation of intracellular Ca^{2+} is thought to be due to an entry of Ca^{2+} through voltage-dependent sarcolemmal channels and an impairment of sarcoplasmatic reticulum function [4, 11].

As abnormalities in Ca^{2+} metabolism of the myocardium might be involved in the pathogenesis of dilated cardiomyopathy, the results described here suggest that the etiology of dilated cardiomyopathy might be related to an antibody-mediated enhancement of Ca current and subsequent Ca^{2+} overload of the myocytes. This could chronically lead to an altered myocardial function by a disturbance of cellular energy metabolism [20] (also Schultheiss et al., this volume).

Summary

Antibodies primarily directed against the ADP/ATP carrier react with the cell surface of cardiac myocytes. The antibodies impair the contractility of cardiac myocytes and exhibit an acute cytotoxic effect which is dependent on extracellular Ca. Ca channel blockers prevent the antibody-mediated cytotoxicity which is also complement independent.

The cross-reacting epitope on the cell surface is the Ca channel. Antibody binding to the α_1-, $\frac{3}{4}_2$-, β-, and γ-subunit of the Ca channel was demonstrated.

Sera of patients with myocarditis contain autoantibodies which bind to the cell surface, induce Ca-dependent cytotoxicity and show immunologically a cross-reactivity between the Ca channel and the ADP/ATP carrier.

Our data suggest a new mechanism of an antibody-mediated cytotoxicity. Acutly, these antibodies showing a cross-reactivity between the Ca channel and the ADP/ATP carrier most likely induce a Ca overload and subsequent cell death. Chronically, those antibodies might alter myocardial function by disturbing cellular energy metabolism.

Acknowledgement. This research was supported by a grant from the Deutsche Forschungsgemeinschaft and the Wilhelm-Sander-Stiftung.

References

1. Bolte HD, Schultheiss HP (1978) Immunological results in myocardial diseases.Postgrad Med J 54: 500–503
2. Campbell KP, Coronado R, Leung AT, Sharp AH, Vilven J (1988) Antibody modulation of Ca^{2+} channel activity. In: The calcium channel: structure, function and implications. Bayer Centenary Syposium, Bayer Forum 1988
3. Dhalla NS, Pierce GN, Panagia V, Singal PK, Beamish RE (1982) Calcium movements in relation to heart functions. Basic Res Cardiol 77: 117–139

4. Gwathmey JK, Copelas L, MacKinnon R, Schoen FJ, Feldman MD, Grossman W, Morgan JP (1987) Abnormal intracellular calcium handling in myocardium from patients with end-stage heart failure. Circ Res 61: 70–76
5. Hohl CM, Altschuld RA, Brierley GP (1983) Effects of calcium on the permeability of isolated adult rat heart cells to sodium. Arch Biochem Biophys 221: 197–205
6. Keraiakes DJ, Parmley WW (1984) Myocarditis and cardiomyopathy. Am Heart J 108: 1318–1326
7. Klingenberg M, Heldt HW (1982) The ADP/ATP translocation in mitochondria and its role in intracellular compartmentation. In: Sies H (ed) Metabolic compartmentation. Academic, London, pp 101–122
8. Laemmli UK (1970) Cleavage of structural proteins during the assembly of the head of bacteriophage T_4. Nature 227: 680–685
9. Leung AT, Imagawa T, Campbell KP (1987) Structural characterization of the 1,4-dihydrophyridine receptor of the voltage-dependent Ca^{2+} channel from rabbit skeletal muscle. J Biol Chem 262: 7943–7946
10. Leung AT, Imagawa T, Block B, Franzini-Armstrong C, Campbell KP (1988) Biochemical and ultrastructural characterization of the 1,4-dihydropyridine receptor from rabbit skeletal muscle. J Biol Chem 263: 994–1001
11. Limas CJ, Olivari MT, Goldenberg IF, Levine TB, Benditt DG, Simon A (1987) Calcium uptake by cardiac sarcoplasmatic reticulum in human dilated cardiomyopathy. Cardiovasc Res 21: 601–605
12. Maisch B, Deeg P, Liebau G, Kochsiek K (1983) Diagnostic relevance of humoral and cytotoxic immune reactions in primary and secondary dilated cardiomyopathy. Am J Cardiol 52: 1072–1078
13. Oakley BR, Kirsch DR, Morris NR (1980) A simplified ultrasensitive silver stain for detecting proteins in polyacrylamide gels. Anal Biochem 105: 361–363
14. Schultheiss HP (1987) The mitochondrium as antigen in inflammatory heart disease. Eur Heart J 8 [Suppl J]: 203–210
15. Schultheiss HP, Bolte HD (1985) The immunochemical characterization of the adenine nucleotide translocator as an organspecific autoantigen in dilated cardiomyopathy. J Mol Cell Cardiol 17: 603–617
16. Schultheiss HP, Klingenberg M (1984) Immunochemical characterization of the adenine nucleotide translocator: organ- and conformation specificity. Eur J Biochem 143: 599–605
17. Schultheiss HP, Berg P, Klingenberg M (1983) The mitochondrial adenine nucleotide translocator is an antigen in primary biliary cirrhosis. Clin Exp Immunol 54: 648–654
18. Schultheiss HP, Schulze K, Kühl U, Ulrich G, Klingenberg M (1987) The ADP/ATP carrier as a mitochondrial auto-antigen — facts and perspectives. Ann NY Acad Sci 488: 44–68
19. Schultheiss HP, Ulrich G, Janda I, Kühl U, Morad M (1988) Antibody-mediated enhancement of calcium permeablity in cardiac myocytes. J Exp Med (in press)
20. Schulze K, Becker B, Schauer R, Schultheiss HP (1988) Antibodies to the ADP/ATP carrier — an autoantigen in myocarditis and dilated cardiomyopathy — disturb cardiac function in vivo. Circ Res (in press)
21. Sharp AH, Imagawa T, Leung AT, Campbell KP (1987) Identification and characterization of the dihydropyridine-binding subunit of the skeletal muscle dihydropyridine receptor. J Biol Chem 262: 12309–12315
22. Smith VE, Katz AM (1983) Inotropic and lusitropic abnormalities in the genesis of heart failure. Eur Heart J 4 [Suppl A]: 7–17
23. Tanabe T, Takeshima H, Mikami A, Flockerzi V, Takahashi H, Kangawa K, Kojima M, Matsuo H, Hirose T, Numa S (1987) Primary structure of the receptor for calcium channel blockers from skeletal muscle. Nature 328: 313–318
24. Towbin H, Staehelin T, Gordon J (1979) Electrophoretic transfer of proteins from polyacrylamide gels to nitrocellulose sheets: procedure and some applications. Proc Natl Acad Sci USA 76: 4350–4354
25. Ulrich G, Kühl U, Janda I, Schultheiss HP (1987) Antibodies to the ADP/ATP carrier of the inner mitochondrial membrane cross-react with cell surface antigens and induce a cytotoxic effect on isolated adult cardiac myocytes. Eur Heart J 8 [Suppl J]: 215–217
26. Werdan K, Reithmann C, Erdmann E (1985) Cardiac glycoside tolerance in cultered chikken heart muscle cell — a dose dependent phenomenon. Klin Wochenschr 63: 1253–1264

Antibodies and Autoantibodies Against ADP/ATP Carrier Enhance Calcium Current in Isolated Ventricular Myocytes

M. MORAD[1], M. NÄBAUER[1], and H.-P. SCHULTHEISS[2]

[1] Department of Physiology, University of Pennsylvania, Philadelphia, PA 19104, USA
[2] Department of Internal Medicine Klinikum Großhadern, University of Munich, Marchinioninistr. 15, D-8000 München 70, FRG

Introduction

Abnormalities in Ca^{2+} metabolism of myocardium and coronary vessels are thought to be involved in genesis of cardiac myopathies [11]. In Syrian hamster (Bio 14.6), Ca^{2+} influx was enhanced and Dihydropytidine (DHP) binding sites increased in early stages of cardiac myopathy [4, 10, 12]. Prolongation of action potential reported in cardiomyopathic hamsters [9] is also consistent with possible enhancement of Ca^{2+} current. Abnormalities in Ca^{2+} metabolism may also result from heart reactive antibodies. Such antibodies have been identified in myocarditis and cardiac myopathies [1, 3]. Antibodies previously described as inhibiting the adenosine diphosphate (ADP)/adenosine triphosphate (ATP) carrier of the inner mitochondrial membrane [8] have been found in myocarditis and dilated cardiomyopathy [7]. Such antibodies were found to bind specifically to the sarcolemma of isolated cardiac myocytes causing oscillating contractions leading to cell death. Cell deterioration, as judged by the number of cells losing their rectangular shape and clear striation, was delayed or prevented by addition of Ca^{2+} channel blockers to the antibody- containing solutions. These results, consistent with those described in Syrian hamster, suggest that etiology of disease may be related to possible enhancement of the Ca^{2+} channel, causing subsequent Ca^{2+} overload. In this report, therefore, we examine the direct effects of antibodies against ADP/ATP carrier and autoantibodies from patients with myocarditis and dilated cardiomyopathies on the Ca^{2+} current recorded in isolated mammalian and amphibian ventricular myocytes.

Electrophysiological Effects of Antibodies in Frog Heart

Antibodies against ADP/ATP carrier were prepared by immunizing rabbits with the isolated and purified ADP/ATP carrier. The antibody binds specifically to a 30 kilodalton band, the isolated carrier protein of the inner mitochondrial membranes. Addition of immunized sera or IgG fraction of sera prolonged the action potential and potentiated the twitch tension (Fig. 1a). The tension-potentiating effect of antibody was complete within 2 min of the exchange of perfusion solution, and the effect was very slowly reversible. To substantiate that the effect of antibody on the action potential and contraction were mediated by enhancement

H.-P. Schultheiß (Ed.)
New Concepts in Viral Heart Disease

of calcium current (i_{Ca}), we studied the effect of the antibody on the i_{Ca} recorded from enzymatically isolated frog, rat, and guinea pig ventricular myocytes [6]. Cells were dialyzed with solution containing high Cs^+ and high ethyleneglycol Bis-(β-aminoethyl)N, N, N′, N′-tetraaciticacid (EGTA) in order to suppress outward K^+ currents and to effectively buffer the $[Ca^{2+}]_i$, and to minimize the effects of Ca^{2+}-induced inactivation on the Ca^{2+} current. Intracellular dialyzing pipet solutions contained 5 m*M* Mg^{2+}ATP and 10–50 μM cylcic adenosine monophosphate (cAMP) and 10–20 m*M* hydroxyethyl piperazine ethane-sulphonic acid (HEPES) at pH 7.2–7.3 in addition to Cs^+ and EGTA. Extracellular solutions contained 5–10 μM tetrodotoxin (TTX) to suppress the Na^+ current. K^+ was often omitted from the bathing solution, and Ca^{2+} concentration ranged between 1 and 3 m*M*. Cells were voltage-clamped using the whole cell patch clamp method [2].

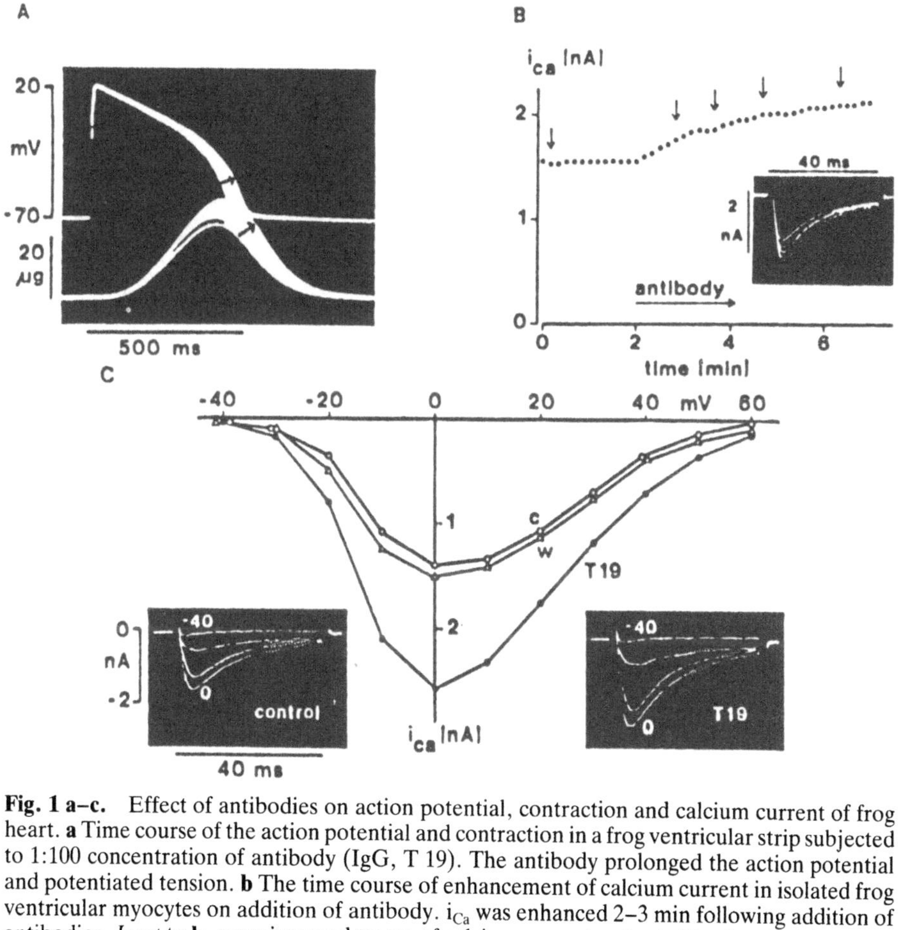

Fig. 1 a–c. Effect of antibodies on action potential, contraction and calcium current of frog heart. **a** Time course of the action potential and contraction in a frog ventricular strip subjected to 1:100 concentration of antibody (IgG, T 19). The antibody prolonged the action potential and potentiated tension. **b** The time course of enhancement of calcium current in isolated frog ventricular myocytes on addition of antibody. i_{Ca} was enhanced 2–3 min following addition of antibodies. *Inset* to **b**, superimposed traces of calcium current activated by depolarizing pulse from −90 to 0 mV at times indicated by the *arrows* on the *graph*. **c** The voltage dependence of i_{Ca} in the presence of antibodies (T 19, 1:100) in control (*c*) and after washout of antibody (*w*). *Inset* to **c** illustrates superimposed traces of i_{Ca}, measured at −40, −30, −20, and 0 mV in control solutions and in the presence of antibody. Patch pipette solution included: 11 m*M* EGTA, 1 m*M* Ca, 10 m*M* HEPES, 25 μM cAMP, 110 m*M* CsCl. $(Ca^{2+})_o$=1 m*M*

In frog ventricular myocytes, i_{Ca} was activated from a holding potential of −90 mV in the presence of TTX. At such holding potentials the myocytes were found to have extremely stable i_{Ca} for at least 30–60 min with little or no "run down". Figure 1b shows the time course of potentiation of i_{Ca} by antibody in a frog ventricular myocyte bathed in Ringer's solution containing 1 m*M* Ca^{2+}. Within 2 min of addition of antibody-containing solution (dilution 1:100), i_{Ca} was significantly increased. Comparison of the antibody at different potentials showed a symmetrical increase of I-V relation (Fig. 1c) suggesting no significant voltage dependence to the enhancing effects of antibody. Removal of antibody was accompanied by reduction of i_{Ca} to control values. The wash-out was much slower than the onset of antibody effect. Frog ventricular myocytes, under these experimental conditions, i.e., high intracellular Ca^{2+}-buffering capacity, tolerated repeated exposures to the antibody showing consistent potentation of i_{Ca}.

Effect of Antibody on i_{Ca} and Developed Tension in Rat and Guinea Pig Myocytes

In order to suppress the Na^+ current completely in rat and guinea pig ventricular myocytes, we used in addition to TTX (10^{-6}–10^{-5} *M*) either holding potentials of

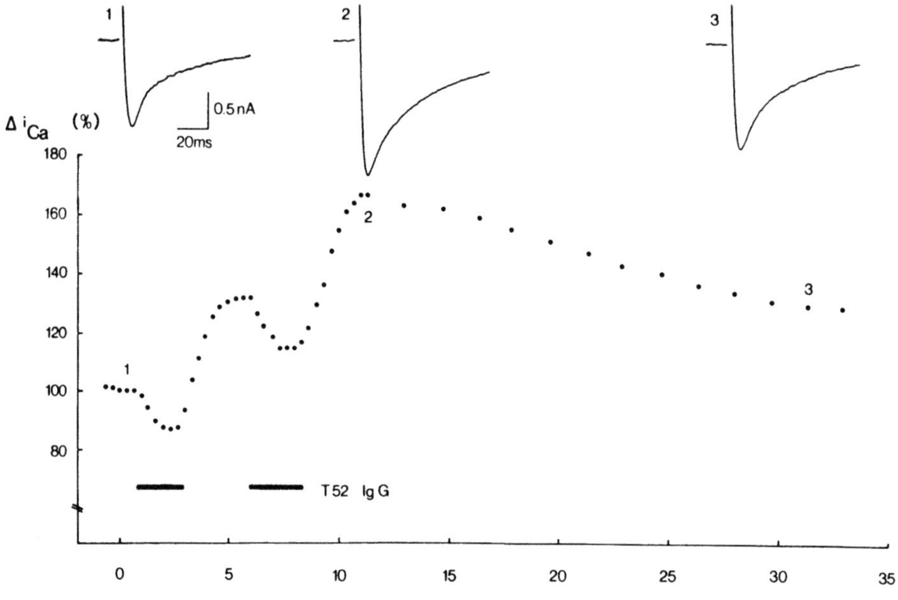

Fig. 2. The time course of enhancement of i_{Ca} by the antibody. The *graph* plots percentage change in the calcium current prior to, during, and following the addition of antibody (1:100) in a guinea pig ventricular myocyte. *Insets* represent the time course of i_{Ca} activated at zero, in control (*1*), during the peak antibody effect (*2*), and following 30 min washout of IgG (T_{52}). The antibody-containing solutions were puffed rapidly on the myocyte. The decrease in i_{Ca} was transient in nature and was not consistently seen. Internal pipette solutions contained 120 m*M* CsCl, 11 m*M* EGTA, 1 m*M* Ca^{2+}, 50 μ*M* cAMP and 5 m*M* Mg-ATP. Extracellular solutions contained 137 m*M* NaCl, 3 μ*M* TTX and 2 m*M* Ca^{2+}. i_{Ca} was measured using a pre-pulse protocol from −80 to −50 mV for 200 ms prior to its activation at 0 mV

-40 mV or conditioning depolarizing pulses to -50 mV for (200–500 ms) prior to activation of i_{Ca}. Addition of antibodies in rat and guinea pig ventricular myocytes consistently increased i_{Ca} and slowed its inactivation. Figure 2 shows the time course of addition of IgG from an immunized rabbit on the i_{Ca} in a guinea pig ventricular myocyte. The antibody-containing solution was puffed directly on the cell, as indicated by the horizontal bar, using a rapid and localized perfusion technique [5]. i_{Ca} increased following exposure of the cell to antibody-containing solutions. Figure 2 also shows that second exposure was accompanied by further increase in i_{Ca}. Not only the peak, but also the maintained component of i_{Ca} appears to be enhanced (trace 2, Fig. 2). As can be seen, the washout of the antibody effect is very slow compared to its onset, so that even after 30 min of continuous washout, i_{Ca} remained significantly potentiated above baseline values (trace 3, Fig. 2). Figure 2 also shows that the onset of antibody effect was accompanied by an initial decrease in i_{Ca}. This effect was highly variable from preparation to preparation and is most likely related to the presence of more than one population of antibodies in the IgG. In support of this idea we found that smaller concentrations of antibody often suppressed i_{Ca}. In frog ventricular myocytes we have not as yet observed the suppressive effects of antibody (Fig. 1), but the effects were seen both in guinea pig and rat ventricular myocytes.

The potentiating effect of antibody was on occasion so large as to make effective voltage control of i_{Ca} difficult at -20 mV. The enhancing effect of antibody on i_{Ca} could be blocked by addition of nifedipine (10^{-6} *M*–10^{-5} *M*). Beta-blockers (propranolol or atenolol, 10^{-6} *M*) did not alter the enhancement of i_{Ca} by the antibody. Figure 3 illustrates the effect of antibody on an isolated ventricular cell in

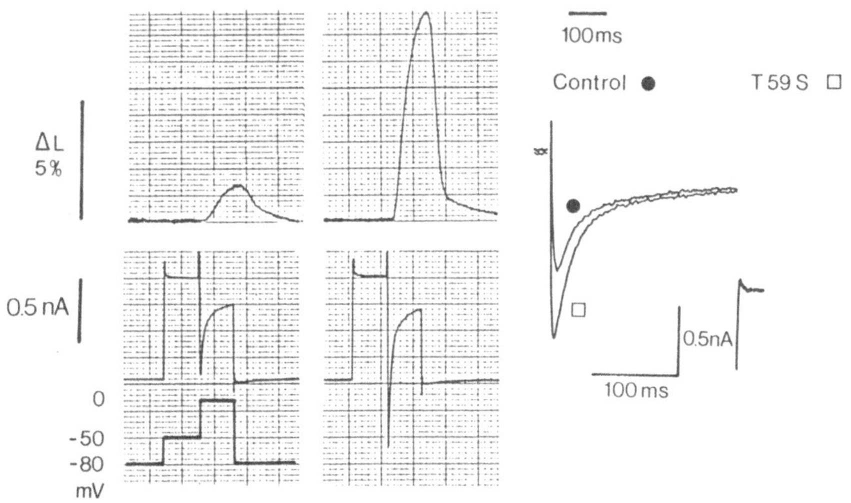

Fig. 3. Effect of antibody on i_{Ca} and contraction in a single guinea pig ventricular mycoyte. Simultaneous recording of i_{Ca} (*lower panels*) and cell shortening (*upper panels*) in control and antibody (T 59,S)-containing solutions. In this example rabbit immunized serum was used. Contraction and i_{Ca} were markedly enhanced. The calcium current in the presence and absence of antibody are shown as superimposed in *traces* on the *far right*. The experimental double pulse procedure is shown *below*. Other experimental conditions were similar to those for Fig. 2

which i_{Ca} and contraction were simultaneously measured. Contraction was monitored by measuring cell shortening using a 256-photodiode array. This cell was dialyzed with only 100 μM EGTA in order to allow it to contract (generally about 1.0 mM EGTA was necessary to completely suppress contraction). Note that both i_{Ca} and contraction are strongly enhanced in the presence of antibody. Thus, antibodies against ADP/ATP carrier are equally effective in enhancing i_{Ca} in amphibian and mammalian myocardium. The enhancement of Ca^{2+} current is reflected in marked potentiation of tension. Since this effect is not mediated through the β-adrenergic receptor and the effect was blocked by nifedipine, we conclude that the antibody against ADP/ATP carrier may cross-react with Ca^{2+} channel, possibly the DHP receptor. This finding is consistent with the observation that exposure of the isolated rat myocytes for 2–3 h to high concentrations of antibody leads to cell deterioration and death which can be prevented with pretreatment of organic Ca^{2+} antagonists.

Effect of Autoantibody from Myocarditis on i_{Ca}

Since the antibodies against ADP/ATP carrier are also found in patients with myocarditis and dilated cardiomyopathies [7], we examined possible effects of sera from such patients on the Ca current in isolated mammalian myocytes. Exposure of whole cell-clamped guinea pig and rat myocytes to sera of such patients (1:100 dilution) caused a marked potentiation of i_{Ca} in two out of five cases. The onset of antibody effect was about ten times faster than its washout. Ca current was potentiated by about 40%, and its inactivation was markedly slowed. Figure 4 shows the voltage dependence of i_{Ca} and the enhancing effect of autoantibodies. Note that the antibody enhanced i_{Ca} at all potentials tested. We also tested the effect of normal human and rabbit sera and found little, or often an inhibitory, effect on Ca current of the isolated rat ventricular myocytes. Thus, we concluded that human sera antibodies may cross-react with the sarcolemmal Ca channel causing an enhancement of the current which may eventually lead to Ca overload and cell toxicity.

Summary

Antibodies against ADP/ATP carrier bind specifically to the sarcolemma of the isolated cardiac myocytes and increase Ca influx by enhancing the Ca current. The enhancement of i_{Ca} results in prolongation of the action potential and potentiation of tension. Autoantibodies from cardiomyopathy also appear to enhance the Ca current but the effect is often variable. The antibody and the autoantibody effects are not mediated through the β-adrenergic receptors. The rapid onset of antibody action and its slow washout suggest tight binding to a receptor site closely associated with the Ca channel. The variability of antibody effect in enhancing the Ca channel may be related in part to the presence of more than one population of antibodies, some of which may inhibit the channel current.

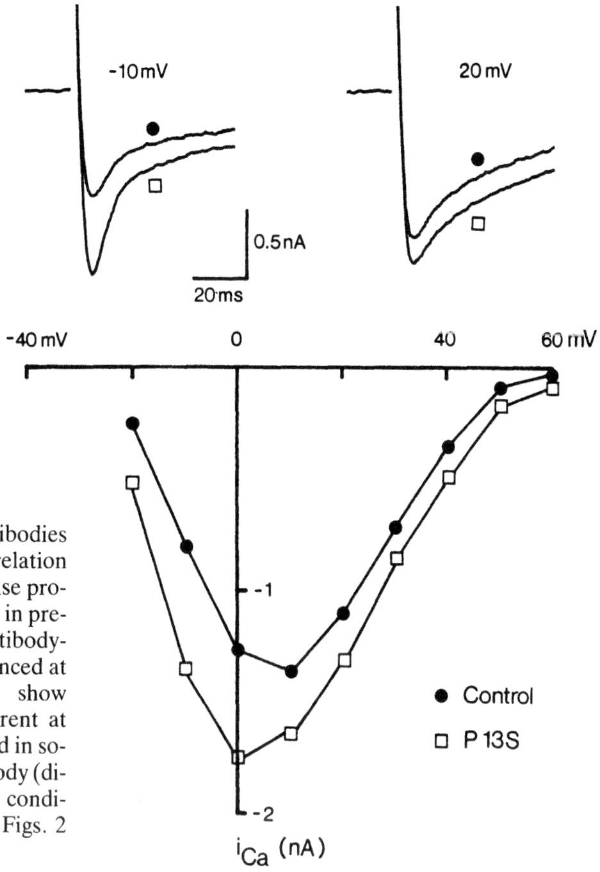

Fig. 4. The effect of autoantibodies on calcium current. The I–V relation for i_{Ca} measured with a two pulse procedure (see protocol for Fig. 3) in presence and absence of autoantibody-containing serum. i_{Ca} was enhanced at all potentials tested. *Insets* show sample tracing of calcium current at −10 and +20 mV in control and in solution containing the autoantibody (dilution 1:100). Experimental conditions were similar to those for Figs. 2 and 3

References

1. Bolte HD, Schultheiss HP (1978) Immunological results in myocardial diseases. Postgrad Med J 54: 500
2. Hamill OP, Marty A, Neher E, Sakmann B, Sigworth F (1981) Improved patch clamp technique for high resolution current recording from cells and cell free membrane patches. Pflügers Arch 391: 85
3. Kereiakes DJ, Parmley WW (1983) Myocarditis and cardiomyopathy. Am Heart J 108: 197
4. Kobayashi A, Yamashita T, Kaneko M, Nishiyama T, Hayashi H, Yamazaki N (1987) Effects of verapamil on experimental cardiomyopathy in the Bio 14.6 Syrian hamster. J Am Coll Cardiol 10: 1128
5. Konnerth A, Lux HD, Morad M (1987) Proton-induced transformation of calcium channel in chick dorsal root ganglion cells. J Physiol 386: 603
6. Mitra R, Morad M (1985) A uniform enzymatic method for dissociation of myocytes from hearts and stomachs of vertebrates. Am J Physiol 249: H1056
7. Schultheiss HP, Bolte HD (1985) The immunological characterization of the adenine nucleotide translocator as an organ specific autoantigen in dilated cardiomyopathy. J Mol Cell Cardiol 17: 603

8. Schultheiss HP, Klingenberg M (1984) Immunological characterization of the adenine nucleotide translocator: organ and conformation specificity. Eur J Biochem 143: 599
8a. Morad M, Davies NW, Ulrich G, Schultheiss (1988) Antibodies against ADP/ATP carrier enhance the Ca^{2+} current in isolated cardiomyocytes. Am J Physiol (Circulation) (in press)
9. Rossner KL, Sachs HG (1978) Electrophysiological study of Syrian hamster hereditary cardiomyopathy. Cardiovasc Res 12: 436
10. Wagner JA, Reynolds IJ, Weissman HF, Dudeck P, Weisfeldt ML, Snyder SH (1986) Calcium antagonist receptors in cardiomyopathic hamster: selective increases in heart, muscle, brain. Science 232: 515
11. Weisman HF, Weisfeldt ML (1987) Toward an understanding of the molecular basis of cardiomyopathies. J Am Coll Cardiol 10: 1135
12. Wrogemann K, Nylen EG (1978) Mitochondrial calcium overloading in cardiomyopathic hamsters. J Mol Cell Cardiol 10: 185

Antibodies Against Adenosine Di-/Triphosphate Carrier Alter Myocardial Function by Disturbing Cellular Energy Metabolism

H. P. SCHULTHEISS[1], U. KÜHL[1], R. SCHAUER[1], K. SCHULZE[1], B. KEMKES[2], and B. F. BECKER[3]

[1] Departments of Internal Medicine
[2] Department of Cardiac Surgery, Klinikum Großhadern, University of Munich, Marchioninistr. 15, D-8000 Munich 70, FRG
[3] Department of Physiology, University of Munich, D-8000 Munich 2, FRG

Introduction

Current concepts of autoimmunity emphasize the role of antibodies in the development and maintenance of most autoimmune disorders. The problem concerning the role of autoantibodies in mediating disease is to show that a given autoantibody is an essential component of a pathological mechanism which leads to tissue damage and organ dysfunction. Besides the understanding of the pathogenesis, the precise characterization of the antibodies involved in an autoimmune process may have diagnostic and prognostic value.

Recently, we showed that the sera of patients with myocarditis or dilated cardiomyopathy contain circulating antibodies directed against the adenosine di-/triphosphate (ADP/ATP) carrier [13, 14, 19]. The ADP/ATP carrier is an intrinsic, hydrophobic protein located in the inner mitochondrial membrane (Fig. 1) [7]. Since this membrane is a priori impermeable to hydrophilic metabolites, the transfer of ATP to the cytosol with its energy-consuming processes and the return of ADP to the inner mitochondrial space for regeneration by oxidative phosphorylation require a particular transport catalysis. The ADP-ATP shuttle, being the only active nucleotide transport system in mitochondria, is highly specific and corresponds exactly to the requirements of ATP production in the eukaryotic aerobic cell [9]. Immunochemically an organ specificity of the ADP/ATP carrier isolated from heart, kidney and liver was demonstrated although a partial cross-reactivity among these three proteins was seen [15, 16]. The organ specificity was observed both with the solubilized and with the membrane-bound translocator protein indicating organ-specific determinants on exposed regions of the carrier [17]. An organ-specific inhibition of the nucleotide transport in heart mitochondria by the heart carboxyatractylate (CAT) protein antiserum leads to the conclusion that the organ specificity is at least partially conditioned by the binding site for the substrate and/or the closely linked gate of the carrier protein [18]. Apart from the organ specificity, the results also demonstrate a specificity of the antibodies for the translocational conformations of the carrier: the "c" conformation stabilized in the CAT protein complex and the "m" conformation present in the bongkrekat-protein complex [13].

H.-P. Schultheiß (Ed.)
New Concepts in Viral Heart Disease

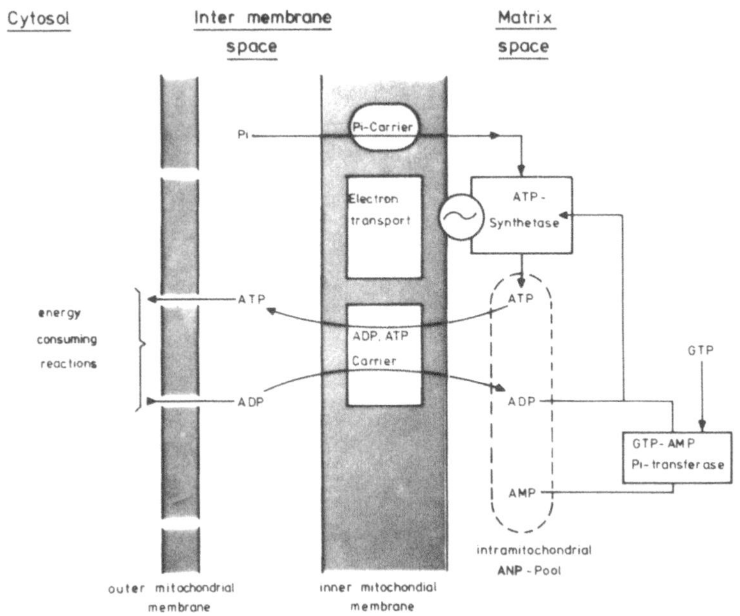

Fig. 1. The major pathways of nucleotide-linked phosphate transfer in the extramitochondrial space. Coupling of the extra- and intramitochondrial nucleotide pools by the ADP/ATP carriers

In this paper we discuss the immunochemical characteristics of the autoantibodies found in patients with myocarditis. To elucidate the role of the autoantibodies against the ADP/ATP carrier in the pathogenesis of heart muscle diseases, we studied the hearts of guinea pigs immunized with the isolated carrier protein. The subcellular distribution of the adenine nucleotides was measured as an inhibition of the carrier should change the intracellular energy balance by diminishing the ATP efflux from the mitochondria into the cytosol. Based on these results we shall discuss our working hypothesis that autoimmunity to intracellular proteins, such as the ADP/ATP carrier, may contribute to the pathophysiology of viral heart disease, a suspected autoimmune disease [20].

Immunochemical Characterization of the Autoantibodies Against the ADP/ATP Carrier

Sera of 66 patients with clinically suspected myocarditis were tested. The clinical diagnosis of myocarditis was established on the basis of anamnestic data, especially with regard to a common cold infection, an accompanying change in viral neutralization titre on follow-up and the typical clinical features like abnormalities in ECG, arrhythmias of recent onset, acute haemodyamic deterioration or elevated serum creatine kinase level. Diagnosis was principally proven by cardiac catheterization data and morphological analysis of endomyocardial biopsies. The biopsy specimens were classified into three major groups according to Olsen [11]: acute myocarditis, healing myocarditis and healed myocarditis. Sera from patients with proven coronary heart disease and from healthy blood donors (controls) were used as controls.

When the sera of patients with clinically suspected myocarditis were tested in the enzyme-linked immunosorbent assay (ELISA) against the ADP/ATP carrier form heart, a significant antibody titre was seen in 20 of 22 patients with acute/healing myocarditis and in 26 of 44 patients with the histological diagnosis "healed myocarditis". In comparison to the controls, the sera from patients with coronary heart disease did not contain autoantibodies against the ADP/ATP carrier (Fig. 2). If the positive sera from patients with suspected myocarditis reacting with the heart-protein were tested against the ADP/ATP carrier from kidney or liver, the binding to the kidney and liver protein was significantly lower than the binding to the heart protein (data not shown) [13, 14]. To exclude the possibility that the suspected organ specificity of the antibodies is only the result of a lower affinity of the antibodies to the antigenic determinants of the liver or kidney protein, immunoadsorption studies were performed. For this purpose, the interaction of the

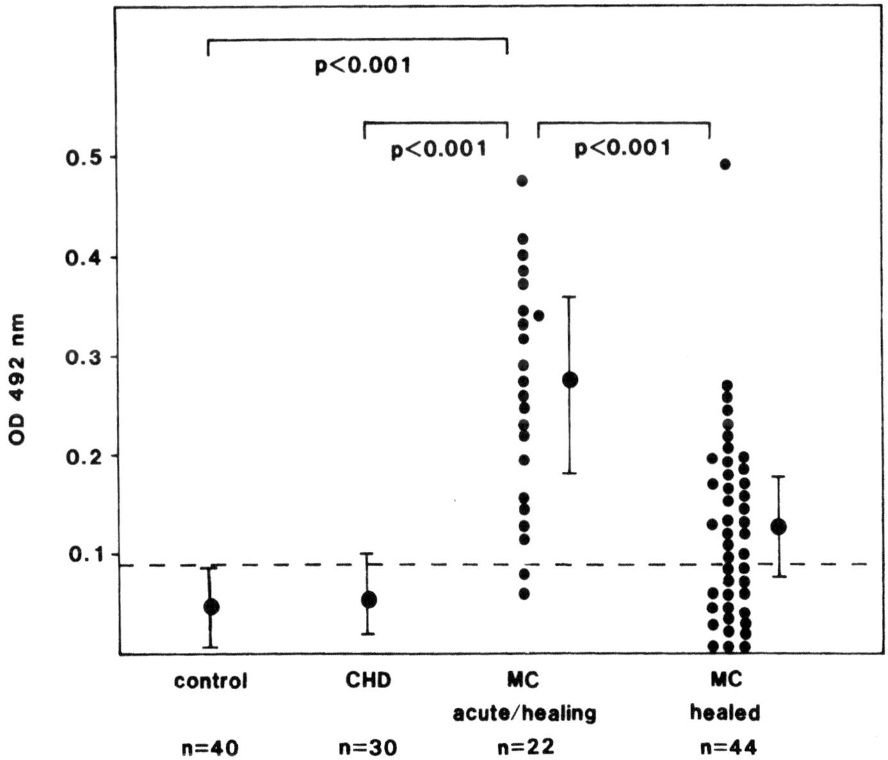

Fig. 2. The binding of autoantibodies to the ADP/ATP carrier from hearts in ELISA from patients with myocarditis (acutehealing, healed), coronary heart disease *(CHD)* and healthy blood donors (controls). For methodological details see legend to Fig. 3

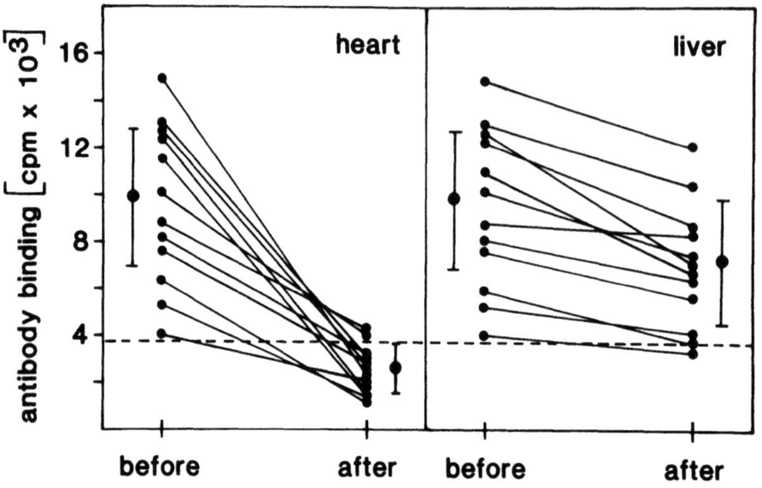

Fig. 3. The binding of antibodies to the ADP/ATP carrier from heart *(before)* and the residual anti-heart protein activity of the antiserum after immunoadsorption on the isolated carrier proteins from heart and liver *(after)*. *Antiserum (90* μl; dilution 1:20) were incubated in microtitre wells, pre-coated with the isolated carrier protein from heart overnight at 4 °C. After incubation with ^{125}I-protein A (80 μl; 100000 cpm/well) the wells were counted in a gamma spectrometer. All samples were tested in duplicates. Values for radioactivity are experimental results minus background (cpm, determined in the presence of equal amounts of non-immune serum). Saturating amounts of the isolated carboxyatractylate protein complexes from heart and liver were incubated with a fixed amount of isolated IgG of antisera. After "precipitation" of the antigen-antibody complexes with protein A-bearing *Staphylococcus aureus* (Cowan I), the residual antibody activity of the supernatant was determined in the immunoradiometric assay with carboxyatractylate-protein heart pre-coated wells

autoantibodies with the isolated ADP/ATP carrier protein from heart and liver or kidney was examined. The maximal amount of an antibody bound by a given immunoadsorbant (plateau binding value) was determined by titrating a fixed amount of antiserum with increasing amounts of antigen [13]. As shown representatively for a group of sera, the ADP/ATP carrier from heart adsorbed most of the autoantibodies, while the residual activity of the antiserum after immunoadsorption on the liver protein was significantly higher ($P < 0.001$) (Fig. 3). These results were also confirmed by immunoadsorption studies on intact mitochondria from heart and liver (data not shown).

To test a possible inhibitory activity of the autoantibodies on the nucleotide exchange rate of mitochondria, we measured the nucleotide transport in vitro. After preincubation of the mitochondria with isolated IgG from myocarditis sera, the exchange rate was determined by the inhibitor-stop method combined with back-exchange [12]. None of the control sera produced a significant inhibition of nucleotide transport. However, about 60% of the positive sera from patients with acute or healing myocarditis immunoinactivated the ADP/ATP carrier from heart. Figure 4 shows a typical example of an immunoinactivation of nucleotide transport by antibodies of a patient with acute myocarditis. To exclude artefactious causes of this inhibitory effect, the tests were performed with a purified IgG

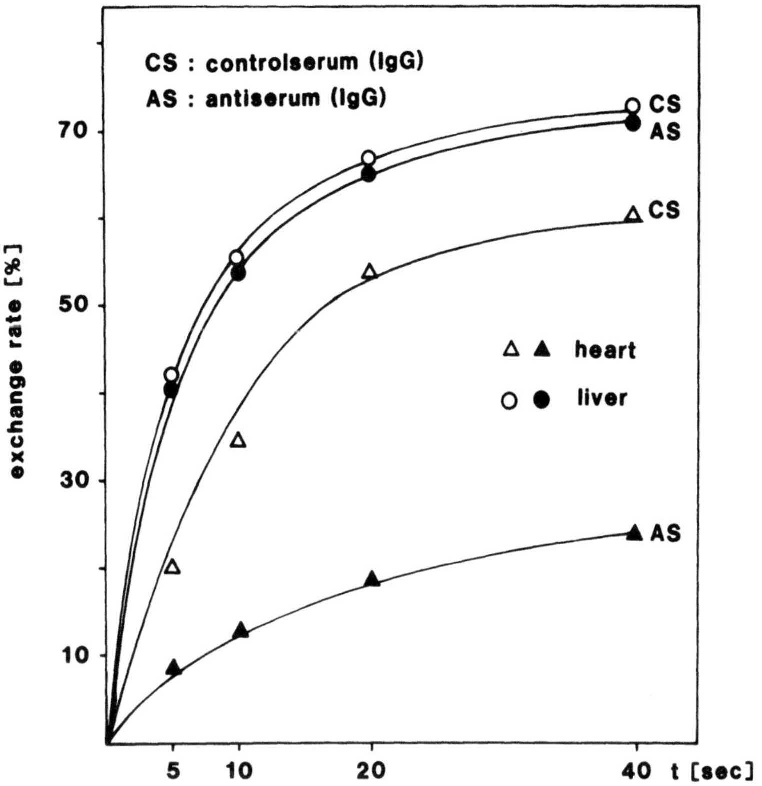

Fig. 4. Influence on the antisera on the nucleotide transport from heart *(triangles)* and liver *(circles)* mitochondria. Isolated and preloaded (^{14}C-ADP) mitochondria were incubated with equilibrated antiserum *(AS;* isolated IgG) for 30 min at 4 °C. Afterwards the ADP/ATP exchange rate was determined by the inhibitor stop method combined with the back-exchange (for methodological details see [12, 15]). Exchange time was 5 s, 10 s, 20 s and 40 s at 4 °C for heart mitochondria and at 8 °C for liver mitochondria. *CS,* control serum (IgG)

fraction. Additionally, in control experiments the inhibition of nucleotide exchange could be abolished by preprecipitation of the antibodies with sepharose protein A or by preadsorption of the antibodies on the isolated heart protein. However, preadsorption of the antibodies on isolated mitochondria from liver did not eliminate the functional activity of the sera, indicating an organ specificity of the immunoinactivating antibodies. This suggestion was supported by results observed in exchange experiments with liver mitochondria. None of the sera which inhibited the nucleotide transport of heart mitochondria inhibited the transport of liver mitochondria (Fig. 4). Again, these results demonstrate the organ specificity not only of the auto-antibodies, but also of the ADP/ATP carrier itself.

It had been postulated that there is a single binding site for the natural substrate (ATP or ADP) and the specific inhibitor CAT [8]. If that is true, the binding of the ligand should be blocked by an immunoinactivating antibody. This was

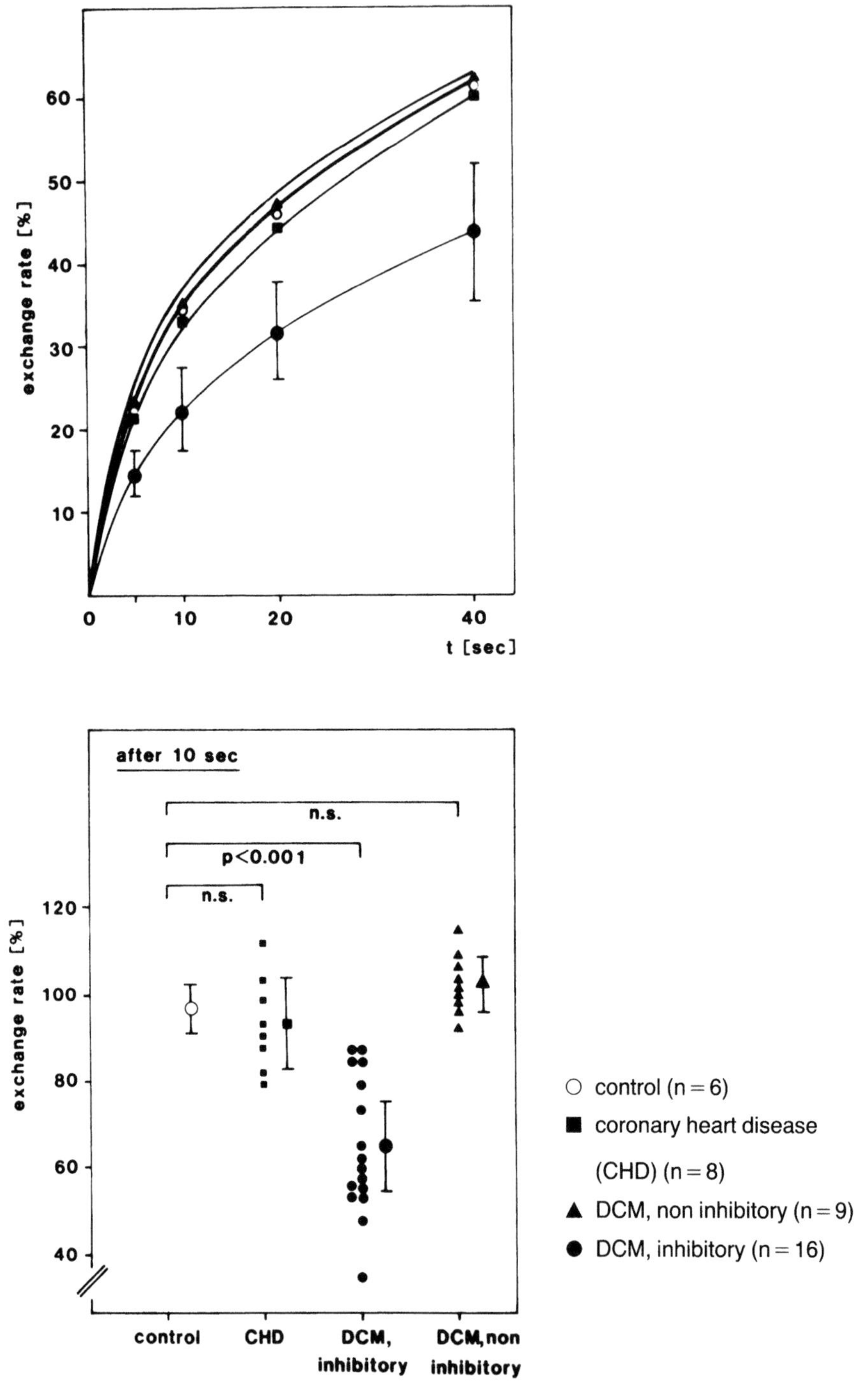

Fig. 5. Nucleotide exchange rates in mitochondria from explanted hearts from patients with coronary heart disease *(CHD; squares;* I = 8), dilated cardiomyopathy *(DCM)* and six controls *(open circles)*. For methodological details see legend to Fig. 4. *Triangles,* DCM, non-inhibitory *(n* = 9); *solid circles,* DCM, inhibitory *(n* = 16)

demonstrated by preincubation of mitochondria with anti-ADP/ATP carrier antibodies (isolated IgG) which decreased the [^{3}H]CAT binding significantly, while the ligand binding was not affected by control IgG. By the fact that the ligand binding on liver or kidney mitochondria was not influenced by the same antibody, the organ specificity of the antibodies could again be shown.

Suggesting a pathogenetic role of the antibodies, one has to ask whether these immunoinactivation antibodies can act as antagonists not only in vitro, but also in vivo, with the consequence of a decreased nucleotide transport across the inner mitochondrial membrane. To answer this question we isolated mitochondria from explanted hearts of patients with dilated cardiomyopathy undergoing heart transplantation. After the isolation of mitochondria, the nucleotide exchange rate was measured without adding antibodies to the mitochondria. In comparison to controls or to coronary heart disease, the nucleotide exchange rate was significantly lowered in 16 out of 25 patients (Fig. 5). These data indicate that the total transport capacity for ATP might be decreased in these patients, implicating a diminished energy supply to the cytosol. This reduced carrier protein activity can depend on several factors. One possibility is that the carrier concentration per milligram of mitochondrial protein is diminished. To answer this question we determined the carrier concentration in the same mitochondrial fraction which was used for the exchange rate measurements by the immunodot technique, using affinity-purified antibodies against the ADP/ATP carrier.

As shown in Fig. 6 the immunochemically determined carrier content in the hearts from patients with coronary heart disease or mitral valve replacement was within the same range as the controls, while the carrier content in dilated cardiomyopathy (DCM) hearts was significantly increased. These data exclude the possibility that the reduced transport capacity for nucleotides is caused by a lowered carrier protein concentraition per milligram of mitochondrial protein, for example, by a diminished de-novo synthesis of the carrier protein or an increased degradation, as shown for several proteins (receptors) in other autoimmune processes. On the contrary, the increased concentration of the carrier protein can be taken as an indication for an upregulation of the ADP/ATP carrier to compensate the diminished transport capacity by an increased protein density. This raises the question of whether the decreased transport capacity is caused by an antibody-mediated inhibition of the ADP/ATP carrier.

Determination of the Subcellular Distribution of Adenine-Nucleotides and the Haemodynamic Function after Immunization with the ADP/ATP carrier

Autoantibodies to cell membrane receptors have been docomented in a number of disease states in man, and there is good evidence for a pathogenic role for at least some of these autoantibodies. However, it is still not clear whether autoantibodies against intracellular antigens may also have a pathophysiological function. To answer the question of whether anti-ADP/ATP carrier antibodies act only in vitro or also influence the adenine nucleotide transport in vivo we established an animal model. Since there are no methods available to determine the function of the ADP/ATP carrier in the living organ, we measured the cytosolic-mitochon-

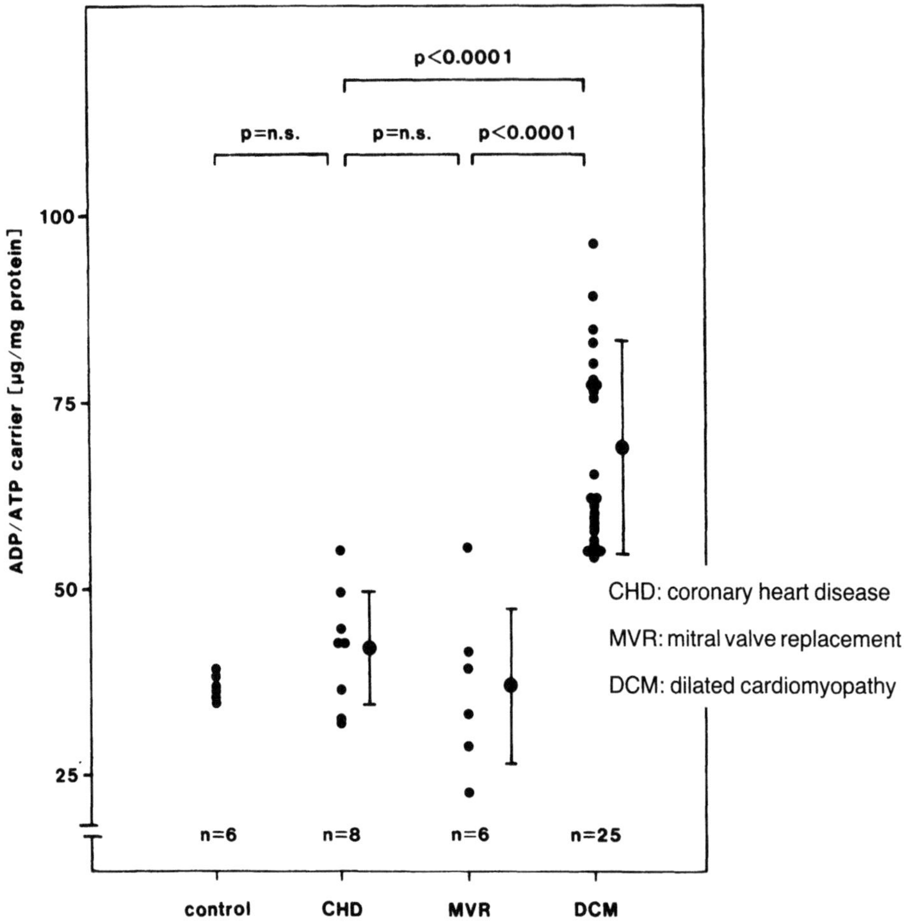

Fig. 6. Quantification of the ADP/ATP carrier content in human mitochondria isolated from explanted hearts from patients suffering from coronary heart disease *(CHD)* or dilated cardiomyopthy *(DCM)* and undergoing heart transplantation. In six patients mitochondria were isolated from the papillary muscle from patients with mitral valve disease undergoing mitral valve replacement *(MVR)*

drial difference of the phosphorylation potential of ATP ($\Delta G_{cyt\text{-}mit}$) to prove whether the transport activity of the ADP/ATP carrier is influenced in the immunized animals. As the energization of the mitochondrial membrane, which is expressed by $\Delta G_{cyt\text{-}mit}$, modulates the ADP/ATP carrier activity so as to operate in the manner required for the energy need of the cell, a decreased phosphorylation potential difference of ATP can be taken as an indicator for a reduced transport capacity of the carrier [9]. Guinea pigs were immunized with the isolated ADP/ATP carrier from heart. After five booster injections at monthly intervals, the sera were tested for their anti-ADP/ATP carrier activity. The antibodies were characterized by the ELISA test, Western blotting and by measuring the nucleotide transport.

In those animals which produced significant titres of immunoinactivating antibodies, an intracellular action of the antibodies in vivo was demonstrated by measurements of the cytosolic and mitochondrial adenine nucleotide concentrations. These were evaluated in haemoglobin-free, perfused working guinea pig hearts using density gradient centrifugation of the myocardial homogenates in non-aqueous media [22, 23]. Perfusion experiments were terminated by stop-freezing the myocardium between aluminium blocks precooled in liquid nitrogen. Since at the time of freeze-clamping the myocardium the heart preparations were in a steady state with respect to haemodynamic function and oxidative metabolism, the obtained compartmentation of adenine nucleotides appears to reflect an equilibrium between the disposal of ATP by extra-mitochondrial sites and the rate of oxidative phosphorylation of ADP in the mitochondria.

The hearts of five control animals exhibited a normal intracellular distribution of ATP (Table 1). These values are in good agreement with those reported by others [4, 23]. In the hearts of guinea pigs which showed immunoinactivating antibodies against the ADP/ATP carrier, the concentration of ATP was slightly lowered in the cytosol but tremendously increased in the mitochondria (Table 1). The ATP/ADP ratio as well as the phosphorylation state of ATP — expressed as the ratio (ATP)/(ADP) x (Pi) — were significantly lower in the cytosol, while in the mitochondria a substantial increase was observed. Consequently, the difference between the cytosolic and mitochondrial phosphorylation potentials was only 0.7 kJ/mol in the hearts of the immunized animals while it was 4.6 kJ/mol in the controls (Fig. 7). These results show a significant decrease of myocardial high-energy phosphates in the immunized guinea pig hearts in comparison to the control hearts. Since it was shown that the ATP/ADP ratio is a suitable parameter for the evaluation of the energy state of the cell or the balance between mitochondrial phosphorylation and ATP consumption in the cytosol, our findings indicate an inhibition of the nucleotide transport in the hearts of the immunized animals. This becomes even more evident by the drastically lowered phosphorylation potential

Table 1. Subcellular distribution of myocardial high-energy phosphates in isolated perfused guinea pig hearts. Ratios for free cytosolic ATP/ADP were calculated from the mass action ratio of the creatine kinase reaction [11, 23]. Values are mean ±

	Controls (n = 5)		Immunized animals (n = 6)	
	cyt	mit	cyt	mit
ATP (m*M*)	13.2 ± 2.0	8.1 ± 2.3	11.3 ± 2.9	18.3 ± 5.9^{b}
ADP (m*M*)	1.9 ± 0.4	1.6 ± 0.3	2.5 ± 0.5^{a}	1.1 ± 0.2^{a}
ATP/ADP$_{(free)}$ (m*M*)	61.8 ± 7.9	5.1 ± 0.5	50.2 ± 8.8^{b}	17.1 ± 4.9^{b}
(ATP) / (ADP) x (Pi) (m*M*)	5487 ± 644	926 ± 215	4105 ± 624	3349 ± 1445^{b}
△ G (kJ/mol ATP)	22.2 ± 0.3	17.6 ± 0.6	21.4 ± 0.4^{b}	20.8 ± 1.2^{b}

^{a} $P < 0.05$
^{b} $P < 0.005$
cyt, cytosol; mit, mitochondria; △ G, phosphorylation potential of ATP

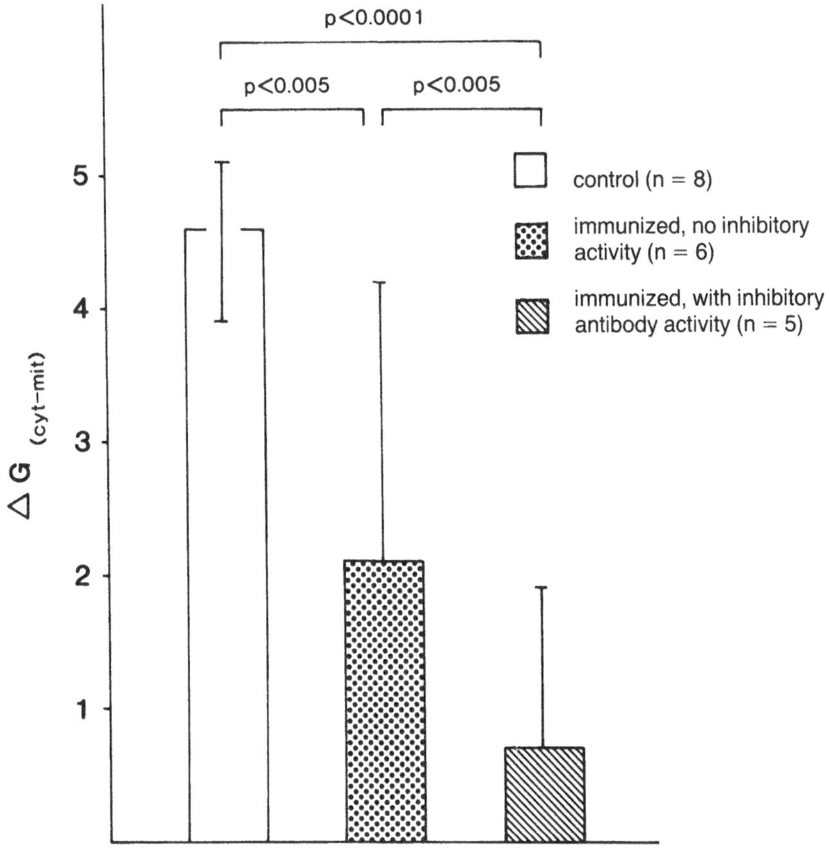

Fig. 7. Calculation of the difference between the cytosolic and mitochondrial phosphorylation potentials in perfused working guinea pig hearts in control animals *(open bars; n* = 5) and in animals immunized with the ADP/ATP carrier. Fractionation of the tissue into cytosol and mitochondria was accomplished utilizing density gradient centrifugation in non-aqueous media according to Schulze et al. [22]. Lyophilized tissue was homogenized in heptane/carbon tetrachloride (d = 1.23 g/ml) and subsequently fractioned in a heptane/carbon tetrachloride density gradient (d = 1.29–1.38 g/ml). Centrifugation (27000 x g, 4 h) yielded eight fractions, each containing different portions of cytosolic and mitochondrial protein. In each fraction the cytosolic and mitochondrial marker enzymes, the high-energy phosphates as well as the protein contents were determined. Subcellular concentrations were obtained assuming 3.8 μl water per milligram of cytosolic protein and 1.8 μl water per milligram of protein in heart mitochondria, respectively. The equilibrium constant of te creatine kinase reaction was $K_{CPK} = 2.04 \times 10^{-9}$. *Dotted bar,* immunized, no inhibitory antibody activity *(n* = 4); *hatched bar,* immunized, with inhibitory antibody activity *(n* = 6)

potential difference ($\Delta G_{cyt\text{-}mit}$) between the cytosolic and mitochondrial ATP and the parallel increase of the mitochondrial ATP/ADP ratio. Similar results were described when carboxyatractyloside was added to the perfused liver [24]. This clearly demonstrates that the diminished difference of the cytosolic and mitochondrial ADP/ATP ratio is indeed caused by the impaired function of the ADP/ATP carrier.

The decreased cytosolic ATP/ADP ratio might hereby stimulate the remaining active ADP/ATP carrier molecules to partly compensate the decreased transport capacity for nucleotides by an increased net transfer of ATP to the cytosol through still normally acting ADP/ATP carrier proteins.

These data raise the question of whether the disturbance of cellular energy balance by diminishing the ATP efflux from mitochondria into the cytosol might alter myocardial function. To prove this hypothesis, we measured the haemodynamic data from immunized guinea pig hearts. The hearts were isolated and perfused as working heart preparations as described previously [1, 3]. The preload of 12 cm H_2O and the mean developed afterload determined the pressure-volume-work of the left ventricle. External heart work was calculated as the sum of pressure-volume-work and acceleration work (½ x ejected volume x mean velocity of flow^2) during ejection. As shown in Fig. 8, exteral heart work was highly reduced in the immunized animals in comparison to the controls. Myocardial oxygen extraction, calculated as the difference of oxygen saturation from the affluent perfusate and the coronary effluent, was found to be lowered (Fig. 9), while the amount of myocardial lactate production, measured as lactate release in the coronary effluent, was found significantly increased in the immunized animals (Fig. 10).

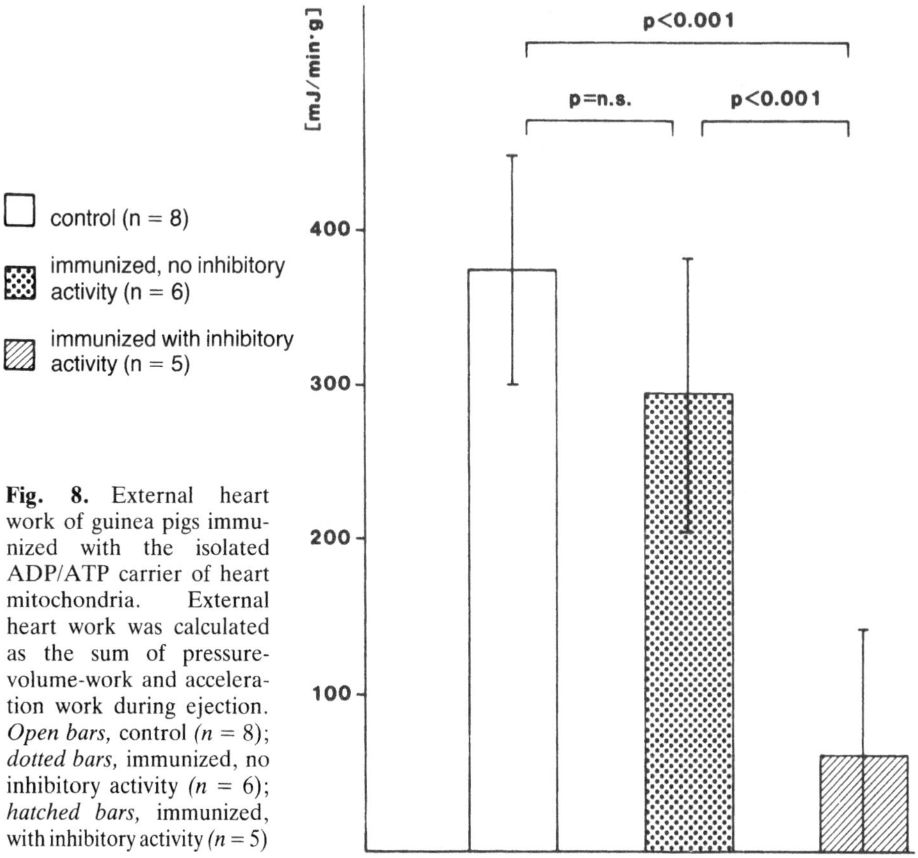

Fig. 8. External heart work of guinea pigs immunized with the isolated ADP/ATP carrier of heart mitochondria. External heart work was calculated as the sum of pressure-volume-work and acceleration work during ejection. *Open bars,* control (*n* = 8); *dotted bars,* immunized, no inhibitory activity (*n* = 6); *hatched bars,* immunized, with inhibitory activity (*n* = 5)

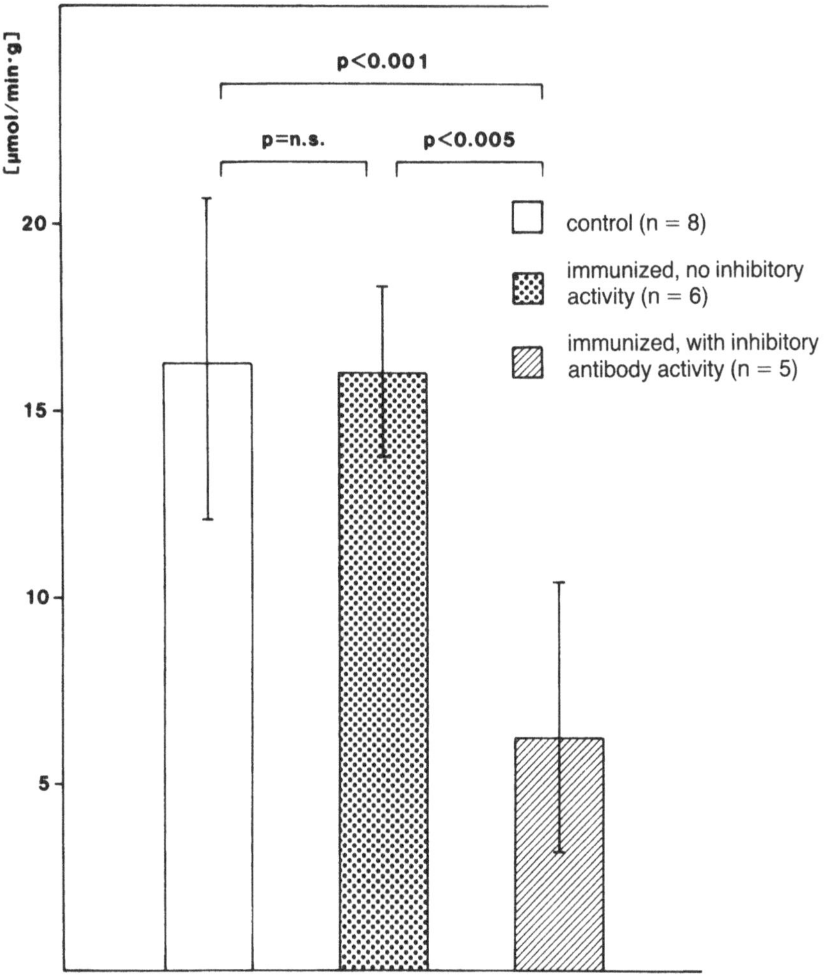

Fig. 9. Myocardial oxygen consumption of isolated hearts from guinea pigs immunized with the ADP/ATP carrier. It was derived from the difference of O_2 tension between aortal perfusate and coronary effluent measured with two Teflon-shielded platinum-silver electrodes (Bachhofer, Reutlingen, FRG)
Open bars, control $(n = 8)$; *dotted bars,* immunized, no inhibitory activity $(n = 6)$; *hatched bars,* immunized, with inhibitory activity $(n = 5)$

Conclusions

After the identification of the ADP/ATP carrier as an autoantigen in myocarditis and dilated cardiomyopathy [13, 14, 19] the precise immunochemical characterization of the antibodies will provide important clues regarding the cause of this suspected autoimmune disease [6]. It has been suggested for a long time that immunopathological mechanisms may contribute, at least in part, to the pathogenesis of viral myocarditis and dilated cardiomyopathy, a possible late state of viral heart disease [6]. Since a viral persistance in the myocardium of patients with

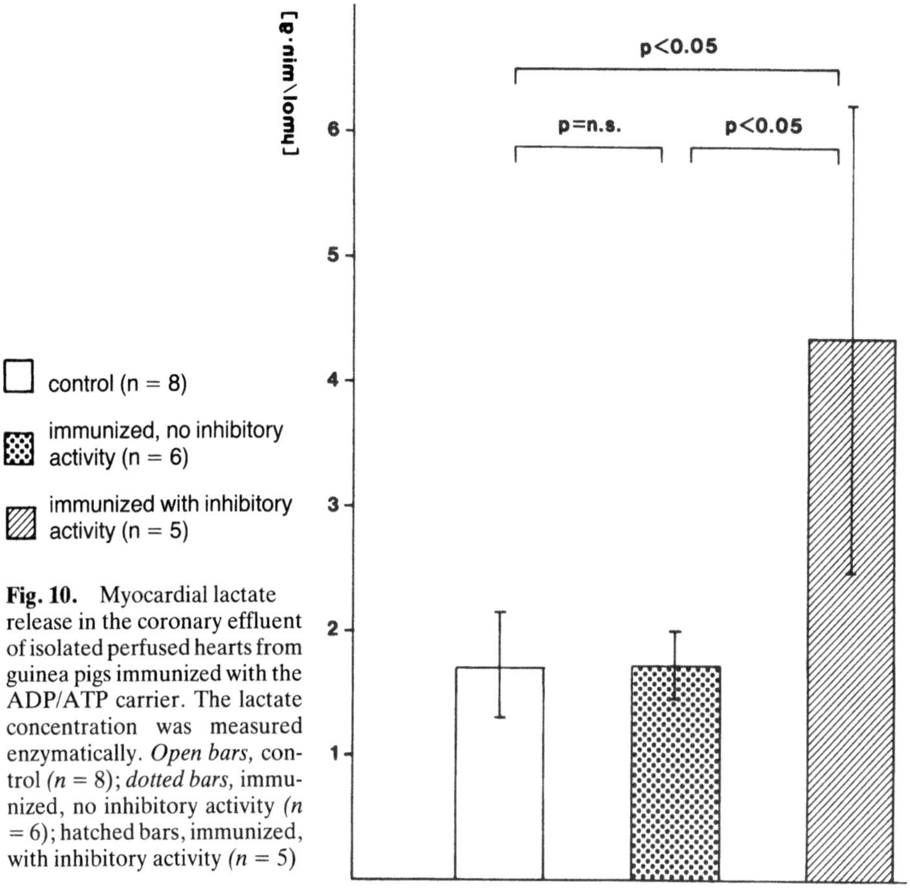

Fig. 10. Myocardial lactate release in the coronary effluent of isolated perfused hearts from guinea pigs immunized with the ADP/ATP carrier. The lactate concentration was measured enzymatically. *Open bars,* control *(n* = 8); *dotted bars,* immunized, no inhibitory activity *(n* = 6); hatched bars, immunized, with inhibitory activity *(n* = 5)

myocarditis and dilated cardiomyopathy has recently been demonstrated [2, 5] the interest is now focused on the link between the persistent viral infection and the alteration of the haemodynamic function in these patients. One possibility, which seems to be quite likely, is that the persistent viral infection triggers the autoimmune reaction. It has been shown recently that mice infected with Coxsackie B3 virus developed an autoimmune reaction against a 30 kilodalton protein, among other proteins [10, 25] which could be the ADP/ATP carrier (K. Beisel, Omaha, personal communication).

The results shown here demonstrate in the sera of patients with myocarditis the existence of organ-specific autoantibodies against the ADP/ATP carrier, a biochemically, immunochemically and functionally well-characterized constituent of the inner mitochondrial membrane [7, 9, 13]. The principal ability of the antibodies to inhibit the nucleotide transport activity was shown by in vitro experiments. It could be shown that this inhibition is caused by an organ-specific inhibition of the substrate/ligand-binding site of the carrier protein through the antibodies [20]. Beside the principal importance of the organspecificity for a better understanding of basic regulatory mechanisms of the energy metabolism in different organs, this aspect has become particular significant for the insight of the ongoing autoimmune process:

1. In relation to the pathogenetic aetiology of the disease one can assume that the cause of the production of these highly specific autoantibodies is limited to the local milieu.
2. The specificity of the autoantibodies and their principal ability to inhibit the nucleotide exchange activity suggests that they might be mediators of the autoimmune disease at the molecular level. This was confirmed by regression analysis which revealed a significant relationship between the immuno-inactiviting autoantibodies and the clinical class and prognosis of the disease [17].

The exact role these autoantibodies play in the pathogenesis of postinfectious heart muscle disease is difficult to evaluate. In principal, there is good evidence for the suspicion that autoantibodies to cell membrane receptors are of importance for several autoimmune diseases. However, up to now it is not clear whether antibodies against intracellular antigens may be of pathophysiological significance. For the first time, the experimental data presented in this paper give evidence that these antibodies can influence cellular function. After immunization of guinea pigs with the isolated ADP/ATP carrier protein, a diminished activity of the nucleotide transport was seen [22]. The decreased transport activity is followed by an increase of the mitochondrial ADP/ATP ratio and a parallel lowering of the cytosolic ADP/ATP ratio. Based on the functional and regulatory characteristics of the ADP/ATP carrier [9] this indicates a stimulation of the remaining active carrier proteins to increase the net transfer of ATP to the cytosol. However, the maximal transport capacity of the remaining active carrier molecules is not sufficient for the energy need of the cytosol. Thus the antibody-mediated inhibition of the ADP/ATP carrier seems to upset the balance between energy delivery and demand in the myocardial cell.

If the energy metabolism of the cell is altered in such a way as to limit the supply of ATP to the organellae responsible for the maintenance of cellular homeostasis, this process should lead to an alteration in the overall myocardial function. As demonstrated by working heart preparations from immunized guinea pigs, these show a marked decrease in cardiac function compared to controls. External heart work fell from 385 mJ $g^{-1}min^{-1}$ to 62 mJ g^{-1} min^{-1} when comparing the performance of the control hearts with the hearts from immunized animals whose sera contained immunoactivating antibodies. Thus the functional data together with the biochemical data clearly support the hypothesis that antibodies against the ADP/ATP carrier cause a dysfunction of the heart by an antibody-mediated disturbance of cellular energy metabolism [20]. This proves the suspicion that antibodies against the ADP/ATP carrier are an essential component of the pathological mechanism involved in viral heart disease [13]. This might be of great importance for a better understanding of the pathophysiology of myocarditis and dilated cardiomyopathy.

There are several ways in which antibodies against the ADP/ATP carrier could alter the function of this protein. For example, the antibodies might inhibit carrier function by antibody-binding to the protein. This inhibition might be due to a direct binding of the antibody to the active site thereby sterically hindering the translocation process. Another possibility to be considered is that the antibodies react with the primary translocation product, hindering the incorporation of the carrier protein into the inner mitochondrial membrane. Both possibilities are suggested by immunohistological and immunoelectron-microscopical results, showing immunoglobulin binding to the mitochondrial membrane in the immunized animals (Schultheiss and Herzog, in preparation). Finally, the antibodies — crossreacting with the calcium chanell (Morad et al., Ulrich et al., this volume) — might cause an alteration of intracellular calcium homeostasis which could also lead to a disturbance of cellular energy metabolism [21].

Regardless of the mechanism of action of the antibodies, the data indicate that autoimmunity to this intracellular protein causes an altered myocardial function by an imbalance between energy delivery and demand. The theoretical implications and clinical consequences following from this new mechanism of an immunologically mediated cellular dysfunction might be of great importance.

Acknowledgements. The excellent technical assistance of E. Ronft, E. Pawels, E. Fiedler, A. von Westphalen is gratefully acknowledged. This research was supported by a grant of the Deutsche Forschungsgemeinschaft and the Wilhelm-Sander-Stiftung.

References

1. Becker BF, Gerlach E (1984) Acute effects of nicotine on hemodynamic and metabolic parameters of isolated, perfused hearts of guinea pigs and rats. Klin Wochenschr 62 [Suppl 2]: 58–66
2. Bowles NE, Richardson PJ, Olsen EGJ, Archard LC (1986) Detection of coxsackie-B-virus-specific RNA sequences in myocardial biopsy samples from patients with myocarditis and dilated cardiomyopathy. Lancet 1: 1120–1122
3. Bünger R, Sommer O, Walter G, Stiegler H, Gerlach E (1979) Functional and metabolic features of an isolated perfused guinea pig heart performing pressure-volume work. Pflugers Arch 380: 259–266
4. Geisbühler T, Altschuld RA, Trewyn RW, Ansel AZ, Lamka K, Brierly GP (1984) Adenine nucleotide metabolism and compartimentalisation in isolated adult rat heart cells. Circ Res 54: 536–546
5. Kandolf R (1988) The impact of recombinant DNA technology on the study of enteroviral heart disease. In: Friedman H, Bendinelli M (eds) Coxsackieviruses — a general update. Plenum, New York, pp 293–318
6. Kereiakes DJ, Parmley WW (1984) Myocarditis and cardiomyopathy. Am Heart J 108: 1318–1326
7. Klingenberg M (1985) The ADP/ATP carrier in mitochondrial membranes. In: Martonosi A (ed) The enzymes of biological membranes. Plenum, New York pp 511–553
8. Klingenberg M, Appel M (1980) Is there a common binding center in the ADP, ATP carrier for substrate and inhibitors? FEBS Lett 119: 195–199
9. Klingenberg M, Heldt HW (1982) The ADP/ATP translocation in mitochondria and its role in intracellular compartmentation. In: Sies H (ed) Metabolic compartmentation. Academic, London, pp 101–122
10. Neu N, Beisel KW, Traystman MD, Rose NR, Craig SW (1987) Autoantibodies specific for the cardiac myosin isoform are found in mice susceptible to coxsackievirus B3-induced myocarditis. J Immunol 138: 2488–2492
11. Olsen EGJ (1983) Myocarditis — a case of mistaken identity? Br Heart J 50: 303–311

12. Palmieri F, Klingenberg M (1979) Direct methods for measuring metabolite transport and distribution in mitochondria. Methods Enzymol 56: 279–290
13. Schultheiss HP (1987) The mitochondrium as antigen in inflammatory heart disease. Eur Heart J 8 [Suppl J]: 203–210
14. Schultheiss HP, Bolte HD (1985) Immunological analysis of autoantibodies against the adenine nucleotide translocator in dilated cardiomyopathy. J Moll Cell Cardiol 17: 603–617
15. Schultheiss HP, Klingenberg M (1984) Immunochemical characterization of the adenine nucleotide translocator — organ specificity and conformation specificity. Eur J Biochem 143: 599–605
16. Schultheiss HP, Klingenberg M (1985) Immunoelectrophoretic characterization of the ADP/ATP carrier from heart, kidney and liver. Arch Biochem Biophys. 239: 273–279
17. Schultheiss HP, Schwimmbeck P (1986) Autoantibodies to the adenine nucleotide translocator (ANT) in myocarditis (MC) — prevalence, clinical correlates and diagnostic value. Circulation 74 [Suppl 2]: 142
18. Schultheiss HP, Schwimmbeck P, Bolte HD (1984) Organ specificity of the inhibitor- and substrate binding site of the adenine nucleotide translocator (ANT). J Mol Cell Cardiol [Suppl 2]: 69
19. Schultheiss HP, Schwimmbeck P, Bolte HD, Klingenberg M (1985) The antigenic characteristics and the significance of the adenine nucleotide translocator as a major autoantigen to antimitochondrial antibodies in dilated cardiomyopathy. Adv Myocardiol 6: 311–327
20. Schultheiss HP, Schulze K, Kühl U, Ulrich G, Klingenberg M (1987) The ADP/ATP carrier as a mitochondrial autoantigen — facts and perspectives. Ann NY Acad Sci 488: 44–64
21. Schultheiss HP, Kühl U, Janda J, Melzner B, Ullrich G, Morad U (1988) Antibody-mediated enhancement of calcium permeability in cardiac myocytes. J Exp Med (in press)
22. Schulze K, Becker BF, Schultheiss HP (1988) Antibodies to the ADP/ATP carrier — an autoantigen in myocarditis and dilated cardiomyopathy — penetrate into myocardial cells and disturb energy metabolism in vivo. Circ Res (in press)
23. Soboll S, Bünger R (1981) compartimentation of adenine nucleotides in the working guinea pig heart stimulated by noradrenaline. Hoppe Seylers Z Physiol Chem 362: 125–132
24. Tager JM, Wanders RJA, Groen AK, Kunz W, Bohnensack R, Küster U, Letko G, Böhme G, Duszynski J, Wojtczak L (1983) Control of mitochondrial respiration. FEBS Lett 151: 1–9
25. Wolfgram LJ, Beisel KW, Rose NR (1985) Heart-specific autoantibodies following murine Coxsackie B3 myocarditis. J Exp Med 161: 1112–1121

Importance of Humoral and Cellular Immunological Parameters for the Pathogenesis of Viral Myocarditis

B. MAISCH, M. HERZUM, T. IZUMI, and S. NUNODA

University Hospital of Internal Medicine, Josef-Schneider-Str. 2, D-8700 Würzburg, FRG

Introduction

Problems in the diagnosis and treatment of myocarditis are not new to medicine. Jean Baptiste Senac, physician in ordinary to the French Emperor Louis XV, wrote some 300 years ago: "The inflammation of our heart is difficult to diagnose, and if we have diagnosed it, can we really treat it more safely?" (J. B. Senac quoted by Schramm [54]). Similarly the WHO/ISCF task force definition of dilated cardiomyopathy [47] as "heart muscle disease of unknown cause" which is postulated to be a common end point of different etiopathogenetic entities, which also include myocarditis, more likely describes our lack of knowledge than a disease entity.

The investigation of immunological regulator [1, 11, 12, 26, 27, 39, 53], humoral [3, 6, 16, 17, 18, 21–38, 40, 42, 44, 48–50, 55; also B. Maisch, E. Bauer, M. Cirsi, and P. Thometzek, to be published; B. Maisch, U. Lotze, T. Schneider, K. Kochsiek, to be published] and cellular effector mechanisms in man [2, 6, 9, 10, 13, 14, 19, 26, 27, 31–33, 41, 55, 56] has broadened both our understanding of the etiopathogenesis of autoreactive mechanisms in myocarditis and has raised new questions. From animal and human studies in myocarditis, it can be deduced that acute and protracted autoreactivity to cardiac epitopes and structures by antibody- and cell-mediated immune reactions may have as one basic mechanism an alteration of regulator mechanisms, be it a modified T helper or T suppressor activity carried out according to established methods [15, 39]. This may ultimately lead to polyclonal or highly selective autoreactivity which could be directed to the triggering virus, to different cardiac epitopes or to both (molecular mimicry) (Fig. 1). This host response, which seems to be initially directed to the virus, may be local, focal, or global. It appears to be also determined by genetic factors although direct evidence for this largely comes from animal data.

Immune Regulation and T Suppressor Cell Activity

Details on methods can be derived from previous work [1, 8, 11, 12, 25, 26]. Enumeration of peripheral T cells in our patients with myocarditis (Fig. 2) revealed an increase in OKIa1-positive B cells or activated T cells [26, 39]. Whereas Fowles et al. [12] and Eckstein et al. [11] demonstrated decreased T suppressor cell acti-

H.-P. Schultheiß (Ed.)
New Concepts in Viral Heart Disease

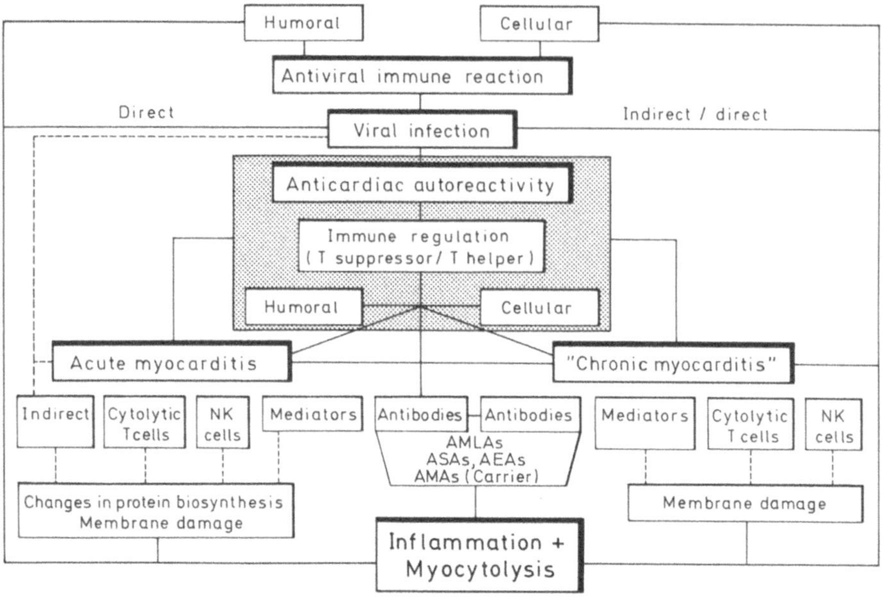

Fig. 1. Hypothesis for the pathogenesis of dilated heart muscle disease following viral myocarditis. *AMLAs,* antimyolemmal antibodies; *AEAs,* antiendothelial antibodies

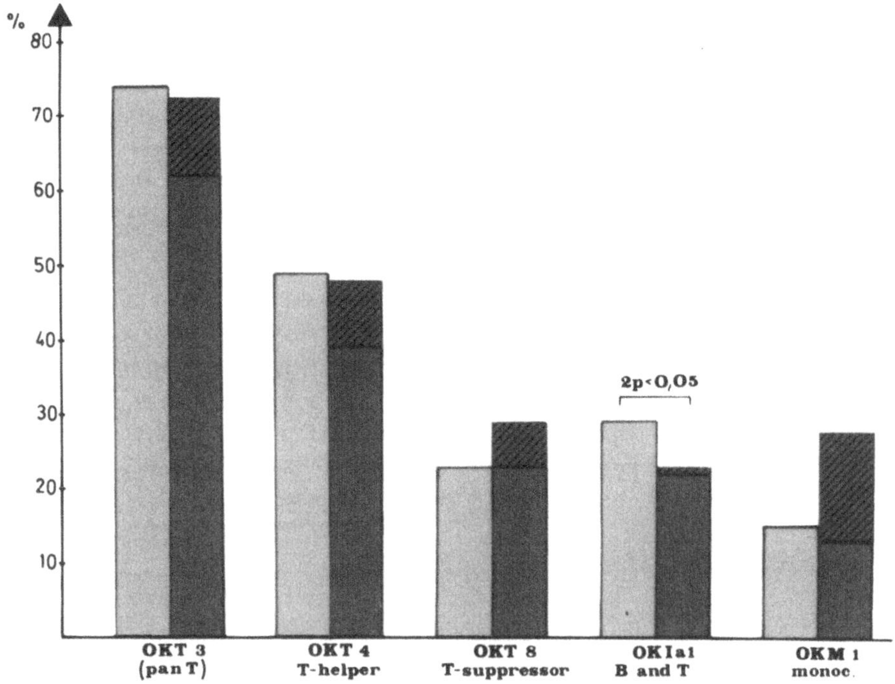

Fig. 2. T cell subpopulations in active myocarditis (fluorescence-activated cell sorter analysis of acute peripheral blood cells in patients with biopsy proven myocarditis). *Red,* myocarditis; *blue,* controls; *hatched areas,* normal range. (From [26] with permission)

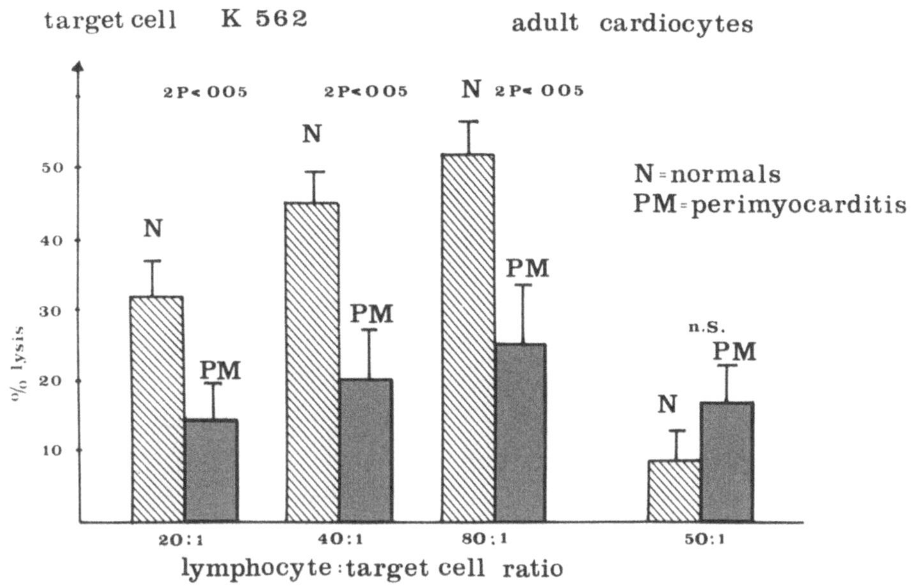

Fig. 3. NK activity *(left)* — target cell-erythroblast cell line K 562 — and specific cytotoxicity *(right)* — target cell-adult rat cardiocytes of periphal blood lymphocytes in acute perimyocarditis *(PM)*. Whereas NK cell activity is reduced in all three lymphocyte to target cell ratios, target cell-specific cytotoxicity remains sustained or even slightly increased. Different lymphocyte to target cell ratios were used for both assays. For target cell-specific non-MHC restricted cardiocytotoxicity the index was not used, but instead the percentage lysis was calculated from data already published [26] to allow a direct comparison with NK cell activity. (From [26] with permission)

vity in dilated cardiomyopathy and speculated on a similarly low reactivity in myocarditis (Table 1), in our studies [26, 39] T suppressor cell activity carried out according to established methods [15, 39] in patients with myocarditis and postmyocarditic heart muscle disease was not different from daily, age-matched controls.

Cellular Effector Mechanisms in Human Myocarditis

At present, data are available only on the effector mechanisms of peripheral blood lymphocytes [2, 13, 26, 27, 33, 41, 55, 56] with the exception of work done by Kurnick et al. [24] on lymphocytes grown from endomyocardial biopsies and expanded by interleukin-2 (IL-2).

Natural Killer Cell Activity

Methods for the assessment of natural killer (NK) cells were first described by Perlmann and Perlmann [46] and these were then adapted to our needs [26]. NK

Table 1. Evidence for and against a discorder of humoral and cellar immunity in perimyocarditis, primary dilated cardiomyopathy, and postmyocarditic cardiomegaly, (Updated from [26, 27]

Discorder/mechanism	Arguments For	Arguments Against	Comment
Regulator cells T suppressor cells (OKT 8)	--	PM/DCM, PMDC unchanged	
B cells (OKIA1)	Increased in PM	--	
monocytes, NK cells,	Increased in DCM	--	
T suppressor cell activity	Increased	Normal	Nonuniform results; global, not tissue-specific activity
T helper cells	No assays done	Subpopulation normal	
Cellular immunity PHA stimulation of peripheral lymphocytes	Decreased in DCM	Also decreased in other heart diseases (CAD); inadequate controls	Sensitive for cardiac discorder but not specific for M or DCM
Lymphocyte stimulation with heart antigen	Slightly increased in DC		
MIF	Positive		
NK cell activity	Decreased in PM, PMDC, DCM	Decrease of NK-cell activity is not cardiospecific	Uniform results, disease characteristic but not cardioselective
Lymphocytotoxicity against heart cells (non-MHC-restricted)	Sustained or increased in PM and PMDC Increased in one-third of patients in DCM	Also sustained or increased in 24% of other heart diseases	Uniform results, cardiospecific but not diseasespecific results for PM or DCM
ADCC	Positive in few patients	Nonuniform for groups of PM, DCM, PMDC	Cardiospecific but not very sensitive
Humoral immunity a. Circulating ASAs b. Bound ASAs (immunohistology)	Present in PM, PMDC; positive in PM, PMDC, DCM here diagnostic if of IgM class or if complement fixation	Except for IgM class and complement fixation, also found in CAD and non cardiac controls	In serum not cardioselective due to similar staining by anticollagen and antilaminin, etc.; if bound cardioselective, not diseasespecific (diagnostic for IgM, and complement fixation)

Table 1. (Continued)

Discorder/mechanism	Arguments For	Arguments Against	Comment
c. Circulating AMLAs, homologous or heterologus type	Diagnostic if IgM (+IgA class)	Not diagnostic if IgG class since also present in non cardiac controls	
d. Circulating anti-M7 antibodies	Diagnostic for part of PM and DCM	Lack of control groups	Membrane expression of the equivalent antigen has to be proven for biological relevance
e. Antiadenine nucleotide translocator antibodies			
f. Circulating immune complexes	Present in most cases	Not cardiospecific since antigen of complex is not characterized	Monitors disease activity
g. AMC in vitro	Diagnostic in CVB, influenza + EBV+ mumps and a large part of undefined PM, M, and PMDC		In vivo relevance is yet to be proven

PM, perimyocarditis; DCM, dilated cardiomyopathy; PMDC, postmyocarditic cardiomegaly; NK, natural killer; PHA, phytohemagglutinin; MIF, migration inhibition factor; CAD, coronary artery disease; AMC, antibody-mediated cytolysis; EBV, Epstein-Barr virus; M, myocarditis.

cells are the first line of a broad but less target-specific immune defense in inflammatory disaese. NK cell activity as measured by the lysis of K562 cells, an erythroblast cell line, was found reduced in active myocarditis (Fig. 3) and postmyocarditic heart muscle disease in our patients [26, 27], as had been previously shown by Anderson et al. [2]. Others supplemented the in vitro assays with interferon gamma [9] or IL-2 [55] in patients with dilated cardiomyopathy and could thus reconstitute a normal NK cell activity (Table 1). All data concerning NK cell activity have to take into account, however, that frozen lymphocytes may give inconsistent and reduced NK cell activity. Thus stringent controls of normal cells frozen in exactly the some way, or even better, immediate assays with freshly isolated effector cells are mandatory [26, 27].

Target Cell-Specific Non-MHC-Restricted Lymphocytotoxicity

Methods for target cell-specific non-MHC-restricted lymphocytotoxicity (LC) have been published in detail [26, 32, 33]. Although the true activity of cytolytic T cells can be assessed only in experimental animals which have autologous myocytes as target cells with an identical MHC determinant available, we could recently present evidence that a heart cell-specific, non-MHC-restricted cytotoxicity of pheripheral blood lymphocytes can be assayed by using heterologous myocytes [26, 27, 31–33, 41]. This lymphocytotoxicity is clearly different from mere NK cell activity; it is directed against vital rat heart cells, and we found it maintained or enhanced in myocarditis (Fig. 3). This is compatible with experiments by Jacobs et al. [19], who used fetal heart cells in their experimental setting for patients with dilated cardiomyopathy (Table 1).

Antibody-Dependent Cellular Cytotoxicity

For antibody-dependent cellular cytotoxicity (ADCC) analysis either autologous lymphocytes ($ADCC_{auto}$) or homologous lymphocytes from a healthy donor ($ADCC_{homo}$) can be tested, the latter system being the one most widely used. Using both systems increased cardiocytotoxicity was present in assays of antibody-dependent cellular cytotoxicity in myocarditis and perimyocarditis [26, 27, 32], and in dilated cardiomyopathy [33] and postmyocarditic heart muscle disease [33] (Table 1).

Function of Macrophages and Neutrophils

Neutrophils and macrophages are vital cellular components of any inflammatory response. They may secrete a number of mediators, e.g., leukotrienes (LT) which could act as a local substrate of the inflammatory response. We have shown in an in vitro setting that isolated heart muscle cells can be damaged primarily by LTB4 but not by LTC4 or LTD4 [38].

The function of macrophages and neutrophils in normal subjects shows a broad intraindividual variance when assessed by in vitro assays measuring the respiratory burst. It can be understood from this that few data are available on their function in myocarditis. From a clinical point of view, imaging methods with In-labeled peripheral blood cells, including granulocytes, have become available in inflammatory heart disease [5]. These imaging methods demonstrate the presence of these cells in diseased myocardium. At present little is known on the function of neutrophils and macrophages in myocarditis: they may be either protective, or detrimental, or both at different times.

Humoral Immune Parameters

Data on humoral immune parameters with antibodies to different cardiac structures, e.g., the membrane [6, 10, 16, 26–37, 40–43; also B. Maisch, E. Bauer, M. Cirsi, P. Thometzek, to be published; B. Maisch, U. Lotze, T. Schneider, K. Kochsiek, to be published], to contractile proteins (see [26, 27, 33] for overview in literature), and to mitchondrial proteins [21, 22, 49, 50] have recently been published.

Table 2. Autoantibodies in perimyocarditis, postmyocarditic cardiomegaly, primary dilated cardiomyopathy, and controls. (From Maisch et al. 1987 [28] with permission)

	CVB ($n = 10$) influenza A/B ($n = 4$) mumps ($n = 4$)	Cytomegalovirus ($n = 15$)	Undefined etiology ($n = 144$)	Tuberculous pericarditis ($n = 12$)	Postmyocarditic cardiomegaly ($n = 18$)	Healthy controls ($n = 200$)	CAD ($n = 51$)
Muscle-specific antibodies							
Antimyolemmal ($AMLAs_a$) (autologous, human cardiocyte)	91[a]	13	62[b]	ND	89[b]	31 (low titer)	43
Antimyolemmal (AMLA) (heterologous, ratcardiocyte	91[a]	6	58[b]	100[b]	89[b]	35 (low titer) 17 (≥ 1:40)	40
Antisarcolemmal (ASAs) (homologous)	100[a]	6	45	100	94	32 (low titer) 20 (≥1:40)	34
Antifibrillary (AFAs)	9	16	31	67[b]	16	4	4
Antiinterfibrillary (IFAs)	18	100[a]	16	12	39	3	6
Nonorganspecific							
Antiendothelial (AEAs)	82[a]	0	73	42	13	25	19
Antismooth muscle (ASMAs)	18	5	16	27	7	11	12

ND, not done; CVB, coxsackie B virus; CAD, coronary artery disease
[a] $P < 0.001$ in x^2 test, when compared to controls (titer 1:40)
[b] $P < 0.01$ in x^2 test, when compared to controls (titer 1:40)

Circulating Antimembrane Antibodies

Since in active (peri)myocarditis DR- and OKIA1-positive cells are increased in the peripheral blood, this could indicate a stimulation of humoral effector organs. Among others, antimyolemmal antibodies (AMLAs) directed against the inner sarcolemmal membrane are of particular interest since they are diagnostic markers (incidence: > 85%). They belong to the IgG, IgM, IgA classes and may fix complement (Tables 1 and 2).

Circulating Antimembrane Antibodies in Coxsackie B Myocarditis in Man

In coxsackie virus type B (CVB) myocarditis (n=10) cross-reactivity of AMLAs to CVB is suggestive since AMLA fluorescence and in vitro cardiocytolytic serum activity in the presence of complement can be absorbed out either with CVB preparations or isolated heterologous myocytes [26, 32; also B. Maisch, E. Bauer, M. Cirsi, P. Thometzek, to be published]. In the Western blot of human sarcolemma — prepared according to Paris et al. [45] — or CVB proteins (Behring) serum antibodies bind to the following bands of identical sites also suggesting cross-reactivity: 110, 48, 35, 31 and 28 Kilodalton (kd). Apart from these, virus- (33 kd, 34 kd) and cardio-specific (90, 78, 72, 67, 45 kd) binding sites were also demonstrated (B. Maisch, E. Bauer, M. Cirsi, P. Thometzek, to be published). Serum antibodies not only bind to myocytes but fix in cryostat sections to different parts of the interstitial tissue and to cytoskeletal proteins as well, which could be demonstrated by using different cells in culture [44] and by an enzyme-linked immunoabsorbent assay (ELISA) with laminin [42]. In addition, evidence from previous publications [26, 32] and very recent data indicate that serum antibodies also bind to the membrane of endothelial cells.

Cytolytic Serum Activity in CVB Myocarditis

Sera of patients with CVB myocarditis demonstrated primarily complement-dependent cardiocytotoxicity (Table 1), which could be eliminated by absorption with CVB preparations and with enriched sarcolemmal proteins [32], indicating that a pathogenetically relevant antibody is both complement fixing and cross-reactive. Circulating anticardiac antibodies are not associated directly with the membrane.

Antifibrillary antibodies directed primarily against myosin are found more often in postmyocarditic heart muscle disease or dilated cardiomyopathy than in perimyocarditis. Anti-interfibrillary antibodies have been observed in cytomegalo- and Epstein-Barr virus infections associated with acute symptoms, segmental wall motion abnormality, and pericardial effusion [26, 27, 32]. Organ-specific antimitochondrial antibodies (anti-M7) were also found in myocarditis [21] and an dilated cardiomyopathy [21,22].

Detailed data on antibodies to specific mitochondrial epitopes, e. g., the adenosine tri-/diphosphate (ATP/ADP) carrier, have been published by Schultheiss et al. [49, 50]. Cross-reactivity to the sarcolemma has been postulated both for antimyosin and antimitochondrial antibodies [43].

Humoral Immune Response in Cytomegalovirus-Associated Myocarditis

In histologically proven myocarditis, antisarcolemmal and antimyolemmal antibodies have also been demonstrated. Remarkably these antimembrane antibodies demonstrate mostly cardiocytotoxicity which is complement independent (unpublished).

Circulating Immune Complexes in Myocarditis

Serial measurements of circulating immune complexes may be helpful in monitoring the progress of inflammatory cardiac diseases. Circulating immune complexes were demonstrated in almost all patients with carditis of viral and undetermined origin [16, 32]. The concentrations and incidences of immune complexes decrease rapidly 4–6 weeks after onset (Table 1). It is not known whether the complexes contain antigens, or whether they only contain aggregated IgG.

Bound Antibodies in the Endomyocardial Biopsy

The bound antisarcolemmal correlative (Fig. 4a) of circulating (Fig. 4b after incubation of the autologous biopsy specimen with the patient's serum) antimembrane antibodies has been detected in more than 90% of cases in the biopsies of patients with histologically proven myocarditis and in more than 65% of patients with postmyocarditic heart muscle disease [26, 27, 29, 33, 34]. They are diagnostic markers of an active autoreactive process only if they belong to the IgM class or if they fix C1q, C3, C4, or C5B9 since in other forms of heart disease (e. g., endocarditis, coronary artery disease) an IgG antibody to sarcolemma and/or extracellular matrix proteins can also be demonstrated.

Cardiac Conduction Disturbances in Myocarditis

In children of mothers with systemic lupus erythematosus, congenital atrioventricular (AV) block was recently demonstrated to be associated to SS-Ro/La antibodies found in the children's sera immediately after birth. The destruction of the AV node involved immunoglobulin binding to the AV node in immunohistological investigations, thus giving indirect evidence of a humoral factor responsible for the destruction of the cardiac conduction system.

In myocarditis AV block may occasionally be found in the acute phase. Our experience of a third-degree AV block in myocarditis is limited to a 43-year-old female pacemaker-dependent patient who died from protracted necrotizing myocarditis that lasted for more than 5 years. Both the initial and the lethal episode of myocarditis were biopsy proven. In the serum distinct antibodies to the human AV node were demonstrated in addition to antisarcolemmal antibodies.

In a more indirect attempt to assess the presence and pathogenetic relevance of antibodies to the human AV- and sinus node, we have analyzed their presence in pacemaker patients many years after a biographic myocarditis [27, 35, 36; also B. Maisch, U. Lotze, T. Schneider, K. Kochsiek, to be published].

In the AV block of these adults, circulating antibodies could be demonstrated that were directed against the human AV node in 20% of patients (B. Maisch, U. Lotze, T. Schneider, K. Kochsiek, to be published). Similarly in the sick sinus syndrome antisinus node antibodies occur in 25% of patients, thus indicating that also in these diseases tissue-specific antibodies may be a diagnostic and perhaps

Fig. 4a

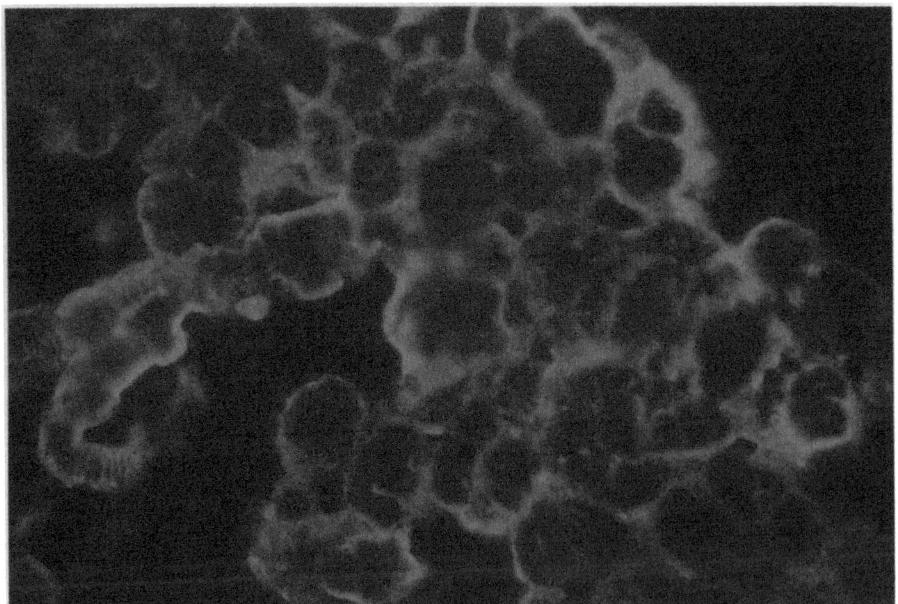

Fig. 4b

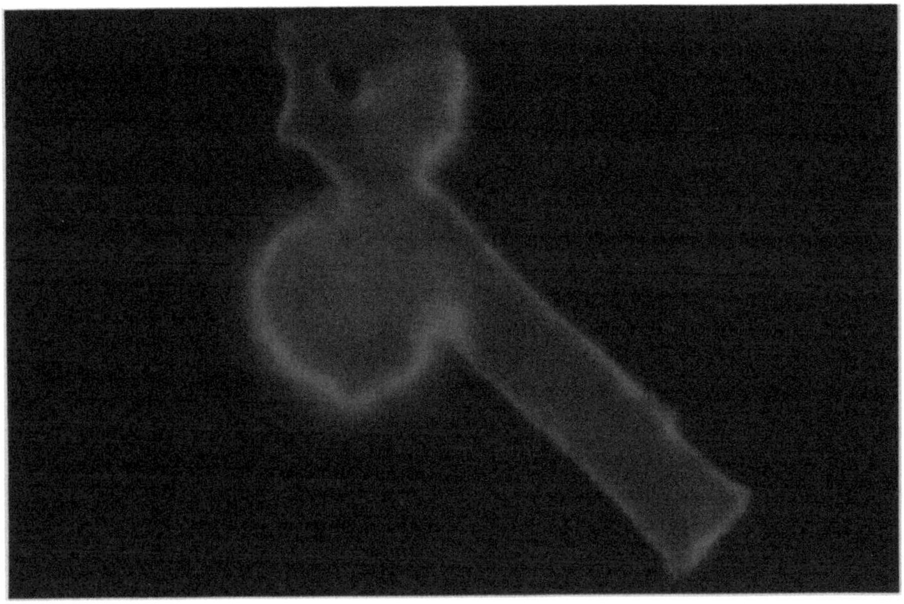

Fig. 5. Antimyolemmal antibodies demonstrated by an indirect FITC-labeled antihuman antiserum [trivalent, $F(ab)_2$ fragments, dilution 1:100]. The antigen used intact human atrial cardiocytes. (From [26]) with permission

even a pathogenic marker [35, 36]. It is remarkable that both in AV block and in the sick sinus syndrome, patients with former myocarditis or rheumatic heart disease have a three-fold risk for this antibody when compared to an AV block or a sick sinus syndrome associated with coronary heart disease, thus providing indirect evidence of patients at risk for cardiac conduction disturbances in myocarditis. It may be possible in the near future to define this risk group for AV block and sick sinus syndrome by the analysis of anticonducting tissue antibodies.

Do Antimyocardial Antibodies belong to "Natural" Antibodies?

Many of the antibodies detected in myocardial diseases are probably "natural" antibodies. They may also be directed against the extracellular matrix (collagen,

---

Fig. 4 a, b. Demonstration of bound immunoglobulin on a cryostat section from a biopsy of a patient with histologically proven myocarditis. In this section immunoglobulin (anti-IgG- and anti-IgM-positive and complement fixation) bound to the sarcolemma and the intersititial space was demonstrated focally using a TRITC (rhodamin)-labeled antiserum (Medac, dilution 1:50). **b** After incubation with unlabeled antihuman immunoglobulin (rabbit) the patient's serum (dilution 1:10) was added. After washing, Fluorescein isothiocyanate(FITC)-labeled goat antihuman immunoglobulin [$F(ab_2)$ fragments, dilution 1:100, Medac] was added, and circulating antimyocardial antibodies directed against the autologous heart tissue were also demonstrated. (From Maisch 1986 [26] with permission)

structural proteins, laminin, microfilaments, etc.) and seem to possess little or no pathogenetic relevance. Recently one has become more and more aware of the fact that in many healthy normals these antibodies may be remnants of former infections without being markers of disease [4, 7, 51, 52]. This may also hold for some of the antimembrane antibodies that are heterophilic antibodies [27, 55] or those directed against archaic, cytoskeletal epitopes [28, 42].

"Natural" antibodies are currently considered to have some regulatory functions, whereas formerly they were referred to as "horror autotoxicus" (Paul Ehrlich) or as being secreted from "forbidden clones" [20]. The current status is that natural antibodies occur in almost all human beings. Their suggested regulatory or pathogenetic relevance is still poorly defined, however. At present natural antibodies are thought to belong primarily to the IgM class. They may be also separated from pathogenetically relevant antibodies by their function: pathogenetic relevance may be primarily attributed to complement-fixing antibodies. From these observations antisarcolemmal antibodies (ASAs) and AMLAs which fix complement and are cytolytic to vital heart cells in vitro can be considered pathogenetically important as they have an autoimmune potential.

Conclusions

Cytolytic antimyolemmal antibodies are a diagnostic marker of a secondary immunopathogenesis in myocarditis. Antigenic mimicry appears to be one principle of "autoimmunity" operating in acute viral heart disease. The membrane of myocytes, of cardiac pacemaker cells, of conducting tissue, of the endothelium, and of the fibroblasts appears to be the target of a polyclonal humoral immune response. Apart from humoral effector mechanisms, cellular parameters are also

Table 3. Immunological classification of dilated heart muscle disease (modified from Maisch et al. 1986 [26] with permission)

Type	Humoral immune response			Cellular immunity	
	AMLAs	ASA (biopsy)	Cytolysis (Serum)	NK activity	cardiospecific lymphocytotoxity
Primary DCM	+/−	IgM−/C3−	−	↓/=	/=
"DCM" with increased cellular cytotoxicity	(+)/−	IgM−/C3−	−	↓/=	↑
"DCM" with increased humoral reactivity	↓	IgM+/C3+/− IgG++, IgA++	↓	↑/=	=
Cardiomegaly with increased humoral and cellular reactivity	↓	↓IgM+/C3+/− IgA++/IgG++	↓	↑/=	↓
Active myocarditis	↓	↓IgG++/IgM++ IgA++/C3+/−	↓	↑	=/↓

altered so that both autoreactive cellular and humoral immune mechanisms are operative in patients with myocarditis and postmyocarditic heart muscle disease. These new findings demand an immunologically based new classification of dilated heart muscle disease as proposed in Table 3.

Acknowledgements. The excellent technical assistance of E. Bauer, M. Cirsi, C. Dienesch, C. Römmelt, U. Schönian, P. Thometzek, and I. Wendl is gratefully acknowledged. The work was supported by grants from the Deutsche Forschungsgemeinschaft for B. Maisch and the A. v. Humboldt Foundation for T. Izumi and S. Nunoda.

References

1. Anderson JL, Greenwood JH, Kawanishi H (1981) Evaluation of suppressor immune regulatory function in idiopathic congestive cardiomyopathy and rheumatic heart disease. Br Heart J 46: 410–414
2. Anderson JL, Carlquist JF, Hammond FH (1982) Deficient natural killer cell activity in patients with idiopathic dilated cardiomyopathy. Lancet 2: 1124–1127
3. Aretz HT, Southern JF, Palacios IF, Dec GW Jr, Howard CA, Fallon JT (1987) Morphological and immunological findings in heart biopsies of patients with suspected or treated myocarditis. Eur Heart J 8 [Suppl J]: 187–190
4. Avrameas S (1986) Natural autoreactive B cells and autoantibodies: the know thyself of the immune system. Ann Inst Pasteur/Immunol [Suppl D] 137: 150–156
5. Becker W, Borst U, Maisch B, Epping J, Börner W, Kochsiek K (1987) ^{111}In-labeled granulocytes in inflammatory heart disease. Eur Heart J 8 [Suppl J]: 307–310
6. Bolte HD, Schultheiss P, Cyran J, Goss F (1980) Binding of immunoglobulins in the myocardium (biopsy). In: Bolte HD (ed) Myocardial biopsy. Springer, Berlin Heidelberg New York, pp 85–92
7. Boyden SV (1963). Natural antibodies and the immune response. Adv Immunol 5: 1–28
8. Breshinhan B, Jasin ME (1977). Suppressor function of peripheral blood mononuclear cells in normal individuals and patients with systemic lupus erythematosus. J Clin Invest 59: 106–116
9. Cambridge G, Campbell-Blay G, Wilmhurst P, Coltan DJ, Stern CMM (1983). Deficient "natural" cytotoxicity in patients with congestive cardiomyopathy (Abstr). Br Heart J 49: 623
10. Das SK, Petty RE, Meengs WA, Tubergen DG (1980) Studies of cell-mediated immunity in cardiomyopathy. In: Sekiguchi M, Olsen EGJ (eds) Cardiomyopathy. University of Tokyo Press, Tokyo, pp 375–377
11. Eckstein R, Mempel W, Bolte HD (1982) Reduced suppressor cell activity in congestive cardiomyopathy and in myocarditis. Circulation 65: 1224–1229
12. Fowles RE, Bieber CP, Stinson EB (1978) Defective in vitro suppressor of cell function in idiopathic congestive cardiomyopathy. Circulation 59: 483–491
13. Frencesini R, Petillo A, Corazza M, Nizzo MC, Azzolini A, Gianrossi R (1983) Lymphocyte response in dilated cardiomyopathy. IRCS Med Sci 11: 1019
14. Gentle TA, Baynham MID, Gammage MD, Lowry PJ, Thompson RA, Littler WA (1987) T cell function in idiopathic congestive cardiomyopathy. Eur Heart J 8 [Suppl J]: 145–146
15. Hallgren HM, Yunis EJ (1977) Suppressor lymphocytes in young and aged humans. J Immunol 118: 2004–2008
16. Herzum M, Maisch B, Kochsiek K (1987) Circulating immune complexes in perimyocarditis and infective endocarditis. Eur Heart J 8 [Suppl J]: 323–325
17. Hufnagel G, Pfeifer U, Maisch B (1987) Immunohistological investigations in suspected cardiac sarcoidosis. Eur Heart J 8 [Suppl J]: 59–62
18. Izumi T, Maisch B, Kochsiek K (1987) Experimental murine myocarditis after immunization with cardiac membranous proteins. Eur Hart J 8 [Suppl J]: 419–424

19. Jacobs B, Matsuda Y, Deodhar S, Shirey S (1979) Cell-mediated cytotoxicity to cardiac cells of lymphocytes from patients with primary myocardial disease. Am J Clin Pathol 72: 1–4
20. Jerne NK (1955) The natural selection theory of antibody formation. Proc Natl Acad Sci USA 41: 849–857
21. Klein R, Maisch B, Kochsiek K, Berg PA (1984) Demonstration of organ specific antibodies against heart mitochondria (anti-M7) in sera from patients with some forms of heart disease. Clin Exp Immunol 58: 283–292
22. Klein R, Spiel L, Kleemann U, Hassenstein P, Berg PA (1987) Relevance of antimitochondrial antibodies (anti-M7) in cardiac disease. Eur Heart J 8 [Suppl J]: 223–226
23. Kühl U, Ulrich G, Schultheiss HP (1987) Cross-reactivity of antibodies to the ADP/ATP translocator of the inner mitochondrial membrane with the cell surface of cardiac myocytes. Eur Heart 8 J [Suppl J]: 219–222
24. Kurnick JT, Leary C, Palacios IF, Fallon JT (1987) Culture and characterization of lymphocytic infiltrates from endomyocardial biopsies of patients with idiopathic myocarditis. Eur Heart J 8 [Suppl J]: 135–144
25. Lowry PJ, Thompson RA, Littler WA (1983) Humoral immunity in cardiomyopathy. Br Heart J 50: 390–394
26. Maisch B (1986) Immunologic regulator and effector functions in perimyocarditis, post-myocarditis heart muscle disease and dilated cardiomyopathy. Basic Res Cardiol 81 [Suppl 1]: 217–242
27. Maisch B (1987) Immunological mechanism in human cardiac injury. In: CJF Spry (ed) Immunology and molecular biology of cardiovascular disease. (Current status of clinical cardiology) MTP, London, pp 225–252
28. Maisch B (1987) The sarcolemma as antigen in the secondary immunopathogenesis of myopericarditis. Eur Heart J 8 [Suppl J]: 155–165
29. Maisch B (1988) The use of myocardial biopsy in heart failure. Eur Heart J [Suppl H] 9: 59–71
30. Maisch B, Kochsiek K (1983) Humoral immune reactions in uremic pericarditis. Am J Nephrol 3: 264–271
31. Maisch B, Maisch S, Kochsiek K (1982) Immune reactions in tuberculous and chronic constrictive pericarditis. Am J Cardiol 50: 1007–1013
32. Maisch B, Trostel-Soeder R, Stechemesser E, Berg PA, Kochsiek K (1982) Diagnostic relevance of humoral and cell-mediated immune reactions in patients with acute viral myocarditis. Clin Exp Immunol 48: 533–545
33. Maisch B, Deeg P, Liebau G, Kochsiek K (1983) Diagnostic relevance of humoral and cytotoxic immune reactions in primary and secondary dilated cardiomyopathy. Am J Cardiol 52: 1071–1078
34. Maisch B, Büschel G, Izumi T, Eigel P, Regitz V, Deeg P, Pfeifer U, Schmaltz A, Herzum M, Liebau G, Kochsiek K (1985) Four years of experience in endomyocardial biopsy — an immunohistologic approach. In: Sekiguchi M, Olsen EGJ, Goodwin JF (eds) Myocarditis and related disorders. Springer, Berlin Heidelberg New York, pp 59–67 and Heart and Vessels [Suppl] 1: 59–67
35. Maisch B, Lotze U, Schneider J, Kochsiek K (1985) Antibodies to human sinus node in sick sinus syndrome. New Trends in Arrhythmias 1: 417–421
36. Maisch B, Lotze U, Schneider J, Kochsiek K (1986) Antibodies to human sinus node in sick sinus syndrome. PACE 9: 1101–1109
37. Maisch B, Cirsi M, Gehrke J, Kirschner A (1987) Lectin binding sites of the sarcolemma and the surrounding interstitial tissue, Eur Heart J 8 [Suppl J]: 175–180
38. Maisch B, Dienesch C, Wendl I (1987) Effect of leukotrienes on isolated vital cardiac cells — an in vitro model of cardiocytotoxic mediators in inflammation? Eur Heart J 8 [Suppl J]: 463–465
39. Maisch B, Hauck H, Köninger U, Endter S, Klopf D, Schmier U, Schmier K (1987) T-suppressor cell activity in (peri)myocarditis and infective endocarditis. Eur Heart J 8 [Suppl J]: 147–153
40. Maisch B, Schwab D, Bauer E, Sandhage K, Schmaltz AA (1987) Antimyolemmal antibodies in myocarditis in children. Eur Heart J 8 [Suppl J]: 167–173

41. Maisch B, Selmayer N, Brugger E, Ertl G, Eilles C, Heinrich J, Gerhards W, Hufnagel G, Schmidt M, Kochsiek K (1987) Cardiac sarcoidosis — clinical and immunoserologic study. Eur Heart J 8 [Suppl J]: 63–71
42. Maisch B, Wedeking U, Kochsiek K (1987) Quantitative assessment of antilaminin antibodies in myocarditis and perimyocarditis. Eur Heart J 8 [Suppl J]: 233–235
43. Maisch B, Wittner B, Wiethölter H (1987) Experimental myocarditis in rabbits — cellular effector mechanisms. Eur Heart J 8 [Suppl J]: 425–432
44. Obermayer U, Scheidler J, Maisch B (1987) Antibodies against micro- and intermediate filaments in carditis and dilated cardiomyopathy — are they a diagnostic marker? Eur Heart J 8 [Suppl J]: 181–186
45. Paris S, Fosset M, Samuel D, Ailhaud G (1977) Chick embryo plasma membrane from cardiac muscle and cultured heart cells: isolation procedures and absence of fatty acid-activating enzymes. J Mol Cell Cardiol 9: 161–173
46. Perlmann P, Perlmann H (1971) Cytotoxic lymphocytes. Mechanisms of activation and target cell destruction. Int Arch Allergy Appl Immunol 41: 36–39
47. Ragosta M, Crabtree J, Sturner WQ, Thrompson PD (1980) Report of the WHO/ISFC task force on the definition and classification of cardiomyopathies. Br Heart J 44: 672
48. Reinauer KM, Klein R, Seipel L, Berg PA (1987) Heart-specific antimitochondrial antibody (anti-M7) in a patient with virus-associated peri-myocarditis. Eur Heart J 8 [Suppl J]: 227–228
49. Schultheiss HP (1987) The mitochondrium as antigen in inflammatory heart disease. Eur Heart J 8 [Suppl J]: 203–210
50. Schultheiss HP, Bolte HD (1985) Immunological analysis of auto-antibodies against the adenine nucleotide translocator in dilated cardiomyopathy. J Mol Cell Cardiol 17: 601–617
51. Schwarz RS (1986) Are natural autoantibodies real? Ann Inst Pasteur/Immunol 137, D, 156–159
52. Seligman M (1986) The origin and nature of autoantibodies. Introduction. Ann Inst Pasteur/Immunol [Suppl D] 137: 149–150
53. Schervish E, O'Connel JB, Kowalczyk D, Robinson JA (1987) Suppressor cell function in dilated cardiomyopathy. Eur Heart J 8 [Suppl J]: 141–143
54. Schramm P (1986) Die Beschwernisse des Herzens und was dagegen dienlich ist. Edition Rarissma, Taunusstein, p 70
55. Yamakawa K, Fukuta S, Yoshinaga T, Umemeto S, Itagaki T, Kusukawa R (1987) Study of immunological mechanisms in dilated cardiomyopathy. Jpn Circ J 51: 665–675
56. Yokoyama A, Aoki S, Aizawa Y, Shibata A (1985) Analysis of peripheral lymphocyte subpopulation in patients with dilated cardiomyopathy. J Dev Med 133(3): 187–188

Characterization of T-Lymphocytes Cultured from Human Endomyocardial Biopsies

J. T. FALLON, I. STAMENKOVIC, D. M. FRISMAN, C. LEARY, I. PALACIOS, and J. T. KURNICK

Department of Pathology, Department of Molecular Biology, Immunopathology Unit and Cardiac Unit, Massachusetts General Hospital, Boston, MA 02114, USA

Introduction

Myocarditis is often characterized by a lymphocytic inflammatory infiltrate in myocardial tissue. However, human myocarditis has neither a definitive etiologic agent nor a well-characterized pathogenesis [1]. Previous investigations of clinical myocarditis have centered around potential viral etiologies and the humoral immune aspects of the acute disease and its possible sequelae, namely dilated cardiomyopathy [2, 3, 4]. A few investigations have examined circulating T lymphocytes in patients with dilated cardiomyopathy [5, 6] while others have applied immunohistochemical techniques to characterize the inflammatory infiltrates in endomyocardial biopsies of myocarditis [7, 8, 9].

Recent work by us [10] has demonstrated that immunologically functional T lymphocytes can be obtained from endomyocardial biopsy samples when cultured in the presence of interleukin-2 (IL-2). During propagation, tissue-derived T lymphocytes retain specificity relevant to the underlying disease [11, 12, 13, 14]. Unfortunately, neither a known antigen nor appropriate target cells are available to study the specificity of myocardially derived T lymphocytes. We therefore elected to examine rearrangements of the T cell receptor (TCR) gene as a possible technique to elucidate T lymphocyte specificity in myocarditis. The TCR is a heterodimeric membrane protein composed of a α chain and a β chain, each of which is encoded by constant (C) and variable (V) region genes which recombine during differentiation and maturation of the T lymphocyte [15]. Added diversity of the receptor is the result of diversity (D) and joining (J) segments between the C and V regions in the TCR gene. The process of combinatorial joining of V-D-J gene regions determines the receptor specificity, which means that a large number of specifications can be produced even though only 20–50 V region genes have been identified for the α and β chains. The use of Southern blot technology allows for detection of a T lymphocyte clone if it represents more than 5% of the total cellular DNA. A stimulus, such as phytohemagglutinin (PHA), activates a large fraction of the T lymphocytes from blood, and the resulting cells are sufficiently diverse that "dominant" clones cannot be distinguished [14]. However, lymphocytes obtained from tissue sites of inflammation [14], rejection [11, 16], and neoplasia [17] often reveal dominant clones.

This report details the results of culturing of endomyocardial biopsy tissue for activated T lymphocytes and the characterization of the phenotype, functionality, and clonal selectivity of cells cultured from patients with myocarditis.

H.-P. Schultheiß (Ed.)
New Concepts in Viral Heart Disease

Materials and Methods

Culture Technique

Endomyocardial biopsy tissues were obtained from the right ventricle using the Caves-Shultz bioptome. Between six and eight tissue fragments were obtained per patient with three to five fragments submitted for routine histopathology and one fragment submitted for culture. Tissue for culture was placed in 15 ml phosphate-buffered saline (PBS) containing 1% human serum, transferred to the tissue culture facility at room temperature, and processed within 24 h as previously reported [10]. Tissue fragments were cut into four to eight pieces and each piece placed in individual 2-ml culture wells containing RPMI 1640 (Gibco) supplemented with 50 μg/ml gentamycin and 5% human serum. Recombinant human IL-2 (a kind gift from Sandoz Research Insitute, Vienna) was added to two-thirds of the wells at 10 ng/ml (10^6 units/mg). Culture plates were then incubated at 37°C in a 5% CO_2 incubator and monitored daily for exudation and proliferation of lymphocytes. All cultures were maintained for 2 weeks with media changed one to two times each week. After 2 weeks, proliferating lymphocytes from positive wells were removed, pooled, counted, and restimulated with PHA in the presence of irradiated mononuclear cells from fresh human blood [12].

Phenotyping

Proliferating lymphocytes were surface phenotyped using flow cytometry [13]. Cells were stained with 5–20 μl fluorescein conjugates of Leu4($CD3^+$), Leu3($CD4^+$), Leu2($CD8^+$), LeuDR, and a γ_2 nonreactive antibody (Becton-Dikkenson). Cells were incubated at 4 °C for 30 min during staining, fixed in 1% paraformaldehyde for 5 min, and stored in the dark at 4 °C until run on an Ortho-Spectrum III flow cytometer.

Functional Assays

IL-2 and macrophage-stimulating factor (ILO) production were measured in supernatants of propagated lymphocytes that were a minimum of 2 weeks post restimulation as previously described [10]. Cytolytic activity of proliferated tissue lymphocytes was determined against three cell types in the presence and absence of a lectin (PHA). The target cells used were: KF — normal human T lymphoblast carried in IL-2; K562 — natural killer (NK) target tumor cell line derived from an erythroleukemia [18]; and ACL-JOG — NK-resistant tumor line obtained from a human adenocarcinoma of the lung [17]. A 4-h, ^{51}Cr-release assay was performed as previously described [10, 12].

TCR Analyses

Cultured, myocardially derived T lymphocytes were expanded with PHA in the presence of irradiated allogenic mononuclear cells as previously described [10, 12]. Cells were harvested for DNA extraction prior to restimulation whenever their number reached 2×10^7. Freshly harvested cells were washed and lysed with 3% lauryl sarcosinate in 3 m*M* ethylenediaminetetra-acetic acid (EDTA). DNA was prepared by proteinase K digestion of the lysate, phenol extraction, and ethanol precipitation.

The TCR β probe was derived from a human cDNA clone, *Jur*β1, isolated and generously provided by T. W. Mak (The Ontario Cancer Institute, University of Toronto). The probe was inserted into an M13 origin PMLC12 plasmid and single-stranded DNA was prepared [19]. To obtain a C region probe, the *Jur*β1 probe was cut with *Pst*I and *Bgl*II yielding a 320-base-pair fragment containing only C region sequences. For hybridizitation, single-stranded cDNA was made and radiolabeled with ^{32}P by primer extension [14]. Southern blots were performed by the method of Reed and Mann [20].

Statistical Analysis

All data were analyzed by parametric and nonparametric analysis in RS/1 (Bolt, Beranek and Newman) using COMPARE.

Results

A total of 261 endomyocardial biopsy fragments from 223 nontransplant patients were obtained for lymphocyte culture between November 1984 and April 1988. Forty-seven biopsies from 40 patients were positive for proliferating lymphocytes. As shown in Table 1, 21 biopsies diagnosed histopathologically as myocarditis resulted in 11 (52%) positive cultures, whereas 226 biopsies that were normal or had nonspecific or specific noninflammatory changes yielded a significantly smaller percentage (14%) of positive cultures *($P = 0.002$)*. Five positive cultures were also obtained from 14 biopsies diagnosed as resolved myocarditis, i.e., from

Table 1. IL-2-positive cultures by histological diagnosis

Histological diagnosis	Biopsies	Positive	
	(n)	*(n)*	(%)
Myocarditis	21	11	52
Resolved Myocarditis	14	5	36
Nonmyocarditis	226	31	14
	261	47	18

patients with a prior histological diagnosis of myocarditis on endomyocardial biopsy.

From most positive cultures, lymphocytes exuded and proliferated slowly over the course of the first 2 weeks in culture. Occasionally, lymphocytes exuded and proliferated rapidly, filling wells with lymphocytes within two to three days. Removal of the tissue from lymphocyte containing wells slowed the further proliferation of cells. Leaving lymphocyte producing tissue fragments in culture for up to eight weeks resulted in continued exudation of cells. No lymphocyte proliferation was seen in the absence of IL-2.

Phenotyping by flow cytometry was performed on cells from 31 biopsies (26 patients). Virtually all cells were $CD3^+$ T lymphocytes. In 19 cultures, $CD4^+$ cells predominated, whereas $CD8^+$ predominance was noted in the remaining 12 cultures. Patients with multiple positive biopsies showed persistent subset predominance in all subsequent biopsies.

Cultures from biopsies of 16 patients produced an average of 67.3 ± 14.2 units of IL-2 per milliliter (range 5.6–195.1 units/ml). The amount of IL-2 produced did not correlate with $CD4^+/CD8^+$ ratio $(P = 0.9)$. An average of 3.5 ± 0.4 units/ml of ILO was produced (range 1.6–7.6 units). As with IL-2, the amount of ILO was not significantly correlated with the predominant phenotype $(P = 0.2)$.

Propagated lymphocytes from 14 biopsies were assayed for cell-mediated cytotoxicity. There was virtually no lysis of normal human T lymphocytes (KF) in the absence of lectin. Greater than 10% NK cell activity was demonstrated against K562 cells in the absence of PHA in only five of the 14 cultures tested. PHA enhanced the lysis of K562 by all of the cultures. Significant lysis of ACL-JOG cells, an NK-resistant tumor cell line, was achieved by only three cultures without PHA but was detected for all cultures when lectin was present. The degree of cytoxicity did not correlate with phenotype $(P > 0.05)$.

Southern blot analysis of the DNA from expanded cultures of 11 biopsies from seven patients with myocarditis showed distinct rearrangements of the $C\beta$ gene. *Eco*RI digests hybridized with $C\beta$ probe, *Jur*β1, revealed loss of the 11-kb germline band and the appearance of one to three rearranged bands in each case (Fig. 1). *Hind*III digests of cultures derived from biopsies of three different patients at different times showed shared rearrangements for two of the three patients and a unique rearrangement in the third (Fig. 2).

Discussion

Myocarditis is a disease of many faces, uncertain pathogenesis, and possibly numerous etiologies. Many investigators feel that myocarditis is a virally caused disease which induces an immunological reaction producing myocyte dysfunction and/or necrosis that eventuates in "sudden death", dilated cardiomyopathy, or complete recovery. While immunological reactions are well established as important pathogenetic mechanisms of experimental viral myocarditis and the subsequent development of dilated cardiomyopathy, relatively few investigations of human myocarditis have examined the cellular immune reaction in the hearts of patients with myocarditis. This study demonstrates that culture of endomyocar-

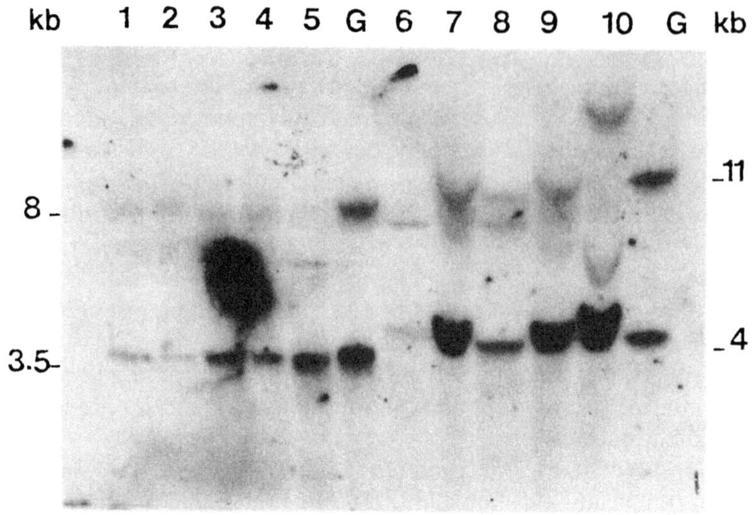

Fig. 1. Southern blot analysis of TCR β-chain rearrangements in DNA digests from uncloned T cells, grown and expanded in IL-2-containing medium, from endomyocardial biopsy fragments of patients with myocarditis. Lanes 1–5 and 6–10 respectively, represent digests of one culture from a single patient. The DNA from lymphocytes and normal human placenta *(G)*, which served as a germ-line control, were digested with *Hind*III *(lanes 1–5)* and *Eco*RI *(lanes 6–10)* and hybridized to the Cβ region probe *(Bgl*II/*Stu*I fragment of the *Jur*β1 probe). One to three rearranged bands are evident in each of *Eco*RI lanes

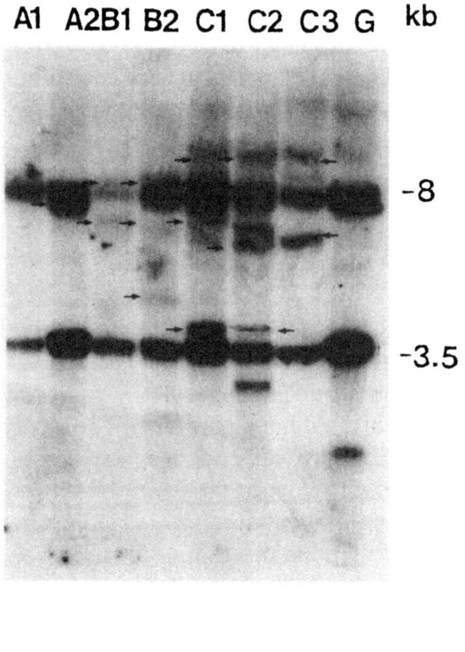

Fig. 2. Southern blot analysis of TCR β chain rearrangement in *Hind*III digests of cultures derived from endomyocardial biopsies of three patients *(A, B,* and *C)* at different times. Shared (B1 and B2; C1, C2, and C3) and unique (A2, B2) rearrangements are present in each set of samples as indicated by the *arrows*

dial biopsy tissue in the presence of IL-2 permits the propagation of activated T lymphocytes which are immunologically functional in that they produce lymphokines and show evidence of cytotoxicity.

Although myocardial and/or viral specificity has not yet been shown, due to the current lack of appropriate target cells, the observations that myocardial tissue fragments continue to exude lymphocytes for up to 8 weeks in culture and that proliferation is enhanced in presence of tissue strongly suggest that the original myocardial tissue continues to provide an "antigenic stimulation" in vitro which was also responsible for activation of T lymphocytes in vivo. Thus, we hypothesize that the lymphocytes derived from myocardial biopsies are enriched for appropriate antigenic specificities. Our hypothesis is further supported by the results of TCR analysis which demonstrate clonal selectivity in all cultures tested from patients with myocarditis. Thus, the inflammatory environment exerts a strong selective force on the immune repertoire of infiltrating lymphocytes.

Our studies provide characterization of infiltrating T cells in myocarditis and hopefully will lead to better rationalization for the use of specific immunosuppressive agents in this disease process. Many patients with "myocarditis" are currently treated with various regimens of immunosuppressive therapy including prednisone, azathioprine and/or cyclosporine. The rationale for using these drugs is based on the presence of a lymphocytic infiltrate similar to that of rejection in endomyocardial biopsies of patients with myocarditis. One possibility is that the immune response in myocarditis represents an example of an immune response to self-antigens. Such autoimmunity may be inferred by finding a limited number of "dominant" clones in the inflammatory infiltrates. Clonal dominance is readily demonstrated by examining the pattern of recombination of the genes encoding the TCR receptor of the cultured T lymphocytes. The demonstration of clonality in myocardially derived T cells suggests the possibility of developing new therapeutic interventions directed at controlling the disease by modulating specific lymphocyte clones present in the tissue.

Summary

To determine the functional characteristics and specificity of the infiltrating T lymphocytes in human myocarditis, we cultured 261 endomyocardial biopsy fragments from 223 patients in the presence of IL-2. IL-2-responsive cells were derived from 47 (18%) of the biopsies. Phenotyping of expanded cultures showed that the cultured cells were almost entirely $CD3^+$ and that $CD4^+$ lymphocytes predominated in 60% and $CD8^+$ cells in 40%. All cultures produced lymphokines and exhibited nonspecific cytotoxicity. In the absence of an identifiable antigen and/or appropriate target cell, we elected to utilize T cell receptor gene rearrangements as a means of detecting selective T lymphocyte accumulation in myocarditis. Cultured lymphocytes from 11 biopsies from seven patients with myocarditis were subjected to restriction fragment length polymorphism analysis of the T cell receptor β gene. DNA was obtained from cultured lymphocytes and subjected to digestion with restriction enzymes — *Eco*R1 and *Hind*III. Southern blots were performed using Jurβ1, a cDNA probe for the β chain of the human T cell receptor. All cultures clearly showed distinct rearrangements of the Cβ gene indicating clonal dominance. In two of three patients with two or more cultures derived from temporally separate biopsies, the rearrangements of the Cβ gene were the same and in the third there were shared rearrangements. These data suggest that the inflammatory infiltrate in the myocardium of patients with myocarditis is dominated by a small number of T lymphocyte clones. Therefore, it may be possible to utilize IL-2-cultured lymphocytes to identify specific antigens involved in the pathogenesis of myocarditis. Furthermore, regulation of specific T lymphocyte clones may represent a potential new approach to therapeutic intervention in patients with myocarditis.

Acknowledgements. The authors would like to thank the many individuals who made this study possible. We are especially thankful to the staff of the cardiac catheterization laboratories and Ms Chris Howard for their assistance in obtaining the specimens for this study. We are grateful to Dr. James Southern and Mr. Bernie Collins for coordinating specimen processing. We are most appreciative to Ms Gail Arbuthnot for her diligent and skillful preparation of this manuscript.

References

1. Abelmann WH (1973) Viral myocarditis and its sequelae. Annu Rev Med 24: 145–150
2. Woodruff J (1980) Viral myocarditis. A review. Am J Pathol 101: 427–479
3. Cambridge G, MacArthur CGC, Waterson AP et al. (1979) Antibodies to Coxsackie B viruses in congestive cardiomyopathy. Br Heart J 41: 692–696
4. Maisch B, Deeg P, Liebau G, Kochsiek K (1983) Diagnostic relevance of humoral and cytotoxic immune reactions in primary and secondary dilated cardiomyopathy. Am J Cardiol 52: 1072–1078
5. Eckstein R, Mempel W, Bolte HD (1982) Reduced suppressor cell activity in congestive cardiomyopathy and myocarditis. Circulation 65: 1224–1229
6. Anderson JL, Carlquist JF, Higashikubo R (1985) Quantitation of lymphocyte subsets by immunofluorescence flow cytometry in idiopathic dilated cardiomyopathy. Am J Cardiol 55: 1550–1554
7. Marboe CC, Knowles DM, Weiss MB et al. (1984) Characterization of the inflammatory infiltrate in human myocarditis — an endomyocardial biopsy study. In: Viral heart disease. Springer, Berlin Heidelberg New York, pp 74–82
8. Cassling RS, Linder J, Sears TD et al. (1985) Quantitative evaluation of inflammation from idiopathically failing or irritable hearts: experience in 80 pediatric and adult patients. Am Heart J 110: 713–720
9. Schnitt SJ, Ciano PD, Schoen FJ (1987) Quantitation of lymphocytes in endomyocardial biopsies: use and limitations of antibodies to leucocyte common antigen. Hum Pathol 18: 796–800
10. Kurnick JT, Leary C, Palacios IF, Fallon JT (1987) Culture and characterization of lymphocytic infiltrates from endomyocardial biopsies of patients with idiopathic myocarditis. Eur Heart J 8 [Suppl J]: 135–139
11. Mayer TG, Fuller AA, Fuller TC et al. (1985) Characterization of in vivo-activated allospecific T lymphocytes propagated from human renal allograft biopsies undergoing rejection. J Immunol 134: 258–264
12. Kurnick JT, Grönvik KO, Kimura AK et al. (1979) Long term growth in vitro of human T cell blasts with maintanence of specificity and function. J Immunol 122: 1255–1260
13. Kurnick JT, Warrens AR, Moscicki RA, Leary CP (1983) T cell subsets in human lymphocytes maintained in IL-2 medium after PHA mixed lymphocyte reaction activation. Clin Immunol Immunopathol 27: 444–451

14. Stamenkovic I, Stegagno M, Wright KA et al. (1988) Clonal dominance among T lymphocyte infiltrates in arthritis. Proc Natl Acad Sci USA 85: 1179–1183
15. Marrack P, Kappler J (1987) The t cell receptor. Science 238: 1073–1079
16. Zeevi A, Fung J, Zerbe TR et al. (1986) Allospecificity of activated T cells grown from endomyocardial biopsies from heart transplant patients. Transplantation 41: 620–626
17. Kurnick JT, Kradin RL, Blumberg R et al. (1986) Functional characterization of T lymphocytes propagated from human lung carcinoma. Clin Immunol Immunopath 38: 367–380
18. Herberman RB, Ortaldo JR (1981) Natural killer cells: their role in defenses against disease. Science 214: 24–30
19. Levinson A, Silver D, Seed B (1984) Minimal size plasmids containing an M13 origin for production of single-strand transducing particles. J Mol Appl Genet 2: 507–517
20. Reed KC, Mann DA (1985) Rapid transfer of DNA from agarose gels to nylon membranes. Nucleic Acids Res 13: 7207–7221
21. Caves PK, Schultz WP, Dong E Jr, Stinson EB, Shumway NE (1974) New instrument for transvenous cardiac biopsy. Am J Cardiol 33: 264–267

V. Diagnostic Criteria

The Role of Biopsy in the Diagnosis and Follow-up of Myocarditis: A Critical Review

E. G. J. OLSEN, B. T. MEANY, and P. J. RICHARDSON*

Department of Histopathology, National Heart Hospital, Westmoreland Street, London WIM 813A
*Department of Cardiology, King's College Hospital, Denmark Hill, London SE5

The techniques of obtaining endomyocardial biopsies have varied over the past decades and have included needle biopsy or surgical intervention, particularly when removal of myocardium formed part of the operation, such as in cases of hypertrophic cardiomyopathy [1]. In 1962 Sakakibara and Konno [2] described and used the bioptome which is still in use. In the early 1970s this technique reached Western Europe, where it has gained widespread acceptance and although such investigations had also been undertaken in the United States at that time, is particularly in the 1980s that investigation by bioptome has been used with increasing frequency in many centres in the United States.

Since the orginal description of the bioptome [2] innumerable variations and adaptations of instruments designed for other purposes have been introduced. The most frequently bioptome employed in Britain is the King's bioptome [3] or that instrument initially described by Caves et al. [4].

The morphological diagnosis of myocarditis has been exceedingly difficult — akin to the clinical diagnosis. At one time several disease processes, including that of hypertension, were ascribed to "chronic myocarditis" [5].

That erroneous attribution was reversed [6], but morphological assessment was largely confined to examination at post mortem, where the incidence varied between 3.5% and 9% in larger autopsy series [7]. The use of the bioptome has reversed the inaccurate assessment of myocarditis, based on clinical or non-invasive techniques, but it was soon found that the interpretation of tissue samples was exceedingly difficult and varied greatly in the reports that appeared in the literature (between 3% and 63%) [8, 9]. This even occurred among experienced cardiac pathologists. [10]. Realisation of this fact led to a meeting of specialist cardiac pathologists in 1984, who defined the condition and suggested a classification of myocarditis by adapting the criteria previously described [11]. Though primarily designed for pathologists participating in a large-scale controlled multi-centre study investigating the efficacy of drug therapy on myocarditis, the suggested criteria will provide guidelines for other workers in the field.

Definition

Myocarditis is defined as "an inflammatory cell infiltrate of the myocardium with necrosis and/or degeneration of adjacent myocytes, not typical of ischaemic

H.-P. Schultheiß (Ed.)
New Concepts in Viral Heart Disease

damage associated with coronary arterial disease". In view of the fact that the natural history of myocarditis is still ill understood, the following sub-classification has been proposed [11].

First Biopsy

Myocarditis with or without fibrosis
Borderline myocarditis (re-biopsy may be indicated)
No myocarditis

Subsequent Biopsies

On-going myocarditis (with or without fibrosis)
Resolving (healing) myocarditis (with or without fibrosis)
Resolved (healed) myocarditis (with or without fibrosis)

Sub-classification of the type of inflammatory cells dominating the morphological picture, as well as the degree and distribution of the inflammatory cell infiltrate and fibrosis, is also advocated. For example, the inflammatory cell infiltrate may be predominantly composed of lymphocytes, eosinophils, neutrophils, giant cells or granulomas. It may also be a mixture of these above mentioned constituents. The distribution of the inflammation and fibrosis may be mild, moderate or servere. It may be focal, confluent or diffuse. The grading of myocarditis is important to monitor the effect of therapy.

The definition describes acute or active myocarditis. The inflammatory cells are in intimate contact with the myocardial fibres, not infrequently resulting in a scalloped contour of the wall the myocytes. In addition, varying degrees of necrosis would also be evident (Fig 1.). Distinction from myocardial infarction can be exceedingly difficult, and reliance is placed on whether or not the nuclei are present in the myocardial fibres adjacent to the inflammatory cell infiltrate. In myocarditis the most frequently encountered cellular infiltrate is lymphocytic. By the time this infiltrate is present in cases of a myocardial infarction, the nuclei would have disappeared from the myocytes. In myocarditis, however, the nuclei are well preserved, that is, in areas without frank necrosis (other changes including eosinophilia are also present in the myocardium of ischaemia). In the acute or active phase of myocarditis, fibrosis is usually (but not invariably) absent.

On-going myocarditis implies that the inflammatory cell infiltrate is as severe or more severe than the most recent previous biopsy. Fibrosis may or may not be present.

In resolving (healing) myocarditis the cellular infiltrate is significantly reduced, and the inflammatory cells are now at a distance from adjacent myocardial fibres, resulting in a smooth contour of these myocytes (Fig. 2). Very occasional foci of necrosis may be evident at the early stage of this phase. An increase in collagen tissue is usally present, initially taking the from of loosely arranged collagen fibres in which the inflammatory cells are embedded. Some focal areas of fibrosis may be recognisable towards the later stage of this phase.

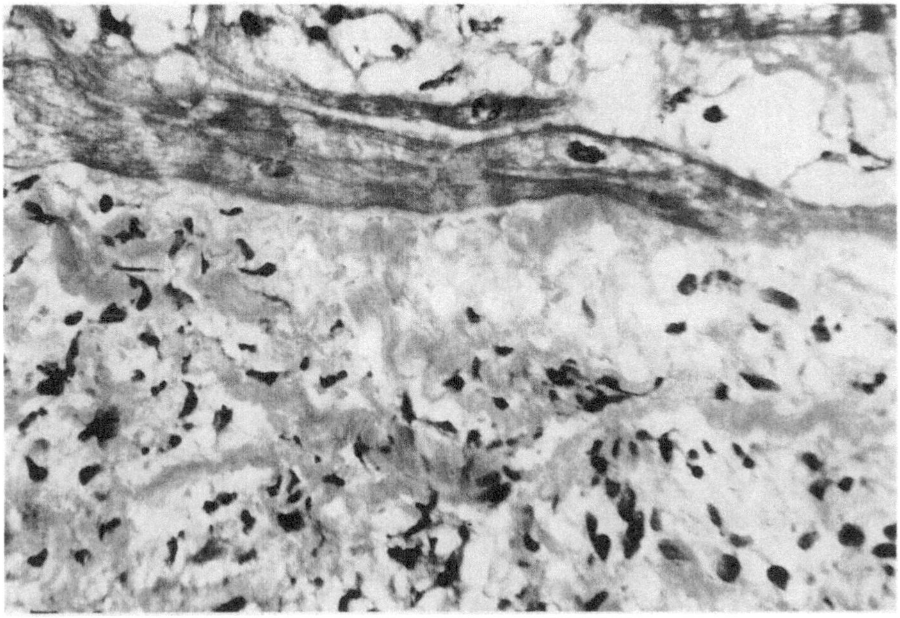

Fig. 1. Acute (active myocarditis). Close-up view of the interstitium showing a predominantly lymphocytic cellular infiltrate. The adjacent myocardial fibre has retained its nucleus. Frozen section, haemotoxylin and eosin, x 550

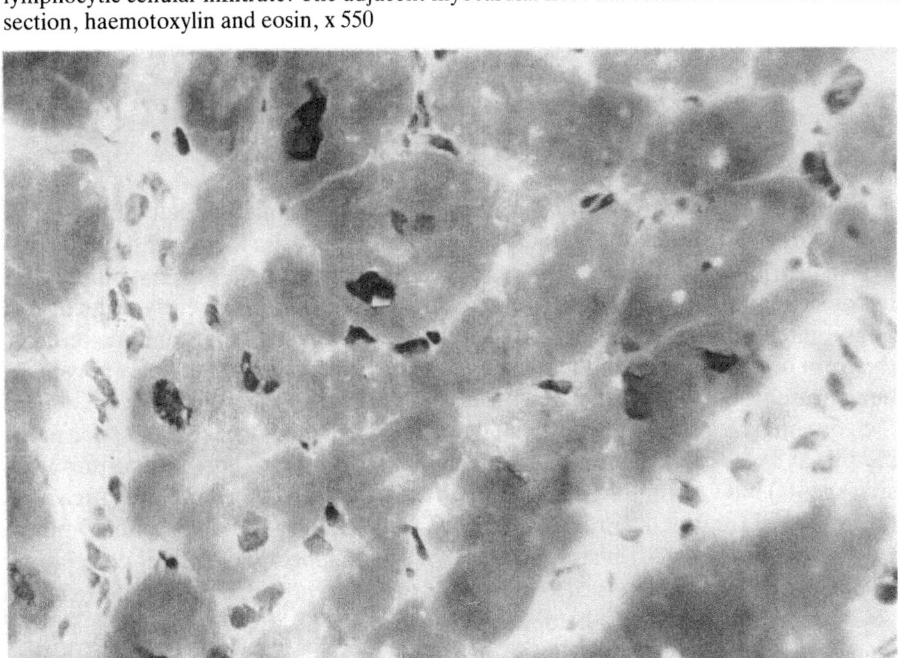

Fig. 2. Resolving (healing) myocarditis. A sparse cellular infiltrate (lymphocytic) can be seen in the widened interstitium. The inflammatory cells are now at a distance from adjacent myocardial fibres. Frozen section, haematoxylin and eosin, x 550

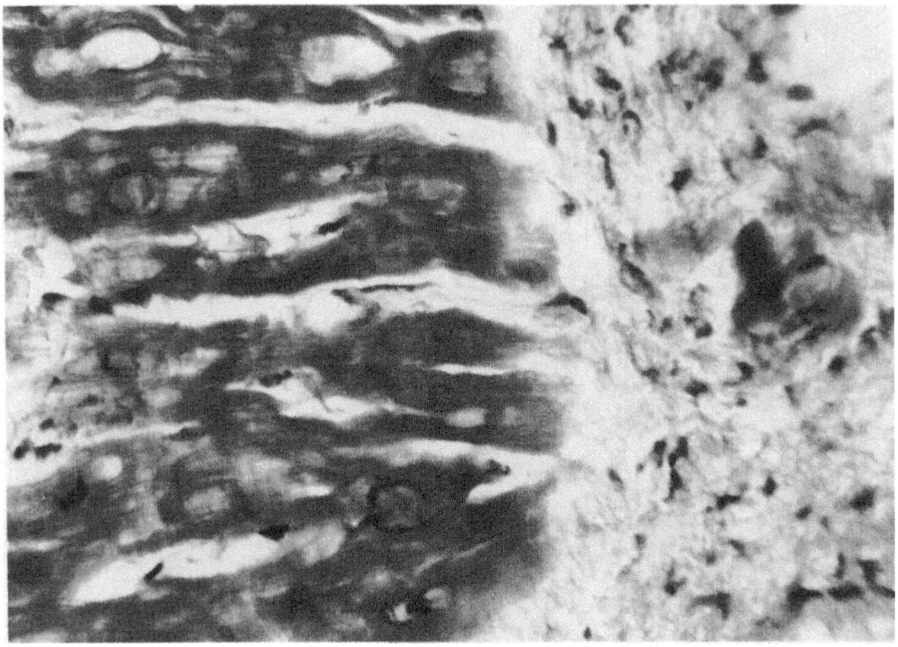

Fig. 3. Resolved (healed) myocarditis. The area on the *right* consists predominantly of collagen tissue, fibrocytes and an occasional lymphocyte. If this had been the patient's first biopsy, myocarditis could not have been diagnosed. Frozen section, haemotoxylin and eosin, x 550

Resolved (healed) myocarditis implies that the inflammatory cell infiltrate has disappeared. Fibrous tissue is usally prominent and may be focal, confluent or diffuse in distribution, often to a severe degree. If this is the patient's first biopsy, the fibrosis cannot be attributed to previous myocarditis. If, in the early stage of this phase, inflammatory cells may still be numerous, at best all that can be offered in the form of a diagnosis would be to suggest that myocarditis in the past may have been present, but a definitive diagnosis is not possible (Fig. 3).

Sequential biopsies to the initial diagnostic biopsy showing acute or active myocarditis at 3 weeks and 6 months (at the end of the course of the immuno-suppressive agents) have permitted monitoring the progress and therefore the efficacy of the results of treatment. Progression from acute to the resolving (healing) phase to the resolved (healed) phase have shown good clinico-pathological correlation.

Unfortunately, no large-scale controlled studies have so far been undertaken to assess the influence drug therapy may have on the disease process. It is well known that acute mycarditis can resolve spontaneously. For this reason, a multicentre controlled trial is in progress in the United States to try and answer this and other questions. Ample evidence has accumulated over recent years that virus-induced myocarditis, particulary Coxsackie B virus, plays a part in the pathogenesis of dilated cardiomyopathy, and this is discussed in other chapters in this volume.

Long-term follow-up of patients whose initial endomyocardial biopsies had shown acute or active myocarditis and who were suspected of suffering from dilated cardiomyopathy has been undertaken in several patients, but this report will concentrate on fifteen patients.

Patients and Methods

The present study includes patients, who have previously been reported [12, 13], with biopsy-proven acute myocarditis at King's College Hospital between January 1980 and December 1984. Of a total of 23 patients in this study, 15 were found to have had serial biopsy performed and to have tissue suitable for histopathological evaluation.

Clinical evaluation included clinical history, physical examination together with routine chest X-ray, electrocardiogram and echocardiography. Routine virology included virus (including Coxsackie B1–5) -neutralising antibody titres. Further investigation included cardiac catheterisation, haemodynamic measurements and, if clinically appropriate, serial endomyocardial biopsy [12]. The latter may have been indicated in the patients receiving immunosuppressive therapy [14].

In those patients who did not have haemodynamic catheter studies, non-invasive evaluation for long-term follow-up included electrocardiogram, chest X-rays and echocardiography: M-mode and two-dimensional (2-D) and radionuclide ventriculography performed as previously described [13, 15].

Results

Fifteen patients had serial biopsy and were divided into two groups as follows: *Group I.* See Table 1 for details representing the ten surviving patients who made

Table 1. Acute myocarditis — serial biopsies — survivors

Patient no.	Age	Sex	Symptom duration (Weeks)	Follow-up (Months)	EF Pre	(%) Last	Outcome
1.	47	F	1	60	29	55	COCM
2.	14	F	1	62	21	67	Normal
3.	33	M	25	70	60	62	Normal
4.	20	F	8	60	26	–	COCM
5.	27	F	1	55	64	74	Normal
6.	21	M	25	58	58	–	Normal
7.	1½	F	6	47	21	51	COCM
8.	43	F	20	18	15	15	COCM
9.	43	M	6	60	43	40	COCM
10.	27	F	4	60	10	70	Normal

EF (%), ejection fraction; COCM, dilated cardiomyopathy

Table 2. Acute myocarditis — serial biopsies — patients who have died or undergone transplantation

Patient no.	Age	Sex	Symptom duration (Weeks)	EF (%)	Outcome
1.	28	F	3	34	Died—arrhythmias
2.	18	M	2	23	Transplanted
3.	2	F	24	17	Died—cardiogenic shock
4.	58	M	16	10	Died—cause unknown
5.	28	F	2	11	Transplant

up group I age was 27 (range: 18 months to 47 years); there were seven females males and three with an average duration of symptoms of 10 weeks with follow-up of 55 months. The ejection fraction at the time of diagnosis was mean 34.7% (n =10) and 54.25% mean on late follow-up (n =8).

Group II. The five patients in group II have either died or undergone cardiac transplantation (see Table 2). The average age was 22.8 years (range 2–38); there were three females and two males. The duration of symptoms before presentation was 10 weeks, and the mean ejection fraction at diagnosis was 19%. Three patients died and two have undergone transplantation.

To illustrate three types of patients let us consider Fig. 4. This shows the long-term serial histological and haemodynamic follow-up in three patients repre-

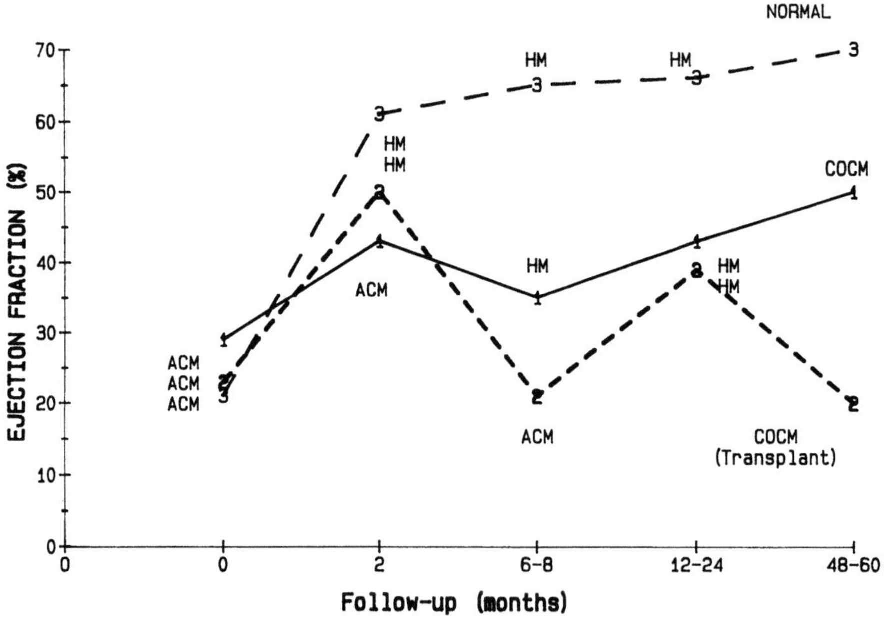

Fig. 4. Ejection fraction and histology features in three patients' follow-up for 60 months. *ACM*, acute myocarditis; *HM*, healing/healed myocarditis; *COCM*, dilated cardiomyopathy

sentative of the natural history of patients with biopsy-proven myocarditis who were initially too severly ill to warrant full haemodynamic investigation. In each case the presenting ejection fractions were very similar, and the biopsy findings at diagnosis were indistinguishable. Whilst active myocarditis may persist for up to 6 months, in the majority of patients a healing or healed pattern will be seen on serial biopsy by 12 months. The haemodynamic response, however, will vary at this stage. In patient 1 there was persistent impairment of left ventricular function and on late follow-up dilated cardiomyopathy was deemed to be present clinically and histologically. In patient 2, whilst there was initial improvement in response to immunosuppression by 2 months, deterioration of cardiac function was associated with persisting or on-going myocarditis. A healed pattern was seen at 1–2 years, but the ejection fraction remained reduced and, indeed, at long-term follow-up and reinvestigation transplantation was thought to be the only possible clinical solution. It is of interest that in patient 2 virus enterovirus RNA was detectable using molecular biological techniques. Patient 1, however, who developed dilated cardiomyopathy at the last biopsy at 60 months follow-up did not have virus RNA detectable. In patient 3, who received immunosuppresive therapy, there was resolution of active myocarditis with improvement of ejection fraction to normal, and this clinical status has persisted to date, and the patient is well and asymptomatic.

Morphological Differential Diagnosis

Myocarditis has also not infrequently been encountered in conditions that are morphologically identical to those of dilated cardiomyopathy, namely heart failure occurring in alcohol abuse and heart failure occurring in the peripartal period. Enzyme analysis in cases of alcohol abuse has been used to distinguish this form of heart failure from dilated cardiomyopathy [16] and in cases of peripartal heart disease though the history suffices to categorise the disease accurately. Studies on peripheral blood T cells and lymphocyte subsets have established a high ratio of helper:suppressor T cells [17].

Myocarditis can occur as a result of many other causes including all viruses, rickettsia, bacteria, fungi, as well as parasites, sensitivity reaction and anaphylaxis. One type of myocarditis in the context of *cardiomyopathy* needs special emphasis and that is *restrictive* cardiomyopathy. The original concept of two different diseases in this category, namely endomyocardial fibrosis [18] which is believed to be confined to the tropical zones and Loeffler's endocarditis parietialis fibroplastica which is associated with eosinophilia and confined to the temperate zones [19], has been superseded by the unitarian approach based on increasingly reliable evidence [20, 21]. It has been shown that morphological distinction between the tropical condition and those examined from the temperate zones is not possible. The origin of the disease process has been traced back to the presence of eosinophils in the myocardium, showing an often intense myocarditis. Further work has shown that eosinophils are morphologically abnormal taking the form of vacuolation and degranulation [21, 22].

Critical Analysis

The interpretation of myocarditis at morphological level can be exceedingly difficult as has been suggested above. The reason for this is that at histological level lymphocytes may be mimicked by other cellular elements of the interstitium, and initially it was sometimes necessary to take recourse to electron-microscopic examination in order to evaluate the cell type accurately. Furthermore, it is well established that lymphocytes can occur in absolutely normal cardiac tissue, and therefore the mere presence of lymphocytes in the myocardium does not necessarily constitute myocarditis. It is the interaction of the interstitium with the cellular infiltrate together with the adjacent myocytes that permit one to make a diagnosis of myocarditis. Reference to the incidence of myocarditis from various reports (3%–63%) has already been made, and the reason, as far as interpretation of endomyocardial biopsy is concerned at morphological level, has also been discussed. Other possible causes for the great ariation includes patient selection; if, for example, a patient is biopsied towards the latter phases of the disease process, resolved or healed myocarditis may well be evident, and this does not permit diagnosis. Experience has, however, shown that if patients in the healing phase are biopsied for the first time, assessment of this stage can be made with a high degree of accuracy at morphological level.

Apart from patient selection, there ist also evidence of seasonal variation, which also points to a virus being responsible for the myocarditis. Regional differences also exist. In the year 1982–1983, from one centre in London the incidence of myocarditis was 56% in patients suspected of suffering from dilated cardiomyopathy, whereas the contribution of myocardial biopsies from all other centres remained between 9% and 30%. All these factors may contribute to the variation of incidence.

What is the evidence that myocarditis leads to dilated cardiomyopathy? Several experimental studies have lent strong support for this assumption. As far as man is concerned, until recently reliance had to be made on neutralisation tests, and in many of these patients high or rising titres to Coxsackie B virus were found in those patients with myocarditis or suspected of suffering from dilated cardiomyopathy. It was not until recently that the virus-specific hybridisation probe has been applied to endomyocardial biopsies, and there is now very good evidence that virus is responsible for myocarditis. Virus has also, however, been found in patients that show the non-specific changes of a hypertrophied dilated myocardium (dilated cardiomyopathy). It is therefore highly likely that progression from myocarditis to dilated cardiomyopathy takes place [13, 23, 24]. Furthermore, it has also been shown that immunological consequences occur as a result of virus infection, and there is now a very good body of evidence that virus infection may or may not result in myocarditis, inducing an immunological reaction which leads to dilated cardiomyopathy. This constitutes the so-called infectious-immune pathogenesis [25].

As far as follow-up studies are concerned, neither the histological findings at diagnosis nor indeed the findings at 2–6 months' follow-up will necessarily predict the outcome. At 8–12 months the finding of a healed myocarditis and a good ejection fraction usually means that these patients will remain stable. Late spontaneous recovery has been seen in our experience and also in other workers' experience [26, 27].

Dec et al. [27] report spontaneous improvement in 40%–50% of patients with acute myocarditis. This is similar to our experience.

The question of whether it is necessary to subject these patients to endomyocardial biopsies rather than using other methods has been explored. In this context the investigation by gallium 67 could be mentioned [28]. Whereas there is no doubt that myocarditis can be diagnosed by this technique, it is unfortunately not possible to identify the phase myocarditis has reached. As this has therapeutic implications it is necessary to subject these patients to bioptome investigation for a correct diagnosis and for monitoring the progression of the disease.

Conclusion

From the above account, there is clear evidence that evaluation of fresh endomyocardial tissue plays an important role in making a diagnosis of myocarditis. It, furthermore, plays an immensely important role in the follow-up studies of patients. Whereas clinical parameters are, of course, the ultimate guideline, morphological changes usually correlate well with the clinical picture. Biopsy examination also permits prognosis to be assessed at 12 months after initial morphological diagnosis, as has been clearly exemplified by the long-term follow-up of the 15 patients who have been discussed.

References

1. Olsen EGJ (1977) Myocardial biopsy. In: Hamer J (ed) Recent advances in cardiology. Churchill Livingstone, Edinburgh, no 7, chap. 13, pp 349–367
2. Sakakibara S, Konno S (1962) Endomyocardial biopsy. Jpn Heart J 3: 537–543
3. Richardson PJ (1974) King's endomyocardial bioptome. Lancet I: 660–661
4. Caves PK, Schultz WP, Dong E Jr, Stinson EB, Shumway NE (1974) New instrument for transvenous cadiac biopsy. Am J Cardiol 33: 264–267
5. Fahr GE (1923) Hypertension heart. The most common form of so-called chronic myocarditis. JAMA 80: 981–984
6. Wood P (1968) Non-rheumatic myocarditis and miscellaneous cardiomyopathes. In Paul Wood's diseases of the heart and circulation, 3rd edn. Eyre and Spottiswoode, London, pp 707–750
7. Olsen EGJ (1980) The pathology of the heart, 2nd edn. Macmillan, London, p 178
8. Fenoglio JJ Jr, Ursell PC, Kellog CF, Drusin RE, Weiss MG (1983) Diagnosis and classification of myocarditis by endomyocardial biopsy. N Engl J Med, 308: 12–8
9. Zee-Cheng C-S, Tsai CC, Palmer DC, Codd JE, Pennington DG, Williams CA (1984) High incidence of myocarditis by endomyocardial biopsy in patients with idiopathic congestive cardiomyopathy. J Am Coll Cardiol, 3: 63–70
10. Shanes JG, Ghali J, Billingham ME, Ferrans VJ, Fenoglio JJ, Edwars WD, Tsai CC, Saffitz JE, Isner J, Furner S, Subramanian R (1987) Interobserver variability in the pathologic interpretation of endomyocardial biopsy results. Circulation 75: 401–405
11. Aretz HT, Billingham ME, Edwards WD, Factor SM, Fallon JT, Fenoglio JJ, Olsen EGJ, Schoen FJ (1987) Myocarditis: a histopathologic definition and classification. Am J Cardiovasc Pathol, 1: 3–13

12. Richardson PJ, Daly K, Gishen P (1984) Haemodynamic findings in biopsy proven acute myocarditis. In: Bolte HD (ed) Viral heart disease. Springer, Berlin Heidelberg New York, pp 165–172
13. Quigley PJ, Richardson PJ, Meany BT, Olsen EGJ, Monaghan MJ, Jackson G, Jewitt DE (1987) Longterm follow up of acute myocarditis. Correlation of ventricular function and outcome. Eur Heart J 8 [Suppl J] 39–42
14. Daly K, Richardson PJ, Olsen EGJ, Morgan-Capner P, McSorley C, Jackson G, Jewitt DE (1984) Acute myocarditis. Role op histological and virological examination in the diagnosis and assessment of immunosuppressive treatment. Br Heart J 51: 30–35
15. Strauss HW, Singleton R, Birow R, Rehn T, Bailey IK, Pitt B (1977) Mutiple gated acquisition (MUGA): an improved noninvasive technique for the evaluation of regional wall motion and left ventricular function. Am J Cardiol 39: 284 (Abstract)
16. Richardson PJ, Wodak AD, Atkinson L, Saunders JB, Jewitt DE (1986) Relation between alcohol intake, myocardial enzyme activity, and myocardial function in dilated cardiomyopathy. Evidence for the concept of alcohol induced heart muscle disease. Br Heart J 56: 165–710
17. Sanderson JE, Olsen EGJ, Gatei D (1986) Peripartum heart disease: an endomyocardial biopsy study. Br Heart J 56: 285–291
18. Davies JNP (1948) Endomyochardial fibrosis in Africans. East Afr Med J 25: 10–14
19. Loeffler W (1936) Endocarditis parietalis fibroplastica mit Bluteosinophilie, ein eigenartiges Krankheitsbild. Schweiz Med Wochenschr 66: 817–820
20. Brockington IF, Olsen EGJ (1973) Loeffler's endocarditis and Davies' endomyocardial fibrosis. Am Heart J 85: 308–322
21. Oakley CM, Olsen EGJ (1977) Eosinophilia and heart disease, Br Heart J 39: 233–237
22. Olsen EGJ, Spry CJF (1979) The pathogenesis of Loeffler's endomyocardial disease, and its relationship to endomyocardial fibrosis. In: Yu PN, Goodwin JF (eds) Progress in cardiology, Lea and Febiger, Philadelphia, pp 281–303
23. Bowles NE, Olsen EGJ, Richardson PJ, Archard LC (1986) Detection of coxsackie-B-virus-specific RNA sequences in myocardial biopsy samples from patients with myocarditis and dilated cardiomyopathy. Lancet I: 1120–1123
24. Archard LC, Bowles NE, Olsen EGJ, Richardson PJ (1987) Detection of persistent coxsackie B virus RNA in dilated cardiomyopathy and myodarditis. Eur Heart J 8: 437–440
25. El-Hagrassy MMO, Banatvala JE, Coltart DJ (1980) Coxsackie B-virus specific IgM responses in patients with cardiac and other diseases. Lancet II: 1160–1162
26. Figulla HR, Rahl FG, Nieger M, Luig H, Kruytzer H (1985) Spontaneous haemodynamic improvement of stabilisation and associated biopsy findings in patients with congestive cardiomyopathy. Circulation 71: 1095–1104
27. Dec WG, Pelakios IF, Fallon JT, et al (1985) Acute myocarditis in the septrum of acute dilated cardiomyopathies. Clinical features histological correlates and clinical outcome. N Engl J Med 312: 885–890
28. O'Connel JB, Robinson JA, Gunnar RM, Scanlon PJ (1985) Clinical aspects of virus/immune myocarditis. In: Sekiguchi M, Olsen EGJ, Goodwin JF (eds) Myocarditis and related disorders. Springer, Berlin Heidelberg New York, pp 101–106

Impaired Myocardial Ultrastructure and Cytoskeleton in Cardiomyopathic Human Myocardium

Jutta SCHAPER[1], R. FROEDE[1], A. BUCK, and N. BLEESE[2]

[1] Max Planck Institute, Department of Experimental Cardiology, Benekestr. 2, D-6350 Bad Nauheim, FRG
[2] Kerckhoff Clinic, Department of Thoracic and Cardiovascular Surgery, Benekestr. 4–6, D-6350 Bad Nauheim, FRG

In dilated cardiomyopathy the principal morphologic changes are hypertrophy of myocytes, increased content of connective tissue, i.e., fibrosis, and degeneration of subcellular myocardial components. In recent years, the ultrastructural changes occurring in various stages of either hypertrophic or dilated cardiomyopathy have been investigated extensively in human patients either in endomyocardial biopsies taken from patients during catheterization or in intraoperatively removed outflow tract muscle [2, 7, 8, 10, 11] (for overview see [1, 6]). With the advent of cardiac transplantation surgery, however, it is possible to investigate changes in end-stage cardiomyopathy. These hearts also offer the possibility of investigating tissue samples larger than biopsy material, so that histologic techniques involving frozen sections and immunocytochemistry using monoclonal antibodies can be applied easily.

We tried to determine the morphologic correlate of severely reduced cardiac function in explanted hearts from patients who had undergone cardiac transplantation surgery because of chronic heart failure in dilated cardiomyopathy. The subcellular composition of the myocardium was studied by electron microscopy, and the state of the cytoskeleton and of associated proteins was investigated by immunfluorescence microscopy using monoclonal antibodies.

Tissue samples were taken immediately upon removal from the left ventricle, snap-frozen in liquid nitrogen and stored at −70 °C. Smaller samples were prepared for electron microscopy by fixation in cold 2% buffered glutaraldehyde and embedded in epon following routine procedures.

Electron Microscopy

The myocardial cells were hypertrophied and the interstitial space widened containing an increased amount of fibrous tissue which consisted of proliferating fibroblasts, macrophages, large amounts of cellular debris and collagen with capillaries interspersed (Fig. 1). The myocardial cell nuclei were often significantly enlarged and abnormally shaped exhibiting an unusual chromatin distribution, multiple nucleoli, and cytoplasmic inclusions (Fig. 2). Mitochondria were of varying size and shape and they were frequently clustered in areas devoid of myofilaments. Glycogen was usually abundant in these areas. The T-tubular system showed proliferation and dilation (Fig. 3). The main ultrastructural alteration,

H.-P. Schultheiß (Ed.)
New Concepts in Viral Heart Disease

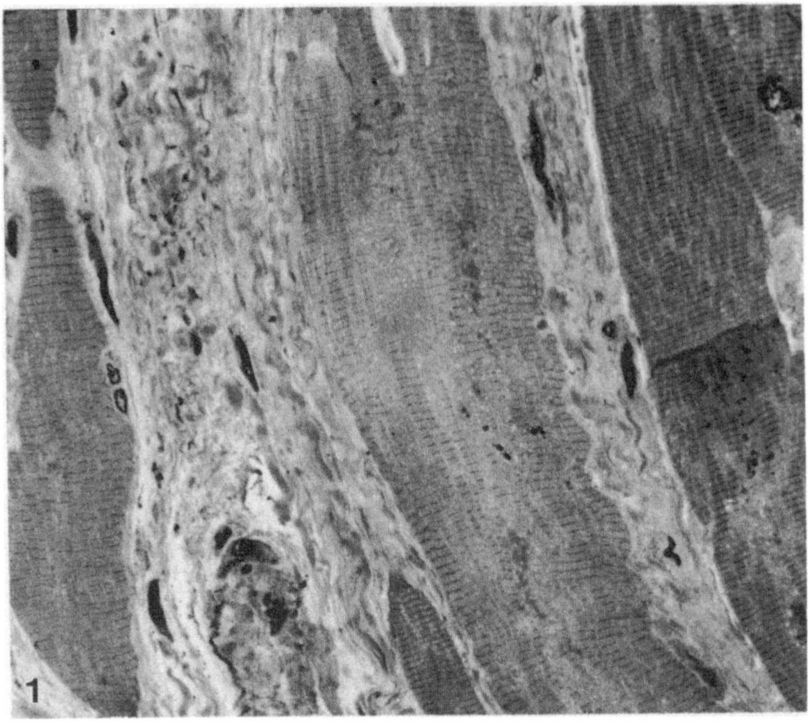

Fig. 1. Light micrograph showing an increased interstitial space that contains collagen, fibroblasts, and cellular debris. Myocardial cells are hypertrophied and show lack of contractile material. x 1000

however, was the lack of contractile material in these hypertrophied myocytes. This change was evident throughout the cells, at the periphery as well as in the central parts around the nuclear poles (Fig. 4).

Immunofluorescence

Monoclonal antibodies against desmin [3, 9, 12], alpha- and beta-tubulin [14], vinculin [4], laminin [13], and fibronectin [5] were used on frozen sections of viable human myocardium. Desmin belongs to the intermediate filaments, it is typical of muscle cells and occurs at the Z-lines of the sarcomeres in normal myocardial cells. In cardiomyopathic hearts it was increased and irregularly distributed in many cells, indicating severe disarrangement within the contractile material (Fig. 5). Desmin was also observed in smooth muscle cells and, to a lesser extent, in endothelial cells. Alpha- and beta-tubulin, another component of the cytoskeleton, was diffusely localized throughout the cells, whereas in normal myocytes from rats it was found to be bound to the microtubular structures. Vinculin, normally present on the inner side of the sarcolemma and at the intercalated discs,

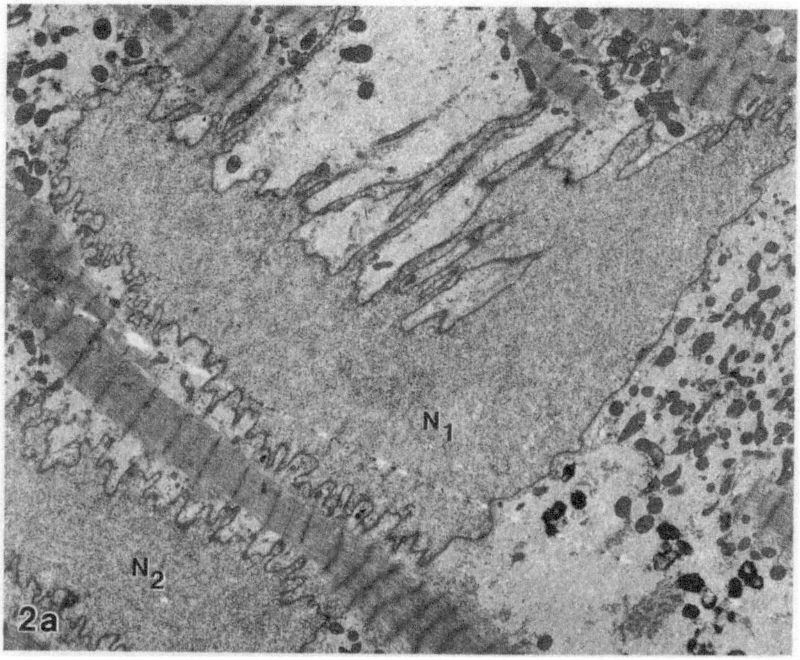

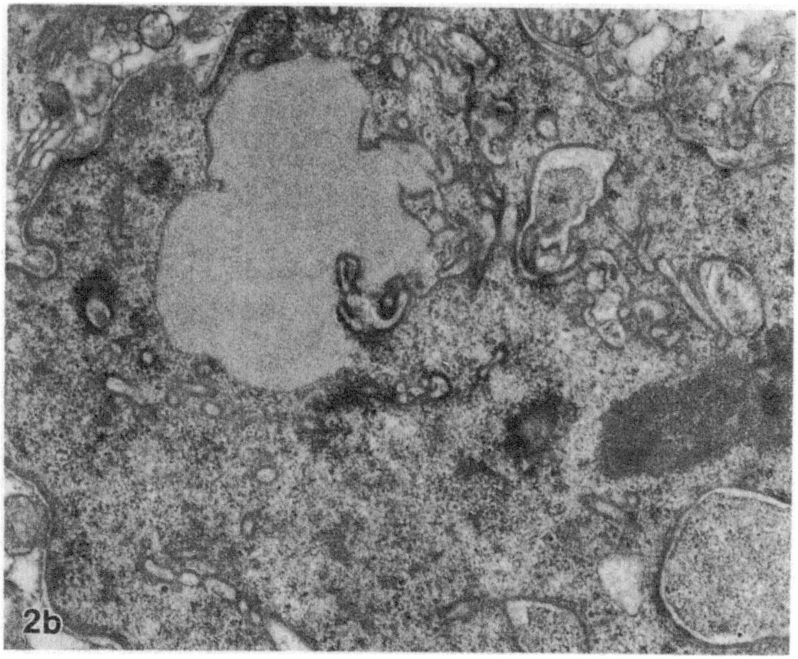

Fig. 2 a, b. Nuclear changes. **a** Nucleus of abnormal size and shape in a myocardial cell *(N_1)*. Next to it there is another nucleus *(N_2)*. **b** Numerous cytoplasmic inclusions. **a** x 4500; **b** x 15000

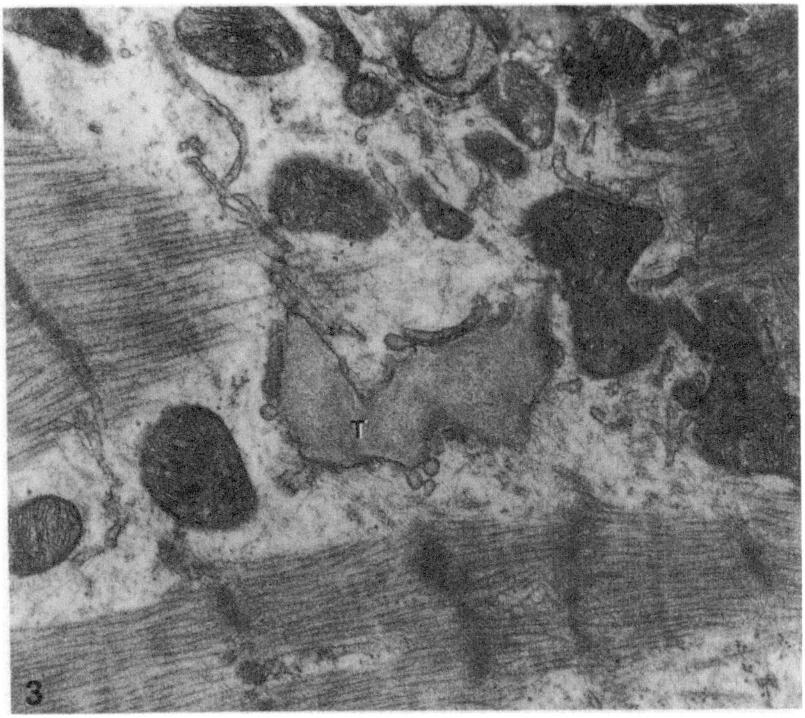

Fig. 3. Dilated T-tubular system *(T)* in a degenerating myocardial cell. x 19500

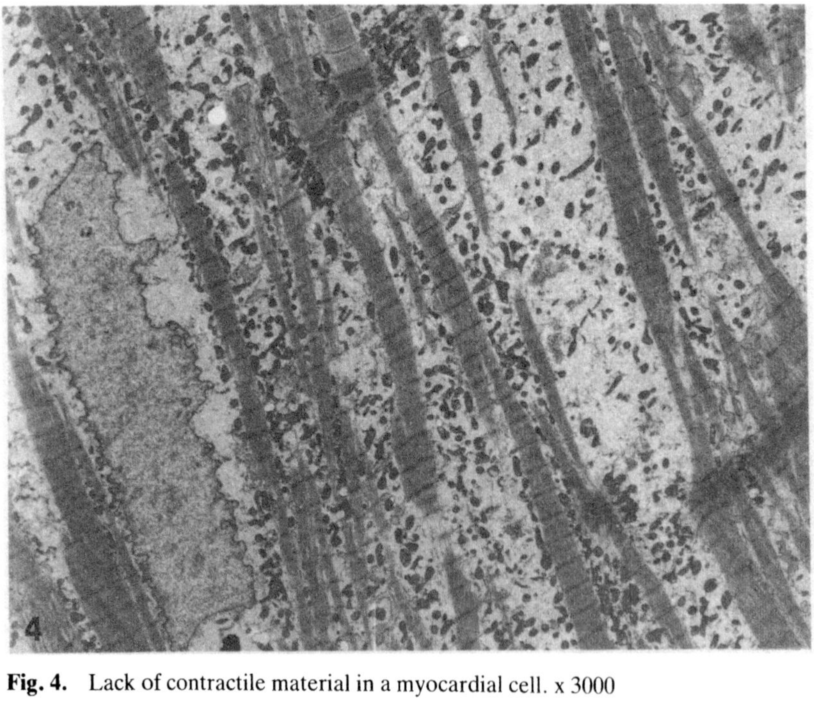

Fig. 4. Lack of contractile material in a myocardial cell. x 3000

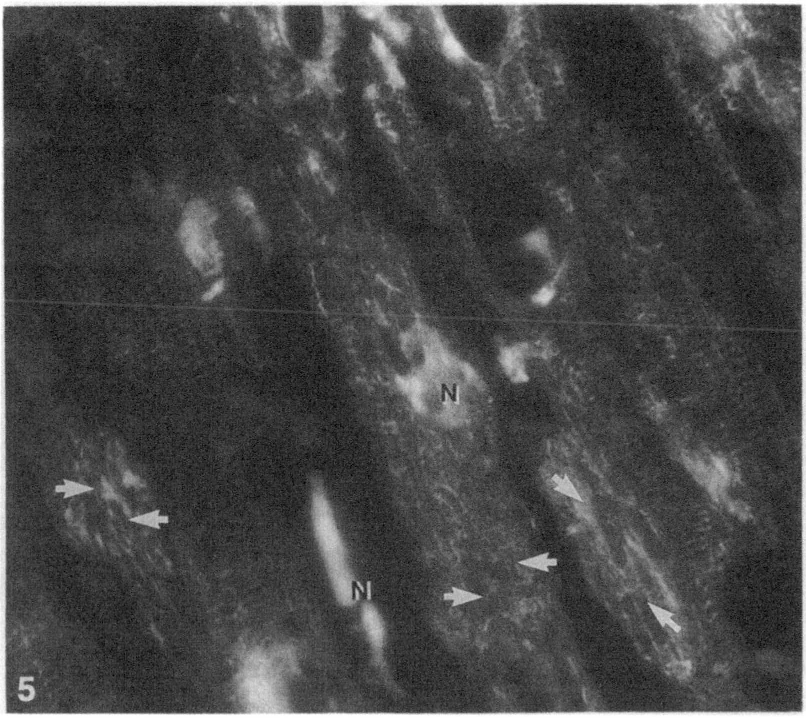

Fig. 5. Irregular distribution of desmin *(arrows)*. Nuclei *(N)* were counterstained with propidium iodide

was greatly increased in amount at these locations and it also occurred within the myocardial cells (Fig. 6). Localization in endothelial and smooth muscle cells appeared to be normal.

Laminin and fibronectin are components of the basal lamina around the myocardial cells. In normal tissue these proteins are localized as a fine rim of fluorescence at the outer side of the sarcolemma. In diseased myocardium, immunofluorescence for both laminin and fibronectin was greatly increased and present in larger proportions in the extracellular space (Fig. 7). Vimentin [12] was present only in vascular endothelium and in cells of the interstitial space, i.e., fibrocytes and fibroblasts. Vimentin-containing cellular elements were increased in number in diseased myocardium corresponding to the increased amount of fibrosis (Fig. 8).

These ultrastructural and immunocytochemical findings indicate that the myocardial morphology in dilated cardiomyopathy is severely impaired. The cells are hypertrophied exhibiting signs of ongoing degeneration of which the decreased content of contractile material seems to be the most severe change. The remaining sarcomeres, even when they appear to be normally organized by electron microscopy, may still be further impaired as seen from the distorted arrangement of an enlarged amount of desmin intermediate filaments or the protein vinculin. Vinculin, a protein anchoring actin filaments at the level of the Z-line and of the

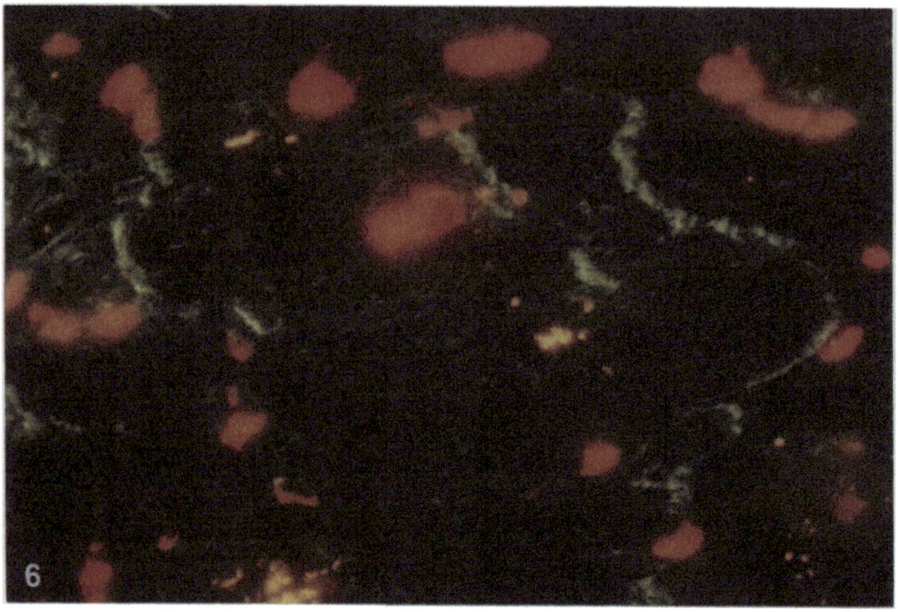

Fig. 6. Vinculin is localized in increased amounts beneath the sarcolemma and at the intercalated disc as well as within the hypertrophied myocardial cells. Nuclei are stained red with propidium iodide

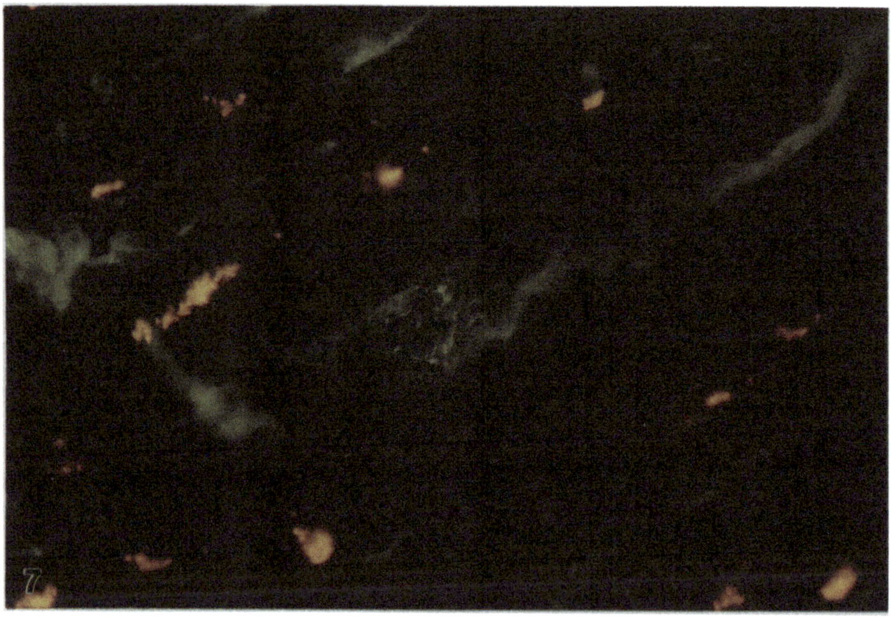

Fig. 7. Laminin fluorescence at the outside of the myocardial cells is greatly enhanced in diseased human myocardium

Fig. 8. Vimentin is localized within the cells of the interstitial space, nuclei are stained red. *Dark areas* correspond to hypertrophied myocardial cells

intercalated disc, was observed to be significantly increased and to occur also in the inner part of the myocardial cell. This may further contribute to the impairment of contractile function of individual sarcomeres. Vinculin and laminin form a close connection across the cell membrane, thereby linking the cytoskeleton (via actin-vinculin-laminin) to the extracellular matrix. An increase in vinculin and laminin content may cause an increased stability of the myocardial cells, but it may also, on the other hand, cause an increase in stiffness and thereby an impairment of contractile function. The augmentation of the fibrous tissue content may further contribute to a reduction of cardiac function.

In conclusion, from these studies it becomes evident that in late-stage dilated cardiomyopathy the morphologic correlate of reduced cardiac function may be: (a) hypertrophy of myocardial cells that show subcellular degeneration but an increase in desmin and vinculin content; (b) increased content of fibrous tissue.

References

1. Fenoglio JJ Jr (ed) (1982) Endomyocardial Biopsy: Techniques and Applications. CRC, Boca Raton, Fla
2. Ferrans VJ, Morrow AG, Roberts WC (1972) Myocardial ultrastructure in idiopathic hypertrophic subaortic stenosis. A study of operatively excised left ventricular outflow tract muscle in 14 patients. Circulation 45: 769–772
3. Franke WW, Schmid E, Osborn M, Weber K (1979) The intermediate sized filaments of human endothelial cells. J Cell Biol 81: 570–580

4. Geiger B, Tokuyasu KT, Dutton AH, Singer SJ (1980) Vinculin, an intracellular protein localized at specialized sites where microfilament bundles terminate at cell membranes. Proc Natl Acad Sci USA 77: 4127–4137
5. Hynes RO, Yamada KM (1982) Fibronectins: multifunctional modular glycoproteins. J Cell Biol 95: 369–377
6. Kaltenbach M, Loogen F, Olsen EGJ (eds) (1978) Cardiomyopathy and Myocardial Biopsy. Springer, Berlin Heidelberg New York
7. Kuhn H, Knieriem HJ, Breithardt G, Hort W, Loogen F (1980) Diagnostic aspects of endomyocardial catheter biopsy (EMCB): clinical and morphological correlations. In: Bolte HD (ed) Myocardial Biopsy. Springer, Berlin Heidelberg New York
8. Kunkel B, Schneider M, Kober G, Bussmann WD, Hopf R (1982) Myocardial biopsy — morphological analysis and clinical correlations. Z Kardiol 71: 787–794

Immunofluorescence Microscopy in the Diagnosis and Follow-up of Patients Suspected of Having Inflammatory Heart Disease

E. H. HAMMOND, R. L. MENLOVE and J. L. ANDERSON

Departments of Pathology, Biophysics, and Cardiolopy, LDS Hospital,
Eighth Avenue and C Street, Salt Lake City, UT 84143, USA
University of Utah College of Medicine Utah, 84132 Salt Lake City, USA

Introduction

In a previously published study, we showed that immunofluorescence (IF) microscopy can help to distiguish myocarditis from idiopathic dilated cardiomyopathy [1]. However, since the previous study was relatively small, consisting only of 79 patients, an evaluation of differing patterns of IF staining as well as serial changes in patients biopsied on several occasions were not reported. We now report our findings in an expanded series of 286 endomyocardial biopsies (208 patients) which were prospectively examined by light and IF microscopy. Our findings suggest that IF microscopy is a complementary technique which serves to define better the presence and type of inflammatory heart disease found in such patients.

Methods

Patient Selection for Endomyocardial Biopsies. Patients selected for endomyocardial biopsies generally presented with dilated hearts and systolic dysfunction leading to unexplained heart failure of variable duration in the absence of coronary artery or valvular heart disease. Rarely, patients presented with arrhythmia without evident heart failure. Tissue examination included in this study were initial and follow-up transvenous endomyocardial biopsies which were of sufficient quantity to be submitted for light microscopy and IF. Right ventricular biopsies were performed transvenously by either the internal jugular approach (method according to Mason [2]) or the femoral venous approach (method of Anderson [3]). Biopsies submitted for assessment of restrictive or hypertrophic cardiomyopathy, transplant rejection, anthracycline cardiotoxicity, or alcoholic cardiomyopathy were excluded from this study.

Sample Preparation. Two to five (median three) endomyocardial samples were obtained from each patient on a saline-soaked filter paper. Each sample measured 2–3 mm in greatest dimension. For IF, a small fragment of at least two pieces was removed and snap frozen in the cryostat and immediately sectioned with ornithine carbamoyltransferase (OCT) as embedding medium. The majority of each portion of the specimen was processed for light microscopy by routine methods.

H.-P. Schultheiß (Ed.)
New Concepts in Viral Heart Disease

Light microscopy. Myocarditis was diagnosed by light microscopy on the basis of the presence of number of mononuclear cells (lymphocytes and histiocytes) found in interstitial regions by use of the Dallas criteria [4]. In addition, the total number of these cells examined under high power (x 400) in the entire specimen was tabulated; only those patients showing an mean number of mononuclear inflammatory cells greater than five per high-power field with associated myocyte injury were considered to have myocarditis [5]. Vasculitis was defined by the presence of infiltrating mononuclear cells in the walls of small aterioles or capillaries (capillaritis) associated with endothelial cell swelling, necrosis of the vessel wall, or thrombosis. Fibrosis in the heart was defined by the use of a trichrome stain. Inflammatory cells found in such fibrotic regions were ignored. Data were recorded prospectively in computer files.

Immunofluorescence. Frozen sections for IF microscopy were incubated with poly-clonal fluorescein-conjugated rabbit antibodies directed against IgG, IgM, IgA, C3, C1q, fibrinogen (BCA-Cappel Laboratories, West Chester, Pennsylvania) and monoclonal mouse antibodies directed against HLA-DR (Ia-like) (Hybritech), pan-B antigen (B-4, Coulter, Hialeah, Florida), pan-T antigen (T-1 lytic, Coulter), and pan-monocyte (M-1, Coulter). T lymphocyte-positive samples were also incubated with monoclonal antibodies to CD4 (helper) and CD8 (suppressor/cytotoxic) (Ortho Pharmaceuticals Corp., Raritan, New Jersey). Frozen sections were air dried, washed in phosphate-buffered saline (PBS), and then incubated with conjugated (IgG, IgM, IgA, C3, C1q, fibrinogen) or unconjugated (HLA-DR, pan-B, pan-T, pan-monocyte) antibodies at optimal dilutions for 30 min at 25 °C. After a wash in PBS, sections were either coverslipped in polyvinyl alcohol (in the case of conjugated anitbodies) or incubated with fluoroscein-labeled IgG fractions of rabbit-antimouse (BCA-Cappel Laboratories or Coulter) for 30 min at 25 °C. Slides were washed after the final incubation and coverslipped in polyvinyl alcohol. Appropriate positive and negative control slides were run with each batch. The IF staining with each antiserum was graded on a scale of to 3+ based upon the amount of antibody deposited in either an interstitial or vascular distribution. The staining was only considered significant (1+) if it was brighter in intensity than or in a different distribution from staining for albumin. Samples containing inflammatory cells were graded according to the infiltrate. Relative amounts of various cell types were expressed as the ratio of one to another. HLA-DR staining of blood vessels was considered strong (bright, generalized), weak (pale, focal), or absent. IF results were recorded before examination by light microscopy. All samples were assessed prospectively by one of us (EHH) who was blinded to patient clinical characteristics and eventual diagnosis and outcome. IF was performed on a Zeiss IF microscope (Carl Weiss Inc., Thornwood, New York). Data were recorded prospectively in computer files.

The presence of circulating (species-nonspecific) serum heart-reactive antibody (HRA) was also routinely evaluated by incubation of serially diluted samples of the patient's serum with sections of normal dog heart and graded by comparison with similar dilutions of normal control serum. The HRA assay was evaluated by indirect IF and was considered positive if HRAs (antisarcolemmal, antimyofibrillary, or antinuclear) were present in a dilution of > 1:16. A dilution of

> 1:16 was graded as 1+; a dilution of > 1:64 was graded 2+; and a dilution > 1:256 was graded 3+.

Tabulation of Data. Results of light microscopy and IF were entered in the LDS Hospital TANDEM computer data base by means of a classification scheme in which the presence and quantity of each features were recorded [1]. In addition, patient clinical characteristics were recorded. Subsequently, the data were downloaded to an IBM PC/XT microcomputer (IBM Corp., Atlanta, Georgia) where the data were manipulated and statistically analyzed.

The strategy of data analysis has been previously described[1] and was largely unchanged for this larger group of patients.In this study, complement (C) had to be present for the IF score to be considered positive. Thus, IgG of 1 was never sufficient for the diagnosis of inflammatory heart disease. A *P* value of < .05 was considered significant for statistical comparison.

Results

Clinical Findings. Clinical features of this patient population did not vary significantly from that previously described [1]. Since no clinical features correlated significantly with the presence or absence of myocarditis or mortality, we did not reanalyze them in the current study.

Light Microscopy. Eighteen (8.7%) of 208 patients were found to have significant cardiac inflammation and myocyte injury diagnostic of "active myocarditis" according to the Dallas criteria [4]. Sixteen of 208 patients (7.7%) showed "borderline myocarditis" on light microscopy. "Borderline myocarditis", as defined by the Dallas criteria, is inflammation present in quantities insufficient to diagnose active myocarditis or that not associated with myocyte necrosis.[4] Cardiac hypertrophy was also found in 120 of 208 patients (58%). Fibrosis was present in 62 patients (30%).

IF Findings. As previously reported, the IF variables that reliably identified patients with cellular infiltrates were IgG and C3. Since IgG was about twice as closely associated with cellular infiltration as was C3, the total IF scale was optimized by the formula 2 (IgG) + C3. This formula was prospectively applied to the present series. The only modification was that C3 was required. This meant that all patients in the current series had values of IF score > 2. This association is seen in Table 1. To construct this table, patients were classified according to IF scale score and the presence or absence of myocarditis by the Dallas criteria. Then, the frequency of patients in each contingent category was tabulated. The receiver operating characteristics (ROC) curve for the IF scale, 2 (IgG) + C3, as previously constructed, showed that the IF scale is a good identifier of an inflammatory infiltrate and that 2 (IgG) + C3 $\geq$ 2 is a reasonable criterion for identification. Thirty-three (16%) of 208 patients were found to have an IF score > 2. Of these 33 patients, 18 (55%) showed active myocarditis; 13 (39%) showed borderline myocarditis; and 2 (6%)

Table 1. Predictive value of immunofluorescence in the diagnosis of myocarditis

Immunofluorescence score	Active myocarditis (*n*)	(%)	Borderline myocarditis (*n*)	(%)	Idiopathic dilated cardiomyopathy (*n*)	(%)
If score ≥ 2 ($n = 33$)	18	55	13	39	2	6
If score < 2 ($n = 175$)	0	0	3	2	172	98

$P \leq 0.0545$ active versus borderline myocarditis $\varkappa^2$
$P \leq 0.00001$ myocarditis versus idiopathic dilated cardiomypathy $\varkappa^2$

showed idiopathic cardiomyopathy ($P \leq 0.001$ for contingency table by $\varkappa^2$). HRA was shown in our previous study not to correlate with myocarditis [1]. Therefore, it was not routinely sought in this larger group of patients.

IF Findings with Vascular Localization of IgG + C. In the current study, ten patients (5%) were found to have vascular IgG + C. Vasculitis was not diagnosed in any patient by light microscopy. Of these patients, three were diagnosed by light microscopy as having active myocarditis; two other patients were thought to have borderline myocarditis on the basis of the Dallas criteria. All of these five patients showed prominent endothelial cell swelling but no infiltration of blood vessels with lymphocytes or polymorphonuclear leukocytes. The remaining five patients showed idiopathic cardiomyopathy by light microscopy.

IF Findings in Other Autoimmune Diseases. During this study, several patients were identified with unusual IF staining patterns. In one patient, diagnosed as having borderline myocarditis according to the Dallas criteria, IF findings diagnostic of mixed connective tissue disease were found [6]. This patient demonstrated a speckled pattern of nuclear staining with IgG as well as interstitial IgG + C3. Two patients with systemic lupus erythematodus were studied; both showed a homogeneous nuclear pattern of staining with IgG suggestive of antinuclear antibody (ANA) activity. In both patients, large amounts of interstitial IgG and C3 were also detected. One patient with scleroderma was studied. This patient showed strong vascular staining with fibrinogen and marked thickening of blood vessel walls by light microscopy. Finally, three patients with acute rheumatic heart dis-ease were examined. In all three patients, only histiocytes were found by light microscopy in interstitial regions. The IF findings were distinctive. All patients showed a predominance of coarse, granular C3 deposits and lesser amounts of IgG, also in a very coarse, granular pattern.

Identification of Infiltracting Cells and Endothelial Cell Activation. Among patients with interstitial infiltration of mononuclear cells, the identity of cells was determined by semiquantitative IF grading with monoclonal antibodies. The current study confirmed our previous study: approximately half of the infiltrating cells in all patients with myocarditis were histiocytes. The majority of the infiltrating lymphocytes in patients with myocarditis were T lymphocytes; these cells were equally divided between T helper and T suppressor cells. Three patients showed only T lymphocytes to be present; two of these three patients showed persistence of the myocarditis on repeat biopsy. Patients with acute rheumatic heart disease, mixed connective tissue disease, and systemic lupus erythematosus showed a predilection for increased numbers of B lymphocytes over T lymphocytes (see Table 2).

Table 2. Lymphocyte subtypes in inflammatory heart disease

Disease type	T only	T + B + H	B > T	Histiocytes only
Myocarditis	3	15	1	2
Acute rheumatic heart disease	0	0	0	3
Systemic lupus erythematodus	0	0	2	0
Mixed connective tissue disease	0	0	1	0

Fisher's exact test $P \leq 0.00128$

An interesting aspect of this study was the investigation of the HLA-DR (Ia-like) staining of vascular endothelium. All types of inflammatory heart disease showed strong vascular HLA-DR staining. By contrast, only two of 78 (3%) patients with idiopathic cardiomyopathy demonstrated vascular HLA-DR staining (Table 3). In both of these patients, the staining was much less intense than in patients with myocarditis or autoimmune disease. Five additional patients with a light-microscopic diagnosis of idiopathic cardiomyopathy showed strong HLA-DR staining; these five patients had strong staining of vessels with IgG + C and are thought to represent a distinctive subgroup.

IF Findings in Serially Studied Patients. Eighteen patients with inflammatory heart disease were followed serially. Nine resolved, but two of these relapsed. Three had persistent myocarditis, and seven progressed to idiopathic cardiomyopathy. HLA-DR was found to predict inflammation accurately, becoming negative in the absence of the light-microscopic evidence of disease. Vascular HLA-DR staining changed within 2–6 months after the initial biopsy; however, the presence of HLA-DR staining did not predict for subsequent relapse or progression to idiopathic cardiomyopathy.

Table 3. HLA-DR in myocarditis and idiopathic dilated cardiomyopathy

	Myocarditis	Idiopathic dilated cardiomyopathy	Autoimmune heart disease
HLA DR +	35	2	17
HLA DR −	0	78	0

$P \leq 0.0001$ for difference in each disease category (χ^2)

Discussion

The current study confirms and extends previous initial observations of the complementary value of IF in the examination of patients suspected of having various types of inflammatory heart disease [1, 7]. IF allows the demonstration of IgG + C3 in the heart, indicators of humoral immune activity which cannot otherwise be detected. It has been proposed that concordant staining for IgG and C3 may indicate deposits of immune complexes; it is also possible that their interstitial distribution is a result of endothelial injury and subsequent permeability, a probable part of the inflammatory response in the heart. In the skin, for example, endothelial injury and permeability is a well-described part of cell-mediated hypersensitivity to diverse antigens [8]. The demonstration of HLA-DR in endothelium from patients with myocarditis is further evidence that the endothelium is an important part of the inflammatory response in the heart. It has been suggested by Burger that the endothelium is acting as an antigen presenting cell [9, 10].

The most important contribution of IF to the examination of endomyocardial biopsies is the confirmatory and complementary role it serves in the diagnosis of myocarditis and other autoimmune disorders affecting the heart. In our hands, the presence of IgG and C3 were significantly predictive that cellular myocarditis would be present on light microscopy. IF provides very important information because cellular infiltrates are often focal and sporadic by light microscopy and could, therefore, be missed due to sampling errors [11, 12]. Confirmation of IgG and C3 staining in such biopsies may permit the diagnosis of myocarditis to be suspected in such cases so that repeat biopsies would be indicated. On the other hand, the absence of significant IgG and/or C3 staining in the presence of a cellular infiltrate may raise the question of a falsely positive diagnosis indicating the need for diagnostic confirmation by electron microscopy or repeat biopsy. IF evaluation also allows for a subclassification of myocarditis into cellular, humoral, and mixed types, each of which may carry a distinct prognostic and therapeutic implication. It should be added that in the current study, insufficient numbers of patients were present to define the subclasses.

Finally, IF allows an accurate diagnosis of other autoimmune disorders to be made. IF permitted accurate distinction of acute rheumatic heart disease, mixed connective tissue disease, systemic lupus erythematosus involving the heart, and scleroderma in the patients in this study. A subset of patients showed vascular localization of immune complexes. The histologic findings in these patients were variable. While half of them had myocarditis on light microscopy, the others showed only idiopathic dilated cardiomyopathy. None showed clear-cut vasculitis by light microscopy. Immune complexes in blood have been shown to produce vasculitis experimentally [13] and clinically [14, 15]. It is common for light-microscopic copic evidence of vasculitis to occur at a later or different time than immunocytochemical demonstration of immune complexes. Thus, the IF findings may provide important data useful in correctly predicting the inflammatory nature of the heart disease in such patients. A recent patient with heart failure found to have such vascular localization of immune complexes without obvious myocarditis or vasenlitis was subsequently proven to have post-streptococcal glomerulonephritis. This patient responded to steroid therapy (E. H. Hammond 1988, unpublished observation). Pulmonary vascular immune complexes, demontrated only by IF without concomitant light-microscopic findings, have been associated with severe interstitial lung disease in a patient with autoimmune disease and urticarial vasculitis [16]. Such anecdotal reports await confirmation in larger series.

Subtyping of lymphocytes did not demonstrate appreciable differences between the patients in the myocarditis category. We suspect that this lack of distinction is due to technical difficulties because, by IF, only semiquantitation of the infiltracting cells is possible. Furthermore, in such small biopsies where only a small number of lymphocytes are present, a significant sampling error is always possible. It will be important to study the distribution of lymphocyte phenotypes in future biopsies to see which phenoytypes of lymphocytes correlate with therapy or prognosis. In this study, two of three patients with only T lymphocytic infiltrates showed progression of myocarditis on serial biopsies. Because the number is so small, it is impossible to make conclusions about its significance.

Conclusions. Our expanded experience demonstrates that IF is an important adjunct to the diagnosis of patients with suspected inflammatory heart disease. The ongoing multicenter cooperative study should further extend these observations as well as determine their precise clinical importance. In the future, more rational diagnostic, prognostic, and therapeutic categories can be formulated.

Summary

A total of 286 consecutive endomycardial biopsies were prospectively examined by light and IF microscopy. Biopsies were graded as 0 to 3+ for immunoglobulin, complement, or fibrinogen by IF. We previously found that the IF score defined as "2 (IgG) + C" correlated in a highly significant manner with myocarditis by light microscopy. Utilizing this IF scoring system, we found 52 patients with IF scores greater than 2. In this new group, the IF findings allowed us to diagnose myocarditis (35 patients), vasculitis (ten patients), systemic lupus erythematosus (two patients), acute rheumatic heart disease (three patients), mixed connective tissue disease (one patient), and scleroderma (one patient). All types of inflammatory heart disease showed strong vascular HLA-DR (Ia-like antigen) staining. Only two of 78 patients with idiopathic cardiomyopathy and any vascular HLA-DR. In both, staining was much less intense than in patients with inflammation. Eighteen patients with inflammatory heart disease were followed serially. Nine resolved, but two of these relapsed. Three had persistent myocarditis, and six progressed to idiopathic cardiomyopathy. HLA-DR was found to accurately reflect inflammation, becoming negative in the absence of light-microscopic evidence of inflammation. It did not predict for recurrence or progression to idiopathic cardiomyopathy in any patients. Lymphocyte typing was useful in discriminating between types of inflammatory heart disease. Acute rheumatic heart disease patients showed mainly histiocytes in their infiltrates; all other patients showed mixed infiltrates of T and B cells in equal proportions except in patients with systemic lupus erythematosus and mixed connective tissue disease in whom B lymphocytes were more common. HRA was not helpful in the distinction of inflammatory heart disease. Four of 17 (24%) patients with inflammatory heart disease had HRA; 15 of 59 patients (25%) with idiopathic cardiomyopathy had HRA. We conclude that IF helps to subclassify patients with inflammatory heart disease. HLA-DR is a useful marker of inflammation regardless of the nature of the underlying inflammatory process; lymphocyte typing and IF score allow accurate recognition and classification of the type of inflammatory disease.

References

1. Hammond EH, Menlove RL, Anderson JL (1987) Predictive value of immunofluorescence and electron microscopic evaluation of endoymocardial biopsies in the diagnosis and prognosis of myocarditis and idiopathic dilated cardiomyopathy. Am Heart J (1978) 114 [5]: 1055–1065
2. Mason JW (1978) Techniques for right and left ventricular endocardial biopsy. Am J Cardiol 41: 887–892
3. Anderson JL, Marshall HW, Allison SB (1984) The femoral venous approach to endomyocardial biopsy: comparsion with internal jugular and transarterial approaches. Am J Cardiol 53: 833–837
4. Aretz HT, Billingham ME, Edwards WD, et al. (1986) Myocarditis: a histopathologic definition and classicfication. Am J Cardiovasc Pathol 1: 3–14
5. Edwards WD, Holmes DR, Reeder GS (1982) Diagnosis of active lymphocytic moycarditis by endomyocarial biopsy. Mayo Clin Proc 57: 419–425
6. Gilbert EM, O'Connell JB, Hammond ME, Renlund DG, Watson FS, Bristow MR (1988) Treatment of myocarditis with OKT3 monoclonal antibody. J Heart Transplant (submitted)
7. Linder J, Cassling RS, Rogler WC, Markin RS, Sears TD, McManus BM (1985) Immunohistochemical characterization of lymphocytes in inflammed ventricular myocardium. Implication for myocarditis. Arch Pathol Lab Med 109: 917–920
8. Clark RAF, Dvorak HF, Colvin RB (1981) Fibronectin in delayed type hypersensitivity reactions: association with vessel permeability and andothelial cell activation. J immunol 126: 787–791
9. Burger DR, Vetto RM (1982) Vascular endothelium as a major participant in T-lymphocyte immunity. Cell Immunol 70: 357–361
10. Wagner CR, Vetto RM, Burger DR (1985) Subcultured human endothelial cells can function independently as fully competent antigen presenting cells. Hum Immunol 13: 33–48
11. Shanes JG, Ghali J, Billingham ME, et al. (1987) Interobserver variability in the pathologic interpretation of endomyocardial biopsy results. Circulation 75: 401–405

12. Tazelaar HD, Billingham ME (1986) Leukocytic infiltrates in indiopathic dilated cardiomyopathy. A source of confusion with active myocarditis. Am J Surg Pathol 10: 405–412
13. Cochrane OG (1971) Mechanisms involved in the deposition of immune complexes in tissues. J Exp Med 134: 875–895
14. Gower RG, Sams WM Jr, Thorne FG, et al. (1977) Leukocytoclastic vasculitis-sequential appearance of immunoreactants and cellular changes in serial biopsies. J Invest Dermatol 69: 477–484
15. Braverman IM, Yen A (1975) Demonstration of immune complexes in spontaneous and histamine induced lesions and in normal skin of patients with leukocytoclastic angiitis. J Invest Dermatol 64: 105–112
16. Fortson JS, Zone JJ, Hammond ME, Groggel GC (1986) Hypocomplementemic urticarial vasculitis syndrome responsive to Dapsone. J Am Acad Dermatol 15: 1137

Induction of Major Histocompatibility Antigens on Myocardial Cells in Patients with Active Myocarditis and Idiopathic Cardiomyopathy

A. HERSKOWITZ, K. L. BAUGHMAN, N. R. ROSE, D. A. NEUMANN, W. E. BESCHORNER, K. W. SELL, and A. AHMED-ANSARI

Johns Hopkins School of Hygiene and Public Health, Department of Immunology and Infectious Diseases, 615 North Wolfe Street, Room 4013, Baltimore, MD 21205, USA

Introduction

The histologic diagnosis of active myocarditis on endomyocardial biopsy has now been standardized and is defined by the presence of interstitial myocardial inflammation associated with myocyte necrosis and degeneration of adjacent myocytes [1]. The use of arbitrary histologic criteria to define a disease whose pathogenesis is not fully unterstood has raised several important diagnostic issues:

1. Defining myocarditis by histology alone may lump together diseases of different etiologies.
2. The histologic criteria may not be sensitve enough to identify all cases of immune-mediated heart disease.
3. The histologic pattern of myocardial injury in most cases of active myocarditis cannot easily explain the severe cardiac dysfunction experienced by these patients.

We have therefore tried to identify nonhistologic markers of myocarditis which may both provide clues to the immunopathogenesis of the disease and potentially serve as markers of immune-mediated myocarditis. The primary focus ot the studies described below will be the potential use of myocardial histocompatibility antigen expression as a clinically useful nonhistologic marker of immune cardiac disease.

Major histocompatibility complex (MHC) molecules play an important role in presenting antigens to the immune system. T lymphocytes can recognize antigens only in the context of self MHC molecules, a phenomenon known as “MHC restriction” [2, 3]. MHC class I molecules primarily activate $CD8^+$ (cytotoxic/suppressor) T cells and are present on virtually all nucleated cells [3, 4]. MHC-class II molecules primarily activate $CD4^+$ T cells and are expressed primarily on antigen-presenting cells of the immune system such as macrophages and B cells [5]. Increased expression of these MHC antigens has been demonstrated in human tissues undergoing autoimmune injury [5–7], allograft rejection [8–11], viral diseases, and other inflammatory states [12, 13].

Normal human fetal and adult cardiac myocytes express low levels of MHC class I (HLA A, B, and C regions) and do not express detectable levels of MHC class II (HLA DP, DQ and DR regions) antigens [4, 5, 14, 15]. We used monoclonal antibodies directed against monomorphic determinants of human MHC class I and II antigens to examine endomyocardial biopsy specimens from select

H.-P. Schultheiß (Ed.)
New Concepts in Viral Heart Disease

patients with active myocarditis and idiopathic cardiomyopathy to determine whether altered expression of MHC antigens could be detected within the myocardium. Patients with no histologic evidence of myocarditis found on endomyocardial biopsy and who were later found to have clear-cut clinical and histopathologic diagnoses explaining their left ventricular dysfunction were used as controls.

Methods

Patients. Between June 1986 and April 1987, 80 patients were suspected on clinical grounds of having myocarditis and underwent endomyocardial biopsy. Histocompatibility antigen expression within endomyocardial biopsies from 13 patients with active myocarditis, 15 with idiopathic cardiomyopathy, and nine nonmyocarditis controls was studied. All patients had some degree of left ventricular dysfunction after an initial presentation of rapid onset congestive heart failure and inadequate clinical response to conventional therapy with diuretics, digoxin, and frequently vasodilators. All patients underwent initial evaluation including history, physical examination, routine hemograms and serum chemistries, screening for collagen-vascular disease, thyroid function testing, and screening for pheochromocytoma.

Endomyocardial Biopsy. Endomyocardial biopsies were all performed via the right internal jugular vein using a modified Caves bioptome. During the procedure, a minimum of three separate endomyocardial biopsy specimens were obtained for examination by routine histology, while one specimen was frozen for immunohistochemistry.

Histology. All biopsy specimens were serially sectioned and examined using hematoxylin-eosin staining. Five serial sections of 5-μm thickness were cut and placed on each of seven glass slides so that the majority of the biopsy tissue was examined. Each set of biopsy specimens was examined by two cardiac pathologists in a blinded fashion. Myocarditis was defined according to the Dallas criteria as inflammation of the myocardium associated with necrosis and degeneration of adjacent myocytes not typical of the ischemic damage seen with coronary artery disease [1].

Semiquantitation of MHC class I/Class II Expression. Endomyocardial biopsy sections from myocarditis and nonmyocarditis controls were assessed for relative levels of MHC claas I and class II expression using standard immunoperoxidase techniques. Frozen section slides were either treated with monclonal anti-human class I (clone W6/32, ATCC, Rockville, MD), anti-human class II (clone L-243, ATCC), or normal mouse serum und incubated with biotin-conjugated rabbit anti-mouse Ig and an avidin-biotinylated horseradish peroxidase complex (Vector Labs, Burlingame, CA). After washing, sections were incubated with a peroxidase substrate, amino ethylcarbazole (AEG), washed, and mounted.

Triplicate slides were coded and the relative reactivity scored semiquantitatively by two independent observers. With each experimental run, tissues from an autopsy normal control and a case of fatal, coxsackievirus type B3 (CVB3) virus culture-positive myocarditis were included. The biopsies were scored for levels of MHC antigen expression on capillary, venular, and arteriolar endothelium as well as for MHC expression on myocytes. The final results represent the consensus opinion of the two individuals who scored the entire panel of slides. Immunoperoxidase staining was scored as follows:

0–1+ Weak, focal staining similar to that seen in negative control samples; myocytes are negative, endothelial cells lining the vessels usually contributing to the weak staining.

2+ Endothelial cells more positive, myocyte sarcolemmal surfaces are negative.

3+ Endothelial cells lining both venous and arteriolar beds are strongly positive. Myocyte surfaces are focally positive, while cytoplasmic structures and intercalated discs have a weak, diffuse staining.

4+ Intense difffuse staining, endothelial cell and multifocal myocyte sarcolemmal staining.

Quantitation of MHC Class I/Class II expression. All specimens were studied. The radioimmunoassay (RIA) was performed by removing the ornithine carbamoyltransferase (OCT) compound from each biopsy specimen and cutting 5-μm sections. Four sections were transferred to glass tubes previously coated overnight with 5% bovine serum albumin (BSA), and each assay was performed in triplicate. Each tube contained 3 ml media (RPMI 1640 supplemented with 100 μg/ml streoptomycin, 100 units/ml penicillin, 2 m*M* *L*-glutamine, and 5% heat inactivated fetal calf serum) containing 0.1 *M* phenylmethylsulphonyl fluoride (Sigma Chem., St. Louis, MO). Following incubation at 4 °C for 20 min, tubes were centrifuged at 650 x *g*, the fluid drained and 3 ml of RPMI media containing 5% normal mouse serum was added. Following incubation at 4 °C for 20 min, to each set of triplicate tubes was added either Normal Mouse Serum (NMS, control), anti-human cardiac myosin (EU-1H1), anti-human monomorphic MHC class I (W6/32) or anti-human monomorphic MHC class II (L-243). The tubes were vortexed, incubated for 60 min at 4 °C and washed twice with phosphate-buffered saline (PBS) pH 7.4. To each tube, RPMI media containing 5 % normal goat serum was added (3 ml/tube) and, following incubation at 4 °C for 20 min, the fluid was drained and affinity-purified ^{125}I-labelled goat anti-mouse Ig [100 μl of 1:50 containing 50000 counts per minute (cpm)] was added. Following gentle mixing and incubation for 30 min at 4 °C, the tubes were washed three times with PBS and then counted in an gamma counter. Autopsy tissue sections from a normal human heart (from a case involving trauma) and tissue sections of heart from a patient who had died of acute myocarditis were run in parallel and served as negative and positive controls, respectively. Mean and standard deviation of cpm for the triplicate samples were calculated. Any assay giving less than a five-fold increase in activity between the negative and positive control was rejected and hence, repeated.

To adjust for differences in the size of each biopsy sample used in the RIA, a monoclonal anticardiac myosin reagent was used. Cardiac myosin is the major intracellular protein within the myocardium and would therefore reflect the amount of tissue used for the RIA. Data are expressed as cpm and as the ratio of class I or class II activity to myosin activity (ratio).

Indirect Immunofluorescence. Standard techniques for indirect immunofluorescence were used [16]. Heart, kidney, liver, skeletal muscle, and stomach were removed from 6-month-old Spargue-Dawley rats (Charles River), frozen, and 4-μm sections were cut in a cryotome. The sections were overlaid with human serum diluted 1:10, 1:20, 1:40, and 1:80 and incubated with goat anti-human IgG, Fc gamma chain-specific (Organon 1202-0121, lot # 27107), 50 μl fluorescein isothiocyanate (FITC) at 1:40 dilution. A rhodamine counterstain was used at 20 μl/ml (Difco Laboratories), the slides were rinsed and washed for 30 min in PBS. The sections were mounted with 90% (vol/vol) glycerol/PBS and examined using a Zeiss fluorescent microscope.

Statistical Analysis. Comparisons between the two patient groups were made using the Student's *t* test. Proportions were analyzed using the $\varkappa^2$ test. Linear and logistic regression analyses were used to determine the relationships among clinical, serologic, and immunohistochemical data. Differences achieved statistical significance when two tailed *P* values were less than 0.05. Group means are expressed as the mean ± standard deviation, unless otherwise noted.

Results

Endomyocardial Biopsies

All biopsy specimens contained at least three separate endomyocardial fragments for routine examination by hematoxylin-eosin. Thirteen patients (16.2%) had biopsy specimens that fulfilled the histologic criteria for active myocarditis. Of the 58 patients with nonspecific biopsies, the first 15 consecutive patients with idiopathic cardiomyopathy were studied. In addition, nine patients (11%) had no evidence of myocarditis, but other clear-cut clinical or histopathologic diagnoses were established. These latter nine patients comprise the nonmyocarditis control group.

Clinical Data

Three of the 13 patients with active myocarditis were postpartum, while the remaining patients within the myocarditis and cardiomyopathy groups remained idiopathic. Of the nine nonmyocarditis control patients, six were found to have severe coronary disease on subsequent left heart catheterization, two were found to have significant amyloid infiltration of the myocardium by endomyocardial biopsy, and one patient had a significant pericardial effusion secondary to metastatic esophageal carcinoma. There were no differences in clinical parameters of left ventricular function between the three groups with the mean ejection fraction less than 30 in all groups.

Immunohistochemistry

Semiquantiatively, using the immunoperoxidase method, the level of MHC class I expression was no greater than 1+ in the control autopsy samples. Myocytes appeared uniformly negative; however, endothelial cells lining the venules appeared to give a faint staining pattern. MHC class II antigen expression was essentially negative except for rare interstitial cells which were faintly positive. These cells were felt to represent interstitial dendritic cells. In the positive controls, the myocytes and microvascular endothelium appeared uniformly and intensely positive for both MHC class I and MHC class II antigens. Three of the nine nonmyocarditis control patients had a grade 2 Immunoperoxidase (IP) score for MHC class I consistent with increased expression of endothelial MHC class I (specificity 67%). Only one of these nine controls had a grade 2 score for MHC class II expression (specificity 89%). None of the nonmyocarditis control samples had myocyte MHC class I or class II expression.

Increased MHC class I capillary endothelial staining (IP score ≥2) was seen in 11 of 13 myocarditis samples (sensitivity 84.6%), while MHC class II capillary endothelial staining was increased in nine of 13 samples (sensitivity 69%) (Fig. 1). Eight of 13 myocarditis samples had myocyte staining for MHC class I (≥ grade 3 by IP) and MHC class II (sensitivity 62%) (Fig. 2). All biopsies with a myocardial pattern of staining also had increased capillary staining (Fig. 3).

RIA results on class I and II MHC expression are expressed as the mean of three separate measurements. Standard deviations for these triplicate measurements did not differ from the raw count by more than 4%–9%. Data is presented as raw counts per minute and is also normalized for variations in biopsy size from patient to patient by expressing the raw counts as a ratio of class I/II to myosin

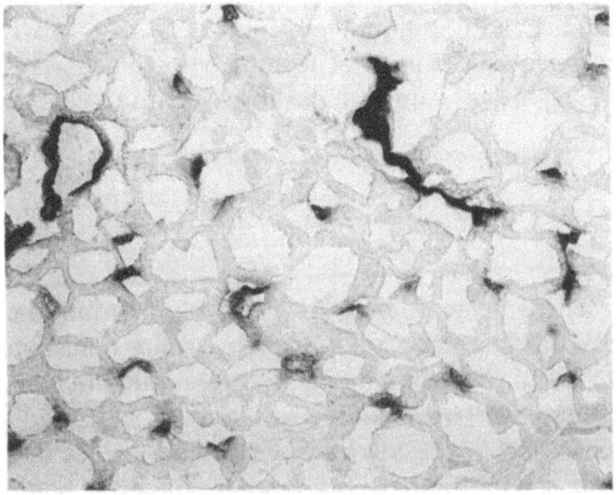

Fig. 1. Immunoperoxidase technique using monoclonal antibody to MHC class I demonstrating grade 2 endothelial capillary staining. The myocardial cells are uniformly negative. x 380

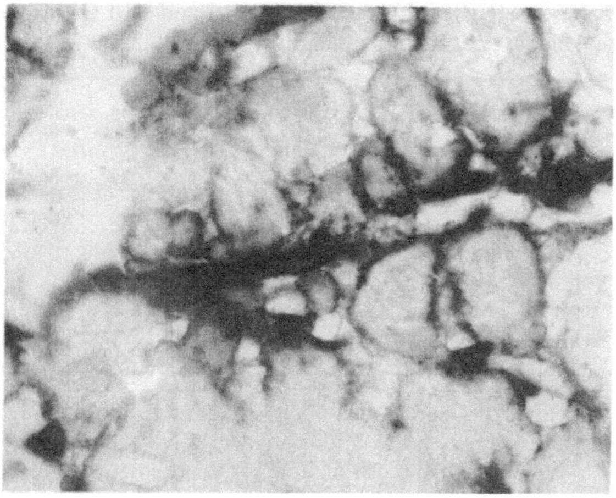

Fig. 2. Immunoperoxidase technique demonstrating MHC class II reaction product staining the sarcolemmal surfaces of a focus of myocytes. Adjacent capillary endothelial cells stain intensely positive as well. x 600

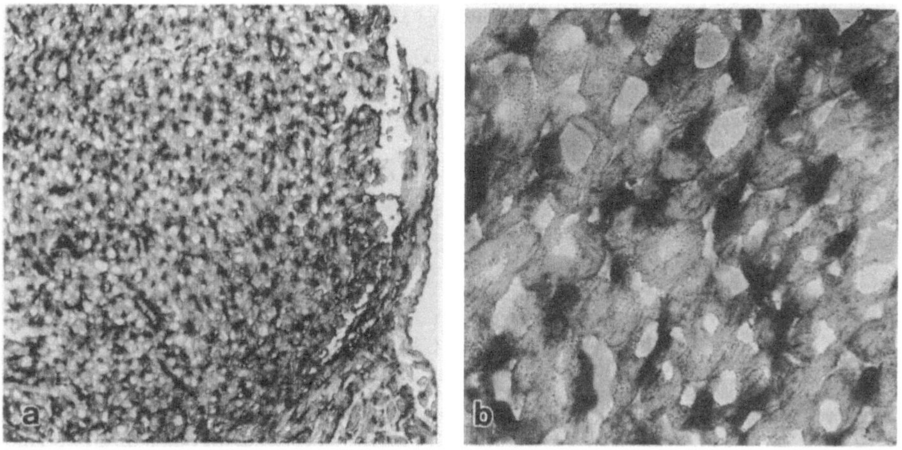

Fig. 3 a, b. Low-power photomicropraph showing intense global capillary staining for MHC class II using the immunoperoxidase technique. **b** Higher-power magnification from same endomyocardial biopsy specimen revealing the intensely positive endothelial staining. There is a faint staining tangentially cut myocytes which probably represents an artifact of spillover of reaction product from endothelium to adjacent myocytes. **a** x 120; **b** x 380

content (Table 1). For the RIA, abnormal expression of MHC class I and MHC class II antigens was defined as any value above 2 standard deviations above the mean of values from nonmyocarditis controls (expressed as rations of cpm to cardiac myosin, class I $\geq$0.156, class II $\geq$0.235). The sensitivity and specificity of the ratio derived from the RIA are listed in Table 2.

Eleven of 13 of the myocarditis biopsies and five of the 15 cardiomyopathy biopsies revealed significant increases in expression of both MHC class I and class II antigens using the RIA (Fig. 4). The subgroup of cardiomyopathy patients with increased MHC expression could not be clinically distinguished from the remai-

Table 1. RIA results for MHD class I and class II antigen expression from one endomyocardial biopsy specimen per patient

	Raito		cpm	
	MHC class I	MHC class II	MHC class I	MHC classII
Myocarditis	0.92 ± 0.61	1.05 ± 0.85	14752 ± 9710	16360 ± 12007
Cardiomyopathy	0.35 ± 0.52	0.31 ± 0.49	5258 ± 6581	4644 ± 5640
Nonmyocarditis Controls	0.06 ± 0.04	0.08 ± 0.07	1137 ± 515	1343 ± 1284
P values	< .001	< .001	< .001	< .001

Data are expressed as means and standard deviation for both raw counts per minute (cpm) and the ratio of raw cpm to cardiac myosin
P values are presented from analysis of variance testing

Table 2. Sensitivity and specificity of the ratio of MHC class I and class II expression to cardiac myosin, derived from the RIA

	Sensitivity (%)	Specificity (%)
Myocarditis		
MHC Class I	85	88
MHC class II	85	88
Cardiomyopathy		
MHC class I	33	
MHC class I	33	

Abnormally increased expression of MHC class I and class II antigen within the myocardium is defined as values greater than 2 standard deviations from the mean for nonmyocarditis controls

ning cardiomyopathy patients. One of nine nonmyocarditis biopsies had minimally increased MHC expression by RIA. Within the three study groups, expression of both MHC class I and class II did not differ significantly.

Circulating Autoantibodies

Significant titers of heart autoantibodies (IgG isotype) were identified in serum from eight of 13 myocarditis patients (61.5%), four of 15 (27%) cardiomyopathy patients, and two of nine (22%) nonmyocarditis control patients. These results are consistent with our other studies where IgG heart autoantibody titers of ≥ 20 were found in 21% of cardiomyopathy patients (n = 85), 60% of patients with a

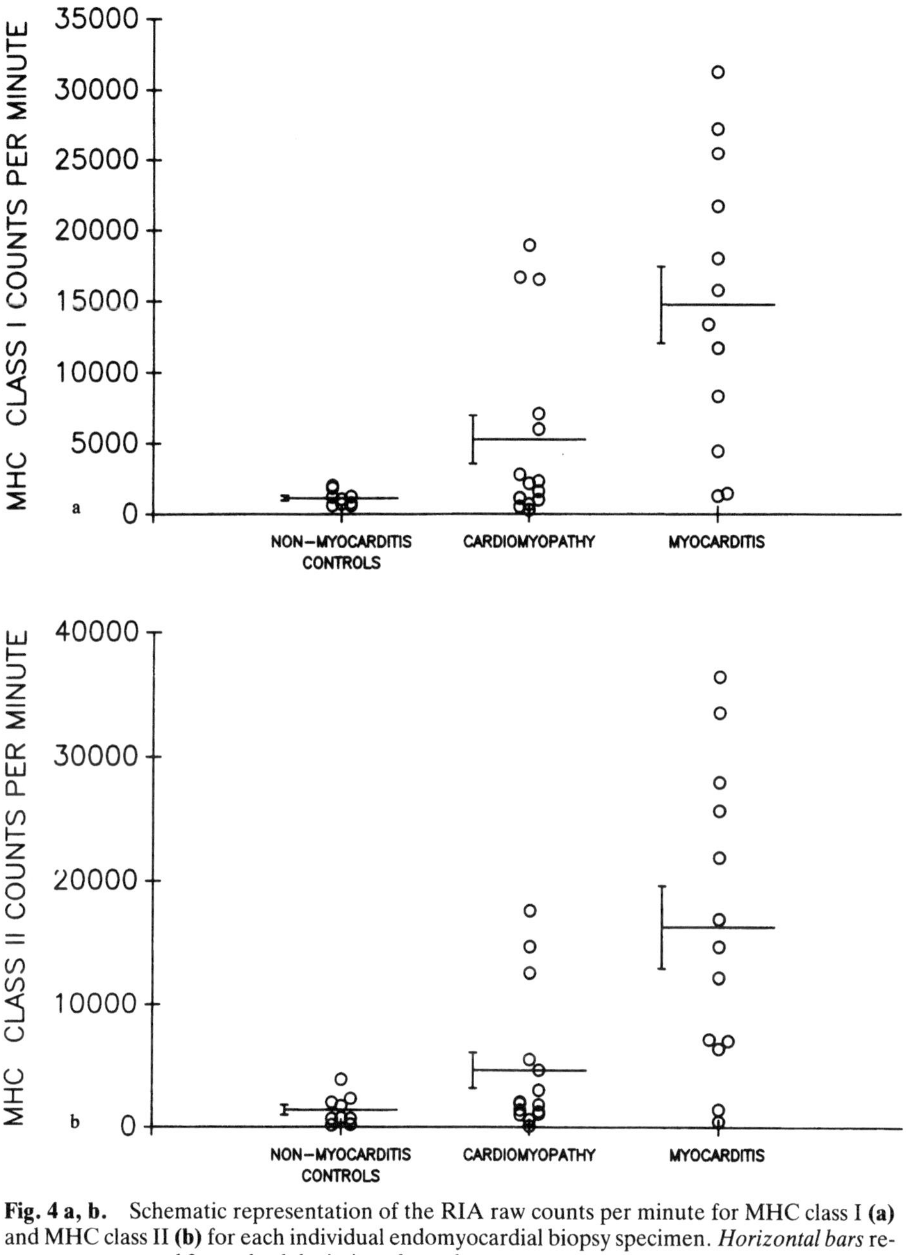

Fig. 4 a, b. Schematic representation of the RIA raw counts per minute for MHC class I **(a)** and MHC class II **(b)** for each individual endomyocardial biopsy specimen. *Horizontal bars* represent means and 2 standard deviations from those means

variety of ischemic heart disease syndromes ($n = 25$), and in 4% of normal volunteer controls ($n = 25$). Preliminary studies using Western immunoblotting suggest that, although ischemic injury to the myocardium may produce a high prevalence of heart autoantibodies, the myocardial antigens may be quite different from those recognized in active myocarditis and cardiomyopathy (data not shown).

Relationship Between Circulating Autoantibodies and Myocapillary MHC Expression

Linear regression analysis revealed a positive correlation between MHC class II expression and circulating IgG heart autoantibody titer ($r = 0.443$, $P = 0.023$). Eighty-three percent of biopsies from patients with serologic evidence of circulating heart autoantibody titers of ≥20 had significantly increased expression of both MHC class I and class II antigens versus only 38% of biopsies from patients without significant titers of heart autoantibodies.

Discussion

The data from this study suggest that abnormal expression of MHC antigens within the myocardium and serologic evidence of circulating heart autoantibodies may be useful nonhistologic markers of immune-mediated heart disease. A significant proportion of endomyocardial biopsies with histologic evidence of active myocarditis have increased expression of MHC antigens within the myocardium (85%), and, interestingly, a subgroup (33%) of biopsies without diagnostic changes of active myocarditis also has similar increases in MHC expression. Serologic evidence of significant titers of circulating heart autoantibodies are present in fewer patients with active myocarditis (61.5%) and cadriomyopathy (27%), but those patients with circulating autoantibodies are more likely to have increased expression of MHC antigens within the myocardium as compared with those patients without circulating autoantibodies (83% vs. 38%). These data support the hypothesis that many cases of human myocarditis and a subgroup of cases of human cardiomyopathy are immune mediated.

The finding of abnormal expression of myocardial MHC antigens is consistent with the leading hypothesis that cases of myocarditis are due to a cardiotropic virus [17–24]. The virus initiates an inflammatory process which is then perpetuated by a persistent immune reaction localized to the heart. Cardiac antigens can only recognized by the immune system in the context of self MHC molecules, a phenomenon known as "MHC restriction" [2, 3]. In a viral infection of the heart, cardiac antigens could be liberated by the direct cytocidal effect on myocytes or as aberrantly expressed antigens on the surface of virally infected cells. This study demonstrates the induction of MHC class I and class II antigens on both microvascular endothelium and myocytes, allowing cardiac antigens to be recognized by the immune system. A chronic state of inflammation characteristic of human myocarditis could thus be generated.

Patterns of Myocardial MHC Expression

Endomyocardial biopsy specimens from myocarditis patients with high RIA counts invariably revealed a strong, diffuse endothelial staining pattern (Fig. 3). Myocyte surface staining was focal or multifocal but frequently had an equally intense staining characteristic as compared to the endothelium (Fig. 2). Of the group 1 myocarditis patients with abnormal expression of MHC class I and class II antigens as measured by RIA, eight had foci where definite myocyte surface staining was seen. Although such sarcolemmal staining could be the result of peroxidase product spilling over from the intensely positive microvasculature, the foci of myocyte staining were rarely adjacent to strongly stained larger venous and arterial vessels. Peroxidase product within the myocyte cytoplasm and within the intercalted discs is more likely to be an artifact of stain spillover (Fig. 3b) and, when present, was seen throughout the entire endomyocardial biopsy specimen.

Similar findings of abnormal espression of MHC antigens in a subgroup of biopsies without characteristic inflammation and necrosis raise important issues. It may reflect a lack of sensitivity of the endomyocardial biopsy technique to identify all cases of myocarditis and possibly indicates that biopsies evolve from active inflammation and necrosis to a chronic, less inflamed state.

Review of MHC Antigen Expression

The well-characterized MHC class I antigens (HLA-A, B, C) are found on virtually all nucleated cells. Dual recognition of MHC class I and antigen is necessary for activating by cytotoxic T cells for lysis of virus-infected cells [13] and cells bearing alloantigens [3]. The MHC class II antigens are involved in communication between cells that regulate the immune response. These antigens are found primarily on B lymphocytes and on antigen-presenting cells, e. g., Kupfer cells in the liver, Langerhans cells in the skin, interstitial dendritic cells in the heart [25]. Low levels of MHC class II antigens may also be found on capillary endothelium. Normal fetal and adult myocytes express very low levels of MHC class I antigens and do not have detectable levels of MHC class II antigens [15]. Endothelial cells lining the microvasculature may express both MHC class I and II antigens, while interstitial dendritic cells (myocardial antigen-presenting cells) express only MHC class II antigens [10, 16]. This study supports previously described patterns of MHC distribution in normal myocardium and demonstrates two significant morphologic patterns of induction of MHC antigens in our study population: (a) intense induction of MHC class I and II antigens within the capillary bed; and (b) abnormal expression of MHC class I and II antigens on myocyte cell surfaces. Our small number of biopsies from patients with ischemic heart disease ($n = 6$) suggets that ischemic heart disease is not associated with abnormal espression of MHC class I and II antigens. The three patients with active myocarditis whose biopsies did not express either MHC class I and class II antigens could not be clinically or histologically distinguished from the other myocarditis patients.

Circulating Heart Autoantibodies

Another marker of immune activation by cardiac antigens is serologic evidence of circulating heart autoantibodies. Circulating heart autoantibodies have been previously described in cardiomyopathy and myocarditis patients [26, 27], post-pericardiotomy syndrome [28], rheumatic carditis, and Chagas' disease [29, 30, 31].

Normal volunteers have a 4% prevalence of circulating heart autoantibodies (unpublished data). Group 1 myocarditis patients had a 62% prevalence of circulating heart autoantibodies, while none of the eight group 2 control patients had significant titers of antibodies. Linear regression analysis indicated that tissue-bound MHC expression and serum circulating heart autoantibodies are indeed related. The quantity of MHC class I and class II expression within an endomyocardial biopsy, as measured by the RIA method correlates well with the titer of circulating IgG heart autoantibodies ($r = 0.593$, $P = 0.006$ and $r = 0.708$, $P < 0.001$, respectively). Seven of the eight myocarditis patients with significant titers of circulating heart autoantibodies had abnormal myocardial MHC class II expression both by immunoperoxidase and by RIA.

Hypothesis

Although any immune hypothesis to explain the immunopathogenesis of human myocarditis is purely conjectural, certain concepts are suggested. An initial injury to the heart, possibly virally mediated, leads to induction of MHC antigens. MHC class I antigen expression on myocytes, endothelial cells, and fibroblasts would favor a $CD8^+$ T cell response. Activated T cells could be directly cytocidal to the myocytes or, by producing lymphokines, could have a detrimental effects on myocardial cell function [32]. Induction of MHC class II antigen expression would preferentially activate $CD4^+$ T cells which could activate B cells to produce immunoglobulins to newly recognized cardiac antigens. Activated T cells producing gamma interferon could induce the diffuse endothelial expression of MHC class I and class II antigens, possibly increasing inflammatory cell traffic within the myocardium.

Although more cases of active myocarditis need to be studied to determine the true prevalence of increased MHC expression in endomyocardial biopsy specimens, this study demonstrates the increased expression of both MHC class I and class II antigens on myocytes and microvascular endothelium in patients with active myocarditis and a subgroup of patients with idiopathic cardiomyopathy. These findings may serve as a useful adjunct to the histologic diagnosis of myocarditis.

References

1. Aretz HT (1987) Myocarditis: the Dallas criteria, Hum Pathol 18: 619–624
2. Bach FH, Bach ML, Sardel PM (1976) Differential function of major histocompatibility complex antigens in T lymphocytic activation. Nature 259: 273–281
3. Zinkernagel RM, Doherty PC (1979) MHC-restricted cytotoxic T cells: studies on the biologic role of polymorphic major transplantation antigens determining T cell restriction, specificity, function and responsiveness. Adv Immunol 27: 51–177
4. Daar AD, Fuggler SV, Fabre JW, Ting A, Morris PJ (1984) The detailed distribution of HLA-ABC antigens in normal human organs. Transplantation 38: 287–292
5. Hanafusa T, Pujol-Borrell R, Chiovato L, Russell RCG, Doniach D, Bottazzo GF (1983) Aberrant expression of HLA-DR antigen on thyrocytes in Graves' disease: relevance for autoimmunity. Lancet 2: 1111–1115

6. Ballardini G, Mirakian R, Bianchi FB, Pisi E, Doniach D, Bottazzo GF (1984) Aberrant expression of HLA-DR antigens on bile duct epithelium in primary biliary cirrhosis: relevance to pathogenesis. Lancet 2: 1009–1013
7. Bottazzo GF, Dean BM, McNally JM, MacKay EH, Swift PGF, Gamble DR (1985) In situ characterization of autoimmune phenomenon and expression of HLA molecules in the pancrease in diabetic insulitis. N Engl J Med 313: 353–360
8. Rose ML, Coles MI, Griffin RJ, Pomerance A, Yacoub MH (1986) Expression of class I and class II major histocompatibility antigens in normal and transplantation human heart. Transplantation 41: 776–780
9. Duquesnoy RJ, Zeevi A, Fung JJ, Kaufman C, Zerbe JR, Griffith B, Trneto A, Kormos R, Hardesty R (1987) Sequential infiltration of class I and class II specific alloreactive T cells in human cardiac allografts. Transpl Proc 19: 2560–2563
10. Hall BM, Bishop GA, Duggin GG, Howath JS, Philips J, Tillen DJ (1984) Increased espression of HLA-DR antigens on renal tubular cells in renal transplantation: relevance to the rejection process. Lancet 1: 247–251
11. Klein J, Hauptfeld V (1980) Ia antigens: their serology, molecular relationships and their role in allograft rejection. Transplant Rev 30: 83–100
12. McMichael AJ, Ting A, Zweevink HF, Askonasi BA (1977) HLA restriction of cell mediated lysis of influenza virus infected human cells. Nature 270: 524–526
13. Appelyard ST, Dunn MJ, Dubowitz V, Rose ML (1985) Increased expression of HLA ABC class I antigens by muscle fibers in Duchenne muscular dystrophy, inflammatory myopathy, and other neuromuscular disorders. Lancet 1: 363–364
14. Rea TH, Shen JY, Modlin RL (1986) Epidermal keratinocyte Ia expression. Langerhans cell hyperplasia and lymphocytic infiltration in skin lesions of leprosy. Clin Exp Immunol 65: 253–259
15. Ahmed A, Leatherbury A, Stone J, Dempsey C, Beisel B, Tadros T, Sell KW (1987) Expression of class I and class II MHC gene products on human and murine cardiac myocytes. Fed Proc 46: 946
16. Rose ML, Coles MI, Griffin RJ, Pomerance A, Yacoub MH (1985) Induction of class I and class II antigens during rejection of allografted human hearts. Adv Exp Med Biol 186: 555–563
17. Lerner AM, Wilson FM (1973) Virus myocardiopathy, Progr Med Virol 15: 63–91
18. Reyes MP, Lerner AM (1985) Coxsackievirus myocarditis—with special reference to acute and chronic effects. Prog Cardiovasc Dis 27: 373–394
19. Burch GE, Giles TD (1979) The role of viruses in the production of heart disease, Am J Cardiol 29: 231–240
20. Dec WG, Palacios IF, Fallon JT, Aretz HT, Mills J, Lee DCS, Johnson RA (1985) Active myocarditis in the spectrum of acute dilated cardiomyopathies. N Engl J Med 342: 885–897
21. Fenoglio JJ Jr, Ursell PC, Kellogg CF, Drusin RE, Wiss MB (1983) Diagnosis and classification of myocarditis by endomyocardial biopsy. N Engl J Med 308: 12–18
22. James TN (1983) Myocarditis and cardiomyopathy, N Engl J Med 308: 39–41
23. Cambridge G, MacArthur CGC, Waterson AP, Goodwin JF, Oakley CM (1979) Antibodies to Coxsackie B viruses in congestive cardiomyopathy, Br Heart J 41: 692–696
24. Herskowitz A, Wolfgram LJ, Rose NR, Beisel KW (1987), Coxsackievirus B_3 murine myocarditis: a pathologic spectrum of myocarditis in genetically defined inbred strains. J Am Coll Card 9: 1311–1319
25. Daar AS, Fuggle SV, Fabre JW, Ting A, Morris PJ (1984) The detailed distribution of MHC class II antigens in normal human organs. Transplantation 38: 293–298
26. Schultheiss HP, Bolte HD (1985) Immunological analysis of autoantibodies against the adenine nucleotide translocator in dilated cardiomyopathy. J Mol Cell Cardiol 17: 603–617
27. Maisch B, Deeg P, Liebau G, Kochsiek (1983) Diagnostic relevance of humoral and cytotoxic immune reactions in primary and secondary dilated cardiomyopathy. Am J Cardiol 52: 1072–1078
28. Maisch B, Berg PA, Kochsiek K (1979) Clinical significance of immunopathological findings in patients with post-pericardiotomy syndrome. Relevance of antibody pattern. Clin Exp Immunol 38: 189–197

29. Acosta AM, Sadigursky M, Santos-Busch CA (1983) Anti-striated muscle antibody activity produced by *Trypanosoma cruzi*. Proc Soc Exp Biol Med 172: 364–369
30 Santos-Busch CA, Scheeter D, Sadigursky M, Acosta AM (1986) Immunopathology of Chagas' disease. In: Milgrom F, Albini BA (eds) Immunopathology. Karger, Basel, pp 262–267
31. Zabriskie JB, Hsu KC, Seegal BC (1970) Heart-reactive antibody associated with rheumatic fever: characterization and diagnostic significance Clin Exp Immunol 7: 147–152
32. Hosenpud JD, Campbell JM, Mendelson DJ (1987) Interleukin-1 (IL-1) induced cardiac dysfunction in an in-vitro isolated perfused heart preparation. J Am Coll Card 9: 143A

Evaluation of Immunohistological Data for the Diagnosis of Myocarditis

U. KÜHL, M. TOUSSAINT, G. ULRICH, D. WAGNER, P. WOLFF, and H.-P. SCHULTHEISS

Department of Internal Medicine, Klinikum Großhadern, University of Munich, Marchioninistraße 15, D-8000 München 70, FRG

Introduction

"Active myocarditis" is defined as the presence of inflammatory cells in the myocardium with evidence of fraying of the adjacent myocardial fibres but without concomitant sequential fibre necrosis. Although endomyocardial biopsies have markedly improved its diagnostic consistency, divergent frequencies of reported positive biopsies for cardiac inflammation have continued to cloud the certainty of morphological diagnosis. This variance stems partially from the different institutional criteria for the morphological diagnosis of myocarditis as well as from interobserver variability [29]. In most cases of clinically suspected active myocarditis, the histopathological diagnosis according to the Dallas classification is consistent with "borderline myocarditis". However, borderline myocarditis may constitute a heterogeneous group clinically.

The purpose of this paper was to find criteria which might help to get a better yield in diagnosis and a better understanding of the mechanisms involved in the pathogenetic process. We provide evidence that immune markers, on the one hand, facilitate the detection of lymphocytes. On the other hand, the identification of the expression of MHC class I and II antigens on myocytes could be helpful for the diagnosis and the evaluation of the activity of the ongoing immunological process.

Methods

Sample Preparation. Five to eight endomyocardial samples were obtained from each patient. For immunofluorescence, the biopsy specimen were frozen in methylbutane at −70 °C.

Direct Immunofluorescence. After air drying, the cryotome sections (3 μm) of the biopsies were incubated with fluorescine (FITC)- labelled second antibodies against IgG, IgM and IgA (Behring) for 30 min at 22 °C. Unbound antibodies were removed by several washes in phosphate-buffered saline (PBS), and sections were coverslipped with 90% glycerol/PBS.

H.-P. Schultheiß (Ed.)
New Concepts in Viral Heart Disease

Indirect Immunofluorescence. The presence of circulating heart reactive antibodies (HRA) was evaluated by incubation of normal guinea pig heart cryotome sections with patient sera, diluted with PBS (1:20), for 30 min at 22 °C. After several washes with PBS, sections were covered with FITC-labelled monospecific goat-anti-human IgG or IgM (Behring), showing no cross-reactivity with guinea pig cardiac tissue. Finally, the sections were washed with PBS, mounted in 90% glycerol/PBS and viewed in a immunofluorescence microscope.

Peroxidase Staining. Cryotome sections (5 μm) were fixed in 10% acetone at 4 °C, air dried and incubated with specific mouse-anti-human antibodies to lymphocyte surface antigens: CD3 (Behring), CD4 (Coulter), CD8 (Behring), pan T 200, and anti-MHC class I and II gene products (gift of Dr. J. Johnson, University of Munich, Dept. of Immunology) for 60 min at 22 °C. After several washes with PBS, peroxidase – labelled anti-mouse gammaglobuline (Behring), diluted 1:100 in 20% fetal calf serum/PBS (30 min, 22 °C) was used to mark bound antibodies. Staining was performed with 3-amino-9-ethylcarbazole and counterstained with haematoxylin.

Enzyme-Linked Immunosorbent Assay. Purification of the matrix proteins laminin, fibronectin and collagen types I, III, IV and V were done as described before [16, 17]. Enzyme-linked immunosorbent (ELISA) microtitre plates (Linbro) were coated overnight with laminin, fibronectin or collagens (50 μg/ml, 4 °C). Plates were washed with PBS and serial dilutions of human sera were added for 24 h at 4 °C. After a thorough washing, biotinylated goat-anti-human immunoglobuline (Amersham International), diluted 1:150, was allowed to bind to the antibodies overnight. Subsequently peroxidase conjugated with streptavidin (Amersham International), diluted 1:1000, was added for 1 h at room temperature followed by a 5-amino-2-hydroxy-benzoic acid substrate. The reaction was stopped with sodium hydroxide, and the plates were read at a wave length of 492 nm using an ELISA reader (Titertek Multiskan MC, Flow Laboratories).

Results

Histological Classification. This study included 39 patients with clinically suspected myocarditis. Pre-biopsy presentations ranged from non-specific systemic symptoms to atypical chest pain, pericarditis, the history of a viral infection preceding onset of cardiac symptoms, electrocardiographic abnormalities, arrhythmias of recent onset, acute hemodynamic deterioration and laboratory changes such as increased erythrocyte sedimentation rate or rise in serum creatine kinase level. Endomyocardial biopsies were undertaken to establish the etiology of these symptoms and to determine the presence or absence of myocarditis as its cause. According to the Dallas criteria, three of 39 biopsies (8%) had features of active myocarditis, 27 (69%) were consistent with borderline myocarditis and nine patients (23%) had no myocarditis.

T lymphocyte Cell Surface Markers. Endomyocardial biopsies were analysed for the presence of inflammatory cell infiltrates by T lymphocyte immune markers (Fig. 1). Only those patients showing a mean number of white blood cells (T 200) greater than five or of lymphocytes (CD 3) greater than three cells per high-power field (x 400) (6) were considered to be positive. In all cases of acute myocarditis, white blood cell and lympocyte infiltrations were seen although leucocyte/lymphocyte ratios varied to a great degree. However, the group of patients classified as borderline myocarditis was not homogeneous. Out of 27 patients, ten showed a significant lymphocytic infiltration (CD 3 positiv). By further subclassification predominantly CD 4-positive cells were seen in seven patients, while CD 8-positive cells were seen in five patients (Table 1). In contrast, only in one biopsy of the patients with the histological diagnosis of "no myocarditis" was a significant infiltrate of lymphocytes detected.

MHC Class I and II Antigens. To gain further knowledge on the immunological state of the disease we studied the MHC antigen expression in biopsy specimen using monoclonal antibodies against class I and II antigens. In controls only MHC class I antigens were seen on interstitials cells and vascular epithelium, while MHC class II could not be detected. In the myocardium from patients with acute myocarditis the expression of MHC I and II antigens was markedly increased in the interstitium. Additionally, in these patients MHC I and II expression was also found on cardiac myocytes in similar frequencies (Table 1). In the borderline group, MHC class I expression on myocytes was still increased in 15 out of 27 patients. Class II gene products were detected (Table 1) in 11 out of 27 patients. In contrast to that in the "no myocarditis group" only one patient showed MHC I and II gene expression. This patient is the same who had a lymphocyte infiltration.

Immunofluorescence Findings in Myocarditis. Sarcolemmal (IgG, IgM) and complement deposits (C3) are a common finding in endomyocardial biopsies from patients with myocarditis [10, 18] (Fig. 2a). Table 2 summarizes the frequencies of IgG and C3 deposits found in these biopsies. However, deposition of IgG was also seen in more than 70% of patients with other heart muscle diseases such as dilated cardiomyopathy (DCM), alcohol toxic cardiomyopathy, coronary heart disease (CHD) or mitral valve disease (MVD). In contrast, C3 could only be detected in

Table 1. Summary of immunohistological findings in endomyocardial biopsies

Diagnosis	*n*	T 200	CD3	CD4	CD8	CD4 > CD8	MHC I	MHC II
Acute MC	3	3/3	3/3	3/3	3/3	2/3	2/3	2/3
Borderline MC	27	8/27	10/27	7/10	5/10	6/10	15/27	11/27
No MC	9	1/9	1/9	1/1	0/1	1/1	1/9	1/9
Control group	25	2/25	1/25	1/25	0/25	0/25	0/25	0/25

MC, myocarditis

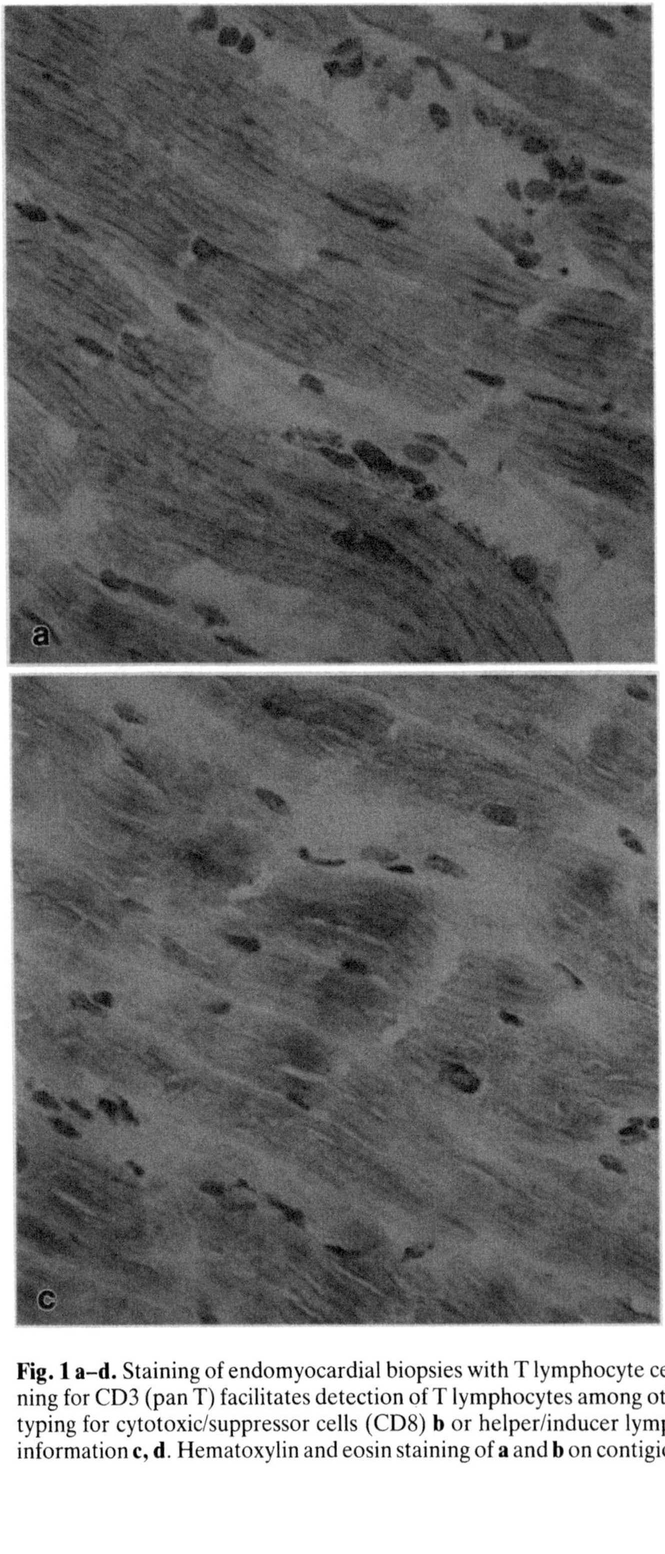

Fig. 1 a–d. Staining of endomyocardial biopsies with T lymphocyte cell surface markers: **a** staining for CD3 (pan T) facilitates detection of T lymphocytes among other infiltrating cells. Subtyping for cytotoxic/suppressor cells (CD8) **b** or helper/inducer lymphocytes provides further information **c, d**. Hematoxylin and eosin staining of **a** and **b** on contigious biopsy sections. x 320

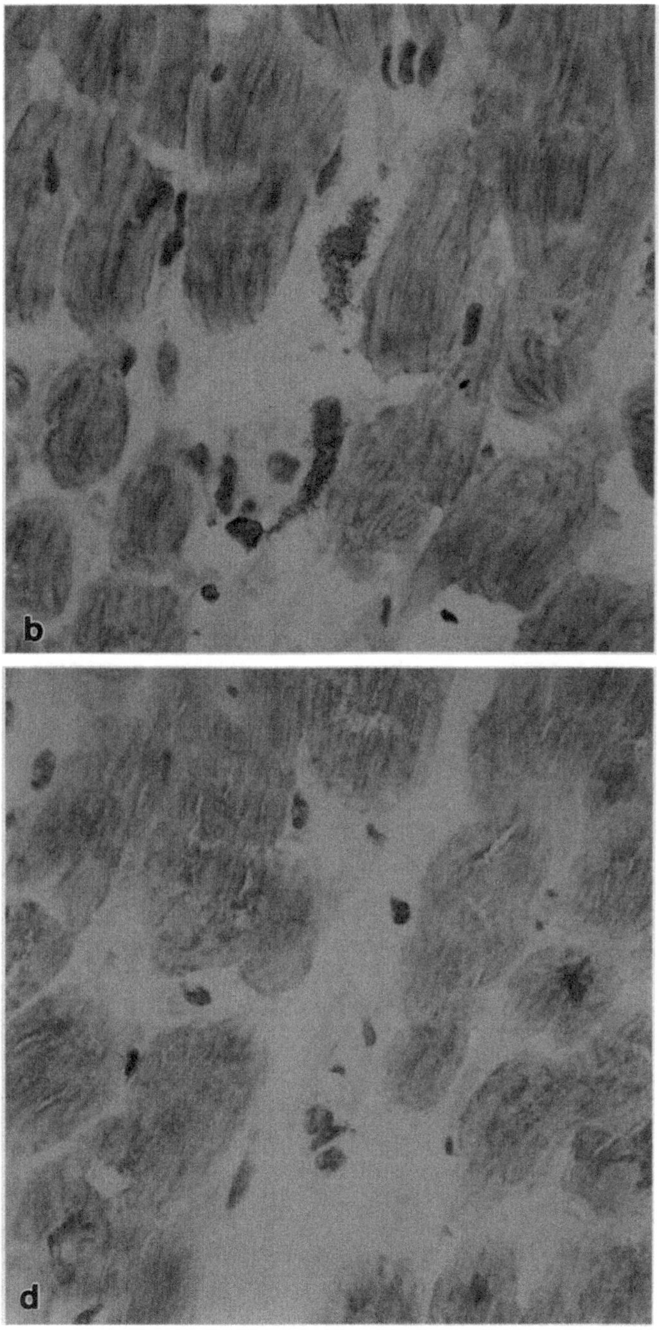
b
d

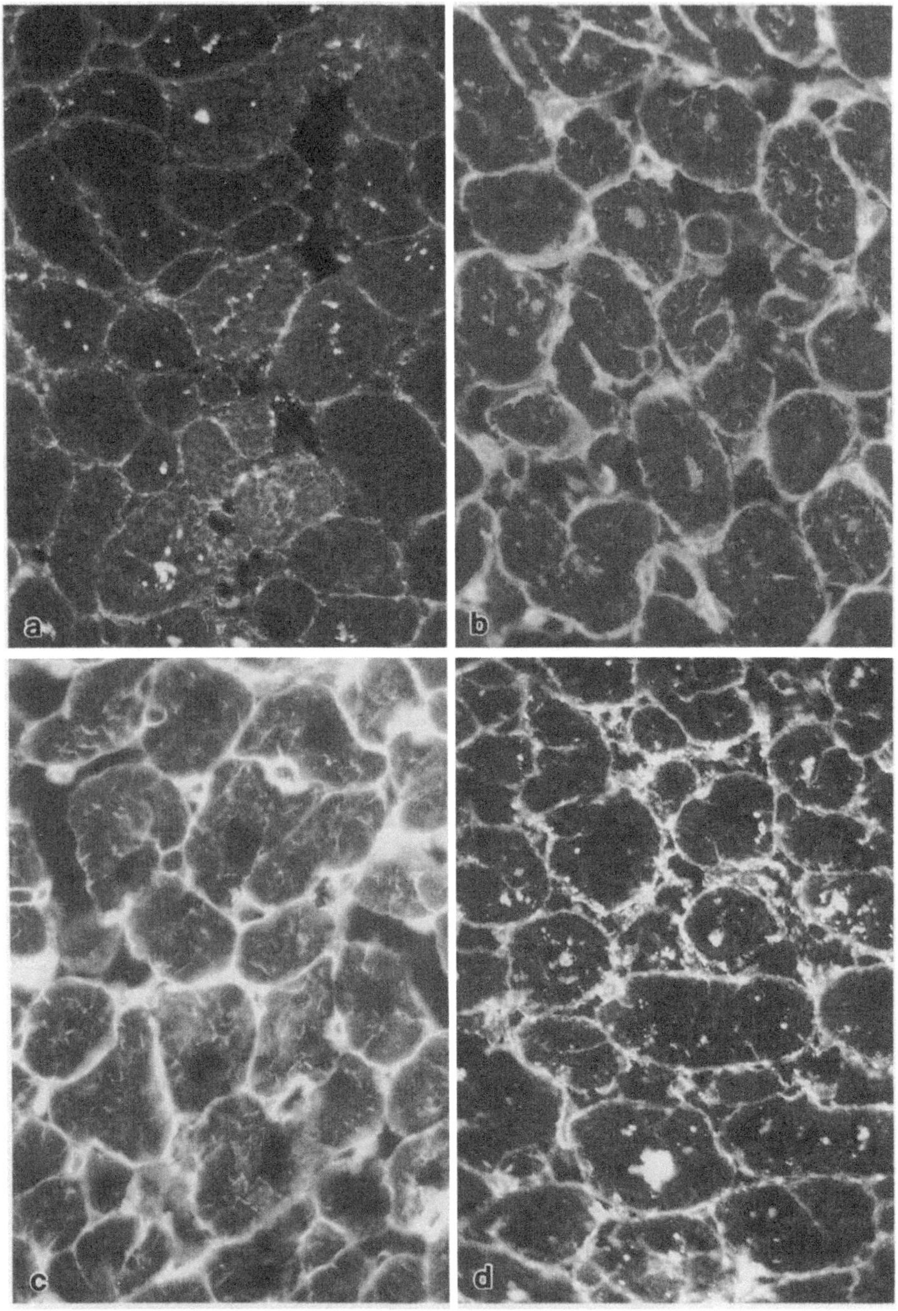

Fig. 2 a–d. Immunofluorescence findings in endomyocardial biopsies: sarcolemmal IgG deposits seen by direct immunofluorescence **a** can hardly be distinguished from staining patterns caused by antibodies against cell matrix proteins such as laminin **b** or type V collagen **c**. They are also found in high frequencies in different heart diseases, for example, CHD **d**. Intracellular IgG deposits **a**, however, are only seen in myocarditis and are not be detected in CHD **d**. x 320

Table 2. Sarcolemmal IgG, complement and intracellular IgG deposits seen in direct immunofluorescence in endomyocardial biopsies

Diagnosis	*n*	ASA (IgG)	C3	Intracellular IgG deposits	anti-ANT antibodies
Acute MC	3	2/3	2/3	2/3	2/3
Borderline MC	27	25/27	13/27	25/27	16/27
No MC	9	6/9	1/9	1/9	1/9
DCM	51	20/51	25/51	39/51	26/51
Alcohol toxic cardiomyopathy	10	10/10	–	–	–
CHD	26	20/26	1/26	1/26	1/26
MVD	16	13/16	1/16	1/16	1/16

ASA, antissarcolemmally bound IgG; C3, complement; ANT, adenine nucleotide translocator; MC, myocarditis; DCM, dilated cardiomyopathy; CHD, coronary heart disease; MVD, mitral valve disease

about 50% of patients with clinically suspected myocarditis (Table 2) in about 50% of patients with DCM and in one patient with CHD or MVD. These data indicate that the detection of bound IgG alone is of no diagnostic value, while the combination of IgG and C3 binding might be significant for myocarditis.

The interpretation of these direct immunofluorescence data is, however, hindered by the fact that resolution of light microscopy is not sufficient to differentiate between sarcolamma of cardiac myocytes and their surrounding connective tissue layers. Autoantibodies against connective tissue components such as collagens, laminin or fibronectin are often seen in many diseases and also found in healthy individuals (Table 3). Affinity-purified antibody staining for laminin, a basement membrane component, gives rise to a cell surface-like staining (Fig. 2b). Similar staining is also seen when carried out for fibronectin or collagens (Fig. 2c). As these staining patterns cannot be discriminated from fluorescence findings normally seen in endomyocardial biopsies (Fig. 2a), the positive staining

Table 3. Autoantibodies against connective tissue laminin in sera of patients with various heart muscle diseases

Diagnosis	*n*	Laminin
MC	30	28/30
No MC	9	8/9
DCM	91	88/91
CHD	20	17/20
Control	20	18/20

MC, myocarditis; DCM, dilated cardiomyopathy; CHD, coronary heart disease

in biopsies does not prove an IgG deposition on the sarcolemma. This is confirmed by the fact that antibodies to these extracellular matrix proteins are present in high frequencies of human sera, tested by the ELISA technique. As shown in Table 3 for laminin, autoantibodies against laminin, fibronectin or collagen types I, III, IV or V are seen in more than 70% of the sera from patients with different myocardial diseases. These data clearly demonstrate that the detection of so-called sarcolemmally bound IgG is of no diagnostic value.

In contrast to these, sarcolemmal-staining, intracellular IgG deposits (Fig. 2a) are only seen in biopsies of patients with myocarditis and, to a lesser extent, in dilated cardiomyopathy (Table 2). These intracellular IgG deposits are often correlated with the presence of autoantibodies against the adenosine di-/triphosphate (ADP/ATP) carrier, a recently characterized organ-specific autoantigen in myocarditis and dilated cardiomyopathy (Table 2) [25]. As intracellular IgG deposits and ADP/ATP carrier antibodies (measured in ELISA) are not found in controls or in other heart muscle diseases (Table 2, Fig. 2d), this combination might have an excellent prognostic meaning for the diagnosis of myocarditis or a postmyocarditic process.

Discussion

Patients with myocarditis may be asymptomatic or may present with acute progressive or chronic symptoms. To overcome the lack of differentiating clinical findings, endomyocardial biopsies were carried out in order to get a more accurate diagnosis. However, histological analysis of endomyocardial biopsies has failed to confirm the clinical diagnosis in most patients, mainly because only 5%–10% of clinically suspected cases show morpholocical criteria of acute myocarditis [2]. This circumstance is attributed to the timing of the biopsy following the viral infection, influencing the ability to detect myocardial necrosis, or to a sampling error [4]. Moreover, the lack of unambigious histological criteria, established quantitative criteria for cellular infiltrations and the difficulty of a histological evaluation by light microscopy to identify an inflammatory process hinder an accurate diagnosis.

Thirty-nine patients with clinical myocarditis were selected for this study. It focused on the value of direct immunofluorescence and the andvantage of additional immune markers for getting a better yield in diagnosis and thus a better understanding of the involved immunological processes. Only three of our patients showed unambigious histological criteria of acute myocarditis. Twenty-seven patients (90%) were consistent with borderline myocarditis according to the Dallas classification, nine patients had no myocarditis.

Nevertheless, diagnosis of myocarditis by light-microscopical evaluation of biopsies still has evident limitations. Cellular infiltrates may be sparse and interstitial cells such as fibroblasts or pericytes, as well as non-lymphocytic inflammatory cells, may mimic lymphocytes. Distinction of myocyte degeneration from necrosis is also difficult and contribution of humoral immune responses is not seen due to the method. In an effort to improve diagnostic and prognostic yield, we tested the added value of different immune markers. As seen in Fig. 1a, detection of lymphocytes is improved by the use of monoclonal antibodies against T lymphocyte surface antigens, allowing their detection even if the lymphocytic infiltrate is sparse. Results of inflammatory cell counts are summarized in Table 1. Although Zee-Cheng et al. [35] regard an even minimal lymphocyte infiltration as reliable criteria for diagnosis of myocarditis, our group, in accordance with other investigators [22, 30], used the presence of more than five chronic inflammatory cells or three lymphocytes per high-power field as positive diagnostic criteria. According to this assumption, 33% of our patients with borderline myocarditis were found to have significant lymphocytic infiltrates.

A subdivision of biopsies classified as borderline myocarditis in cases with or without a lymphocytic infiltrate might provide additional information. This fact is also taken into account by a new classification of myocarditis recently described by Olsen [23]. This classification will perhaps allow the description of different activities of the involved immunological process in future. Present results suggest that the more acute state is represented by CD4-positive cells, while in chronic myocarditis CD8-positive cells might be predominant.

To gain a better understanding of the pathogenetic role of cellular immune events in the onset and course of this heart muscle disease, a further characterization of infiltrating cells could be helpful.

Subtyping of lymphocytes in CD4-positive helper/inducer and CD8-positive cytotoxic/suppressor cells showed the presence of both lymphocyte subtypes in biopsy specimen. In seven out of ten cases, CD4-positive lymphocytes were the predominant cell type, while in two biopsy specimen mainly CD8-positive cells were found. One biopsy contained both subtypes in similar amounts. Moreover, first results from lymphocyte cultures, obtained from biopsy specimen indicate that the predominant cell types found by immunohistological methods will also be preserved in raised cell cultures (unpublished results).

This suspicion would be in accordance with studied animal models of virally infected mice, showing mainly CD4-positive cells during the early phase of the disease, while cytotoxic/suppressor T lymphocytes (CD8) are found in the late stage [9, 12–14]. Present evidence suggests that T helper cells indirectly mediate the disease. Probably these cells are also involved in humoral mechanisms, whereas cytotoxic/suppressor T lymphocytes may directly cause myocarditis indicating progression or a resolved immunological process.

It is important, however, that timing of biopsies greatly influences the ability to detect a cellular infiltration. If an acute lymphocytic infiltrate is to be found with myocyte damage, then it is most likely to occur within 3–6 weeks of the acute onset [4]. Later only reparative changes may be seen, which are not diagnostic. Thus the detection of MHC class I and II antigen expression might provide additional advantage in diagnosing an acute or resolved immunological process, even in the absence of infiltrating lymphocytes.

Recent data indicate that expression of MHC gene products, which are involved in humoral and cellular defence mechanisms [7], is increased in myocarditis [1, 11] (also Hammond et al., Herskowitz et al., this volume). Both antigens are normally not expressed on cardiac myocytes [5, 31], although MHC I can be detected on interstitial cells and vascular epithelium, the latter of which is thought to be involved in antigen presentation [33]. Studies on cardiac and kidney transplant rejections [21, 24] have shown that MHC antigen expression is induced during organ rejection and associated with an increase of leucocyte infiltrate, although not necessarily of T cells [8]. The presence of MHC II may persist for several weeks before return to normal [8, 31], thus enabling the diagnosis of a resolved cellular or humoral immunological event even in the absence of other immunological markers.

Anti-cardiac-specific humoral autoreactivity has been shown to occur in the biopsies of patients with myocarditis by several investigators [15, 18–20, 26, 34], however, its significance is not clear at present. Suggesting a major contribution of humoral events in the onset and course of myocarditis, [19, 27, 28] several groups have strengthened the hypothesis that deposits of IgG and complement (C3), easily detected by direct immunofluorescence, are of diagnostic relevance [10, 18, 19, 26]. Sarcolemmally bound antibodies, however, are not specific for myocarditis. They are also detected in the myocardium of patients with various other heart diseases (Table 3). In addition, it is well known that autoantibodies directed against connective tissue components are a common feature in many diseases [32] and are even found in healthy controls [3]. Resolution of light microscopy as shown in Fig. 2, is too low to discriminate between the sarcolemma of cardiac myocytes and the extracellular matrix. The high incidence of circulating autoantibodies against laminin, fibronectin and collagens in different heart diseases (Table 3) therefore attenuates the diagnostic value of direct immunofluorescence even if specific antibodies are present. Depositions of complement (C3), focally or at the myocyte sarcolemma, however, in contrast to IgG deposition, is only found in myocarditis. Its detection in about 50% of cases with clinically suspected myocarditis indicates its high specificity for diagnosis of myocarditis (Table 2) although it is of low sensitivity.

In contrast, intracellular IgG deposits are only associated with myocarditis and not found in other cardiac diseases. The pathophysiological relevance of these findings, however, has to be established. Nevertheless, in contrast to sarcolemmally bound antibodies, the close correlation of intracellular IgG deposits with myocarditis and, to a lesser extent, with dilated cardiomyopathy, which is thought to result from viral myocarditis at least in some instances, might provide a helpful criterion for the diagnosis of myocarditis.

In summary, sarcolemmally bound antibodies (IgG), seen by the direct immunofluorescence technique, are not helpful for the diagnosis of myocarditis. Specific cross-reactive antibodies doubtlessly are present in the concert of a polyclonal immune response to cell surface and matrix proteins. Until the cross-reacting myocyte surface antigens are known, the diagnostic relevance, however, is hindered by the high frequencies of autoantibodies to these extracellular matrix components. In contrast, intracellular IgG deposits in combination with autoantibodies against the ADP/ATP carrier [27], as well as complement deposits, are found to be specific for myocarditis or a postmyocarditic process. Moreover, the identification of infiltrating T lymphycytes and their subsets by immune markers facilitates the recognition of myocardial inflammation. In addition, detection of MHC antigen expression allows the characterization of an immunological process even in the absence of cellular or humoral immune markers. These observations suggest that functional abnormalities are not only caused by an inflammatory cell response but also by non-cell-mediated immune processes. The use of immune markers in endomyocardial biopsies will thus develop into a helpful tool for diagnosis and prognosis of myocarditis.

Acknowledgements. The excellent technical assistance of E. Ronft, E. Pawels, E. Fiedler and A. von Westphalen is gratefully acknowledged. This research was supported by a grant from the Deutsche Forschungsgemeinschaft and the Wilhelm-Sander-Stiftung.

References

1. Ahmed A, Leatherbury A, Stone J, Dempsey C, Beisel K, Tadros T, Sell KW (1987) Expression of class I and class II MHC gene products on human and murine cardiac myocytes. Fed Proc 46: 946–957
2. Aretz T (1986) Diagnosis of myocarditis by endomyocardial biopsy. Med Clin North Am 70: 1215–1226
3. Bernard A, Passelecq B, Foidart JM, Mahieu P, Lauwerys R (1987) Mise en evidence d'anticorps antilaminine dans le serum de sujets sains. Biomed Pharmacother 41: 183–186
4. Billingham E (1985) The diagnostic criteria of myocarditis by endomyocardial biopsy. Heart and Vessels [Suppl] 1: 133–137
5. Burger DR, Vetto RM (1982) Vascular endothelium as a major participant in T-lymphocytes immunity. Cell Immunol 70: 357–361
6. Edwards WD, Holmes DH, Reeder GS (1982) Diagnosis of active lymphocytic myocarditis by endomyocardial biopsy. Mayo Clin Proc 57: 419–425
7. Foulis AK (1986) Class II major histocompatibility complex and organ specific autoimmunity in man. J Pathol 150: 5–11
8. Fuggle SV, McWhinnie DL, Chapman JR, Taylor HM, Morris PJ (1986) Sequential analysis of HLA class II antigen expression in human renal allografts: induction of tubula class II antigens and correlation with clinical parameters. Transplantation 42: 144–154
9. Guthrie M, Lodge PA, Huber SA (1984) Cardiac injury in myocarditis induced by coxsackievirus group B, Type 3 in BALB/c mice is mediated by lyt 2^+ cytolytic lymphocytes. Cell Immunol 88: 558–567
10. Hammond EH, Menlove RL, Anderson JL (1987) Predictive value of immunofluorescence and electron microscopic evaluation of endomyocardial biopsies in the diagnosis and prognosis of myocarditis and idiopathic dilated cardiomyopathy. Am Heart J 114: 1055–1065
11. Hewan-Lowe K, Ansari AA, Leatherbury A (1988) Induction and increased expression of MHC class I and class II antigens on human cardiac myocytes in cardiac tissue of patients with viral myocarditis. Am J Pathol, (in press)
12. Huber SA, Lodge PA (1984) Coxsackievirus B-3 myocarditis in BALB/c mice. Evidence for autoimmunity to myocyte antigens. Am J Pathol 116: 21–29
13. Huber SA, Lodge PA (1986) Coxsackievirus B-3 myocarditis. Identification of pathogenetic mechanisms in DBA/2 and BALB/c mice. Am J Pathol 122: 284–291
14. Huber SA, Leyden DC, Lodge PA (1985) Immunopathogenesis of experimental coxsackievirus induced myocarditis: role of autoimmunity. Herz 10: 1–7
15. Klein R, Seipel L, Kleeman V (1987) Relevance of antimitochondrial antibodies (anti m-7) in cardiac disease. Europ H J [Suppl J] 8: 223–226
16. Kühl U, Öcalan M, Timpl R, Mayne R, Hay E, von der Mark K (1984) Role of muscle fibroblasts in the deposition of type IV collagen in the basal lamina of myotubes. Differentiation 28: 164–172
17. Kühl U, Timpl R, von der Mark K (1982) Synthesis of type IV collagen and laminin in cultures of skeletal muscle cells and their assembly on the surface of myotubes. Dev Biol 93: 344–354

18. Lowry PJ, Thompson RA, Littler WA (1983) Humoral imunity in cardiomypathy. Br Heart J 50: 390–394
19. Maisch B, Deeg P, Liebau G (1983) Diagnostic relevance of humoral and cytotoxic immune reactions in primary and secondary dilated cardiomyopathy. Am J Cardiol 52: 1072–1078
20. Maisch B (1987) The sarcolemma as antigen in the secondary immunopathogenesis of myocarditis Eur H J [Suppl J] 8: 155–165
21. Milton AD, Fabre JW (1985) Massive induction of donor-type class I and class II major histocompatibility complex antigens in rejecting cardiac allografts in the rat. J Exp Med 161: 98–104
22. Nippoldt TB, Edwards WD, Holmes DR, Reeder GS, Hartzler Go, Smith HC (1982) Right ventricular endomyocardial biopsy. Clinicopathologic correlates in 100 consecutive patients. Mayo Clin Proc 57: 407–418
23. Olsen EGJ (1985) The role of biopsy in diagnosis of myocarditis. Herz 10: 21–26
24. Rose NL, Coles MI, Griffin RJ, Pomerance A, Yacoub MH (1986) Expression of class I and class I major histocompatibility antigens in normal and transplanted human heart. Transplantation 41: 776–787
25. Schultheiss HP, Bolte HD (1985) Immunological analysis of autoantibodies against the adenine nucleotide translocator in dilated cardiomyopathy. J Mol Cell Cardiol 17: 603–617
26. Schultheiss HP, Kühl U, Ulrich G (1987) The mitochondrium as antigen in inflammatory heart disease. Eur Heart J 8: 203–210
27. Schultheiss HP, Schulze K, Kühl U, Ulrich G, Klingenberg M () The ADP/ATP carrier as a mitochondrial auto-antigen — facts and perspectives. Ann NY Acad Sci 488: 44–68
28. Schultheiss HP, Ulrich G, Janda I, Kühl U, Morad M (1988) Antibodymediated enhancement of calcium permeability in cardiac myocates. J Exp Med (in press)
29. Shanes JG, Ghali J, Billingham ME, Ferrans VJ, Fenoglio JJ, Edwards WD, Tsai CC, Saffitz JE, Isner J, Furner S, Subramanian R (1987) Interobserver variability in the pathologic interpretation of endomyocardial biopsy results. Circulation 75: 401–405
30. Steenbergen C, Kolbeck PC, Wolfe JA, Anthony RM, Sanfilippo FP, Jennings RB (1986) Detection of lymphocytes in endomyocardium using immunohistochemical techniques. Relevance to evaluation of endomyocardial biopsies in suspected cases of lymphocytic myocarditis. J Appl Cardiol 1: 63–73
31. Suitters A, Rose M, Higgins A (1987) MHC antigen expression in sequential biopsies from cardiac transplant patients — correlation with rejection. Clin Exp Immunol 69: 575–583
32. Tonietti G, Giacomeli R, Accini L, Caruso L (1986) Evidence of antimitochondrial autoantibodies in patients affected by chronik myocardial ischemia. IRCS Med Sci 14: 1075–1076
33. Wagner CR, Veto RM, Burger DR (1985) Subcultured human endothelial cells can function independently as fully competent antigen-presenting cells. Hum Immunol 13: 33–47
34. Yamakawa K, Fukuta S, Kimura Y (1983) Circulating antiheart antibodies in heart disease detected using immunofluorescent technique. Jpn Circ J 47: 1173–1178
35. Zee-Cheng CS, Tsai CC, Palmer DC, Codd JE, Pennington DG, Williams GA (1984) High incidence of myocarditis by endomyocardial biopsy in patients with idiopathic congestive cardiomyopathy. Am Coll Cardiol 3: 63–70

Enteroviral Heart Disease: Diagnosis by In Situ Hybridization

R. KANDOLF[1,2], P. KIRSCHNER[2], D. AMEIS[2], A. CANU[2], E. ERDMANN[1], H. P. SCHULTHEISS[1], B. KEMKES[3], and P. H. HOFSCHNEIDER[2]

[1] Department of Internal Medicine I, University of Munich, D-8000 München 70, FRG
[2] Department of Virus Research, Max-Planck-Institut für Biochemie, D-8033 Martinsried, FRG
[3] Department of Cardiac Surgery, University of Munich, D-8000 München 70, FRG

Introduction

Enteroviruses of the human Picornaviridae such as the group B coxsackieviruses (types 1 to 5) are generally considered to be the most common agents of viral myocarditis [1, 2, 11, 25, 29, 35]. Other members of the human enteroviruses comprising at present over 70 serotypes (e.g., various coxackie A viruses and echoviruses) have also been associated with human viral heart disease. These agents appear to be capable of producing dilated cardiomyopathy of acute onset or lead to life-threatening arrhythmias. Particularly intriguing is the concept of chronic dilated cardiomyopathy evolving from acute or subacute infections of the human heart.

The structure and molecular genetics of the enteroviruses are well understood, chiefly by analogy with the thoroughly studied poliovirus [25, 32]. The genomic single-stranded RNA of enteroviruses consists of about 7500 nucleotides of positive polarity, covalently linked at the 5' end to a small virus-encoded protein. The viral RNA, which is polyadenylylated at the 3' end, functions as a messenger in the cytoplasm of infected cells. It is translated into a polyprotein which is subsequently cleaved by viral proteases into the mature viral proteins.

Diagnosing enteroviral heart disease by clinical features is an exercise of uncertain validity. In practice, the measurement of serum antibody titers in suspected enteroviral heart disease seldom proves useful diagnostically because of the high number of enterovirus serotypes that might be implicated [5, 18]. Thus, we lack information on how often enteroviruses are actually the cause of acute dilated cardiomyopathy, what the prognosis is in such cases, and whether specific therapy can affect prognosis. The reason for this fundamental lack of knowledge is the difficulty of establishing an unequivocal diagnosis of viral heart disease. Confirmation of the clinical suspicion of enterovirus-induced cardiomyopathy demands the demonstration of replicating virus inside myocardial cells, which is exceedingly difficult by conventional methods [18, 24]. In addition, the increasing availability of potential antiviral agents — e.g., interferons [13,17] — has accentuated the need for methods by which endomyocardial biopsy specimens of patients with suspected viral heart disease can be diagnosed conclusively.

Molecular genetic techniques have now provided investigators with powerful tools to define the role of enteroviruses in the induction of the disease [3, 8, 14, 16, 30] as well as to study the molecular basis of pathogenicity. This chapter will

H.-P. Schultheiß (Ed.)
New Concepts in Viral Heart Disease

focus on the development of an enterovirus group-specific *in situ* hybridization approach [16, 18], which at present provides the method of choice for diagnosing enteroviral heart disease using endomyocardial biopsy samples. Owing to the high sensitivity of the molecular *in situ* hybridization approach, not only extensive virus replication is detected, but also more elusive virus-cell interactions — e. g., in persistent infections where permissive viral replication is replaced by restricted replication.

Molecular Cloning and Characterization of Full-Length Reverse-Transcribed Coxsackievirus B3 cDNA

One major prerequisite for the introduction of *in situ* hybridization as a diagnostic tool in suspected enterovirus-induced cardiomyopathy was the molecular cloning and characterization of the single-stranded genomic RNA of the cardiotropic coxsackievirus B3 (CVB3) [14]. The polyadenylylated viral RNA isolated from a large plaque variant of the Nancy strain of CVB3 that had been propagated in cultured human heart cells [12, 13] served as a template for first-strand cDNA synthesis by using reverse transcriptase. Making use of the RNase H- and DNA polymerase I-mediated second-strand cDNA synthesis [6, 27], an efficient cloning strategy for large RNA molecules was designed and followed (Fig. 1), representing an improvement on earlier methods used in the molecular cloning of viral genomes. After second-strand cDNA synthesis, double-stranded CVB3 cDNA was directly tailed with dGMP by using terminal deoxynucleotidyl transferase. Oligo $(dG)_{10-15}$-tailed CVB3 double-stranded cDNA was annealed to oligo $(dC)_{10-15}$-tailed plasmid vector p2732B, previously linearized with *Cla*I and *Bgl*II in its cloning site (Fig. 1). On transformation of competent *Escherichia coli* BJ 5183, at least 4×10^4 bacterial colonies were obtained per microgram of viral RNA used in the initial reverse transcription.

The efficiency of the cloning strategy is given by the fact that full-length reverse-transcribed CVB3 cDNA of about 7500 nucleotides was molecularly cloned [14]. Moreover, full-length reverse-transcribed CVB3 cDNA generated infectious virus antigenically identical to CVB3 upon transfection of recombinant viral cDNA into mammalian cells, demonstrating the molecular cloning of a faithful transcript of the viral RNA. Because there is no DNA intermediate in the life cycle of enteroviruses, the availability of the infectious CVB3 cDNA copy offers unique opportunities for the genetic analysis of this cardiotropic virus. From the diagnostic point of view, cloned CVB3 cDNA offers the unique possibility for a group-specific diagnosis of enterovirus infections.

Cloned Coxsackievirus B3 cDNA as a Diagnostic Reagent for the Detection of Enteroviruses

Because of the high degree of nucleic acid sequence identity shared among the numerous serotypes of the human enterovirus group, detection of the most commonly associated agents of human viral heart disease, including the group A and

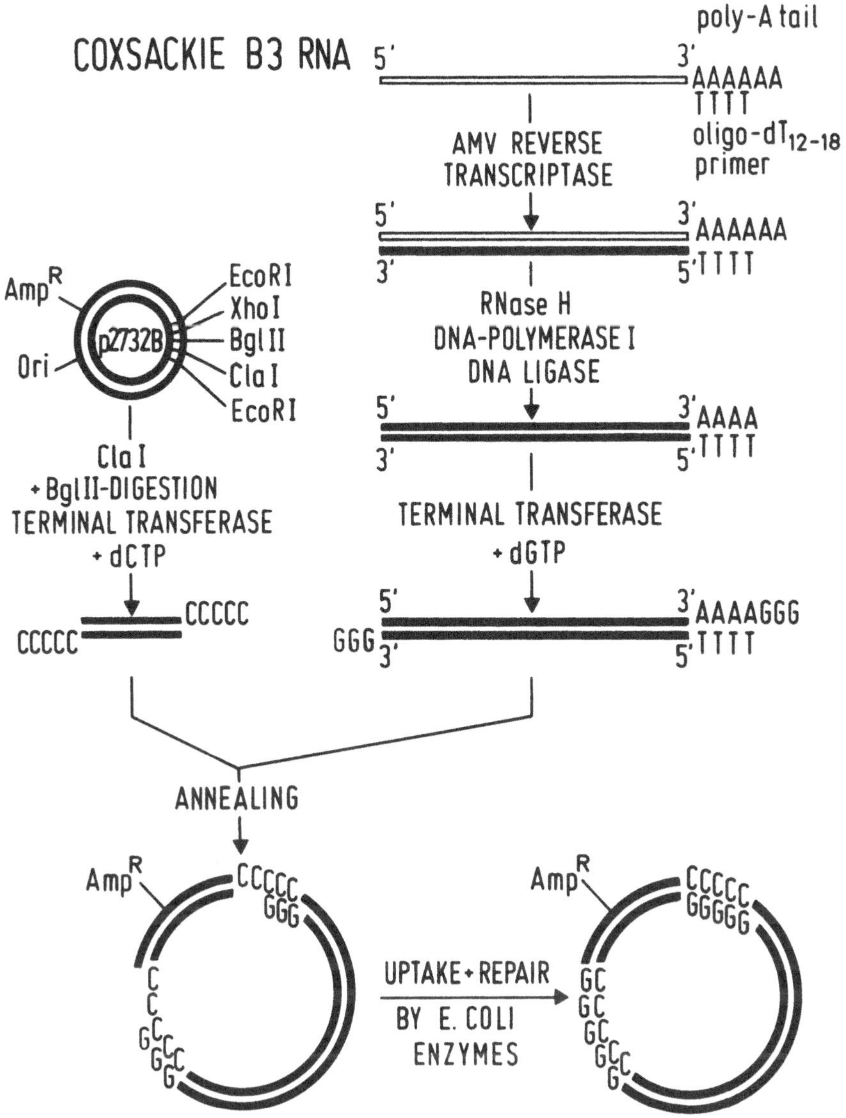

Fig. 1. Cloning strategy of CVB3 cDNA. The polyadenylylated purified CVB3 RNA served as a template for oligo(dT)$_{12-18}$ primed first-strand complementary DNA (cDNA) synthesis, using reverse transcriptase as previously described [14]. The RNA strand of the RNA-cDNA hybrid was replaced by RNase H and DNA polymerase I-mediated second-strand synthesis and homopolymer tails of 10–15 residues of dGMP were added directly to double-stranded cDNA by using terminal deoxynucleotidyl transferase. Plasmid vector p2732B, a derivate of pBR322 containing the ampicillin-resistance gene, the plasmid origin of replication and a cloning site array between two *Eco*RI sites, was kindly provided by J. D. Monahan (Roche Institute, Nutley, New Jersey). Plasmid vector p2732B was linearized at its cloning site by digestion with *Cla*I and *Bgl*II, tailed with dCMP, hybridized to the oligo (dG)-tailed double-stranded cDNA and used to transform *Escherichia coli* BJ 5183 as previously described [14]

B coxsackieviruses as well as the echoviruses, is possible in a single hybridization assay. To estimate the degree of nucleotide sequence homology between different enteroviral serotypes by a slot-blot hybridization assay, Vero cells were infected with various enteroviruses. Total RNA of infected cultures, which was isolated when about 50% of the cells exhibited cytopathologic effects, was blotted on nitrocellulose and hybridized with the ^{32}P-labeled CVB3 cDNA probe corresponding to the region form 0.066 to 7.128 kilobases of the viral genome. The results indicate a remarkably high degree of nucleotide sequence identity among different human enterovirus strains, as shown in Fig. 2 for CVB (types 1 to 6), coxsackievirus A9 (CVA9), echoviruses 11 and 12, and poliovirus type 1. Because comparative nucleotide sequence data for CVB3 [22, 23, 31] and poliovirus type 1 [21, 28] show on overall homology of about 70% within functional genes (e.g., the replicase gene) or within the noncoding 5' region, these cross-hybridization data (Fig. 2) appear to predict an even higher degree of nucleotide sequence

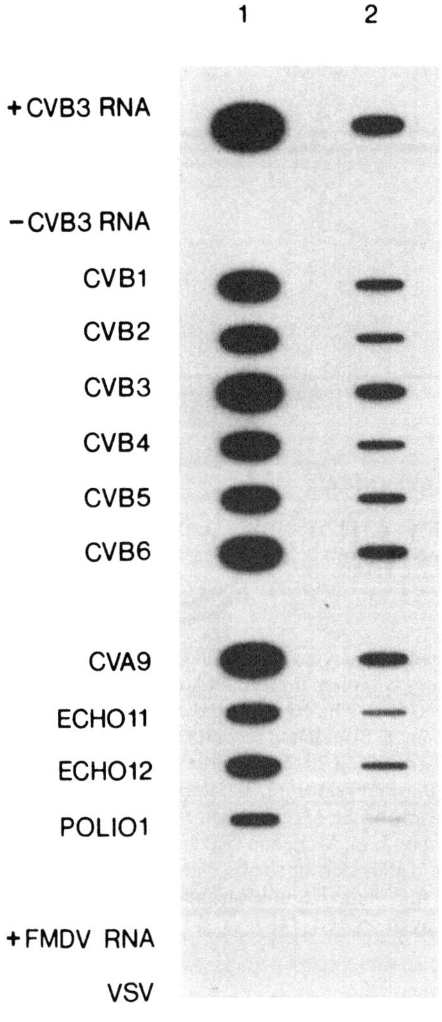

Fig. 2. Hybridization of cloned CVB3 cDNA to total RNA from Vero cells infected with various enteroviruses. Cells were infected with CVB1, CVB2, CVB3, CVB4, CVB5, CVB6, CVA9, echovirus 11, echovirus 12, or poliovirus type 1. Total RNA was isolated when about 50% of the cells exhibited cytopathologic effects. One microgram *(column 1)* and 0.1 microgram *(column 2)* of total cellular RNA were applied to nitrocellulose paper and hybridized with ^{32}p-labeled cDNA fragments corresponding to the region from 0.066 to 7.128 kilobases of the CVB3 genome. Reference slots *(+CVB3 RNA)* consisted of 5 ng *(column 1)* and 0.5 ng *(column 2)* of purified CVB3 RNA and were loaded in the presence of 1 μg *(column 1)* and 0.1 μg *(column 2)* total uninfected Vero cell RNA. Analogously, purified RNA of foot-and-mouth disease virus *(+FMDV RNA)* was applied in the presence of Vero cell RNA. Uninfected Vero cell RNA *(−CVB3 RNA)* served as a negative control. Vesicular stomatitis virus *(VSV)* RNA was obtained by isolating total RNA from infected Vero cells. Autoradiographic exposure was at −80 °C for 18 h. (From [16])

homology between the different serotypes of CVA and CVB as well as echoviruses. With respect to CVB, the extent of cross-hybridization is in agreement with the recently published sequences of CVB1 [9] and CVB4 [10] which indicated a homology of 85% and 82% for regions such as the replicase gene of these viruses. In contrast, foot-and-mouth disease virus, a nonhuman picornavirus, or the genetically unrelated vesicular stomatitis virus do not hybridize (compare Fig. 2). These results indicate that the cloned CVB3 cDNA can be used as an enterovirus group-specific probe.

This broad detection spectrum of cloned CVB3 cDNA greatly facilitates diagnosis of enteroviral heart disease since, from the clinical point of view, the antigenic typing of an etiologically implicated enterovirus serotype is of secondary importance and can be carried out later — e.g., by standard virologic techniques or by hybridization with serotype-specific cDNA fragments, available to date only for a limited number of enterovirus serotypes.

Highly specific hybridization conditions have been established for the detection of enteroviruses [14]. When radioactively labeled cloned CVB3 cDNA corresponding to 95.4% of the viral genome (nucleotides 66—7128) was hybridized to electrophoretically resolved CVB3 RNA and to total RNA from cultured human heart cells, specific hybridization was found for the viral RNA and not for human RNA.

In Situ Detection of Enterovirus RNA in Infected Cells by Nucleic Acid Hybridization

The feasibility of the *in situ* hybridization technique [7, 34] for the detection of enterovirus RNA was first established in cell culture systems and then applied to myocardial tissue of athymic mice persistently infected with CVB3 [16]. Uninfected Vero cells exhibited essentially no grains when hybridized to the ^{3}H-labeled or ^{35}S-labeled CVB3 cDNA probe. By contrast, highly significant labeling was achieved in Vero cells infected with various enteroviruses — e.g., CVB1, CVB3, CVB5, or echovirus 11.

Using myocardial tissue sections of CVB3-infected mice, *in situ* hybridization proved to be a powerful tool not only with respect to establishing an unequivocal diagnosis of myocardial infection but also with respect to understanding pathogenesis (Fig. 3). The autoradiographic silver grains, which indicate hybridization between viral RNA and the radiolabeled CVB3 cDNA probe are clearly related to distinct infected myocytes (Fig. 3A). These cells are easily identified by interference contrast microscopy in unstained sections because of their characteristic size and morphology. In this model system of enteroviral heart disease, myocardial involvement was found to be multifocal and randomly distributed in the heart muscle (Fig. 3B). Myocardial cross-sections revealed a transmural disseminated infection of the myocardium, as demonstrated in Fig. 3C for the left ventricle. Infected myocytes were often found in clusters within areas of severe myocardial lesions and fibrosis (Fig. 3D). Furthermore, progression of the infection could be observed from areas with myocardial fibrosis to as yet uninfected myocytes (Fig. 3E and F), indicating the possible cell-to-cell spread of the virus. However, viral

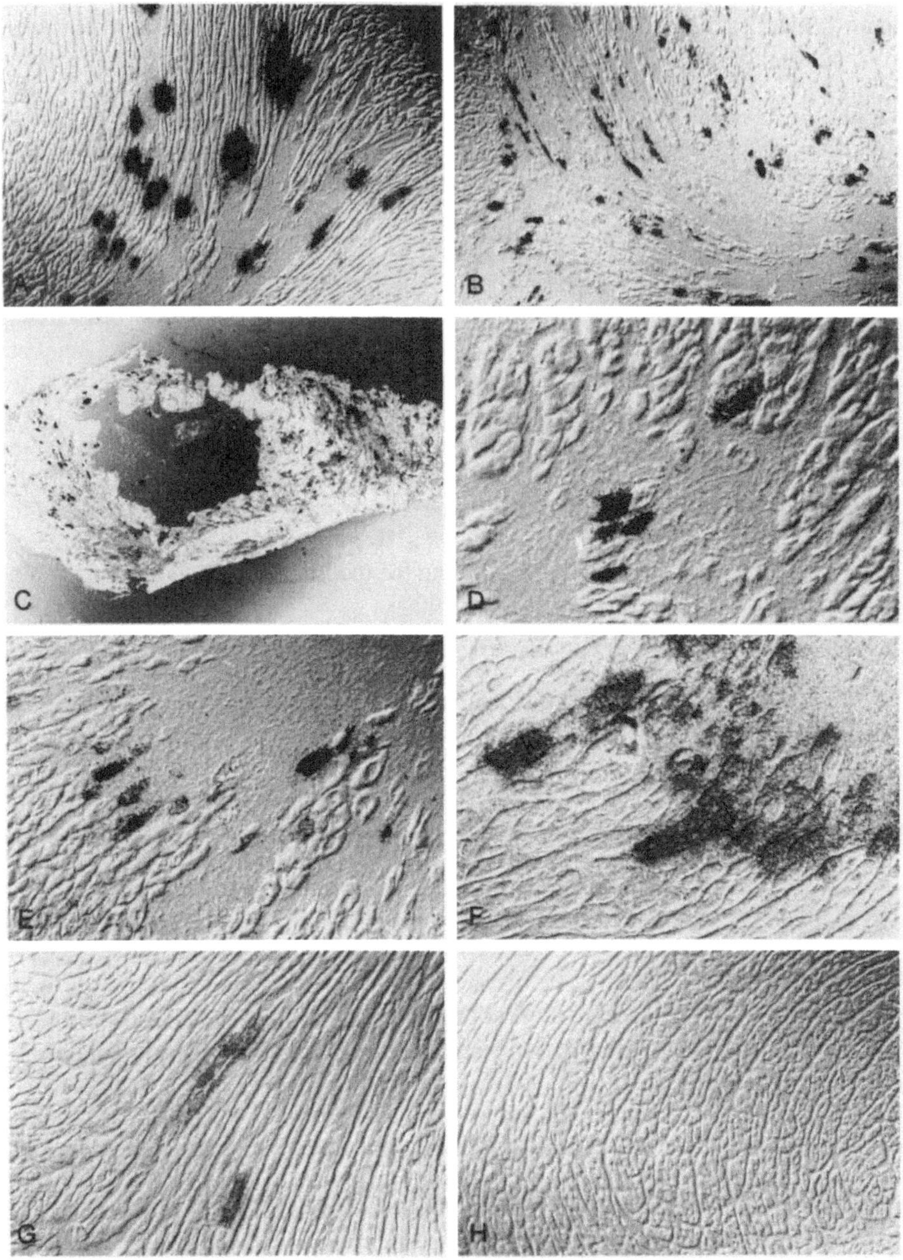

Fig. 3 A–H. Autoradiographs of CVB3-infected **(A–G)** and uninfected **(H)** mouse myocardial tissue hybridized *in situ* with ^{3}H-labeled **(B–E)** or ^{35}S-labeled **(A, F–H)** cloned CVB3 cDNA. The number of days after infection of athymic mice was 56 for **A**, 23 for **B–E**, 42 for **F**, and 4 for **G**. Exposure times were 9 days for **A** and **H**, 6 weeks for **B–E**, 4 days for **F**, and 2 days for **G**. Note that since silver grains are positioned at various levels within the photoemulsion, some grains are not observed and appear out of focus in photography. (From [16]) Interference contrast microscopy of unstained sections; x 200 **(A, G,** and **H)**, x 100 **(B)**, x 25 **(C)** x 400 **(D** and **E)**, x 650 **(F)**

RNA was also found in isolated myocytes in apparently normal myocardial tissue, which was primarily observed in the early stage of the infection (Fig. 3G). In addition, viral RNA also appeared to be located within the small interstitial myocardial cells (Fig. 3B and E). No labeling of myocardial cells was found when myocardial sections were probed with the radiolabeled plasmid p2732B control DNA [14] or with the cloned *Eco*RI J fragment of the genetically unrelated cytomegalovirus [26]. In addition, no labeling of myocardial cells was found when myocardial tissue sections of uninfected mice were probed with the radiolabeled CVB3 cDNA probe (Fig. 3H).

High sensitivity of detection was made possible by several improvements in methodology, including optimized hybridization conditions to prevent nonspecific binding and use of radiolabeled cloned cDNA fragments of about 100 nucleotides in length [16]. Quantification of CVB3 copy numbers in infected Vero cells by RNA blot analysis and comparison with *in situ* hybridization of cells from the same culture indicated that as few as 20 viral copies are easily detectable within 2 weeks of autoradiographic exposure. Clearly positive hybridization signals with infected mouse myocardial tissue were already observed after 2 days of exposure with the ^{35}S-labeled cDNA probe, indicating a high copy number of replicating viral genomes in myocardial cells (Fig. 3G). Furthermore, overexposed myocardial slide preparations showed extremely low background signals, demonstrating the high specificity of *in situ* hybridization for the detection of enteroviral RNA.

In Situ Detection of Enterovirus RNA in the Myocardium of Patients with Myocarditis and Dilated Cardiomyopathy

To date, *in situ* hybridization has already proved to be a valuable tool in assessing the presence of enterovirus RNA in endomyocardial biopsy samples obtained from patients with etiologically unexplained heart failure [15, 16, 18, 19]. Replicating enterovirus RNA was found to be present in 19 of 81 patients with a clinical suspicion of acute myocarditis, including eight of 27 patients with dilated cardiomyopathy of recent onset. Thirty patients with other specific heart muscle diseases — e.g., ischemic, hypertrophic, or metabolic cardiomyopathies — were used as an internal control group and were all negative when myocardial tissue was examined by the *in situ* hybridization approach.

Moreover, replicating enterovirus RNA was not only detected at an early stage of clinically obvious myocarditis, but also in chronic dilated cardiomyopathy, indicating persistence of the virus in the human heart. Seven of 39 patients, with chronic dilated cardiomyopathy, including four of 19 patients who had undergone heart transplantation because of end-stage dilated cardiomyopathy, were found to have enteroviral heart disease. In addition, the concept of enterovirus persistence in chronic dilated cardiomyopathy is further substantiated by the presence of enterovirus RNA in subsequent biopsies obtained from patients with ongoing disease. Figure 4 shows as an example *in situ* detection of replicating enterovirus RNA in the myocardium of a 49-year-old male patient with chronic dilated cardiomyopathy who underwent heart transplantation. The autoradiographic silver grains (Fig. 4), which indicate hybridization between viral RNA and the radio-

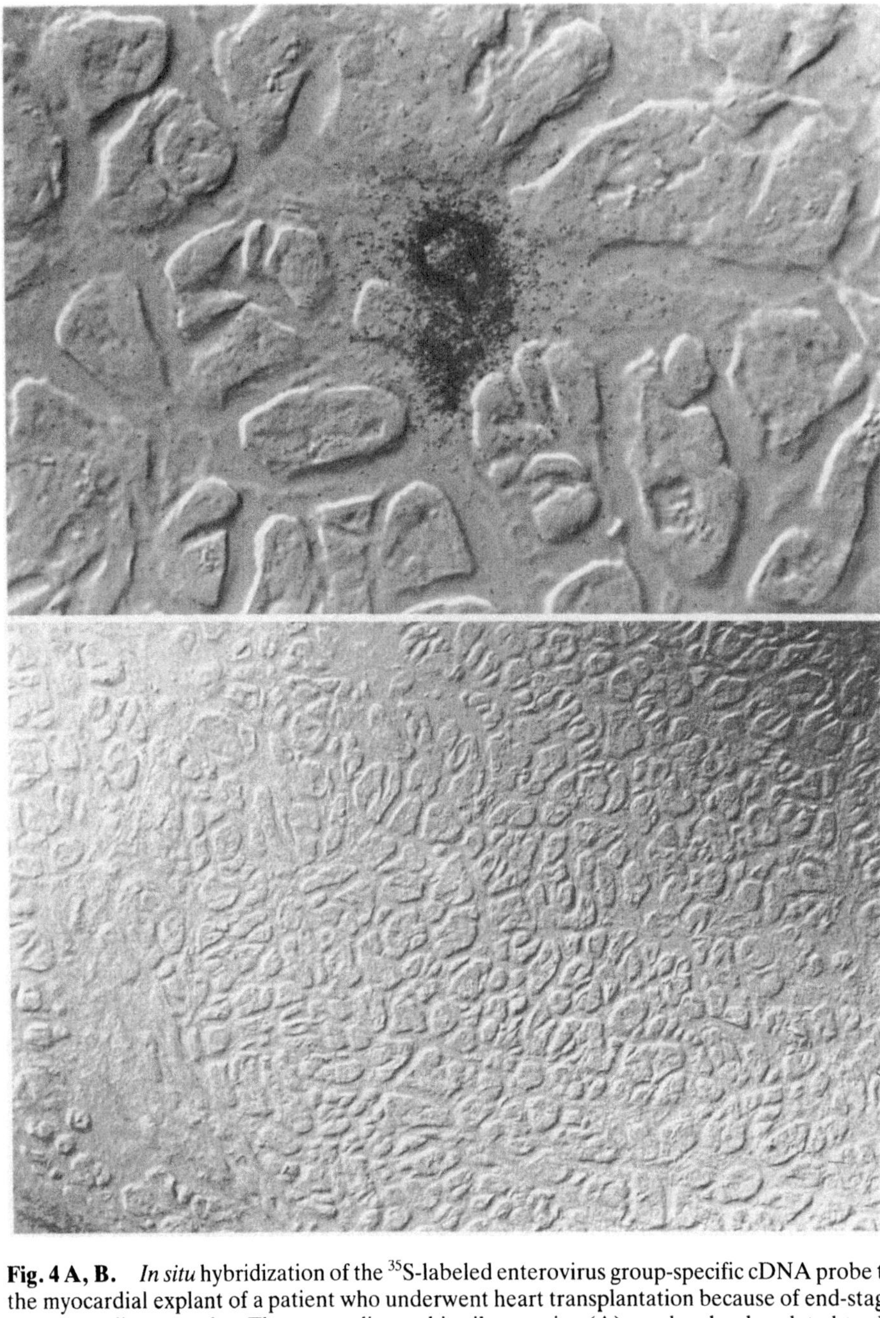

Fig. 4 A, B. *In situ* hybridization of the ^{35}S-labeled enterovirus group-specific cDNA probe to the myocardial explant of a patient who underwent heart transplantation because of end-stage dilated cardiomyopathy. The autoradiographic silver grains **(A)** can be clearly related to distinct infected myocytes, thereby providing the possibility for an unequivocal diagnosis of the enterovirus infection. No labeling of myocardial cells was observed when myocardial cryocuts were hybridized with the ^{35}S-labeled plasmid vector p2732B (**B**, control). Interference contrast of unstained sections

labeled cloned cDNA probe, are clearly related to single myocytes, thereby providing a unique possibility for an unequivocal diagnosis of the disease at the cellular level. Infected myocytes are easily identified by interference contrast microscopy in unstained sections because of their characteristic size and morphology.

Another important finding is our observation that replicating enterovirus RNA was not only found to be present in myocytes, but also in small interstitial cells, presumably fibroblasts, which agrees with the previous *in vitro* findings in cultured human heart cells and persistently infected human myocardial fibroblasts [13, 17]. Moreover, the number of infected myocardial cells appears to be related to the severity of clinical symptoms. Whereas in patients with benign perimyocarditis or healing myocarditis only a few myocardial cells were found to express enterovirus RNA, numerous infected cells were found in patients with severe dilated cardiomyopathy.

Summary and Outlook

Enterovirus research is currently in a very dynamic phase. Recombinant DNA techniques have not only allowed an important breakthrough in the study of enterovirus genetics but have already contributed significantly to an improved diagnosis of enterovirus infections. To date, there is firm evidence from the *in situ* nucleic acid hybridization approach that enterovirus infections of the human heart result in a significant number of patients presenting with clinical symptoms and signs of myocarditis or dilated cardiomyopathy. In patients with dilated cardiomyopathy of recent onset, the most dramatic manifestation of myocarditis, the incidence of enterovirus infections was found to be approximately 30%. The concept of enterovirus persistence in chronic dilated cardiomyopathy is further substantiated by the presence of enterovirus RNA in follow-up biopsies of patients with ongoing cardiac disease.

Presumably, enterovirus persistence in myocardial cells is based on continued synthesis of the viral precursor polyprotein and correct processing at least of the virus-encoded polymerase. However, lack of correct processing of other enterovirus gene products, e.g., coat proteins, could result in restricted replication with a limited production of infectious progeny virus. Restricted or defective replication could explain the common failure to isolate infectious virus from patients after the acute phase of virus replication. In addition, reactivation of infectious virus replication could explain the phenomenon of recurrence or recurrences, and progression to long-term cardiac disease.

To investigate further the molecular mechanisms of enterovirus persistence, the infectious CVB3 cDNA is used as a source for the generation of distinct viral proteins in different expression systems. In a first approach, fusion proteins containing distinct CVB3 coat proteins or the RNA-dependent RNA polymerase of CVB3 were expressed in *Escherichia coli.* Polyclonal antisera raised in rabbits against the purfied expression products offer the possibility of enterovirus group-specific identification of the various commonly implicated agents of viral heart disease, including CVA and CVB as well as the echoviruses [33]. In addition to the *in situ* hybridization approach, this broad detection spectrum greatly facilitates identification of enterovirus infections by antigen detection. The application of these antibodies in combination with *in situ* hybridization is currently being pursued. Using a double-labeling assay [4], the simultaneous *in situ* detection of enterovirus RNA and the demonstration of distinct viral proteins is now possible at the single-cell level. This approach will allow the question of whether restricted virus replication is implicated in persistent forms of enterovirus-induced cardiomyopathy to be resolved. Moreover, new constructions involving eucaryotic expression vectors are being designed in order to study the interaction between distinct CVB3 proteins and cellular mechanisms.

In addition to providing an unequivocal diagnosis of myocardial infection, there are therapeutic implications from the demonstration of replicating enterovirus RNA in myocardial cells. Corticosteroids appear to be clearly proscribed in the presence of replicating viral RNA since increased viral replication [20] and inhibition of the endogenous interferon system may follow their use [17]. The most appealing treatment for enteroviral heart disease, which in some cases may prevent the development of chronic dilated cardiomyopathy, appears to be antiviral therapy. We have previously reported on the high antiviral activity of natural fibroblast interferon (IFN-β) in CVB3-infected cultured human heart cells [13, 17]. Although the relationship between the *in vitro* effects of IFN-β and its potential in vivo activities remains to be determined, our *in vitro* observations are an important prerequisite for correct use of exogenous IFN-β in enterovirus infections.

Acknowledgments. We thank Ilona Dahmen and Sabine Höner for excellent technical assistance. This work was supported in part by Grant Ka 593/2-1 from the Deutsche Forschungsgemeinschaft and by Grant 321-7291-BCT-0370 from the German Ministry for Research and Technolgy. R. K. is a Hermann and Lilly Schilling Professor for Experimental Medicine.

References

1. Abelmann WH (1973) Viral myocarditis and its sequelae. Annu Rev Med 24: 145–152
2. Bolte HD (ed) (1984) Viral heart disease. Springer, Berlin Heidelberg New York
3. Bowles NE, Richardson PJ, Olsen EGJ, Archard LC (1986) Detection of coxsackie-B-virus-specific RNA sequences in myocardial biopsy samples from patients with myocarditis and dilated cardiomyopathy. Lancet 1: 1120–1123
4. Brahic M, Haase AT, Cash E (1984) Simultaneous in situ detection of viral RNA and antigens. Proc Natl Acad Sci USA 81: 5445–5448
5. Dec GW, Palacios IF, Fallon JT, Aretz HT, Mills J, Lee DC-S, Johnson RA (1985) Active myocarditis in the spectrum of acute dilated cardiomyopathy. N Engl J Med 312: 885–890
6. Gubler U, Hoffmann BJ (1983) A simple and very efficient method for generating cDNA libraries. Gene 25: 263–269
7. Haase A, Brahic M, Stowring L, Blum H (1984) Detection of viral nucleic acids by in situ hybridization. In: Maramorosh K, Koprowski H (eds) Methods in Virology, vol 7, pp 189–226
8. Hyypiä T, Stålhandske P, Vainionpää R, Petterson U (1984) Detection of enteroviruses by spot hybridization. J Clin Microbiol 19: 436–438
9. Iizuka N, Kuge S, Nomoto A (1987) Complete nucleotide sequence of the genome of coxsackievirus B1. Virology 156: 64–73

10. Jenkins O, Booth JD, Minor PH, Almond JW (1987) The complete nucleotide sequence of coxsackievirus B4 and its comparison to other members of picornaviridae. J Gen Virol 68: 1835–1848
11. Johnson RA, Palacios I (1982) Dilated cardiomyopathies of the adult. N Engl J Med 307: 1119–1126
12. Kandolf R, Hofschneider PH (1984) Effect of interferon on the replication of Coxsackie B3 virus in cultured human foetal heart cells. In: Bolte HD (ed) Viral heart disease. Springer, Berlin Heidelberg New York, pp 57–63
13. Kandolf R, Canu A, Hofschneider PH (1985) Coxsackie B3 virus can replicate in cultured human feotal heart cells and is inhibited by interferon. J Mol Cell Cardiol 17, 167–181
14. Kandolf R, Hofschneider PH (1985) Molecular cloning of the genome of a cardiotropic Coxsackie B3 virus: Full-length reverse-transcribed recombinant cDNA generates infectious virus in mammalian cells. Proc Natl Acad Sci USA 82: 4818–4822
15. Kandolf R, Kirschner P, Ameis D, Canu A, Hofschneider PH (1986) Virale Herzerkrankung: Etablierung einer neuen hochsensitiven Diagnostik der Virusinfektion des Myokards durch in situ DNA/RNA Hybridisierung. Z Kardiol 75, Supp 1, 57
16. Kandolf R, Ameis D, Kirschner P, Canu A, Hofschneider PH (1987) *In situ* detection of enteroviral genomes in myocardial cells: An approach to the diagnosis of viral heart disease. Proc Natl Acad Sci USA 84: 6272–6276
17. Kandolf R, Kirschner P, Ameis D, Canu A, Hofschneider PH (1987) Cultured human heart cells: a model system for the study of the antiviral activity of interferons. Eur Heart J 8: Supp J, 453–456
18. Kandolf R (1988) The impact of recombinant DNA technology on the study of enterovirus heart disease. In: Bendinelli M, Friedman H (eds) Coxsackieviruses — A General Update. Plenum Press, New York, pp 293–318
19. Kandolf R, Hofschneider PH (1988) Enterovirus-induced cardiomyopathy. In: Notkins AL, Oldstone MBA (eds) Concepts in Viral Pathogenesis III. Springer, New York, in press
20. Kilbourne E, Horsfall F (1951) Lethal infections with Coxsackievirus of adult mice given cortison. Proc Soc Exp Biol Med 77: 135–138
21. Kitamura N, Semler BL, Rothberg PG, Larsen GR, Adler CJ, Dorner AJ, Emini EA, Hanecack R, Lee JJ, van der Werf S, Anderson CW, Wimmer E (1981) Primary structure, gene organization and polypeptide expression of poliovirus RNA. Nature 291: 547–553
22. Klump WM, Bergmann I, Müller B, Ameis D, Hofschneider PH, Kandolf R (to be published) Complete nucleotide sequence of infectious coxsackievirus B3 cDNA: aspects of infectivity and RNA replication
23. Lindberg AM, Stålhandske POK, Petterson U (1987) Genome of coxsackievirus B3. Virology 156: 50–63
24. Melnick JL, Wenner HA, Phillips CA (1979) Enteroviruses. In: Lennette EH, Schmidt NJ (eds) Diagnostic Procedures for Viral, Rickettsial and Chlamydial Infections, 5th edn. American Public Health Assoc., Washington, D.C., pp 471–534
25. Melnick JL (1985) Enteroviruses: polioviruses, coxsackieviruses, echoviruses, and newer enteroviruses. In: Fields BN (ed) Virology. Raven Press, New York, pp 739–794
26. Nelson JA, Fleckenstein B, Galloway DA, McDougall JK (1982) Transformation of NIH 3T3 cells with cloned fragments of human cytomegalovirus strain AD169. J Virol 43: 83–91
27. Okayama H, Berg P (1982) High-efficiency cloning of full-length cDNA. Mol Cell Biol 2: 161–170
28. Racaniello VR, Baltimore D (1981) Molecular cloning of poliovirus cDNA and determination of the complete nucleotide sequence of the viral genome. Proc Natl Acad Sci USA 78: 4887–4891
29. Reyes MP, Lerner AM (1985) Coxsackievirus myocarditis — with special reference to acute and chronic effects. Prog Cardiovasc Dis 27: 373–394
30. Tracy S (1984) A comparison of genomic homologies among the coxsackievirus B group: Use of fragments of the cloned coxsackievirus B3 genome as probes. J Gen Virol 65: 2167–2172
31. Tracy S, Liu HL, Chapman NM (1985) Coxsackievirus B3: Primary structure of the 5' noncoding region and capsid protein coding regions of the genome. Virus Res 3, 263–270

32. Wimmer E, Kuhn RJ, Pincus S, Yang CF, Toyoda H, Nicklin MJH, Takeda N (1987) Molecular events leading to picornavirus genome replication. J Cell Sci, Suppl 7:251–276
33. Werner S, Klump WM, Schönke H, Hofschneider PH, Kandolf R (1988) Expression of coxsackievirus B3 proteins in *Escherichia coli* and generation of virus-specific antisera. DNA 7: 307–316
34. Wolf H, zur Hausen H, Becker V (1973) EB viral genomes in epithelial nasopharyngeal carcinoma cells, Nature New Biol 244: 245–247
35. Woodruff JF (1980) Viral myocarditis — a review. Am J Pathol 101: 427–479

Persistence of Enterovirus RNA in Dilated Cardiomyopathy: A Progression from Myocarditis

L. C. Archard[1], C. A. Freeke[1], P. J. Richardson[2], B. Meany[2], E. G. J. Olsen[3], P. Morgan-Capner[4], M. L. Rose[5], P. Taylor[5], N. R. Banner[5], M. H. Yacoub[5], and N. E. Bowles[1]

[1] Department of Biochemistry, Charing Cross and Westminster Medical School, Fulham Palace Rd. London, W6 8RP, UK
[2] Kings College Hospital, Cardiac Department, London SE 5 9RS, UK
[3] National Heart Hospital, Histology Department, London W1M 8BA, UK
[4] Preston Infirmary, District Laboratory, Preston, PR1 6PS, UK
[5] Harefield Hospital, Harefield, Middlesex UB9 6JH, UK

Introduction

We have investigated the role of enterovirus infection in the pathogenesis of myocarditis and dilated cardiomyopathy. In this article we will present evidence (from molecular hybridizations with virus-specific probes) that enteroviruses are major aetiological agents of myocarditis and dilated cardiomyopathy and that virus persistence in myocardium is implicated in the progression from myocarditis to dilated cardiomyopathy.

Myocarditis

While a few cases of inflammatory disease of cardiac muscle may be caused by toxins or parasites, the majority are believed to be caused by virus infection [35]. Although a range of viruses has been implicated [24], the enteroviruses, and particularly the Coxsackie group B viruses, have been associated most frequently.

Other than infants who have developed a multisystem infection in which myocarditis often predominates, or adult patients with acute myocarditis of short history, virus isolation from heart muscle or pericardial fluid is generally unsuccessful [15]. Similarly, immunofluorescent detection of enterovirus-specific antigens in myocardium has been unsuccessful with the exception of a single report [7]. Thus, in a detailed study of 19 patients with histologically confirmed active myocarditis and 16 with healed or healing myocarditis no virus was isolated from, or virus antigens detected in, the myocardium of any patient [31].

Evidence for the role of Coxsackie B viruses in myocarditis has been derived principally from animal model systems and human serological studies. Several groups have described murine models of myocarditis in BALB/c mice infected with a cardiotropic strain of Coxsackie B3 virus [41]. In this system, virus can be isolated from the myocardium up to 14 days post-infection although pathological changes do not develop until the 5th day when virus titres are falling. However, inflammation persists for weeks after virus is apparently eliminated [18], and tissue

H.-P. Schultheiß (Ed.)
New Concepts in Viral Heart Disease

damage occurs through immune-mediated mechanisms. Such observations have been assumed to apply to patients with myocarditis since biopsies are unlikely to be taken so soon after infection.

Studies of the humoral immune response to Coxsackie B viruses by neutralization tests have revealed elevated titres in patients with myocarditis [16]. More significantly, 36% of myocarditis patients were shown to be positive for Coxsackie B virus-specific IgM [10], and these IgM responses can persist for up to 6 months [30]. These data suggest that a pool of infectious virus persists within the body and appear to contradict the animal studies.

A number of explanations are possible for this apparent discrepancy between the animal and human studies. First, the virus may have been eliminated from cardiac muscle but persists elsewhere, such as gut epithelium, with the possibility of mediating a remote toxic effect or antigenic mimicry. Secondly, the virus may persist within the myocardium, but its infectivity may be reduced by immunological interactions or by defective synthesis or processing of virus-specific proteins. We will show the latter to occur.

Molecular cloning strategies have facilitated the derivation of sequences complementary to regions of Coxsackie B virus genomes [6, 21]. Nucleotide sequence analysis [19, 26], and hybridization experiments [42] have shown that the 5' and 3' non-translated regions and sequences encording virus-specific enzymes (e.g. the RNA-dependent RNA polymerase of protease) are highly conserved. In contrast, regions of the virus genome encoding structural proteins defining serotype are less conserved. This has enabled probes to be chosen as either enterovirus-group specific or serotype-specific in hybridization experiments to detect the presence of virus RNA. The availability of such probes and the development of endomyocardial biopsy [36] facilitated investigations of the persistence of enteroviruses in the myocardium of patients with heart muscle disease [1, 6].

RNA is isolated from left ventricular endomyocardial biopsy specimens from patients with histologically confirmed myocarditis or dilated cardiomyopathy (see below) or from patients with specific heart muscle diseases unrelated to virus infection. This is immobilized on duplicate nitrocellulose or nylon membranes using a slot blot apparatus and hybridized with either an enterovirus group-specific probe or a control probe (beta-tubulin [17] or 7B6 [20]) to quantitate the RNA immobilized. Hybridization probes are isolated from vector sequences by restriction endonuclease digestion and gel electrophoresis and are labelled with 32_P by random oligonucleotide primer extension [11]. After stringent washing and autoradiography (illustrated in Fig. 1), the signals for each sample are quantitated by densitometry and expressed as a ratio of the signal with the enterovirus probe to that of the control probe (hybridization index: HI). The pathological control group with other specific heart muscle diseases, e. g. amyloid give low background signals (for example, Fig. 2: group C; HI = 0.07 ± 0.017) similar to RNA from normal tissue (for example, pig heart; HI = 0.08). Some of the samples from the myocarditis/dilated cardiomyopathy group give low signals which are statistically indistinguishable from the control group (for example, Fig. 2: group B; HI = 0.082 ± 0.021). However, others are positive for the presence of enterovirus RNA (group A: HI greater than mean HI of controls plus 3 x standard deviation).

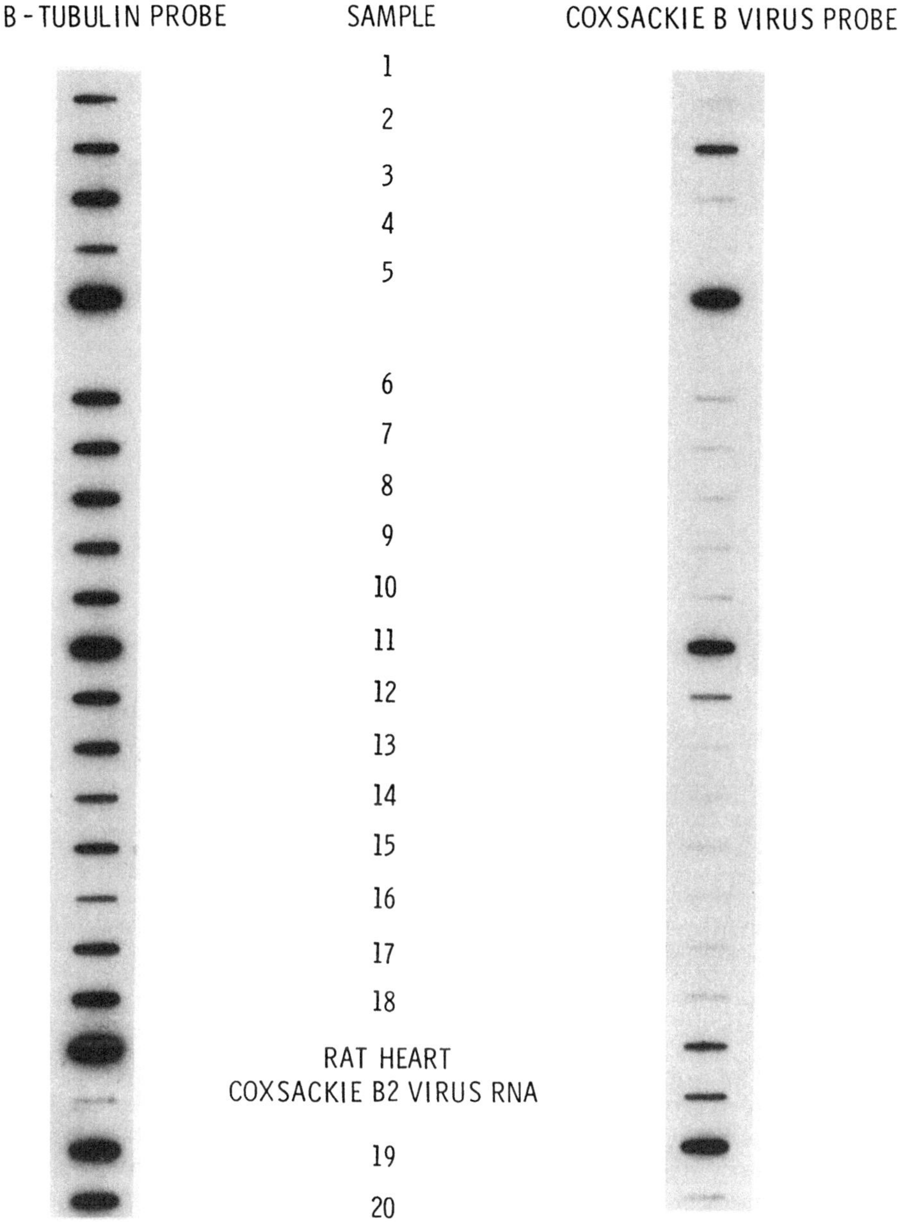

Fig. 1. Representative autoradiograph of RNA isolated from biopsy specimens from patients with myocarditis or dilated cardiomyopathy or from pathiologically irrelevant controls after hybridization with the enterovirus group-specific probe (derived from a conserved region of coxsackievirus B2 RNA) or the control (beta-tubulin-specific) probe

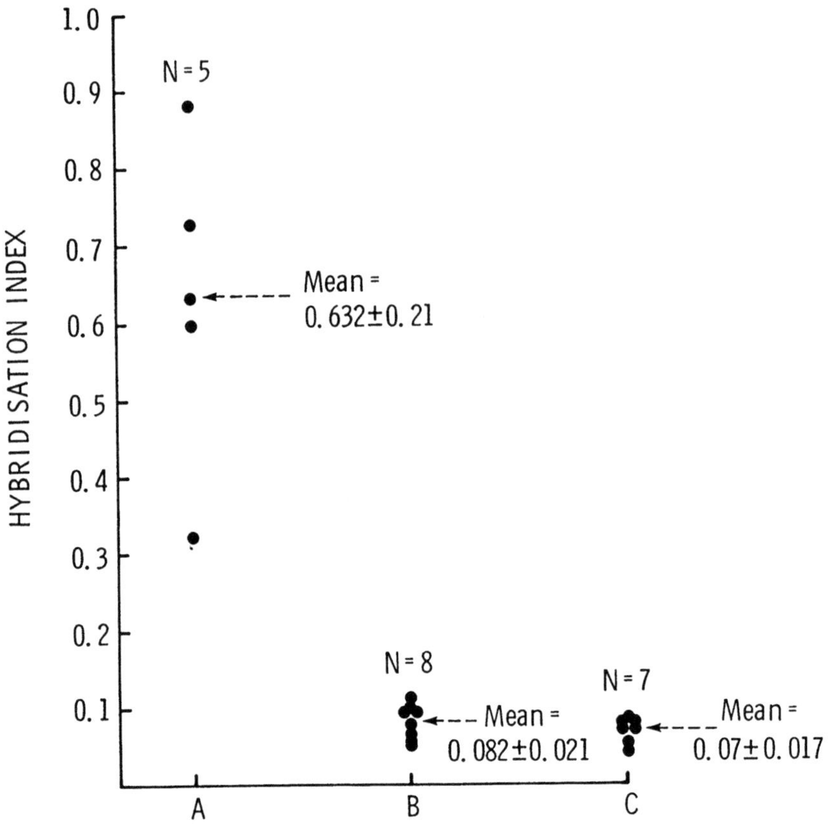

Fig. 2. Detection of enterovirus RNA in endomyocardial biopsy specimen. In this illustrative experiment, samples in groups *A* and *B* are from patients with myocarditis or dilated cardiomyopathy. Samples in group C are from patients with heart diseases unrelated to virus infection. Samples in group A are positive (greater than mean HI of controls plus 3 x standard deviation) and samples in group B negative for enterovirus RNA

To date, a total of 44 patients with myocarditis have been studied. These were divided into two groups based upon the histological assessment of sections from endomyocardial biopsy specimens [33]: first, active myocarditis, characterized by an inflammatory infiltrate in the interstitium and in close proximity to myocardial fibres with injury to adjacent fibres that is not typical of infarction: the contours of some myocardial fibres are irregular, indicating necrosis; secondly, healing myocarditis, indicated by reduced myocyte necrosis, but some remaining inflammatory cells. Of the 44 myocarditis patients studied, ten were diagnosed histologically as active myocarditis and, of these, six (60%) were positive for enterovirus RNA. Eleven of the remaining 34 with healing myocarditis (32%) were positive for virus RNA (Table 1).

One of the histological features of myocarditis is the focal nature of the pathology [4]. This observation is reflected by our experience of probing for enterovirus RNA: where multiple specimens taken on the same occasion have been investigated,

Table 1. Frequency of detection of enterovirus RNA in endomyocardial biopsy specimens (134 patients)

Histological Diagnosis (Dallas Criteria)			
Acute Myocarditis	Healing Myocarditis	Healed Myocarditis/ Dilated Cardiomyopathy	Other Specific Heart Diseases
6/10 = 60%	11/34 = 32%	22/54 = 41%	0/36 = 0%

virus RNA is sometimes detected only in a proportion (for example, Fig. 3, groups A and B).

Coxsackie B virus-specific neutralizing antibody [31] and IgM [22] titres have been determined in 16 of the 44 myocarditis patients. Seven of these 16 were positive for enterovirus RNA. Neutralization titres (IgG) reveal no significant differences between patients positive or negative for enterovirus RNA. Of the 16, three were positive for Coxsackie B virus-specific IgM and these were positive also for enterovirus RNA, evidence for recent persisting virus infection. However, two of ten controls in the series also had Coxsackie-specific IgM. Two possible explanations may account for failure to detect virus-specific IgM antibodies in a high proportion of these patients. First, the virus RNA detected by hybridization may be an enterovirus serotype other than a Coxackie B virus, the humoral immune response against which would not be detected. Secondly, it may be due to the patient failing to mount a continuing immune response to normal antigens: this in turn could be due to immunosuppression (although these patients are apparently immunocompetent) or altered virus antigen expression.

Dilated Cardiomyopathy

Dilated cardiomyopathy presents as left ventricular dysfunction which progress to biventricular disease and congestive cardiac failure. The relationship between enterovirus infection and dilated cardiomyopathy is based on the proposal that dilated cardiomyopathy is a long-term sequela of viral myocarditis [14]. The histological features of dilated cardiomyopathy are non-specific including myocardial fibre degeneration, irregular hypertrophy of myocardial fibres, myocardial fibrosis and occasional inflammatory cells [32] making differentiation from healed myocarditis difficult.

Animal studies and retrospective serology both provide some evidence for a role of enteroviruses in the pathogenesis of dilated cardiomyopathy.

Mice infected with a cardiotropic strain of Coxsackie B3 virus initially developed myocarditis and, after recovery, were observed over a 15-month period. Over 90% of the animals subsequently developed pathological changes of the left ventricle characteristic of dilated cardiomyopathy with almost 50% mortality [35].

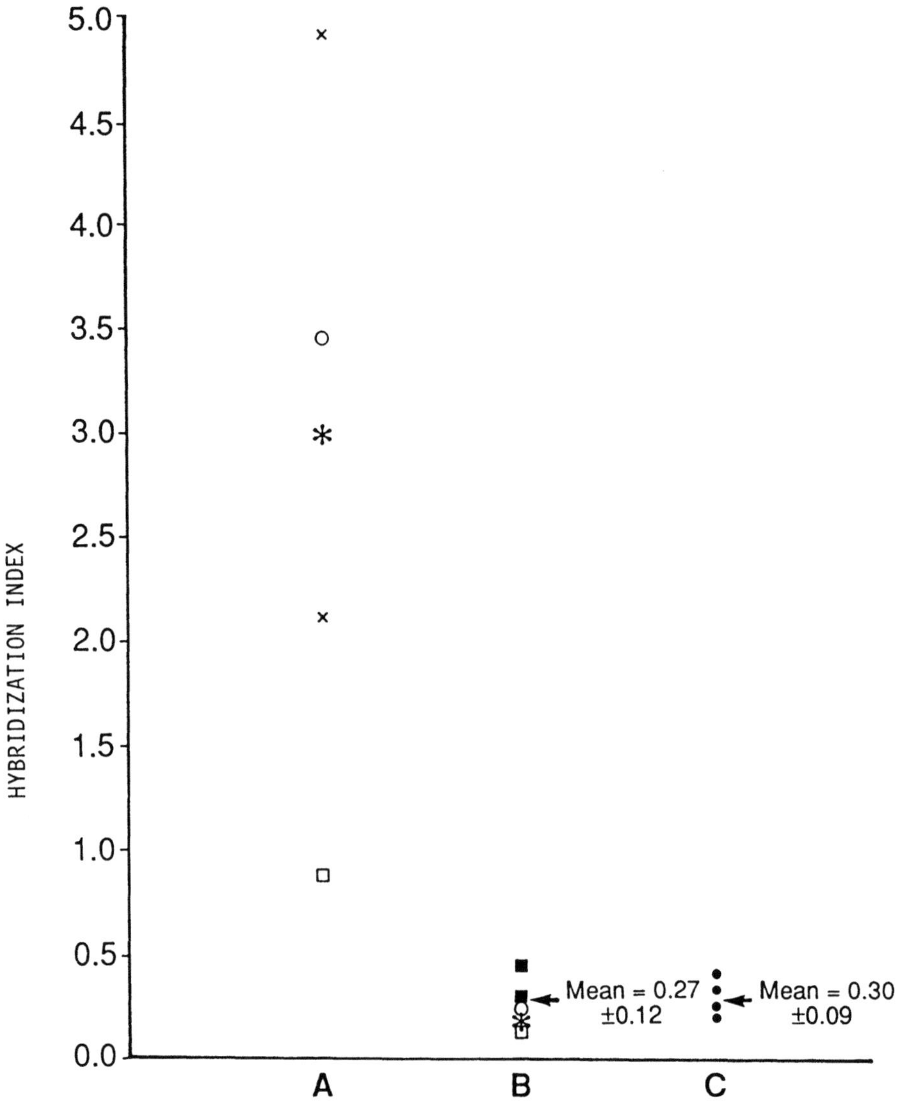

Fig. 3. Focal distribution of enterovirus RNA in endomyocardial biopsy specimens. Groups *A*, *B* and *C* in Fig. 2. *Crosses*, previous myocarditis/dilated cardiomyopathy; *asterisks* and *open circles*, healing myocarditis; *open squares*, healed myocarditis; *solid squares*, unconfirmed myocarditis; *solid circles*, normal or pathologically irrelevent controls

Although attempts to isolate infectious virus or demonstrate virus-specific antigens in biopsy samples failed [31], increased titres of Coxsackie B virus-specific antibody were reported: 15 of 50 (30%) patients with dilated cardiomyopathy had neutralization titres of at least 1:1024, compared with one of 50 (2%) controls [8]. High titres were found particularly in patients with short history of disease. Coxsackie B virus-specific IgM responses in dilated cardiomyopathy patients have also been reported: two of 28 (7%) patients were positive whilst all 46 controls were negative [9].

We have investigated left ventricular endomyocardial biopsy specimens from patients with dilated cardiomyopathy by molecular hybridization for the presence of enterovirus RNA.

Biopsy samples from 54 patients were studied and of these, 22 (41%) were positive for enterovirus RNA (Table 1). As with specimens from myocarditis patients, the involvement of virus appears to be focal. Seven of these 54 patients, including three positive for enterovirus RNA, were tested for Coxsackie B virus-specific antibodies. As in myocarditis, neutralizing titres revealed no differences between patients positive or negative for virus RNA. Only one patient was positive for Coxsackie B virus-specific IgM and this patient was also positive for enterovirus RNA. The possible reasons for the proportion of patients positive for virus RNA but without virus-specific IgM have been outlined in the previous section.

We also investigated tissue from the explanted hearts of 40 patients with end-stage disease who had undergone orthotopic cardiac transplantation. Twenty-one patients had dilated cardiomyopathy whilst the other 19 had cardiac failure of specific cause (17 had angiographically proven coronary artery disease, one had congential heart disease and one had amyloid heart disease). Up to four pieces of tissue were cut from a sample of right ventricular myocardium obtained during the transplant procedure. RNA was isolated from these and hybridized with an enterovirus group-specific probe as described for endomyocardial biopsy samples. The signals were quantitated and expressed as before (Fig. 4).

The majority of the samples gave background hybridization signals statistically indistinguishable from the normal controls (Fig. 4: group C; HI = 0.135 ± 0.046). However, eight samples were positive for enterovirus RNA. Seven of these were samples from the explanted hearts of six patients with dilated cardiomyopathy (i.e. one patient had two samples positive: Fig. 4; group A). The eighth was from a patient with ischaemic heart disease (Fig. 4; group B).

Eleven of the 21 cardiomyopathy patients had a history of 1 year or longer and this subset included all of the six positive for enterovirus RNA. This would not have been expected given the data of Cambridge of al. [8] and may be a statistical abberation due to the fact that the group is relatively small.

The proportion of the cardiomyopathy transplant patients found to be positive for virus RNA (29%) was lower than the corresponding figure from endomyocardial biopsy specimens (41%), described above. However, our patients from whom endomyocardial biopsy specimens were studied were of relatively recent diagnosis compared to the transplant group who were end-stage. While enteroviruses clearly can persist for years in a proportion of patients, virus may be eliminated from some during progression of the disease.

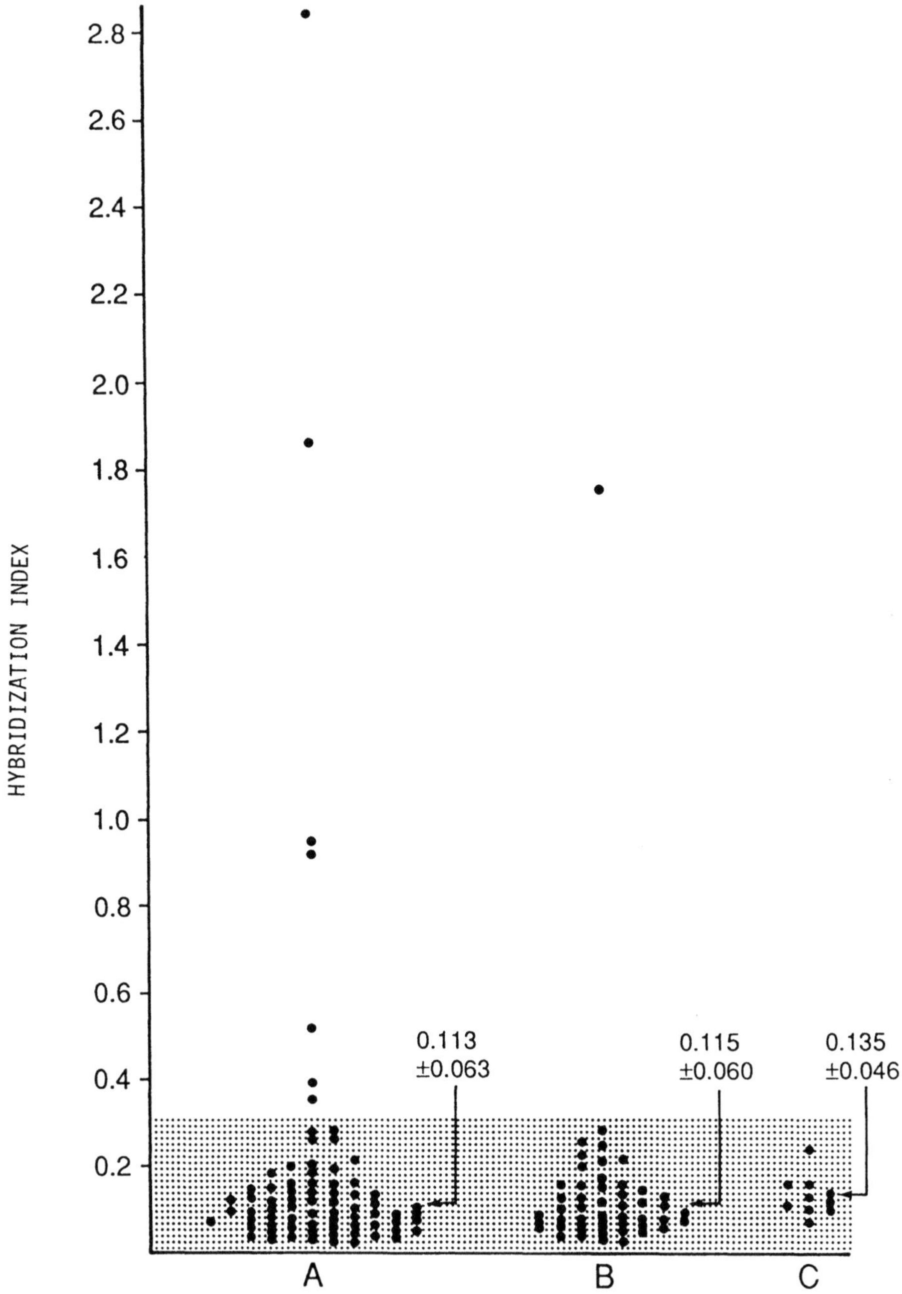

Fig. 4. Detection of enterovirus RNA in right ventricular samples of explanted hearts from patients with end-stage disease. Group *A* is tissue from patients with dilated cardiomyopathy. Group *B* is tissue from patients with heart disease of specific cause. Group *C* is normal tissue. The samples in the *shaded area* are negative (less than 3 x standard deviations greater than mean HI of group *C)* and the remainder are positive for enterovirus RNA

In our transplant patient study, serological investigations again revealed no significant differences between the cardiomyopathy or control groups in the frequency of either Coxsackie B virus-specific neutralizing antibody or IgM titres. The frequency of significant neutralizing titres (greater than 1:320) in both groups of patients was greather than reported for controls in previous studies [31]. Only one patient, with dilated cardiomyopathy but negative for enterovirus RNA, was positive for Coxsackie B virus-specific IgM.

Leucocyte infiltration of tissue from the explanted hearts of these patients was investigated by immunostaining [39]. Thin sections were stained with monoclonal antibody against the leucocyte common antigen F10, or the T cell antigen Leu 4 (Fig. 5). This revealed no significant difference between the cardiomyopathy and ischaemic heart disease groups or between the cardiomyopathy patients positive or negative for enterovirus RNA. Clearly, in contrast to viral myocarditis, persistence of virus in dilated cardiomyopathy does not attract a cellular immune response. The single patient with ischaemic heart disease and positive for enterovirus RNA had no obvious history of an acute enteroviral illness but had a significant number of infiltrating T lymphocytes. While we do not conclude that enteroviruses are a pathogenetic factor of ischaemia, some serological evidence for such a link has been described [9.25]. The presence of enterovirus RNA in the myocardium of this patient may be due to an undetected cardiomyopathy or viral myocarditis as indicated by the T cell infiltrate. The fact that dilated cardiomyopathy is diagnosed principally by exclusion in these end-stage patients means that disease other than ischaemia may be undetected.

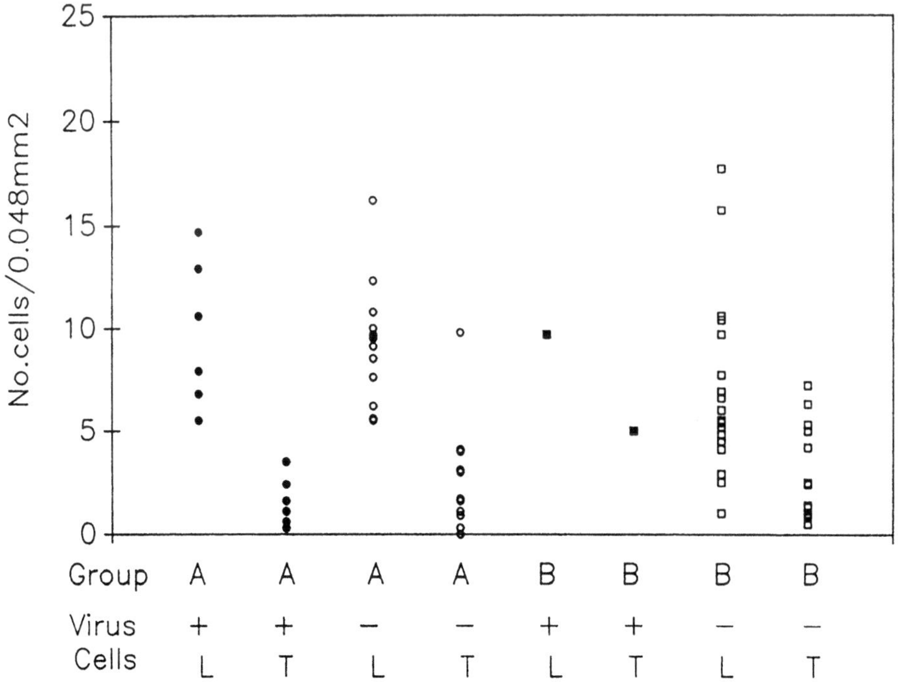

Fig. 5. Immunocytochemical detection of leucocytes *(L)* or T cells *(T)* in heart tissue from patients with dilated cardiomyopathy (group *A)* or other heart muscle disease (group *B)*

Pathogenesis of Dilated Cardiomyopathy

Whilst many studies have implicated cellular immune-mediated damage in the pathogenesis of myocarditis [13, 41], the lack of a chronic infiltrate in the myocardium of patients with dilated cardiomyopathy suggests an alternative mechanism.

There are two possibilities worth considering. The first is that a virus-initiated immune response causes the damage observed in myocarditis. The subsequent healing process may fail to restore the full function of the myocardium and results in ventricular dilatation.

The second is that persistent virus infection perturbs the function of myocytes. Persistent infection of specialized cells has been shown to result in alterations of non-essential gene functions [40], and a virus-induced metabolic myopathy has been proposed in patients with non-inflammatory, post-viral muscle fatigue syndrome [2]. Evidence for dysfunction of respiratory metabolism in such patients has been derived from nuclear magnetic resonance (NMR) studies of exercising muscle [3]. This technique is being used to investigate some of our dilated cardiomyopathy patients who developed this disease after myocarditis, and in whom persisting enterovirus infection of myocardium has been demonstrated by detection of virus RNA.

It is worth considering whether there is a pathogenetic difference between patients with disease of rapid or slow progression. Our failure to detect virus RNA in the explanted hearts of patients with dilated cardiomyopathy that progressed rapidly, combined with the serological studies of Cambridge et al. [8], suggests that virus-induced cytolysis may be followed by elimination of the virus and subsequent antibody-mediated cell damage. The presence of autoantibodies directed against myolemma in patients developing dilated cardiomyopathy post-myocarditically supports the role of humorally mediated cell damage [28]. Lymphocyte-mediated cytolysis was not observed in these patients. By contrast, patients with a slowly progressing disease may have a persistent infection leading to metabolic dysfunction.

Progression of Myocarditis to Dilated Cardiomyopathy

Several studies demonstrate a link between histologically confirmed myocarditis and the subsequent development of dilated cardiomyopathy. Endomyocardial biopsy specimens from four of 15 dilated cardiomyopathy patients with raised enterovirus antibody titres had histological evidence of active or healed myocarditis [8]. Two further small series have shown a progression from histologically confirmed myocarditis to dilated cardiomyopathy [27, 37]. In particular, patients with a short history of dilated cardiomyopathy had evidence active or healed myocarditis.

A long-term follow-up of a larger series of myocarditis patients has been reported by some of us [34]. Of 23 patients who presented with clinical and histological signs of myocarditis, 12 (52%) subsequently developed dilated cardiomyopathy during a mean follow-up period of 43 months. Four of these died, two underwent cardiac transplantation, one recovered and the rest remained ill at the end of the study.

We examined biopsy samples from three such dilated cardiomyopathy patients with a previous history of myocarditis: all were positive for enterovirus RNA. One of these patients was biopsied six times in a 5-year period. Histologically, this series of biopsies showed a progression from first focal and then healing myocarditis through healed myocarditis to the non-specific changes compatible with the clinical diagnosis of dilated cardiomyopathy. Enterovirus RNA was detected in this final biopsy before cardiac transplantation: previous biopsy samples were no longer available for hybridization studies. This patient had unremarkable neutralizing antibody titres to Coxsackie B virus serotypes and, like the majority of dilated cardiomyopathy patients with persisting enterovirus RNA in myocardium (above), no Coxsackie B virus-specific IgM.

The consistent detection of enterovirus RNA in biopsy samples from cases of myocarditis and dilated cardiomyopathy supports the hypothesis of a role for persistent virus infection in the progression from myocarditis to dilated cardiomyopathy. We are now investigating patients with active myocarditis prospectively with the aim of following the progression to dilated cardiomyopathy and of monitoring the presence of virus in serial biopsies.

It was suggested previously that histological assessment of the grade of myocarditis and the degree of fibrosis in subsequent biopsy specimens could be useful in predicting the long-term outcome of the disease [38] although in the patients described by Quigley et al. [34] this appeared not to be the case. It is possible that the presence of enterovirus may be an alternative prognostic marker.

The presence of virus RNA in biopsy material may be an important factor in considering treatment regimens. Myocarditis is frequently treated with steroids and transplant patients are iatrogenically immunosuppressed. Systematic study will show whether presence of virus contra-indicates these treatments.

Virus Persistence and Antigen Expression

Various mechanisms of persistence of usually cytopathic viruses have been proposed [12]. These include the generation of defective interfering (DI) particles, temperature-sensitive mutants and the effects of interferon on virus replication. Various in vitro systems have been described for persistent infection with Coxsackie B viruses including carrier state infections of heart cells [21a] and lymphocytes [29]. However, at present there is little evidence from studies of pathological tissue.

The well-documented failure of conventional virological techniques to demonstrate virus in the affected tissue may be explained by either immunological concealment of virus or altered virus antigens with a concomitant reduction of infectivity. The observation that, in the patients studied here and found to be positive for enterovirus RNA, virus-specific IgM is mainly detected in patients with short history is compatible with a progressive alteration in virus-specific antigens.

Reduction in the infectivity of the virus could provide mechanisms for both virus persistence and the evasion of host immune response. The selection of DI mutants could result in alteration of antigen expression. In polioviruses, a large deletion of the genome in the region encoding the capsid proteins has been described [23] resulting in a reduction of infectivity. Also, an in vitro model for persistent echovirus infection of WISH cells has been described where virus infectivity is reduced by a failure to cleave the precursor normally generating two of the capsid proteins [5].

Conclusions

Enterovirus RNA is present in the myocardium of a sustantial proportion of patients in whom a diagnosis of either dilated cardiomyopathy of myocarditis is supported by histological examination of biopsy material. Examination of multiple biopsy specimens supports the histopathological observation that distribution of virus is focal resulting in false negatives through sampling error. Our detection of enterovirus RNA in myocardium suggests a pathogenetic role, and our understimate of the frequency of involvement of virus implies that enteroviruses are the major aetiological agent of viral heart disease.

It is clear that enterovirus RNA can persist within the dilated myocardium for many years, even until end-stage disease requiring cardiac transplantation. We detected virus RNA less frequently in these end-stage cardiomyopathy patients than in those with recently diagnosed disease, suggesting that some patients may eliminate virus during progression of the disease.

Although direct evidence is difficult to obtain, we feel there are now sufficient data to implicate persistence of enteroviruses in the progression from myocarditis to dilated cardiomyopathy. We have detected enterovirus RNA in a substantial proportion of all grades of histologically proven myocarditis or dilated cardiomyopathy and in dilated cardiomyopathy patients with a proven history of myocarditis. Our prospective study directed at identifying patients with acute viral myocarditis and correlating their clinical progress with histopathological changes and persistence of virus RNA will provide the definitive link.

It is surprising that the persistence of enteroviruses in dilated cardiomyopathy does not attract a detectable IgM immune response in a higher proportion of patients, although this may be due to the expression of altered virus antigens leading to a response not detected in the standard assay. Alternatively, the synthesis or display of normal virus antigens may be reduced. It will be interesting to determine if such factors are related to the molecular mechanism of virus persistence.

Acknowledgements. We wish to thank Mr. R. Florio and Ms. D. Hornall for their invaluable assistance.

References

1. Archard LC, Bowles NE, Olsen EGJ, Richardson PJ (1987) Detection of persistent Coxsackie B virus RNA in dilated cardiomyopathy and myocarditis. Eur Heart J 8 [suppl J]: 437–440
2. Archard LC, Behan PO, Bell EJ, Doyle D, Bowles NE (1988) Postviral fatigue syndrome: persistence of enterovirus RNA in muscle and elevated creatine kinase. J R Soc Med 81: 326–329

3. Arnold DL, Bore PJ, Radda GK, Styles P, Taylor DJ (1984) Excessive intracellular acidosis of skeletal muscle on exercise in a patient with post-viral exhaustion/fatigue syndrome. Lancet 1: 1367–1369
4. Baandrup U, Florio RA, Olsen EGJ (1982) Do endomyocardial biopsies represent the morphology of the rest of the myocardium? A quantitative light microscopic study of single v. multiple biopsies with the King's bioptome. Eur Heart J 3: 171–178
5. Blackburn RV, Righthand F (1987) Viral protein profiles of echovirus 6 recovered from lytic and non-lytic steady state infections. Seventh International Congress of Virology, Edmonton, Canada (Abstr R16.13) 9–14 August 1987
6. Bowles NE, Richardson PJ, Olsen EGJ, Archard LC (1986) Detection of Coxsackie B virus-specific RNA sequences in myocardial biopsy samples from cases of myocarditis and dilated cardiomyopathy. Lancet 1: 1120–1123
7. Burch GE, Sun SC, Colclough HL, Sohal S, DePasquale NP (1967) Coxsackie B viral myocarditis and valvulitis in routine autopsy specimens by immunofluorescent techniques. Am Heart J 74: 13–23
8. Cambridge G, MacArthur CGC, Waterson AP, Goodwin JF, Oakley CM (1979) Antibodies to Coxsackie B viruses in primary congestive cardiomyopathy. Br Heart J 41: 692–696
9. Coltart DJ, El-Hagrassy MMO, Banatvala JE (1984) Coxsackie B virus-specific IgM responses in patients with cardiac and other diseases. In: Bolte HD (ed) Viral heart disease. Springer, Berlin Heidelberg New York, pp 89–94
10. El-Hagrassy MMO, Banatvala JE, Coltart DJ (1980) Coxsackie B virus-specific IgM responses in patients with cardiac and other diseases. Lancet 2: 1160–1162
11. Feinberg AP, Vogelstein B (1984) A technique for radiolabelling DNA restriction fragments to high specific activity. Anal Biochem 137: 266–267
12. Friedman RM, Ramseur JM (1979) Mechanisms of persistent infections by cytopathic viruses in tissue culture. Arch Virol 60: 83–103
13. Godeny EK, Gauntt CJ (1987) In situ immune autoradiographic identification of cells in heart tissues of mice with Coxsackievirus B3-induced myocarditis. Am J Pathol 129: 267–276
14. Goodwin JF (1983) Myocarditis as a possible cause of cardiomyopathy. In: Just H, Schuster HP (eds) Myocarditis — cardiomyopathy. Springer, Berlin Heidelberg New York, pp 7–11
15. Grist NR (1978) Coxsackie virus infections of the heart. In: Waterson AP (ed) Recent advances in clinical virology. Churchill Livingstone, Edingburgh, pp 141–150
16. Grist NR, Bell EJ (1974) A six year study of Coxsackie virus B infections in heart disease. J Hyg 73: 165–172
17. Hall JL, Dudley L, Dobner PR, Lewis SA, Cowan NJ (1983) Identification of two human beta-tubulin isotypes. Mol Cell Biol 3: 854–862
18. Huber SA, Lodge PA, Job LP (1984) The role of virus-, and immune-mediated cardiocyte injury in Coxsackievirus B3-induced myocarditis. In: Bolte HD (ed) Viral heart disease. Springer, Berlin Heidelberg New York, pp 64–73
19. Iizuka N, Kuge S, Nomoto A (1987) Complete nucleotide sequence of the genome of Coxsackievirus Bl. Virology 156: 64–73
20. Kaczmarek L, Calabrett B, Baserga R (1985) Expression of cell-cycle-dependent genes in phytohemagglutinin stimulated human lymphocytes. Proc Natl Acad Sci USA 82: 5372–5379
21. Kandolf R, Hofschneider PH (1985) Molecular cloning of the genome of a cardiotropic Coxsackie B3 virus: full-length reverse-transcribed recombinant cDNA generates infectious virus in mammalian cells. Proc Natl Acad Sci USA 82: 4818–4822
21.a Kandolf R, Canu A, Hofschneider PH (1985) Coxsackie B3 virus can replicate in cultured human foetal heart cells and is inhibited by interferon. J Mol Cell Cardiol 17: 167–181
22. King ML, Shaikh A, Bidwell D, Voller A, Banatvala JE (1983) Coxsackie B virus-specific IgM responses in children with insulin-dependent (juvenile-onset; type I) diabetes mellitus. Lancet 1: 1397–1399
23. Kuge S, Saito I, Nomoto A (1986) Primary structure of poliovirus defective-interfering particle genomes and possible generation mechanisms of the particles, J Mol Biol 192: 473–487

24. Lansdown ABG (1978) Viral infections and diseases of the heart. Prog Med Virol 24: 70–113
25. Lau RCH (1986) Coxsackie B virus-specific IgM responses in coronary care unit patients. J Med Virol 18: 193–198
26. Lindberg AM, Stalhandske POK, Pettersson U (1987) Genome of Coxsackievirus B3. Virology 156: 50–63
27. MacArthur CGC, Tarin G, Goodwin JF, Hallidie-Smith KA (1984) The relationship of myocarditis to dilated cardiomyopathy. Eur Heart J 5: 1023–1035
28. Maisch B, Deeg P, Liebau G, Kochsiek K (1983) Diagnostic relevance of humoral and cytotoxic immune reactions in primary and secondary dilated cardiomyopathy. Am J Cardiol 52: 1072–1078
29. Matteucci D, Paglianti M, Giangregorio AM, Capobianchi MR, Dianzini F, Bendinelli M (1985) Group B Coxsackieviruses readily establish persistent infections in human lymphoid cell lines. J Virol 56: 651–654
30. McCartney RA, Banatvala JE, Bell EJ (1986) Routine use of μ-antibody-capture ELISA for the serological diagnosis of Coxsackie B virus infections. J Med Virol 19: 205–212
31. Morgan-Capner P, Richardson PJ, McSorley C, Daly K, Pattison JR (1984) Virus investigations in heart muscle disease. In: Bolte HD (ed) Viral heart disease. Springer, Berlin Heidelberg New York, pp 95–115
32. Olsen EGJ (1979) Pathology of cardiomyopathies. A critical analysis. Am Heart J 98: 385–392
33. Olsen EGJ (1985) Histologic aspects of viral myocarditis and its diagnostic criteria. In: Sekiguchi M, Olsen EGJ, Goodwin JF (eds) Myocarditis and related disorders. Springer Berlin Heidelberg New York, pp 130–132
34. Quigley PJ, Richardson PJ, Meany BT, Olsen EGJ, Monaghan MJ, Jackson G, Jewitt JE (1987) Long-term follow-up of acute myocarditis. Correlation of ventricular function and outcome. Eur Heart J 8 (Suppl J): 39–42
35. Reyes MP, Ho K-L, Smith FE, Lerner AM (1981) A mouse model of dilated-type cardiomyopathy due to coxsackievirus B3. J Infect Dis 144: 232–236
36. Richardson PJ (1984) Endomyocardial biopsy: technique and evaluation of a new disposable forceps and catheter sheath system. In: Bolte HD (ed) Viral heart disease. Springer, Berlin Heidelberg New York, pp 173–176.
37. Richardson PJ (1985) Clinical aspects of myocarditis. In: Sekiguchi M, Olsen EGJ, Goodwin JF (eds) Myocarditis and related disorders. Springer, Berlin Heidelberg New York, pp 97–100
38. Richardson PJ, Daly K, Gishen P (1984) Haemodynamic findings in biopsy proven acute myocarditis. In: Bolte HD (ed) Viral heart disease. Springer, Berlin Heidelberg New York, pp 165–172
39. Rose ML, Cotes MI, Griffin RJ, Pomerance A, Yacoub MH (1986) Expression of class I and class II major histocompatibility antigens in normal and transplanted human heart. Transplantation 41: 776–780
40. Southern P, Oldstone MBA (1986) Medical consequences of persistent viral infection. N Engl J Med 314: 359–367
41. Woodruff JF (1980) Viral myocarditis, a review. Am J Pathol 101: 425–484
42. Zhang HY, Yousef GE, Bowles NE, Archard LC, Mann GF, Mowbray JF (1988) Detection of enterovirus RNA in experimentally infected mice by molecular hybridization: specificity of sub-genomic probes in quantitative slot blot and in situ hybridization. J Med Virol (in press)

Diagnosis of Acute Myocarditis with Radiolabeled Monoclonal Antimyosin Antibody: Immunoscintigraphic Evaluation

B. A. KHAW, T. YASUDA, I. F. PALACIOS, J. T. FALLON, G.W. DEC, P.D. NICOL, A. J. FISCHMAN, H. W. STRAUSS, and E. HABER

Cardiac Unit, Jackson 13, Massachusetts General Hospital Boston, MA 02114, USA

Introduction

Myocarditis may be difficult to diagnose due to the paucity of common clinical and laboratory findings. The gold standard for diagnosis of acute myocarditis is the histopathological examination of endomyocardial biopsy samples for evidence of myocyte necrosis with concurrent cellular infiltrates [1]. Although this method is highly specific, endomyocardial biopsy is relatively insensitive because of the limited number of biopsy samples which can be obtained from each patient, and the focal nature of the disease in some patients. The disease may only cause necrosis in a portion of one ventricle. Since the biopsy can only sample the endocardium, necrosis involving only the epicardium or mid-portion of the myocardium cannot be appreciated from histological examination of the specimen. A method of evaluating necrosis at any location in the myocardium would help to identify patients with myocarditis. To determine whether a radioimmunoscintigraphic approach that is highly specific for delineation and visualization of acute myocardial necrosis in myocardial infarction [2–10] could also serve as a noninvasive method for the diagnosis of acute myocarditis, we studied a series of patients scheduled for ventricular biopsy who were suspected of having acute myocarditis.

This immunoscintigraphic approach is based on the principle that myocytes with an intact cell membrane do not permit macromolecules to enter or exit these cells. Following any disruption of the cell membrane, intracellular macromolecules such as actin, myosin, and other contractile proteins are exposed to the extracellular environment. When this occurs, soluble intracellular substances such as CK [11], LDH [12], SGOT [13] and myoglobin [14] are washed out into the systemic circulation. However, myosin, a major component of the fibrous contractile element of the muscle cell, is insoluble and, therefore, remains in situ after disruption of the cell membrane. Administration of a radiolabeled antibody raised against an epitope of cardiac myosin can then bind to the exposed myosin antigen, and delineate the areas of myocardial necrosis.

The specificity of antimyosin for the identification of myocyte necrosis has been shown by various methods. Figure 1b shows a scanning electron micrograph of a necrotic myocyte in culture [15]. This cell shows a region of membrane disruption with extrusion of the intracellular myofibrils which had bound specifically with antimyosin-coated 1-μm diameter beads. Normal myocyte with intact sarcolemma did not concentrate the antimyosin beads (Fig. 1a). A closer examination of a

H.-P. Schultheiß (Ed.)
New Concepts in Viral Heart Disease

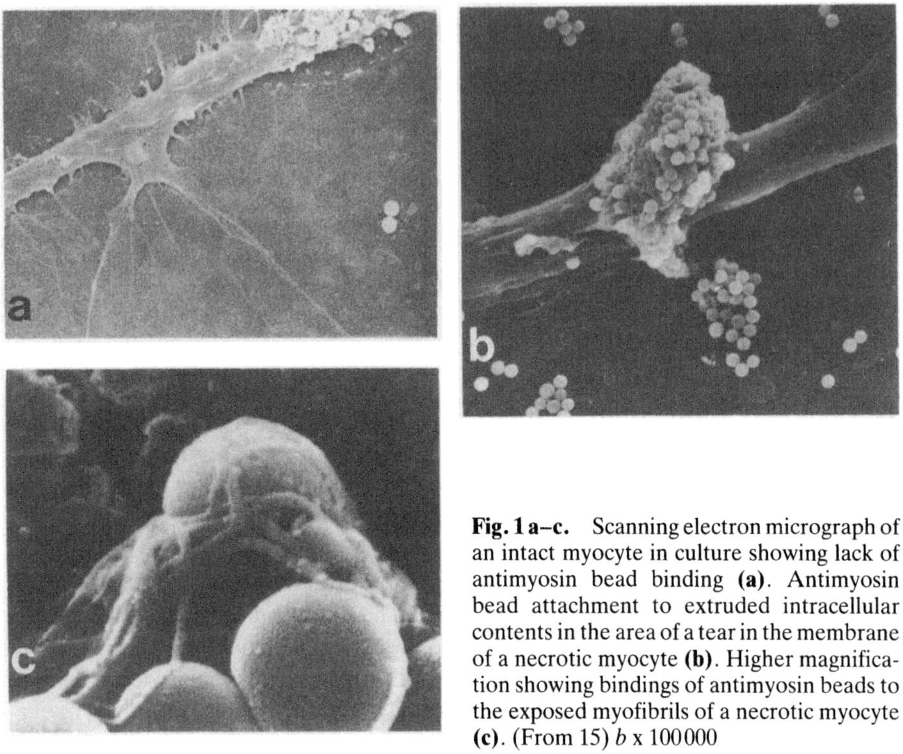

Fig. 1 a–c. Scanning electron micrograph of an intact myocyte in culture showing lack of antimyosin bead binding **(a)**. Antimyosin bead attachment to extruded intracellular contents in the area of a tear in the membrane of a necrotic myocyte **(b)**. Higher magnification showing bindings of antimyosin beads to the exposed myofibrils of a necrotic myocyte **(c)**. (From 15) *b* x 100000

region of membrane disruption showed the myofilaments which contained myosin to be wrapped around the antimyosin beads (Fig. 1c). Similarly, specificity of antimyosin for necrotic myocytes has been shown in fixed tissue sections by micro-autoradiography [4], in gross specimens of the hearts [4] and in experimental [6, 7] and clinical [9, 10] situations involving myocyte necrosis due to ischemic injury.

Localization of radiolabeled monoclonal antimyosin Fab in acute experimental myocardial infarction is shown in Fig. 2. Although the blood pool activity at 5 h

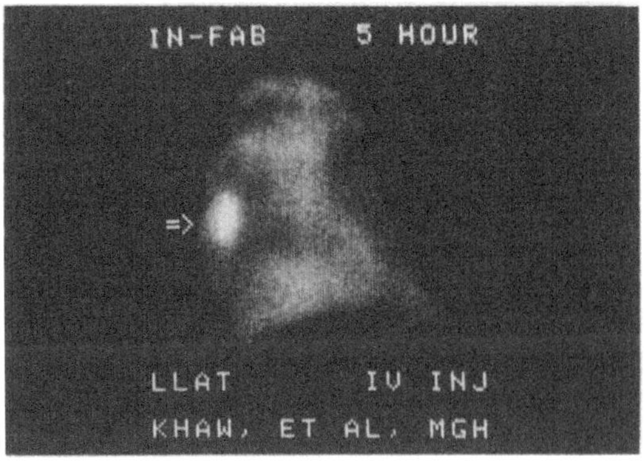

Fig. 2. Left lateral gamma scintigram of a dog obtained 5 h after intravenous administration of In-111-labeled antimyosin Fab, showing localization of the radioactivity in the anteroapical region of the heart, the site of experimental myocardial infarction

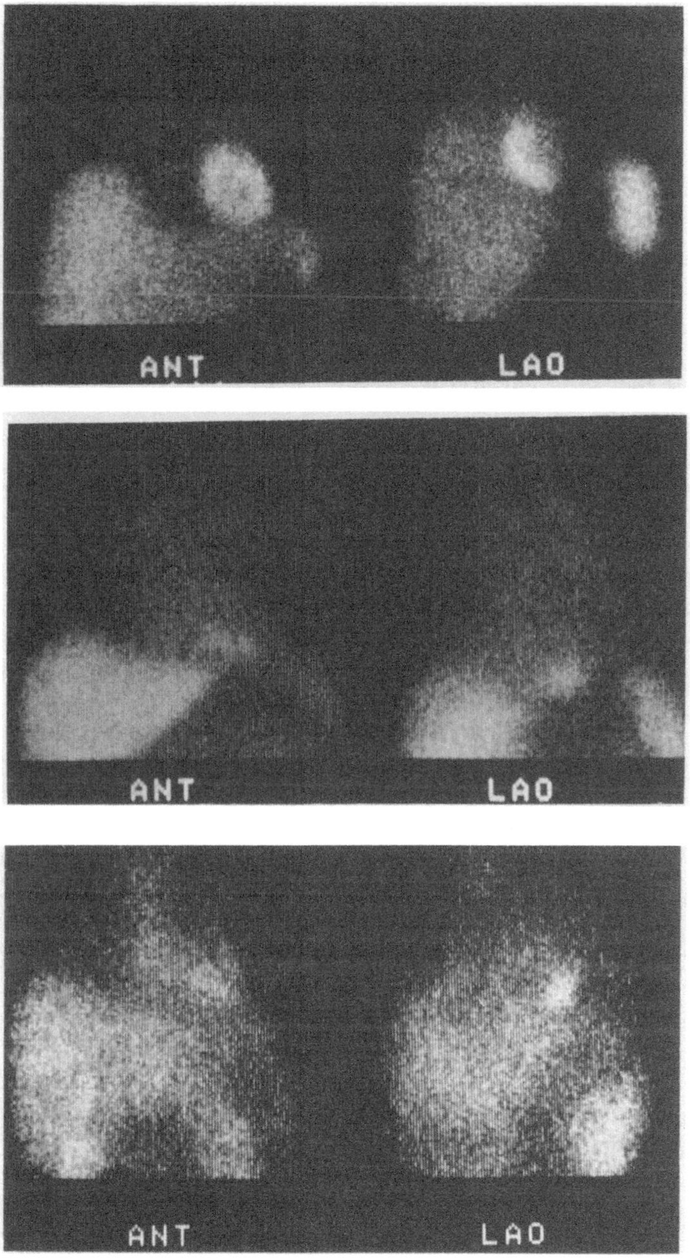

Fig. 3 a–c. Planar anterior and 45° left anterior oblique gamma scintigraphic images of three patients with an anterior **(a)**, inferior **(b)**, and a posterior **(c)** acute myocardial infarction. The images were obtained approximately 24 h after intravenous administration of the radiolabeled monoclonal antimyosin Fab

after intravenous administration of the antibody is apparent, the delineation of necrotic myocardium is unequivocally seen at the anteroapical region of the heart. Similarly, localization of the necrotic myocardium in patients with anterior, inferior, and posterior acute myocardial infarction is demonstrated in Fig. 3 [16].

Radiolabeled antimyosin can delineate myocardial necrosis regardless of the cause of myocardial injury. For example, myocardial necrosis resulting from heart transplant rejection in a murine heterotopic heart transplant model can readily be defined with radiolabeled antimyosin [17]. Similarly, Addonizio et al. [18] showed that acute rejection of heterotopically transplanted hearts in dogs can be visualized by gamma scintigraphy. In our laboratory, we have been able to visualize heterotopic heart transplant rejection in a murine model [19], and Frist et al. [20] have reported the application of In-111 antimyosin imaging for diagnosis of human cardiac transplant rejection.

Animal Studies

The feasibility of applying antimyosin imaging to the detection of acute myocarditis was tested in a murine viral autoimmune myocarditis model. In these studies, a monoclonal antimyosin, 2G42D7, IgG1 isotype, with no isomyosin specificity was used. Fab of 2G42D7 were used in the study by Rezkalla et al. [21] to demonstrate localization of I-125-labeled Fab in a murine viral myocarditis model. These investigators used microautoradiography with I-125-labeled antimyosin Fab to demonstrate specific localization in necrotic myocytes with concurrent white cell infiltration. Recently, Neu et al. [22] reported the induction of autoimmune myocarditis in a series of inbred strains of mice following immunization of the animals with cardiac myosin.

It has also been reported that immunization of mice with sarcoplasmic reticular ATPase containing the cross-reacting antigen SRA elicited autoimmune myocarditis similar to myocarditis induced by *Trypanosoma cruzi* infection [23]. Furthermore, Cossio et al. [24] were able to induce experimental myocarditis in Swiss mice by immunizing them with homologous heart homogenates. Therefore, we attempted to determine whether inbred CD rats immunized with homologous heart homogenates would develop autoimmune myocarditis, and whether radiolabeled antimyosin antibody could be used as a noninvasive diagnostic method for visualization of myocardial necrosis associated with experimental autoimmune myocarditis.

Histological examination of the hearts from the rats immunized with heart homogenates showed focal myocyte necrosis with white cell infiltration. A group of these rats was imaged with two monoclonal antimyosin antibodies. One antimyosin, R11D10, is specific for the V3 isomyosin, and the other, 2G42D7, has no isomyosin specificity reacting with both V1 and V3 isoforms. Figure 4 shows a set of posteroanterior images of the same rat that had been immunized with myocardial extracts, injected with In-111-labeled R11D10, then 1 week later injected with In-111-labeled 2G42D7. The predominant isomyosin in adult rats is V1, with minimal V3. Therefore, R11D10 antimyosin is not expected to show intense uptake. However, 2G42D4, which reacts with all isoforms of myosin, showed

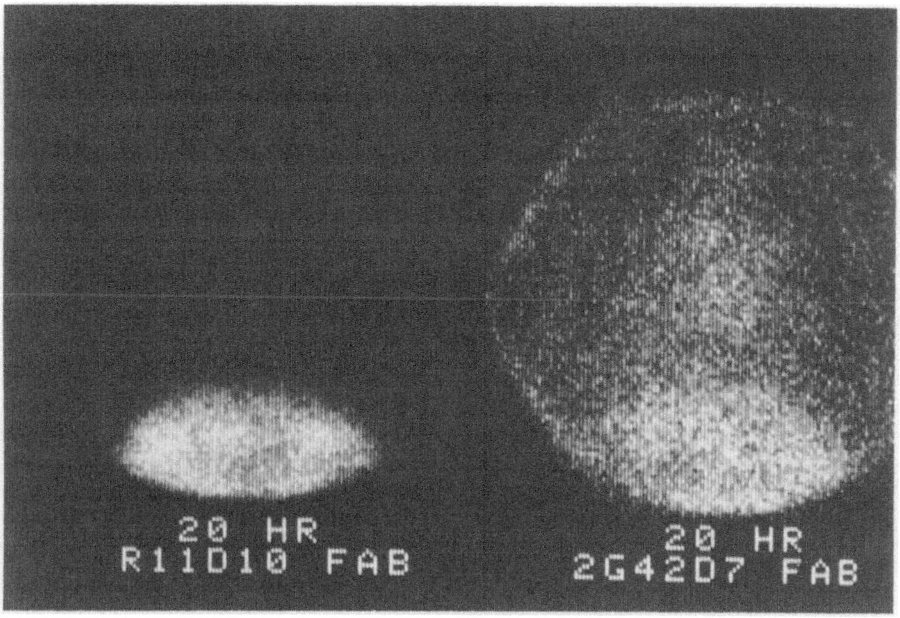

Fig. 4. Posteroanterior gamma scintigrams of a rat immunized with isologous heart extract to elicit autoimmune myocarditis. The gamma image on the left was obtained 20 h after intravenous administration of In-111-labeled antimyosin Fab specific for the V1 isomyosin. The image on the left was obtained 1 week later in the same animal 20 h after intravenous administration of In-111-labeled antimyosin Fab which reacts equally with both isoenzymes of cardiac myosin. The activity toward the bottom of the images corresponds to the liver, and only the image on the *left* showed distinct myocardial activity *(center)*

unequivocal but diffused delineation of global myocardial necrosis associated with experimental autoimmune myocarditis.

Human Studies

While clearance of radiolabeled antimyosin Fab in the rodent models is rapid, the clearance in patients is more prolonged. Figure 5 shows the clearance of In-111-labeled antimyosin Fab in patients with acute myocardial infarction [10]. Even at this clearance rate, the percentage of injected dose remaining at 24 h after intravenous administration was substantial. However, by 48 h only about 4% of the injected dose remained in the circulation. Because myocardial necrosis associated with acute myocarditis is not as extensive, focal, or homogenous as in acute myocardial infarction, residual blood pool activity can easily be misinterpreted as necrosis in the 24-h antimyosin images. To avoid this confusion, patient studies were interpreted from anterior and left anterior oblique planar images, as well as from the transverse, sagittal, and coronal tomograms obtained at 48 h post intravenous administration of the In-111-labeled antimyosin Fab. Figure 6 shows a set

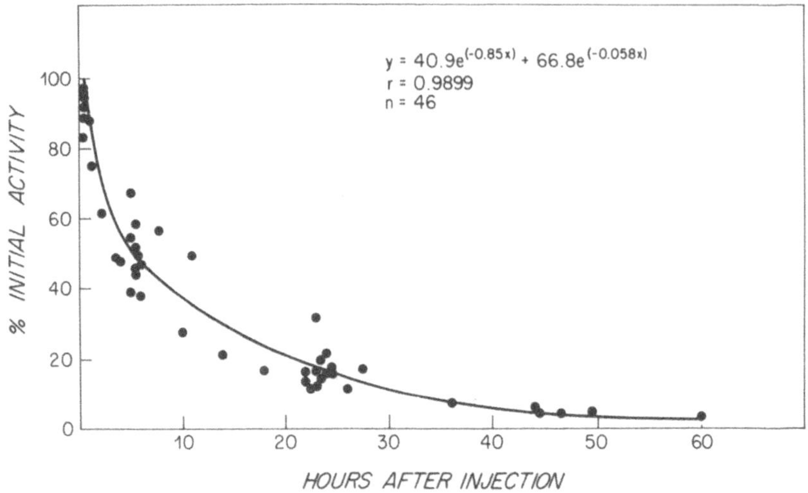

Fig. 5. Blood clearance of intravenously administered In-111-labeled monoclonal antimyosin Fab in 14 patients. The $T_{1/2}$ of the fast component was 0.8 h *($P < 0.001$)*, that of the slow component was 12 h *($P < 0.001$)*

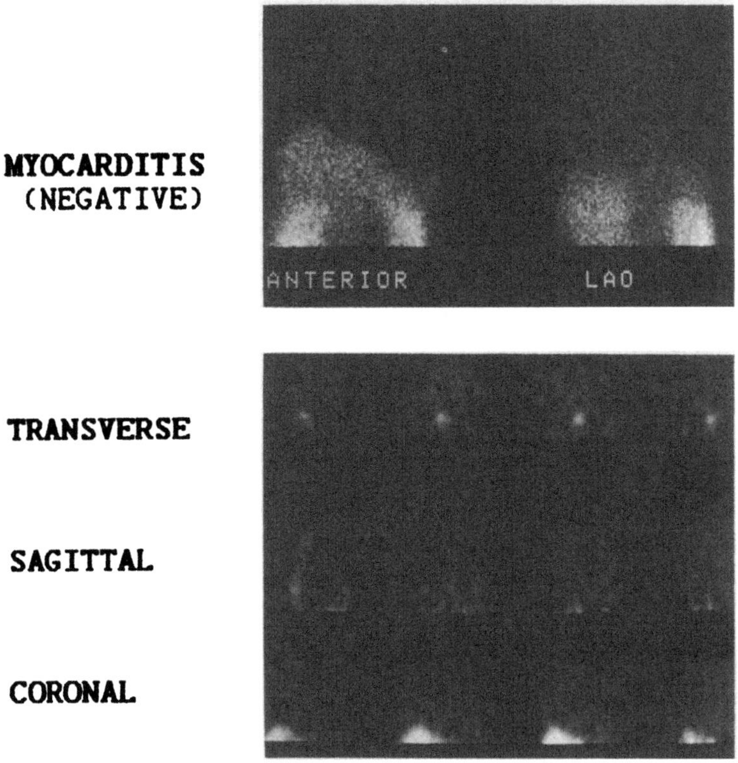

Fig. 6. A negative antimyosin image. No tracer uptake in the myocardial region is seen either in the planar (anterior and left anterior oblique; *LAO)* or the tomographically reconstructed images (transverse, sagittal, or coronal)

of negative images of a patient suspected of having acute myocarditis, but who had a negative biopsy for myocarditis. To avoid misconstruing residual blood pool activity as myocardial activity, it is imperative that both planar and tomographic images be evaluated, and that myocardial activity be present in the planar and at least two of the three tomographic reconstructions [25]. Figure 7 shows a set of positive antimyosin scans. In this case, localization of In-111-labeled antimyosin can be visualized in all of the views.

Patients who were available for a 6-month follow-up study with thorough clinical evaluation, endomyocardial biopsies and determination of left ventricular ejection fraction (LVEF) were reexamined with antimyosin scans [26]. Figure 8 shows gamma scintigrams of a patient who was positive for acute myocarditis by histopathological criteria, had an LVEF of 34%, and initial positive antimyosin imaging. The 6-month follow-up scintigrams showed minimal myocardial antimyosin activity in the planar images, probably representing residual blood pool activity. No unequivocal myocardial activity could be discerned in the transverse, sagittal, and coronal tomographic reconstructions. The biopsy was negative and

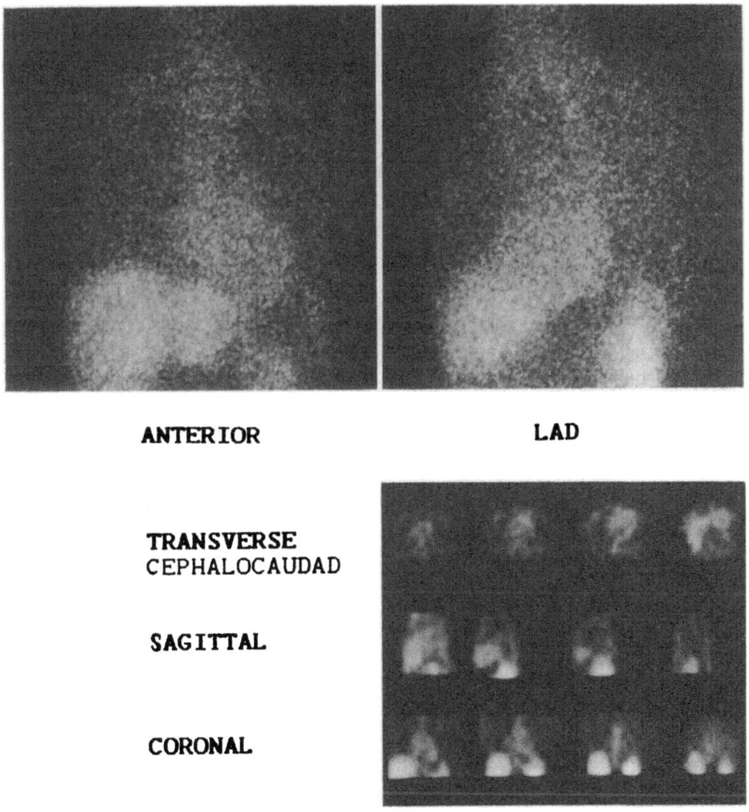

Fig. 7 a, b. A positive antimyosin image for unequivocal diagnosis of acute myocarditis. Diffuse uptake in the cardiac region is seen in the planar images **(a)** and in the tomographic reconstructions **(b)**

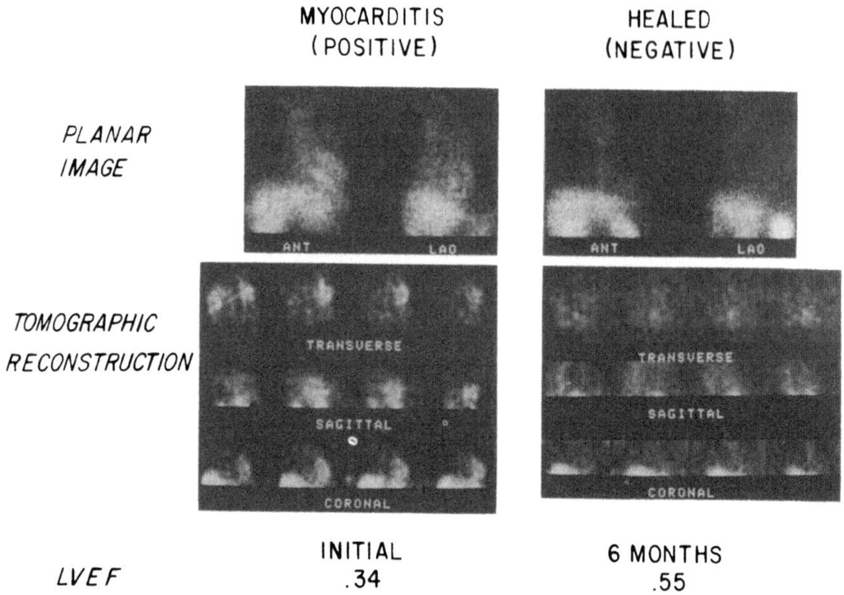

Fig. 8. Initial and follow-up In-111 antimyosin planar and tomographic scintigrams from a patient. See text

the LVEF increased to 55% at the time of the follow-up study. To date, we have analyzed antimyosin gamma scintigraphic studies in 57 patients with suspected acute myocarditis [27]. The scintigraphic studies were performed within 1 year from the onset of a dilated cardiomyopathy. The left ventricular ejection fraction was less than 45% (mean 24% ± 10%) in every patient. Figure 9 shows the correlation of antimyosin to histological studies.

The patients who are of most concern are those with positive antimyosin scans and negative biopsy evidence for acute myocarditis. While these results could be attributed to actual false-positive scans, a more likely cause is the poor sensitivity

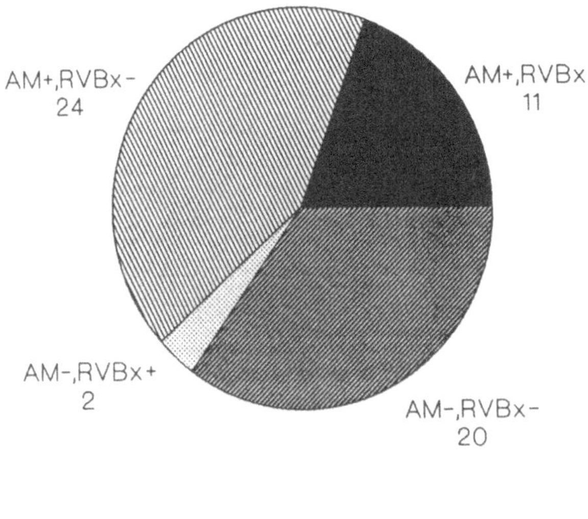

Fig. 9. Correlation of antimyosin *(AM)* to right ventricular biopsy data (RVBx) for diagnosis of acute myocarditis. Utilizing RVBx as the standard for comparison, sensitivity of antimyosin for detection of myocarditis was 85% and specificity was 45%. Substantial spontaneous rise in LVEF (> 10%) and decrease in NYHA class within 6 month was seen in nine of the 24 patients with positive antimyosin and negative RVBx findings

of right ventricular biopsy for the detection of myocarditis (only five to seven right ventricular biopsy samples were obtained for histological studies), combined with antimyosin's ability to delineate myocardial injury anywhere in the heart with higher sensitivity.

At this time we cannot ascertain which method is more reliable for diagnosis of acute myocarditis. However, of the 24 patients with positive antimyosin scans and negative biopsies, there was a substantial spontaneous rise in LVEF (> 10%) and a decrease in New York Heart Association (NYHA) class within 6 months. Repeat antimyosin scans in five of these patients were negative at the time of LVEF improvement. These findings suggest that the focal myocarditis detected by antimyosin immunoscintigraphy was missed by right ventricular biopsy.

The importance of reinjection of antimyosin for follow-up assessment of myocarditis necessitated our investigation of possible adverse immunological effects of multiple murine monoclonal antimyosin administrations. Use of murine intact monoclonal antibodies or their fragments in higher concentrations in oncological immunoscintigraphic and therapeutic studies, as well as in immunosuppression of kidney transplant rejections with OK-t antibodies, have elicited human antimurine antibody (HAMA) responses. Presence of HAMA in the circulation of patients at the time of follow-up antimyosin scans would cause rapid clearance of the radiolabeled antimyosin Fab and may cause complications due to immune complexes. Therefore, we obtained blood samples from patients before the first antimyosin injection, and usually within 1–2 weeks before the second antimyosin administration. The sera from these patients were assessed for HAMA activity using the molecular sieve chromatographic technique. Positive control rabbit antimurine IgG antiserum was mixed with an aliquot of In-111-labeled antimyosin Fab, and the mixture was chromatographed on an Ultrogel ACA-34 column (45 x 0.5 cm). Free In-111 antimyosin Fab eluted in the 50-kilodalton (kd) molecular weight region, whereas the antimyosin Fab antimurine antibody complexes eluted in the 200-kd and larger molecular size region. To date, none of the sera from patients with multiple antimyosin antibody administrations showed the presence of HAMA activity. Thus, it appears that murine monoclonal antibody Fab fragments at a dosage of 0.5 mg do not elicit HAMA response after intravenous administration [28].

Conclusion

Radiolabeled antimyosin may be useful in the screening of patients with suspected myocarditis. A positive antimyosin scan indicates the need for endomyocardial biopsy to establish the histological diagnosis. This need was exemplified by the study of one of the 57 patients studied thus far. This patient had a negative right ventricular biopsy on his initial study. However, his antimyosin scan showed marked antimyosin accumulation in the left ventricle. A left ventricular biopsy performed subsequently was strongly positive for acute myocarditis by histological criteria. Furthermore, the disappearance of antimyosin positivity in the follow-up scans in some of the patients who initially had positive antimyosin scans, low LVEF, but who, at the time of the follow-up study, had a spontaneous increase in LVEF and marked improvement in clinical symptoms, indicates a potential noninvasive method which may be capable of identifying the disease in those patients whose lesions escape diagnosis due to sampling error inherent in endomyocardial biopsy.

References

1. Aretz HT, Billingham ME, Edwards WD, Factor SM, Fallon JT, Fenoglio JJ, Olsen EGJ, Schoen FJ (1986) Myocarditis: a histopathologic definition and classification. Am J Cardiovasc Pathol 1: 3–14
2. Khaw BA, Beller GA, Haber E, Smith TW (1976) Localization of cardiac myosin-specific antibody in myocardial infarction. J Clin Invest 58: 439–446
3. Khaw BA, Fallon JT, Strauss HW, Haber E (1980) Myocardial infarct imaging of antibodies to canine cardiac myosin with Indium-111-diethylenetriamine pentaacetic acid. Science 209: 295–297
4. Khaw BA, Fallon JT, Beller GA, Haber E (1979) Specificity of localization of myosin-specific antibody fragments in experimental myocardial infarction: histologic, histochemical, autoradiographic and scintigraphic studies. Circulation 60: 1527–1531
5. Khaw BA, Mattis JA, Melincoff G, Strauss HW, Gold HK, Haber E (1984) Monoclonal antibody to cardiac myosin: imaging of experimental myocardial infarction. Hybridoma 3: 11–23
6. Khaw BA, Strauss HW, Moore R, Fallon JT, Yasuda T, Gold HK, Haber E (1987) Myocardial damage delineated by In-111 antimyosin Fab and Tc-99m pyrophosphate. J Nucl Med 28: 76–82
7. Frame LH, Lopez JA, Khaw BA, Fallon JT, Haber E, Powell J Jr (1983) Early membrane damage during coronary reperfusion in dogs: detection by radiolabeled anticardiac myosin $(Fab')_2$. J Clin Invest 72: 535–544
8. Khaw BA, Strauss HW, Carvalho A, Locke E, Gold HK, Haber E (1982) Technetium-99m labeling of antibodies to anticardiac myosin Fab and to human fibrinogen. J Nucl Med 23: 1011–1019
9. Khaw BA, Gold HK, Yasuda T, Leinbach RC, Kanke M, Fallon JT, Barlai-Kovach M, Strauss HW, Sheehan F, Haber E (1986) Scintigraphic quantification of myocardial necrosis in patients after intravenous injection of myosin-specific antibody. Circulation 74: 501–508
10. Khaw BA, Yasuda T, Gold HK, Leinbach RC, Johns JA, Kanke M, Barlai-Kovach M, Strauss HW, Haber E (1987) Acute myocardial infarct imaging with Indium-111-labeled monoclonal antimyosin Fab. J Nucl Med 28: 1671–1678
11. Shell WE, Kjekshus JK, Sobel BE (1971) Quantitative assessment of the extent of myocardial infarction in the conscious dog by means of analysis of serial changes in serum creatine phosphokinase activity. J Clin Invest 50: 2614–2652
12. Sobel BE, Shell WE (1972) Serum enzyme determinations in the diagnosis and assessment of myocardial infarction. Circulation 45: 471–482
13. Nydick I, Wroblewski F, La Due JS (1955) Evidence for increased serum glutamic oxalacetic transaminase (SGO-T) activity following graded myocardial infarcts in dogs. Circulation 12: 161–168
14. Kagen L, Scheidt S, Butt A (1977) Serum myoglobin in myocardial infarction: the "staccato phenomenon". Am J Med 62: 86–92
15. Khaw BA, Scott J, Fallon JT, Cahill SL, Haber E, Homcy C (1982) Myocardial injury: quantitation by cell sorting initiated with antimyosin fluorescent spheres. Science 217: 1050–1053
16. Khaw BA, Yasuda T, Strauss HW (1988) Monoclonal antibodies for imaging myocardial necrosis. In: Early PJ, Sodee DB (eds) Principles and practice of nuclear medicine, 2nd edn. Mosby, Lanham (in press)

17. Khaw BA, Fallon JT, Russell PS, Rosseel J, Ferguson P, Haber E (1981) New approach to determination of heart transplant rejection by I-125 antimyosin antibody and Tc-99m-fibrinogen. Clin Res 29: 496A (Abstract)
18. Addonizio LJ, Seldin DW, Esser PD, Johnson LL, Michler RE, Marboe C, Rose EA, Alderson PO, Cannon PJ (1986) SPECT imaging of cardiac transplant rejection using In-111 antimyosin antibody. J Nucl Med 27:746 (Abstract)
19. Isobe M, Nossiff ND, Chase C, Strauss HW, Haber E, Khaw BA (1988) Early detection with indium-111 monoclonal antimyosin imaging of cardiac rejection in heterotopic heart transplantation in mice. J Nucl Med 29 (5, Abstr.): 746
20. Frist W, Yasuda T, Segall G, Khaw BA, Strauss HW, Gold H, Stinson E, Oyer P, Baldwin J, Billingham M, McDougall IR, Haber E (1987) Noninvasive detection of human cardiac transplant rejection with indium-111 antimyosin (Fab) imaging. Circulation 76: V81–V85
21. Rezkalla S, Khatib R, Khaw BA, Haber E, Smith FE, Kloner RA (1987) Monoclonal antimyosin antibody in coxsackievirus B3 murine myocarditis. Clin Res 35: 319A
22. Neu N, Rose NR, Beisel KW, Herskowitz A, Gurri-Glass G, Craig SW (1987) Cardiac myosin induces myocarditis in genetically predisposed mice. J Immunol 139: 3630–3636
23. Acosta AM, Santos-Buch CA (1985) Autoimmune myocarditis induced by Trypanosoma cruzi. Circulation 71: 1255–1261
24. Cossio PM, Bustuoabad O, Paterno E, Iotti R, Casanova MB, Podesta MR, Bolomo N, Arana RM, de Pasqualini CD (1984) Experimental myocarditis induced in Swiss mice by homologous heart immunization resembles chronic experimental Chagas' heart disease. Clin Immunol Immunopathol 33: 165–175
25. Yasuda T, Palacios IF, Dec GW, Fallon JT, Gold HK, Leinbach RC, Strauss HW, Khaw BA, Haber E (1987) Indium 111-Monoclonal antimyosin antibody imaging in the diagnosis of acute myocarditis. Circulation 76: 306–311
26. Haber E, Yasuda T, Palacios IF, Dec GW, Khaw BA (1987) Antimyosin scintigraphy in the diagnosis of acute myocarditis. In: Kawai C, Abelmann WH (eds) Cardiomyopathy update 1. Pathogenesis of myocarditis and cardiomyopathy: recent experimental and clinical studies. University of Tokyo Press, Tokyo, pp 277–280
27. Dec GW, Yasuda T, Strauss HW, Khaw BA, Southern J, Fallon JT, Palacios I Antimyosin antibody imagings vs. right ventricular biopsy in patients with dilated cardiomyopathy of recent onset and suspected myocarditis. International Symposium in Inflammatory Heart Disease at Snowmass, Colorado (in press)
28. Nicol PD, Yasuda T, Locke E, Strauss HW, Haber E, Khaw BA (1988) Multiple intravenous administration of In-111 labeled antimyosin Fab: determination of anti-murine Fab response in patients with myocarditis. J Nucl Med 29: 939 (Abstract)

VI. Immuno-Modulating Therapy: Experimental and Clinical Approaches

Animal Models for Therapeutic Trials of Viral Myocarditis: Effect of Ribavirin and Alpha Interferon on Coxsackievirus B3 and Encephalomyocarditis Virus Myocarditis

A. MATSUMORI[1], I. OKADA[1], C. KAWAI[1], C. S. CRUMPACKER[2], and W. H. ABELMANN[2]

[1] The Third Division, Department of Internal Medicine, Kyoto University, Kyoto, Japan
[2] Divisons of Cardiology and Infectious Diseases, The Charles A. Dana Research Institute and the Harvard-Thorndike Laboratory of Beth Israel Hospital and Harvard Medical School, Boston, Massachusetts, USA

Introduction

Currently, the treatment of viral myocarditis is symptomatic and supportive, no specific therapy having yet been found. Therefore, we investigated the effects of antiviral agents, ribavirin and recombinant alpha interferon (IFN-α) on viral myocarditis using animal models of encephalomyocarditis (EMC) virus and coxsackievirus B3 myocarditis in mice (Matsumori and Kawai 1982, a, b).

EMC Virus Myocarditis

Effect of Recombinant IFN-α on EMC Virus Myocarditis

Recombinant human leukocyte IFN-α A/D was supplied by Hoffmann-La-Roche, Inc., Nutley, New Jersey, USA or Kamakura, Japan. Antiviral activity in vitro was assayed by a plaque reduction method using FL (human amnion) cells. Treatment of cells with increasing concentrations of IFN resulted in progressive reduction of plaque formation. Linear regression analysis showed good negative correlation of $y = 91.9–48.5 \log x$ (r = -0.938, $P < 0.001$) between plaque forming units (PFU) and the logarithm of IFN concentration. The 50% inhibitory dose (ID_{50}) was 7.2 units/ml.

IFN was administered subcutaneously once daily, beginning 1 day before, the same day, or 1 day after virus inoculation. Animals were sacrificed on day 4. Myocardial virus titers were significantly lower in mice treated with 10^4 units/g IFN daily when treatment was started 1 day before or on the day of infection. Myocardial cell necrosis, cellular infiltration, and calcification were scored on a scale of 1+ to 4+ in terms of severity (Matsumori et al. 1985). The histologic scores of cellular infiltration and myocardial necrosis were significantly less severe in the treated groups (Matsumori et al. 1987). Prevention was dependent upon dosage and the time when treatment was started. When treatment was started before or simultaneously with virus inoculation, IFN-α A/D, at a dose of 10^4 units/g daily, effectively reduced the inflammatory response and myocardial damage.

H.-P. Schultheiß (Ed.)
New Concepts in Viral Heart Disease

Effect of Ribavirin on EMC Virus Myocarditis

Ribavirin (Virazole, 1-β-*D*-ribofuranosyl-1,2,4-triazole-3-carboxamide) is a synthetic nucleoside analogue, structurally related to inosine and guanosine, and has broad antiviral activity against RNA and DNA viruses. Clinically, its efficacy has been demonstrated in measles, influenza, and respiratory syncytial virus.

The in vitro effect of ribavirin was assayed by a plaque reduction method in FL cells. Linear regression analysis showed good negative correlation of $y = 116–55.5 \log x (r = -0.893, P < 0.001)$ between PFU and logarithm of drug concentration. The median inhibitory dose was 15 μg/ml.

Ribavirin was administered subcutaneously beginning immediately after 10 PFU EMC virus inoculation. EMC virus infection was lethal in these 4-week-old DBA/2 mice (Jackson Lab.) and mean survival time of control mice was 5.2 days. On the other hand, mean survival time was 6.6 days in mice treated with 100 mg/kg ribavirin daily, 7.4 days in the group with 200 mg/kg daily, and 7.7 days in the group with 400 mg/kg daily. All treated groups survived significantly longer than controls.

Myocardial virus titers were determined in the hearts of mice that had died on days 6–8. The degree of inhibition of viral replication in the heart was influenced by drug dosage, and 200 mg/kg ribavirin significantly inhibited viral replication. Cellular infiltration was significantly less marked in mice treated with 200 mg/kg ribavirin, and myocardial necrosis was significantly less severe in mice treated with 400 mg/kg compared with that in controls (Matsumori et al. 1985). In order to study the effect of later administration, ribavirin was administered subcutaneously at a dose of 400 mg/kg beginning 1 day or 3 days after inoculation with 10 PFU EMC virus. Three of ten control mice died before day 6, and seven mice died before day 8. In contrast to control animals, no mice in which treatment was started 1 day after infection died before day 6, but there was no significant difference in the number of deaths before day 8 compared with controls. Myocardial necrosis and cellular infiltration were significantly less severe than in controls when treatment was started 1 day after virus inoculation.

Synergistic Treatment with IFN and Ribavirin for EMC Virus Myocarditis

As described above, the use of IFN-α A/D and ribavirin, given simutaneously with inoculation with EMC virus, inhibited multiplication of virus in the heart and protected mice from developing myocarditis. Large amounts of these agents, however, were necessary to obtain beneficial effects. The side effects of these drugs are usually dose related, and synergistic combination of drugs may allow effective treatment at lower drug concentrations. This strategy has been highly successful in the treatment of some bacterial diseases and cancers. Since the mode of action of IFN may differ from that of ribavirin, we investigated the possibility that recombinant IFN-α A/D might be synergistic with ribavirin.

Plaque reduction assays in tissue cultures indicated that ribavirin and IFN-α A/D inhibit EMC virus replication in a synergistic manner (Matsumori et al. 1986).

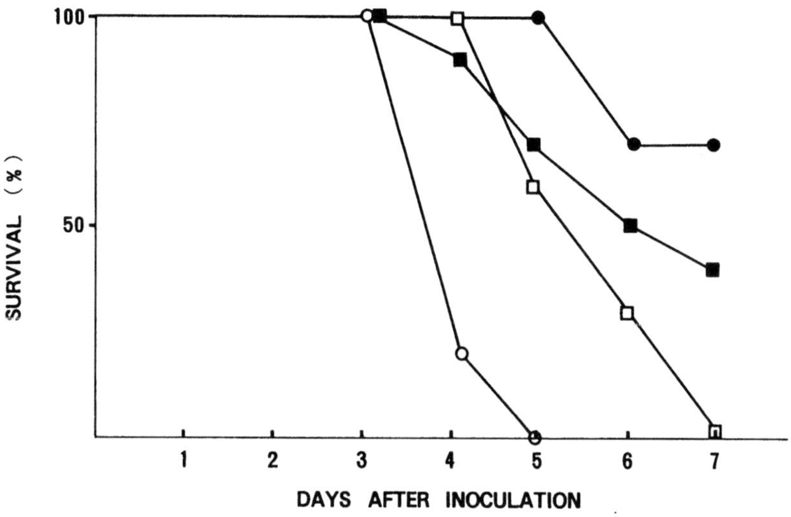

Fig. 1. Synergistic combination treatment with IFN-α A/D and ribavirin in EMC virus myocarditis produced enhanced survival in mice. *Open circles,* infected control; *solid squares,* IFN 1000 units/g; *open squares,* ribavirin 0.1 mg/g; *solid circles,* IFN + ribavirin (each group, $n = 10$)

The use of the synergistic combination of IFN 10^3 units/g daily and ribavirin (0.1 mg/g daily) to treat mice infected with EMC virus produced an enhanced survival as only three of ten mice died at day 8. This was compared to all ten of the untreated mice who died before day 6, and all ten who died before day 8 with ribavirin treatment alone. In the group treated with IFN alone, six of ten died before day 8 (Fig. 1).

The myocardial virus titer was also effectively reduced in mice treated with the combination of ribavirin and IFN. Groups of five mice each were sacrificed at day 4 following infection, and the myocardial virus titer determined and histologic evaluation carried out. The mean myocardial virus titer of control mice was 4.86 log PFU/mg. The titer of mice hearts treated with IFN alone was 4.79 log PFU/mg, and ribavirin alone was 4.89. In mice treated with the combination of IFN and ribavirin the myocardial virus titer was 2.19 log PFU/mg and this was a statistically significant decrease when compared to all the other regimens. The combined treatment with IFN and ribavirin produced a highly significant reduction in the inflammatory response and myocardial damage. This combination achieves the effects at lower concentrations than either drug used alone, and the frequency of undesirable side effects of either drug may be able to be reduced.

In a previous study, ribavirin, 0.1 mg/g, or IFN, 10^3 units/g, alone did not inhibit EMC virus replication in the heart. When used together, 0.1 mg/g ribavirin daily and 10^3 units/g IFN daily achieved a striking effect. This study provides convincing evidence that the demonstration of synergy between two antiviral drugs in tissue culture can be predictive of impressive biologic effect when these drugs are used together to treat animals with a life-threatening viral infection. The use of combinations of antivirals deserves further careful study for other serious viral infections.

Coxsackievirus B3 Myocarditis

Effect of Recombinant IFN-α on Coxsackievirus B3 Myocarditis

The Nancy strain of coxsackievirus B3 was used. Virus stock was prepared as described previously (Matsumori and Kawai 1980) and was stored at −70 °C until it was diluted for use. Four-week-old male C3H/He (Shizuoka Agricultural Coop., Shizuoka, Japan) were inoculated intraperitoneally with 0.1 ml coxsackievirus B3, diluted in Eagle's minimum essential medium to a concentration of $10^{6.5}$ TCD_{50}/ml. Severe myocarditis with extensive myocardial necrosis and prominent cellular infiltration was induced in this model (Fig. 2). IFN-α A/D was administered subcutaneously at a dose of 10^4 units/g once daily, beginning 1 day before or on the same day as virus inoculation. Control mice were injected with saline. Each group consisted of ten mice. Animals were sacrificed on day 7.

Two mice without IFN treatment died, but no mice died in the treated groups. The myocardial virus titer of the controls was 4.4 ± 0.9 log TCD_{50}/mg (mean ± SD). Titers of treated group were 0.7 ± 0.1 log TCD_{50}/mg in mice with prior treatment, and 2.5 ± 1.2 log TCD_{50}/mg in mice with simultaneous treatment. Myocardial virus titers were significantly lower in both groups compared with controls (Fig. 3, $P < 0.005$).

Cellular infiltration and myocardial necrosis were found in all hearts of untreated mice. On the other hand, nine of ten hearts in mice with prior treatment and

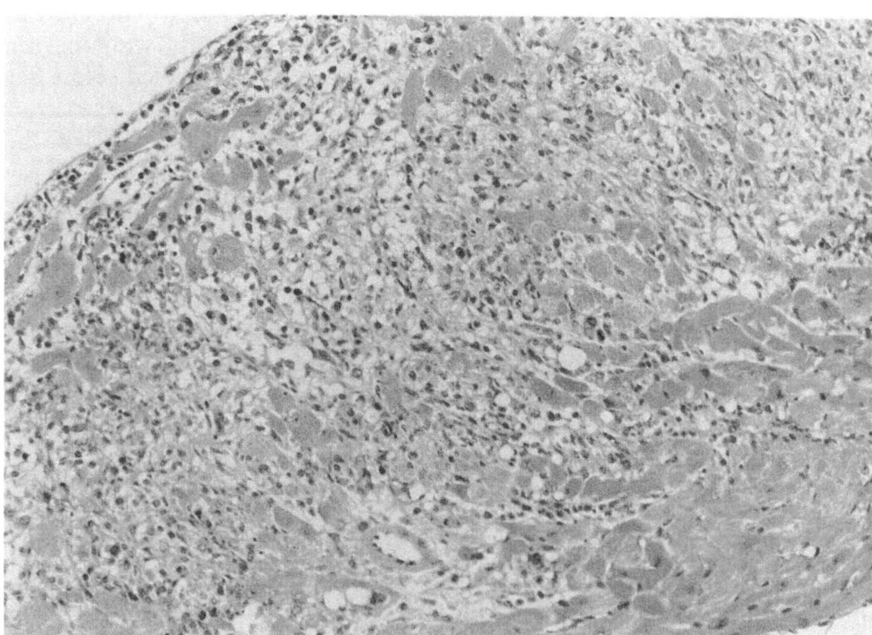

Fig. 2. Histologic findings in the heart of a C3H/He mouse inoculated with coxsackievirus B3 — 7 days after virus inoculation. Marked inflammatory cell infiltration and extensive myocardial necrosis are seen. Hematoxylin-eosin stain; x 180

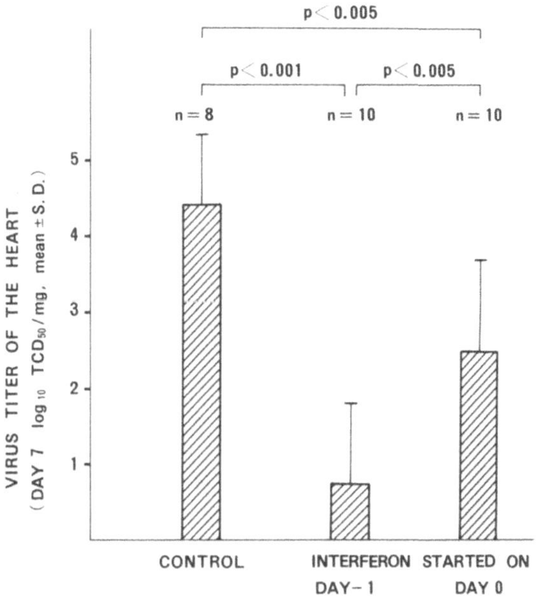

Fig. 3. Effect of IFN-α A/D on coxsackievirus B3 replication in the heart. Viral replication was inhibited when 10^4 units/g IFN was administered daily starting 1 day before or on the same day as virus inoculation. (From Matsumori et al. 1988)

two hearts in mice with simultaneous treatment showed neither cellular infiltration nor myocardial necrosis. The scores of cellular infiltration were 0.1 ± 0.3 and 0.9 ± 0.6, respectively, and were significantly less severe than those of controls (3.6 ± 0.7, $P < 0.005$). The scores of myocardial necrosis were 0.1 ± 0.3, 1.1 ± 0.7, and 3.8 ± 0.6, respectively ($P < 0.005$, Fig. 4). Cellular infiltration and myocardial necrosis were significantly less severe in the treated groups (Matsumori et al. 1988).

Coxsackieviruses are known to be the most common causative agents in viral myocarditis in humans. IFN-β has been shown to inhibit replication of coxsackievirus B3 in cultured human heart cells (Kandolf et al. 1985), Lutton and Gauntt

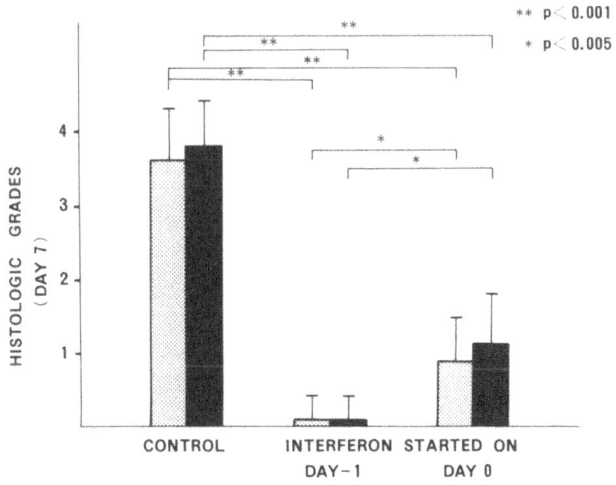

Fig. 4. Effect of IFN-α A/D on cardiac histopathology in coxsackievirus B3 myocarditis. Cellular infiltration *(screened bars)* and myocardial necrosis *(solid bars)* in the treated groups were significantly less severe than in controls. (From Matsumori et al. 1988)

(1985) reported that murine IFN-β reduced the number of coxsackievirus B3-induced myocardial lesions in CD-1 mice but did not reduce virus titers in heart tissues. Murine IFN-α/β has been shown to be effective against lethal infection of coxsackievirus A16 and enterovirus 71 in mice when administered simultaneously with or early after infection (Sasaki et al. 1986). In this study, in an animal model of coxsackievirus myocarditis, we demonstrated that human leukocyte IFN-α A/D given 1 day before and simultaneously with inoculation with virus inhibited multiplication of virus in the heart and protected mice from developing myocarditis.

Effect of Ribavirin on Coxsackievirus B3 Myocarditis

Four-week-old C3H/He mice were inoculated intraperitoneally with 3 x 10^4 PFU coxsackievirus B3. Ribavirin was administered subcutaneously at a dose of 50, 100, or 200 mg/kg daily beginning immediately after virus inoculation. All mice survived and were sacrificed on day 7. Myocardial virus titer on day 7 was significantly lower in mice treated with 100 mg/kg of ribavirin (0.5 $\pm$ 0.9 $\log_{10}$ PFU/mg; $P < 0.05$) compared with controls (2.9 $\pm$ 0.9 $\log_{10}$ PFU/mg, Fig. 5). Histologic examination of myocardium showed that there were less cellular infiltration and necrosis in mice treated with 100 or 200 mg/kg (Fig. 6). Thus, ribavirin effectively inhibited myocardial virus replication and reduced the inflammatory response and myocardial damage in this experimental model of coxsackievirus B3 myocarditis. The minimal effective dose for coxsackievirus myocarditis was smaller than that for EMC virus myocarditis (Okada et al. 1988).

Treatment with ribavirin and IFN has advantages in that they are broadly active antiviral agents and therefore should be of great advantage in diseases of multiple etiology such as viral myocarditis.

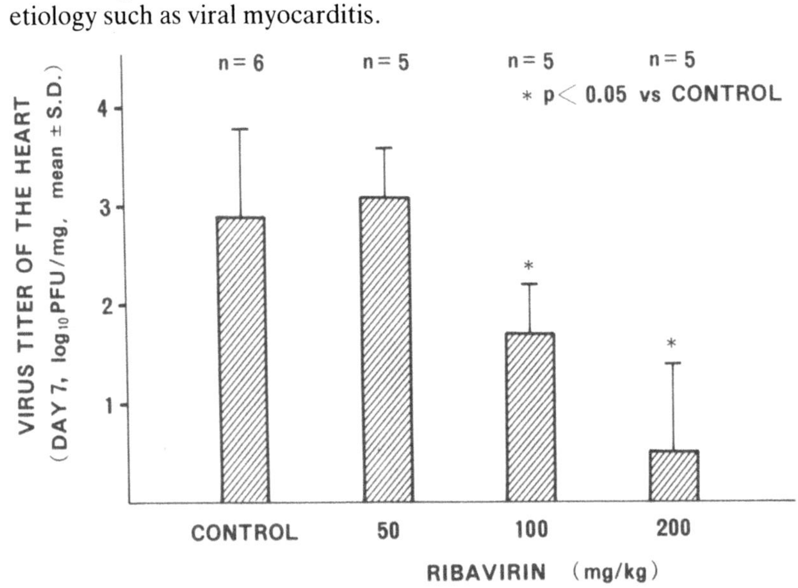

Fig. 5. Effect of ribavirin on coxsackievirus B3 replication in the heart. Viral replication was inhibited by 100 and 200 mg/kg ribavirin daily

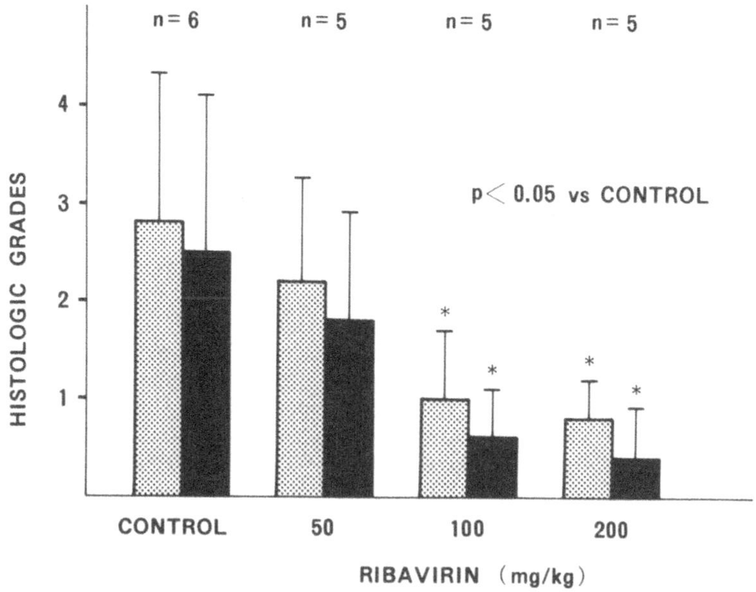

Fig. 6. Effect of ribavirin on cardiac histopathology in coxsackievirus B3 myocarditis. Cellular infiltration *(screened bars)* and myocardial necrosis *(solid bars)* were significantly less severe in mice treated with 100 and 200 mg/kg ribavirin daily than in controls

Summary

Murine models of myocarditis induced by EMC and coxsackievirus B3 were used for the evaluation of the effects of ribavirin and/or recombinant IFN-α A/D. Ribavirin, a synthetic nucleoside analogue, inhibited EMC virus replication in the heart when given simultaneously with infection, and this inhibitory effect correlated well with the decrease of severity of myocardial damage and inflammatory cell infiltration. Myocardial damage and cellular infiltration were also decreased when ribavirin was administered early after virus inoculation. Recombinant human leukocyte IFN-α A/D, given 1 day before or simultaneously with EMC virus, inhibited multiplication of virus in the heart and protected mice from developing myocarditis. Ribavirin was synergistic with IFN-α A/D when used in combination against EMC virus infection. This combination may be able to reduce the frequency of unfavorable effects of ribavirin or IFN by lowering the effective dose of both agents. The effect of ribavirin and IFN was also investigated in an animal model of coxsackievirus B3 myocarditis. Four-week-old C3H/He mice were inoculated intraperitoneally with coxsackievirus B3. IFN-α A/D, 10^4 units/g daily, inhibited viral replication in the heart and reduced the inflammatory response and myocardial damage when treatment was started before or simultaneously with virus inoculation. Ribavirin was also effective on coxsackievirus B3 myocarditis, at a dose of 100 mg/kg daily which is smaller dose than the minimal effective dose for EMC myocarditis.

Acknowledgements. This work was supported, in part, by a Research Grant for Cardiomyopathy from the Ministry of Health and Welfare of Japan, Grant-in-Aid for Scientific Research from the Ministry of Education, Science and Culture and Kanazawa Research Fund.

References

1. Kandolf R, Canu A, Hofschneider PH (1985) Coxsackie B3 virus can replicate in cultured human foetal heart cells and is inhibited by interferon. J Mol Cell Cardiol 17: 167–181
2. Lutton CW, Gauntt CJ (1985) Ameliorating effect of IFN-β and anti-IFN-β on coxsackievirus B3-induced myocarditis in mice. J Interferon Res 5: 137–146
3. Matsumori A, Kawai C (1980) Coxsackievirus B3 perimyocarditis in BALB/c mice: experimental model of chronic perimyocarditis in the right ventricle. J Path 131: 97–106
4. Matsumori A, Kawai C (1982a) An experimental model for congestive heart failure following encephalomyocarditis virus myocarditis in mice. Circulation 65: 1230–1235
5. Matsumori A, Kawai C (1982b) An animal model of congestive (dilated) cardiomyopathy: dilatation and hypertrophy of the heart in the chronic stage in DBA/2 mice with myocarditis caused by encephalomyocarditis virus. Circulation 66: 355–360
6. Matsumori A, Wang H, Abelmann WH, Crumpacker CS (1985) Treatment of viral myocarditis with ribavirin in an animal preparation. Circulation 71: 834–839
7. Matsumori A, Abelmann WH, Crumpacker CS (1986) Combination treatment of interferon and ribavirin for viral myocarditis in an animal model. Tenth World Congress of Cardiology Sept. 14–19, Washington D.C. (Abstr) p 123
8. Matsumori A, Crumpacker CS, Abelmann WH (1987) prevention of viral myocarditis with recombinant human leukocyte interferon α A/D in a murine model. J Am Coll Cardiol 9: 1320–1325
9. Matsumori A, Tomioka N, Kawai C (1988) Protective effect of recombinant alpha interferon on coxsackievirus B3 myocarditis in mice. Am Heart J 115: 1229–1232
10. Okada I, Matsumori A, Matoba Y, Ohkusa T, Kawai C (1988) The effect of ribavirin on experimental coxsackievirus B3 myocarditis in mice. Jpn Circ J (in press) (Abstr)
11. Sasaki O, Karaki T, Imanishi J (1986) Protective effect of interferons with hand, foot and mouth disease virus in newborn mice. J Inf Dis 153: 498–502

Antiviral Activity of Human Fibroblast Interferon in Coxsackievirus B3-Infected Cultured Human Heart Cells

P. KIRSCHNER[1], R. KANDOLF[1,2], A. CANU[3], A. HEIM[1,2], and P. H. HOFSCHNEIDER[1]

[1] Department of Virus Research, Max-Planck-Institut für Biochemie, D-8033 Martinsried, FRG
[2] Department of Internal Medicine I, University of Munich, D-8000 München 70, FRG
[3] Département de Microbiologie, Unité de Formation et de Recherche des Sciences Pharmaceutiques, F-14000 Caen, France

Introduction

Effective antiviral chemotherapy has not yet been established in the treatment of viral heart disease. The optimal goal of antiviral treatment — to restore function to the infected cells — usually appears unattainable. A realistic goal would be to inhibit virus replication and thus prevent virus spread to as yet uninfected cells. This is achieved, in part, in many natural virus infections by the endogenous interferon (IFN) system [1, 2].

With respect to the potential clinical application of exogenous IFN in viral heart disease, one prerequisite is the demonstration that the virus to be treated is susceptible *in vitro* to the action of a given type of IFN in the specific host cell. For this reason we have established tissue cultures of enriched human fetal myocytes and human myocardial fibroblasts [3]. Cultured human heart cells offer the unique possibility to simulate an infectious situation *in vitro* as well as to study the effect of antiviral agents in specific host cells. We chose coxsackievirus B3 (CVB3) as a model because of the obvious significance of coxsackieviruses in clinical cardiology [4].

Materials and Methods

Preparation and Culture of Human Fetal Heart Cells

Human fetuses, approximately 12–16 weeks of age, were obtained within 2 h of hysterectomy. Hearts were removed aseptically, minced and subjected to six to eight disaggregation cycles with 0.1% trypsin in calcium- and magnesium-free phosphate-buffered saline (PBS) as described previously [3, 5]. Myocytes were depleted of endothelial cells and fibroblasts by two sequential 1-h adsorptions [6] in petri dishes (Falcon) at 37 °C. Enriched contractile myocytes as well as fibroblasts were maintained in Dulbecco's modified Eagle's medium (DMEM) (Gibco) supplemented with 10%–20% fetal bovine serum (FBS) (Boehringer Mannheim) and penicillin at 100 μg/ml.

The established tissue culture conditions allow the cultivation of spontaneously beating, functional intact myocytes up to 6 weeks. Spontaneous contractility of

H.-P. Schultheiß (Ed.)
New Concepts in Viral Heart Disease

cultured heart cells was followed at 37 °C by recording the change of light intensity from the edge of beating cells with the aid of a video system.

Virus Propagation and Virus Titration

Human heart cells were infected using a purified large plaque variant of CVB3 (Nancy strain; American Type Culture Collection) which was isolated by three successive plaque purification cycles [3, 7]. Virus titration was carried out by the plaque method on confluent monolayers of Vero cells which were overlaid with 1% Seaplaque agarose in DMEM supplemented with 7.5% FBS, 10 m*M* $MgCl_2$ and kanamycin at 50 μg/ml as previously described [3].

Interferon

Human fibroblast interferon (IFN-β) was prepared from the human fibroblast cell line FS-4 by Bioferon (supported by the German Ministry for Research and Technology). The specific activity was approximately 1.5×10^8 IU/mg protein.

Results

CVB3 Infection of Beating Human Fetal Myocytes

A useful approach to the study of the interaction between CVB3 and human myocardial cells is to simulate coxsackievirus myocarditis in tissue culture. When beating fetal myocytes were infected with a high dose of 50 plaque-forming units (PFU) of CVB3 per cell, titers of infectious progeny virus as high as 10^8 PFU/ml culture medium were found within 2 days [3]. Virus replication was accompanied by loss of spontaneous contractility of myocytes within 9 h, followed by the CVB3-produced classic cytopathic effects with cell retraction and severe destruction of formerly beating networks of myocytes within 20 h of infection. Productive infection of human fetal myocytes could also be demonstrated by the presence of crystalline arrays of virus particles inside myocytes [3]. If the rate of virus assembly is rapid and many viral particles are formed nearly simultaneously at a circumscribed site, crystallization may occur. The predominant features of ultrastructural pathologic changes observed in CVB3-infected human fetal myocytes were disintegration or total dissolution of myofibrils and myofilaments, together with the formation of vesiculated regions which appeared morphologically as the sites of the cytocidal action of the virus [3]. Certain of the observed ultrastructural lesions are in agreement with the findings *in vivo,* both in humans and in animal models, e.g., necrosis of myofibers in the absence of inflammatory cells [4].

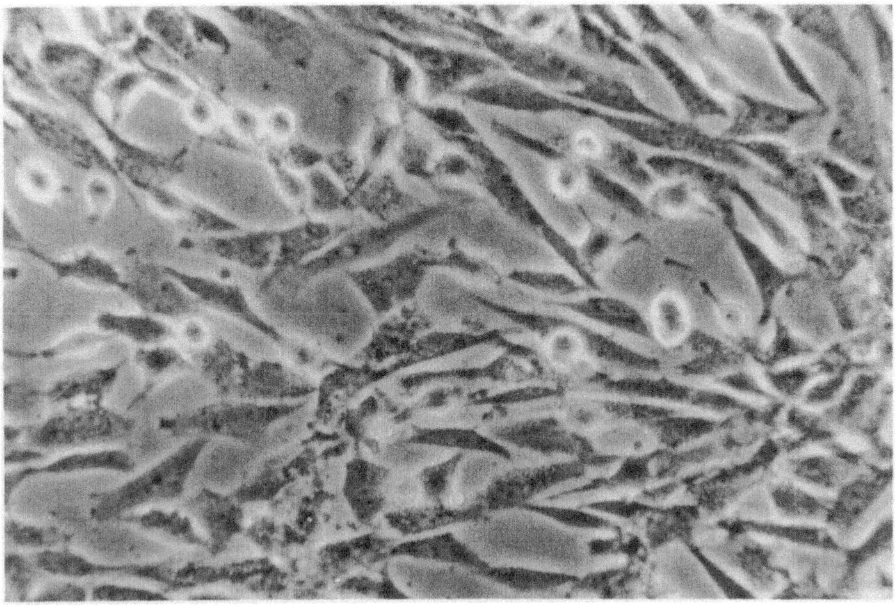

Fig. 1. Myocardial fibroblasts were isolated and grown in tissue culture as described under Materials and Methods. Upon infection of these cells with CVB3, a persistent carrier-state infection is induced. These persistently infected cells can easily be maintained in tissue cultures over periods of many months. Only a few cells show the classic cytopathic effect with cell retraction and rounding. Phase contrast, x 250

Carrier-State Infection of Cultured Human Fetal Myocardial Fibroblasts by CVB3

During these experiments, we made the important observation that lysed areas of formerly beating networks of myocytes were replaced by overgrowing myocardial fibroblasts which apparently survived the infection with CVB3 [3]. Continued propagation of these cells surviving the initial infection with CVB3 yielded cultures of myocardial fibroblasts which produce CVB3 continuously, as observed for periods of up to 9 months, indicating a persistent carrier-state infection [8] of human myocardial fibroblasts by CVB3. These cultures consisted mainly of myocardial fibroblasts (Fig. 1). However, a few heart muscle cells were detected by electron microscopy, which might indicate decreased susceptibility of the differentiated human myocyte, as previously shown for cultured differentiated mouse skeletal muscle cells infected with coxsackievirus A13 [9].

As shown by indirect immunofluorescence studies and molecular *in situ* hybridization [10], viral persistence is based on a carrier-state infection with production of infectious virus in about 10% of the total cell population (Fig. 2). In these productively infected cells a regular viral multiplication cycle appears to take place, usually terminating in cell death, giving rise to virus titers of 10^7 PFU of infectious CVB3 per 1 ml culture medium.

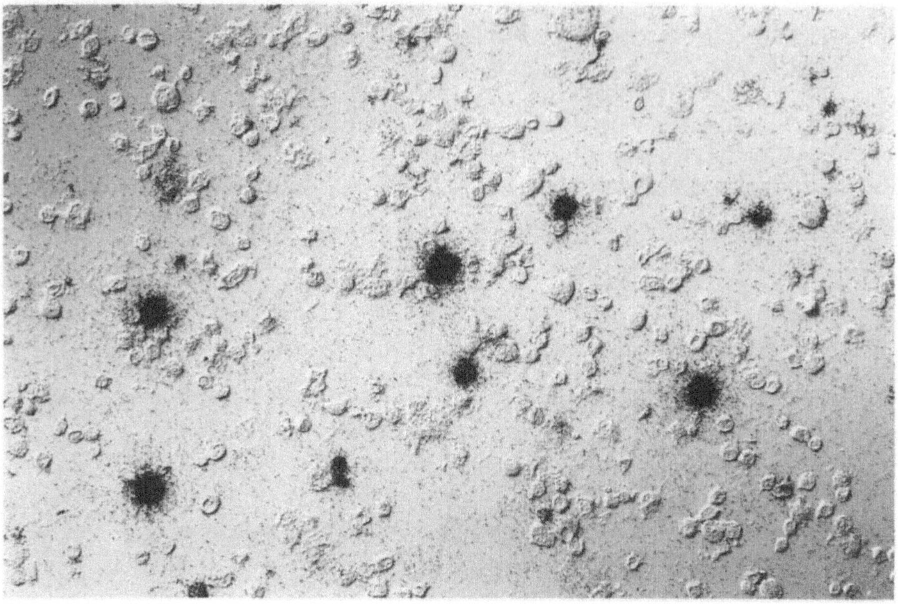

Fig. 2. *In situ* hybridization of ^{35}S-labeled CVB3 complementary DNA (cDNA) probe to persistently infected CVB3 carrier cultures of human myocardial fibroblasts, indicating an asynchronous course of virus replication in this type of infection. The autoradiographic silver grains can be clearly related to distinct fibroblasts which replicate large amounts of viral RNA. Interference contrast, x 150

Protection of Cultured Human Beating Heart Cells by IFN-β

We have previously reported on the major protective role of natural human IFN-β in CVB3-infected enriched human fetal myocytes [3]. To determine the antiviral potency of human IFN-β, we proceeded as follows. Since IFN causes antiviral resistance not directly but by activating cellular genes for antiviral proteins [1, 2], permissive cells must be exposed to IFN for several hours before infection. IFN-mediated antiviral resistance was achieved by exposing 7-day-old cultures of human fetal myocytes to IFN-β 20 h before CVB3 infection at a multiplicity of 50 PFU/cell. Myocytes protected by IFN-β (10^3 IU/ml) continued to beat rhythmically as long as IFN-β was given with changes of medium every 2 days. By contrast, virus replication in unprotected cultures was accompanied by loss of spontaneous contractility, followed by complete lysis of myocytes within 20 h of infection. In agreement with other virus-cell systems, we found no significant effect when IFN-β was given after viral replication had already begun in most of the highly permissive myocytes, e.g., 9 h post infection.

Stimulation of the Beat Frequency of Cultured Human Myocytes by Human IFN-β

Cultured human heart cells offer the unique possibility to study potential side effects of IFNs, which may have an influence on the clinical usage of IFNs in cardiac patients. For example, a norepinephrine-like stimulation of the beat frequency caused by IFN has been described for cultured mouse myocardial cells [11]. Using beating human heart cells as a model system, we also observed a dose-dependent increase in the beat frequency caused by IFN-β, which was not found after preincubation of IFN-β with monoclonal anti-IFN-β antibodies. When beating human fetal myocytes were incubated with IFN-β at 10^3 IU/ml supernatant medium, we observed an increase in the beat frequency of about 50%, e.g., from 50 per minute to 78 per minute within 2 min. However, no significant increase in the beat frequency was observed at the lower dose of 300 IU IFN-β per 1 ml, indicating that the stimulation of the beating rate is dose dependent.

To examine whether IFN-β might have a direct cardiotoxic effect, we used the relatively high dose of 10^3 IU/ml of IFN-β, although lower levels of 100–300 IU/ml of IFN-β were found to be sufficient for the induction of the antiviral state. In agreement with other reports [12], so far no morphologic differences have been observed between myocardial cells treated with 10^3 IU IFN-β per 1 ml culture medium and untreated cells. At the electron-microscopic level, myofibrillar structures had the same morphologic appearance as those in control cultures [3].

Persistently Infected Human Myocardial Fibroblasts as a Test System for the Study of the Antiviral Activity of IFN

Our finding of persistently CVB3-infected human myocardial fibroblasts offers a particularly useful *in vitro* test system for the study of the antiviral activity of IFNs and other agents [13]. In contrast to beating myocytes, infected myocardial fibroblasts can easily be grown and maintained in tissue culture by repeated passages over a period of many months. In this model system, virus-cell interactions can be followed and the effect of antiviral agents analysed on well-defined cell and virus populations. In this type of infection only a small proportion of cells appears to be productively CVB3 infected.

When persistently infected carrier cultures of human myocardial fibroblasts were treated with human IFN-β at 300 IU/ml supernatant medium every 24 h, elimination of infectious CVB3 from the cultures was achieved within 11–13 days (Fig. 3). This clearly demonstrates the potent *in vitro* activity of IFN-β in human myocardial fibroblasts at a dose of 300 IU/ml, a concentration comparable with therapeutically achievable plasma levels [14].

Using molecular hybridization methods, the quantitative impact of therapeutic interventions on viral replication can now be measured at the viral nucleic acid level. Preliminary results obtained with the *in situ* hybridization approach [10] indicate that, following IFN-β treatment, small amounts of CVB3 RNA are still detectable within some fibroblasts although infectious virus is no longer assayable by biologic tests. The molecular basis of this type of enterovirus persistence with reduced or even undetectable infectious virus remains of focal interest.

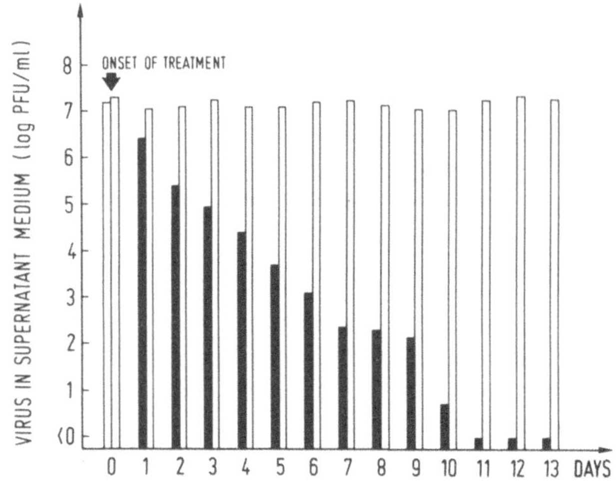

Fig. 3. Effect of human IFN-β on the replication of CVB3. Persistently infected CVB3 carrier cultures of human fetal myocardial fibroblasts were grown in tissue culture. The culture medium was changed every 24 h, and the concentration of infectious progeny virus in the supernatant medium was determined as the number of plaque-forming units *(PFU)* per 1 ml. IFN-β (300 IU/ml) was given with medium changes beginning on day 0 up to day 13 *(solid bars; open bars,* without IFN-β). Elimination of infectious CVB3 from the supernatant medium was achieved within 11 days. Values shown are the averages from duplicate cultures

Inhibition of the Antiviral Activity of IFN-β by Prednisolone

Immunosuppressive therapy of patients with myocarditis still remains controversial. However, it is generally accepted that immunosuppressive agents are contraindicated in acute viral myocarditis. We have therefore investigated the interaction of exogenous fibroblast IFN with clinically used immunosuppressive agents such as prednisolone, azathioprine, or cyclosporine. Whereas azathioprine as well as cyclosporine do not interfere with the potent *in vitro* activity of IFN-β, the high antiviral activity of IFN-β is *in vitro* completely blocked in the presence of prednisolone. Whereas infectious CVB3 is eliminated from persistently infected myocardial fibroblasts by IFN-β at a dose of 300 IU/ml every 24 h within 11–13 days, no significant reduction of virus titers is found when cells are treated with IFN-β in the presence of prednisolone (10 *μM)*. This clearly demonstrates that the antiviral activity of IFN-β is completely antagonized in the presence of corticosteroids.

Downregulation of IFN Receptors

Another important finding obtained in this model system is the observation of an IFN-β-mediated induction of a refractory state to IFN-β at the lower dose of 100 IU/ml. When persistently infected CVB3 carrier cultures of human myocardial fibroblasts were treated with IFN-β at 100 IU/ml every 24 h, a reduction of infectious virus was observed from $3x10^7$ to $9x10^3$ PFU/ml culture medium within 5 days. However, initial virus titers were reached again later on, in spite of continued IFN-β treatment at 100 IU/ml. This decreased antiviral action of IFN-β at a dose of 100 IU/ml is presumably mediated by an IFN-β-induced downregulation of IFN-β receptors within 5 days, which could be overcome by a higher dosage of IFN-β, e.g., 300 IU/ml. The finding of IFN-β-induced downregulation of IFN-β receptors is worth further investigation with respect to a possible role of IFN-β receptor downregulation *in vivo*.

Discussion

From the clinical point of view, it is hoped that antiviral agents, e.g., IFNs, may afford protection against the acute and chronic effects of viral myocarditis. In a first attempt we chose human IFN-β as the antiviral agent since clinical trials have shown the efficacy of IFN-β in serious virus diseases such as generalized herpes zoster or viral encephalitis [15, 16]. Our *in vitro* studies clearly demonstrate that IFN-β acts as a potent inhibitor of the replication of CVB3 in cultured human myocytes, with the consequence that myocytes protected by IFN-β could be kept alive and beating. Moreover, treatment of persistently infected CVB3 carrier cultures of human myocardial fibroblasts with IFN-β at a dose of 300 IU/ml culture medium every 24 h resulted in elimination of infectious CVB3 from the culture medium within 11–13 days. Although the relationship between the described *in vitro* effects of IFN-β and its potential *in vivo* activities remains to be determined, our *in vitro* observations are an important prerequisite for correct therapeutic use of exogenous IFN in severe enterovirus infections.

When applied at an early stage of this disease, IFN would probably reduce the spread of the virus in the myocardium by protecting uninfected cells and thereby limit the disease. But we have to be careful in patients with severely impaired hemodynamic functions who might not tolerate the increase in oxygen demand induced by fever and tachycardia which are common side effects of IFN therapy [16]. Our *in vitro* results with cultured human beating myocytes show an IFN-β-mediated dose-dependent increase of the beat frequency at the high dose of 1000 IU/ml culture medium, which was not found at a dose of 300 IU/ml. Nonetheless, this finding indicates the necessity of a careful evaluation of potential *in vivo* activities of IFN-β in cardiac patients.

The observation of an IFN-β-mediated induction of a refractory state to IFN-β at a dose of 100 IU/ml after 5 days, which was not found at a dose of 300 IU/ml, should be further investigated with respect to a possible role of IFN receptor downregulation *in vivo,* analogous to some forms of insulin resistance *in vivo* [17, 18]. Although clinically relevant conclusions cannot directly be drawn from these *in vitro* findings, an interval treatment schedule for IFN therapy appears to be more appropriate than continuous treatment.

Finally, attention should be drawn to the finding that the antiviral activity of IFN-β is completely inhibited in the presence of prednisolone. By contrast, azathioprine as well as cyclosporine do not interfere with the antiviral activity of IFN-β. Because endogenous IFNs play an important role within the natural host defense mechanisms against viral infections, our data imply that corticosteroids are proscribed in enteroviral heart disease.

Acknowledgements. We thank Mrs. I. Dick and Mrs. B. Schwaiger for excellent technical assistance. This work was supported in part by Grant Ka 593/2-1 from the Deutsche Forschungsgemeinschaft. P. K. received support from a research fellowship from the Deutsche Forschungsgemeinschaft.

References

1. Joklik WK (1986) Interferons. In: Fields BN, Knipe DM (eds) Fundamental Virology. Raven Press, New York, pp 281–307
2. Gresser I (1984) Role of interferon in resistance to viral infection *in vivo*. In: Vilcek J, De Maeyer E (eds) Interferon Vol 2, Interferons and the Immune System. Elsevier, Amsterdam, pp 221–247
3. Kandolf R, Canu A, Hofschneider PH (1985) Coxsackie B3 virus can replicate in cultured human foetal heart cells and is inhibited by interferon. J Mol Cell Cardiol 17: 167–181
4. Wynne J, Braunwald E (1980) The cardiomyopathies and myocarditides. In: Braunwald E (ed) Heart Disease. A Textbook of Cardiovascular Medicine. W. B. Saunders, Philadelphia, pp 1437–1498
5. Kandolf R, Hofschneider PH (1984) Effect of interferon on the replication of Coxsackie B3 virus in cultured human fetal heart cells. In: Bolte HD (ed) Viral Heart Disease. Springer, Berlin, pp 57–63
6. Kasten FH (1973) Mammalian myocardial cells. In: Kruse PFJ, Patterson MKJ (eds) Tissue Culture, Methods and Applications. Academic Press, New York, pp 72–81
7. Kandolf R, Hofschneider PH (1985) Molecular cloning of the genome of a cardiotropic Coxsackie B3 virus: Full-length reverse-transcribed recombinant cDNA generates infectious virus in mammalian cells. Proc Natl Acad Sci USA 82: 4818–4822
8. Takemoto KK, Habel K (1959) Virus-cell relationship in a carrier culture of HeLa cells and Coxsackie A9 virus. Virology 7: 28–44
9. Goldberg RJ, Crowell RL (1971) Susceptibility of differentiating muscle cells of fetal mouse in culture to coxsackievirus A13. J Virol 7: 759–769
10. Kandolf R, Ameis D, Kirschner P, Canu A, Hofschneider PH (1987) *In situ* detection of enteroviral genomes in myocardial cells by nucleic acid hybridization: An approach to the diagnosis of viral heart disease. Proc Natl Acad Sci USA 84: 6272–6276
11. Blalock JE, Stanton JD (1980) Common pathways of interferon and hormonal action. Nature 283: 406–408
12. Lampidis TJ, Brouty-Boyé D (1981) Interferon inhibits cardiac cell function *in vitro*. Proc Soc Exp Biol Med 166: 181–185
13. Kandolf R, Kirschner P, Ameis D, Canu A, Hofschneider PH (1987) Cultured human heart cells: a model system for the study of the antiviral activity of interferons. Eur Heart J 8, Supp J: 453–456
14. Treuner J, Dannecker G, Joester KE, Hettinger A, Niethammer D (1981) Pharmacological aspects of clinical stage I/II trials with human beta interferon in children. J Interferon Res 1: 373–380
15. Heidemann E, Wilms K, Treuner J, Niethammer D (1982) Fibroblasten-Interferon zur Behandlung des Herpes Zoster. Dtsche Med Wochenschr 107: 695–697
16. Heidemann E, Obert HJ (1985) Clinical trials and pilot studies with natural interferons in Germany. In: Kirchner H, Schellekens H (eds) The Biology of the Interferon System 1984. Elsevier, Amsterdam, pp 557–567
17. Aguet M, Groebke M, Dreiding P (1983) Down-regulation of interferon receptors. In: De Maeyer E, Schellekens H (eds) The Biology of the Interferon System 1983. Elsevier, Amsterdam, pp 161–170
18. Kahn CR (1979) The role of insulin receptors and receptor antibodies in states of altered insulin action. Proc Soc Exp Biol Med 162: 13–21

Immunosuppressive Therapy in Myocarditis: Acute Viral Versus Chronic Idiopathic Myocarditis

J. D. HOSENPUD

Division of Cardiology, Department of Medicine, The Oregon Health Sciences University, 3181 SW Sam Jackson Park Road, Portland, Oregon 97201, USA

Introduction

The myocardium is involved during a variety of acute viral illnesses [41]. This acute myocarditis is most commonly an incidental clinical finding but more rarely presents as acute myocardial dysfunction and congestive heart failure, or a life threatening arrhythmia [16, 13]. The acute inflammatory process involving lymphocytes and other round cells is usually present in a diffuse distribution and associated with myocardial cell necrosis. Contrasting acute viral myocarditis is a more chronic, indolent disorder, again characterized by round cell infiltration into the myocardium, but the distribution can be focal or diffuse [2]. The intensity of the inflammatory process is usually less impressive, and viral neutralizing titers are less consistent. The clinical presentation is one of a dilated cardiomyopathy of recent onset, with or without preceding viral illness, and the diagnosis is made by screening endomyocardial biopsy or at autopsy [33]. Because of the presence of myocardial inflammation, manipulation of the immune system or use of anti-inflammatory agents has been attempted in both of these scenarios, with inconsistent results. This paper reviews both the experimental and clinical data available on immunosuppressive (and immunomanipulative) treatment of myocarditis.

Acute Viral Versus Chronic Idiopathic Myocarditis

To best review results of therapy in myocarditis, first it is necessary to define and differentiate between the acute myocardial inflammation accompanying an acute viral infection and the more chronic indolent form of myocardial inflammation. As previously noted, acute viral myocarditis presents during or shortly after an acute viral syndrome, usually respiratory [16]. The incidence of myocardial involvement with an acute viral illness is difficult to ascertain from the available literature. In one study of military recruits with acute respiratory illnesses following the outbreak on an influenza A epidemic, approximately 8% of recruits had electrocardiographic evidence of myocarditis [19]. In all of these patients, echocardiographic evidence of wall motion abnormalities were also present.

The manifestations of acute myocarditis have been best studied in patients with Coxsackie B viral infections; however, most series of acute myocarditis present a mixture of etiologic agents [29]. In one older study of 60 patients with "myoperi-

H.-P. Schultheiß (Ed.)
New Concepts in Viral Heart Disease

carditis," myocardial involvement was manifest by chest pain. Pericardial pain represented 32% of all cases and "coronary mimic" or angina-like pain represented an additional 40%. Only 5% of the patients presented with congestive heart failure [13]. In another study of 185 patients with respiratory infection and abnormal electrocardiograms, 104 patients were felt to have these ECG abnormalities on the basis of acute myopericarditis, and of these, approximately one-half had elevations in neutralizing antibody confirming the acute infection. Again, myocardial dysfunction as manifest by a third heart sound or an enlarged cardiac silhouette, although present in a minority of patients (32% and 22% respectively), was not associated with overt clinical congestive heart failure in any of the patients studied [16].

That myocardial inflammation can be a chronic process was first recognized by Kelle in 1892 [20]. Later Boikan [4] termed this entity "pernicious myocarditis" to emphasize the chronic indolent nature of this disease. The most complete early series of chronic myocarditis which included a review of the literature as well as new cases was presented by Kline and Saphir in 1960 [23]. In contrast to the myocardial disease described above, the etiology, pathophysiology, and natural history of chronic idiopathic myocarditis is less well understood. The most popular hypothesis is that following an initial insult to the myocardium, most likely from an acute viral infection, a more chronic inflammatory process develops depending on a variety of host factors. The chronic inflammatory process ultimately progresses to myocardial cell loss, fibrosis, and dilation. This schema is shown in Fig. 1.

In contrast to acute viral myocarditis, chronic idiopathic myocarditis is usually not associated with signs of systemic inflammation and most commonly presents as unexplained congestive heart failure of recent onset in a younger individual

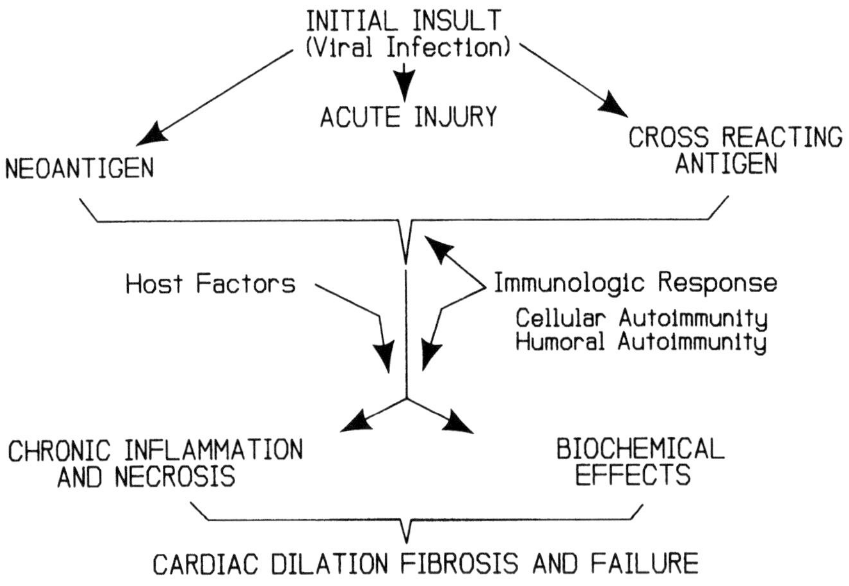

Fig. 1. The hypothesis for the development of chronic myocarditis and subsequent dilated cardiomyopathy. The initial insult, presumably a viral infection, results in an antigenically changed myocardium which elicits an autoimmunologic response to the heart

Table 1. Clinical charateristics of patients with dilated cardiomyopathy and myocarditis

Parameter	DCM		MYO		All		P^a
Age (years)	45 ± 11	(60)	31 ± 15	(22)	43 ± 12	(82)	.0004
M/F *(n)*	43/17		10/12		53/29		.0274
Duration of symptoms (months)	25 ± 56	(60)	4 ± 9	(22)	19 ± 48	(82)	.0081
Heart rate (bpm)	90 ± 18	(44)	102 ± 19	(19)	93 ± 19	(63)	.0226
Pulmonary wedge (mmHg)	20 ± 10	(42)	20 ± 8	(20)	20 ± 10	(62)	NS
Cardiac Index (liters per min per m^2)	2.5 ± .9	(29)	2.7 ± .7	(17)	2.6 ± .8	(46)	NS
Left ventricular end diastolic dimension (mm)	67 ± 13	(24)	55 ± 15	(17)	62 ± 15	(41)	.011
Ejection fraction (%)	27 ± 14	(27)	31 ± 15	(12)	29 ± 15	(39)	NS

[a] Probability comparing myocarditis to dilated cardiomyopathy.
DCM, dilated cardiomyopathy; MYO, myocarditis; All, both gropus combined; NS, not significant; ( ), number of patients with data available

[17]. Table 1 presents data from the Oregon Health Sciences University comparing patients with dilated cardiomyopathy to those with the histologic diagnosis of chronic idiopathic myocarditis. Although several features such as age, duration of illness, and left ventricular chamber size are different among groups, the considerable overlap between groups make histologic documentation necessary for the diagnosis of myocarditis in these patients.

Immunosuppressive Therapy for Acute Viral Myocarditis

More than 30 years ago, Kilbourne et al. [21] reported that treating mice infected with Coxsackie B3 virus with cortisone resulted in a marked increase in the severity of the inflammation and myocardial necrosis seen on histology. This increased pathogenicity was confirmed by Woodruff [42] who demonstrated that corticosteroid therapy increased mortality in a dose-response fashion. A defect in mononuclear cell migration was associated with higher viral titers and a delay in clearing the virus. This increase in mortality and histopathologic myocardial changes in animal models of Coxsackie viral myocarditis is not unique to corticosteroids. Rager-Zisman and Allison [36] reported increased mortality, prolongation of the ability to isolate virus, and increased levels of interferon production when Coxsackie B3-infected mice were pretreated with cyclophosphamide. Most recently, cyclosporine has been shown to increase mortality in both Coxsackie B3 murine models of myocarditis [32] and in myocarditis produced by the encephalomyocarditis virus in DBA-2 mice [28].

In addition to the more conventional immunosuppressive agents exacerbating viral myocarditis, a recent report suggests that nonsteroidal anti-inflammatory agents can produce similar effects. Costanzo-Nordin et al. [6] treated BALB/c mice with buprofin 7 days following infection with Coxsackie B3 virus. This group demonstrated that, although viral cultures were negative in both the buprofin-treated and control animals, the degree of necrosis was significantly greater in the animals treated with the nonsteroidal agent. They postulated that the mechanism of immunomodulation may be a block in mononuclear cell release of prostaglandin which may increase interleukin-2 elaboration and intensify the immunologic response.

Despite the almost uniformly negative results using immunosuppressant or anti-inflammatory therapy in animal models of viral myocarditis, this has not dissuaded investigators from attempting other manipulations of the immune system in this disease. Matsumori et al. [26] demonstrated in a recent study that therapy with intravenous interferon reduced the mortality in mice infected with Coxsackie B3. Most recently Kishimoto and Abelmann [22] demonstrated that treating C3H/He mice with monoclonal antibodies to both helper/inducer (Lyt1) and suppressor/cytotoxic (Lyt2) T lymphocytes reduced Coxsackie B3-induced myocardial inflammatory lesions. This protective effect was seen when the monoclonal antibodies were given either during the viremic stage or the postviremic stage. Of further interest is that the protective effect was lost if either anti-Lyt1 or anti-Lyt2 were used alone.

The use of immunosuppressive therapy in patients with acute myocarditis has little experience to draw from. There are case reports and small series of patients with acute viral myocarditis who have been treated with immunosuppression. These patients are summarized in Table 2, and in all cases were treated with either corticosteroids or adrenocorticotrophic hormone (ACTH), [1, 14, 15, 37, 38]. In contrast to the uniformily negative results in animals treated with corticosteroids, the results in patients were more heterogeneous with several of the patients

Table 2. Immunosuppressive therapy for acute viral myocarditis

First author	Age	Sex	Virus	Therapy	Outcome
Sanyai[37]	13	F	Coxsackie B4	Prednisone	Died
Ainger[1]	3	M	Influenza A2	Prednisolone	Survived, improved
	.5	M	Influenza A2	Prednisolone	Survived, improved
	1.5	M	Influenza A	Prednisolone	Survived, improved
	3.5	M	Influenza A	Prednisolone	Survived, improved
	4	F	Influenza B	Prednisone	Survived, improved
	2	F	Influenza B	Prednisolone	Survived, improved
	2.5	F	Influenza B	Prednisolone	Survived, improved
	6	F	Influenza B	Meth-Prednisolone	Survived, improved
Garrison[14]	7	M	Unknown	ACTH	Survived, improved
Glajchen[15]	24	F	Coxsackie B1	Prednisone	Deteriorated
Sutinen[38]	15	F	Coxsackie B3	"Steroids"	Died

appearing to have responded to therapy. However, primarily based on the previously presented animal data, clinicians have been appropriately conservative when approaching these patients, and immunosuppressive therapy in general has been avoided, especially during the presumed viremic stage.

Immunosuppressive Therapy for Chronic Idiopathic Myocarditis

Mason et al. [25] reported a small series of patients with a clinical presentation of dilated cardiomyopathy who on endomyocardial biopsy demonstrated a chronic inflammatory picture consisting of a round cell infiltrate and myocardial necrosis. Because of the histologic similarity to cardiac allograft rejection, these patients were treated with immunosuppression consisting of prednisone and azathioprine or prednisone alone. Of the ten patients treated, five patients improved, and four patients stabilized on therapy. Following this report, the enthusiasm for finding and treating chronic myocarditis in patients with dilated cardiomyopathy increased significantly as evidenced by the multiple reports of incidence and therapy which appeared in the medical literature in the subsequent 5 years [3, 5, 7, 8, 10, 18, 24, 30, 31, 35, 40, 43]. Based on these studies, the overall incidence of myocarditis in patients with dilated cardiomyopathy appears to be approximately 10%–15%; however, this incidence varies widely. A summary of immunosuppression trials are displayed in Table 3 [7–10, 18, 25, 34, 43]. Although an average 46% improvement rate is reported, there are several problems with these data. First, as evidenced by the incidence differing from 0.5% [31] to 63% [43], there are clearly differences in patient selection, epidemiologic factors, or the criteria utilized for the histologic diagnosis. Most easily identified are the differences in histologic definition of myocarditis ranging from any lymphocytes in the myocardial interstitium [43] to a prescribed number of lymphocytes per high-power field [9], to the

Table 3. Immunosuppressive therapy in chronic moycarditis

First author	Total treated	Improved (*n*)	(%)
Mason[25]	10	5	50
Edwards[9]	4	2	50
Fenoglio[10]	19	8	42
Daly[7]	9	7	78
O'Connell[34]	15	6	40
Zee-Cheng[43]	11	5	45
Dec[8]	9	4	44
Hosenpud[18]	6	1	17
	84	38	46

Table 4. Spontaneous improvement in dilated cardiomyopathy myocarditis

First author	Total studied	Improved (*n*)	(%)
McDonald[27]	31	19	61
Fuster[12]	104	18	17
O'Connell[34]	20	6	30
Dec[8]	18	6	33
Figulla[11]	18	7	39
Taliercio[39]	24	6	25
	215	62	29

most conservative definition requiring evidence of myocardial damage associated with an inflammatory infiltrate [2].

Other problems include the small numbers of patients studied, the nonuniformity of immunosuppressive protocols and the inconsistent definition of "improvement" ranging from clinical symptomatic improvement to documentation of systolic performance. Furthermore, as none of the studies were controlled or randomized, the incidence of spontaneous improvement or improvement from conventional therapy could not be taken into account. Table 4 summarizes data which demonstrate that a substantial incidence of spontaneous improvement is present both in patients with dilated cardiomyopathy with or without myocardial inflammation demonstrated on myocardial biopsy [11, 12, 27, 34, 39].

To try to better address the issue of immunosuppression in chronic idiopathic myocarditis, a multicenter trial, organized through the University of Utah and funded by the National Heart, Lung, and Blood Institute of the National Institutes of Health, is currently in progress. This study is designed in two major parts, a clinical therapeutic trial of immunosuppression in biopsy-proven myocarditis and an immunologic study investigating mechanisms of the disease.

In order to have uniform criteria for the histologic diagnosis of myocarditis, a pathology panel was convened to develop specific histologic criteria for diagnosing myocarditis for the purposes of the study. These criteria have been subsequently published [2]. Patients with congestive heart failure of recent onset and of unknown etiology (no coronary or significant valvular disease) are screened for histologic evidence of myocarditis. Those with myocarditis are randomly assigned to receive conventional therapy consisting of digitalis, diuretics, and vasodilators, or conventional therapy plus immunosuppression. The current immunosuppression protocol used in the study is a combination of cyclosporine and prednisone. Patients are treated for a total of 6 month and followed for a total of 1 year. Specific end points of the study are mortality, left ventricular systolic function (radionuclide angiography), and treadmill exercise duration.

The second goal of the study is to investigate the pathogenic mechanisms in chronic myocarditis. In addition to extensive virologic surveillance, both cellular and humoral immunity are being studied. Circulating lymphocyte subset analysis and characterization of the inflammatory cells within the myocardium are being performed. Studies of humoral immunity involve the isolation and definition of circulating antibodies to myocardium and a determination of specific myocardial antigens that serve as immunologic targets.

It is hoped that the results of this ongoing study, now 2 years in progress, will define both pathogenic mechanisms of chronic idiopathic myocarditis as well as the role of immunosuppressive therapy in this disease.

Summary

Between 10% and 15% of patients presenting with a recent onset of congestive heart failure and a clinical dilated cardiomyopathy will have evidence of a chronic inflammatory process involving the myocardium. Although a history of a "viral syndrome" is frequently elicited, this historical data is difficult to separate from symptoms solely related to the onset of heart failure. There is, however, evidence that this chronic myocarditis is an autoimmune sequela of an acute viral myocarditis.

Immunosuppressive therapy in acute viral myocarditis in patients is of unclear benefit. In murine Coxsackie viral myocarditis there is evidence that immunosuppressive or anti-inflammatory therapy (corticosteroids, cyclosporine, nonsteroidal anti-inflammatory agents) may prolong viral infection and increase the severity of myocardial necrosis.

Approximately 100 patients with chronic myocarditis have thus far been treated with varying regimens of corticosteroids, azathioprine, and cyclosporine. The overall response rate is approximately 50%. The difficulty with this grouped data is that the series reported to date have been small, uncontrolled, and have not defined improvement in a uniform way. Moreover, the spontaneous improvement in patients with myocarditis may be as high as 30%. In a small series of patients studied at the Oregon Health Sciences University, a 6-month course of azathioprine and prednisone resulted in no significant change in echocardiographic indices of cardiac function. There were, however, significant side effects from the immunosuppressive therapy in half of the patients treated.

Based on the lack of conclusive evidence supporting immunosuppressive therapy in chronic myocarditis, a randomized multicenter study is currently in progress. This study will compare the efficacy of prednisone/cyclosporine to conventional therapy in biopsy-proven myocardits. End points include mortality, left ventricular ejection fraction, and clinical status including exercise tolerance. Results of this study should clarify the role of immunosuppressive therapy in myocarditis.

References

1. Ainger LE (1964) Acute aseptic myocarditis: corticosteroid therapy. J Pediatr 64: 716–723
2. Aretz HT, Billingham ME, Edwards WD, Factor SM, Fallon JT, Fenoglio JJ, Olsen EGJ, Schoen FJ (1986) Myocarditis: a histopathologic definition and classification. Am J Cardiovasc Pathol 1: 3–14
3. Baandrup U, Florio RA, Rehahn M, Richardson PJ, Olsen EGJ (1981) Critical analysis of endomyocardial biopsies from patients suspected of having cardiomyopathy. II: Comparison of histology and clinical/hemodynamic information. Br Heart J 45: 487–493
4. Boikan WS (1931) Myocarditis perniciosa. Virchows Arch 282: 46–66
5. Bolte HD, Ludwig B (1984) Viral myocarditis: symptomatology, clinical diagnostics and haemodynamics. In: Bolte HD (ed) Viral heart disease. Springer, Berlin Heidelberg New York, pp 177–187
6. Costanzo-Nordin MR, Reap EA, O'Connell JB, Robinson JA, Scanlon PJ (1985) A non-steroidal anti-inflammatory drug exacerbates Coxsackie B3 murine myocarditis. J Am Coll Cardiol 6: 1078–1082
7. Daly K, Richardson PJ, Olsen EGJ, Morgan-Chapner P (1984) Acute myocarditis: role of histological and virological examination in the diagnosis and assessment of immunosuppressive treatment. Br Heart J 51: 30–35
8. Dec GW Jr, Palacios IF, Fallon JT, Aretz HT, Mills J, Lee DCS, Johnson RA (1985) Active myocarditis in the spectrum of acute dilated cardiomyopathies. N Engl J Med 312: 885–890
9. Edwards WD, Holmes DR, Reeder GS (1982) Diagnosis of active lymphocytic myocarditis by endomyocardial biopsy. Quantitative criteria for light microscopy. Mayo Clin Proc 57: 419–425
10. Fenoglio JJ, Ursell PC, Kellogg CF, Drusin RE, Weiss MB (1983) Diagnosis and classification of myocarditis by endomyocardial biopsy. N Engl J Med 308: 12–18
11. Figulla HR, Rahlf G, Nieger M, Luig H, Kreuzer H (1985) Spontaneous hemodynamic improvement or stabilization and associated biopsy findings in patients with congestive cardiomyopathy. Circulation 71: 1095–1104
12. Fuster V, Gersh BJ, Giuliani ER, Tajik AJ, Brandenburg RO, Frye RL (1981) The natural history of idiopathic dilated cardiomyopathy. Am J Cardiol 47: 525–531
13. Gardiner AJS, Short D (1973) Four faces of acute myopericarditis. Br Heart J 35: 433–442
14. Garrison RF, Swisher RC (1953) Myocarditis of unknown etiology (Fiedler's?) treated with ACTH. J Pediatr 42: 591–599
15. Glajchen D (1961) Myocarditis due to coxsackie virus infection in an adult. Br Med J 2: 870–871
16. Heikkila J, Karjalainen J (1982) Evaluation of mild acute infectous myocarditis. Br Heart J 47: 381–391
17. Hosenpud JD (1988) Chronic idiopathic myocarditis: controversies in causes and therapy. Cardiovasc Rev Rep (in press)
18. Hosenpud JD, McAnulty JH, Niles NR (1985) Lack of objective improvement in systolic function in patients with myocarditis treated with azathioprine and prednisone. J Am Coll Cardiol 6: 797–801
19. Karjalainen J, Nieminen MS, Heikkila J (1980) Influenza A1 myocarditis in conscripts. Acta Med Scand 207: 27–31
20. Kelle K (1892) Über primäre chronische Myokarditis. Dtsh Arch Klin Med 49: 442–456
21. Kilbourne ED, Willson CB, Perrier D (1956) The induction of gross myocardial lesions by a coxsackie (pleurodynia) virus and cortisone. J Clin Invest 35: 362–365
22. Kishimoto C, Abelmann WH (1988) Monoclonal antibody therapy for prevention of acute coxsackie myocarditis in mice. J Am coll Cardiol 11: 196A (Abstract)
23. Kline IK, Saphir O (1960) Chronic pernicious myocarditis. Am Heart J 59: 681–697
24. Kunkel B, Lapp H, Kober G, Kaltenbach M (1978) Correlations between clinical and morphologic findings and natural history in congestive cardiomyopathy. In: Kaltenbach M, Loogen R, Olsen EGJ (eds) Cardiomyopathy and myocardial biopsy. Springer, Berlin Heidelberg New York, pp 771–783

25. Mason JW, Billingham ME, Ricci DR (1980) Treatment of acute inflammatory myocarditis assisted by endomyocardial biopsy. Am J Cardiol 45: 1037–1044
26. Matsumori A, Crumpacker CS, Abelmann WH (1987) Prevention of viral myocarditis with recombinant human leukocyte interferon alpha A/D in a murine model. J Am Coll Cardiol 9: 1320–1325
27. McDonald CD, Burch GE, Walsh JJ 1972) Prolongend bed rest in the treatment of idiopathic cardiomyopathy. Am J Med 52: 41–50
28. Monrad ES, Matsumori A, Murphey JC, Fox JG, Crumpacker CS, Abelmann WH (1986) Therapy with cyclosporine in experimental murine myocarditis with encephalomyocarditis virus. Circulation 73: 1058–1064
29. Nieminen MS, Heikkila J, Karjalainen J (1984) Echocardiography in acute infectious myocarditis: relation to clinical and electrocardiographic findings. Am J Cardiol 53: 1331–1337
30. Nippoldt TB, Edwards WD, Holmes DR, Reeder GS, Hartzler GO, Smith HC (1982) Right ventricular endomyocardial biopsy. Clinicopathologic correlates in 100 consecutive patients. Mayo Clin Proc 57: 407–418
31. Noda S (1980) Histopathology of endomyocardial biopsies from patients with idiopathic cardiomyopathy: quantitative evaluation based on multivariate statistical analysis. Jp Circ J 44: 95–116
32. O'Connell JB, Reap EA, Robinson JA (1986) The effects of cyclosporine on acute murine coxsackie B3 myocarditis. Circulation 73: 353–359
33. O'Connell JB, Robinson JA, Gunnar RM, Scanlon PJ (1985) Clinical aspects of virus/immune myocarditis. Heart Vessels 1: 102–106
34. O'Connell JB, Robinson JA, Henkin RE, Gunnar RM (1981) Immunosuppressive therapy in patients with congestive cardiomyopathy and myocardial uptake of gallium-67. Circulation 64: 780–786
35. Parillo JE, Aretz HT, Palacios I, Fallon JT, Block PC (1984) The results of transvenous endomyocardial biopsy can frequently be used to diagnose myocardial disease in patients with idiopathic heart failure: endomyocardial biopsies in 100 consecutive patients revealed a subtantial incidence of myocarditis. Circulation 69: 93–101
36. Rager-Zisman B, Allison AC (1973) Effects of immunosuppression on coxsackie B-3 virus infection in mice, and passive protection by circulating antibody. J Gen Virol 19: 339–351
37. Sanyai SK, Mahdavy M, Gabrielson MO, Vidone RA, Browne MJ (1965) Fatal myocarditis in an adolescent caused by coxsackie virus group B type 4. Pediatrics 35: 36–41
38. Sutinen S, Pohjonen KR, Vastamaki R (1971) Fatal generalized coxsackie B3 virus infection in an adolescent with successful isolation of the virus from pericardial fluid. Ann Clin Res 3: 241–246
39. Taliercio CP, Seward JB, Driscoll DJ, Fisher LD, Gersh BJ, Tajik AJ (1985) Idiopathic dilated cardiomyopathy in the young: clinical profile and natural history. J Am Coll Cardiol 6: 1126–1131
40. Unverferth DV, Fetters JK, Unverferth BJ, Leier CV, Magorien RD, Arn AR, Baker PB (1983) Human myocardial histologic characteristics in congestive heart failure. Circulation 68: 1194–1200
41. Woodruff JF (1980) Viral myocarditis: a review. Am J Pathol 101: 426–478
42. Woodruff JF (1979) Lack of correlation between neutralizing antibody production and suppression of coxsackie B-3 replication in target organs: evidence for involvement of mononuclear inflammatory cells in host defence. J Immunol 123: 31–36
43. Zee-Cheng C, Tsai CC, Palmer DC, Codd JE, Pennington DG, Williams GA (1984) High incidence of myocarditis by endomyocardial biopsy in patients with idiopathic congestive cardiomyopathy. J Am Coll Cardiol 3: 63–70

Immunosuppressive Treatment in Myocarditis

F. Camerini[1], A. Salvi[1], R. Bussani[2], L. Dreas[1], A. Di Lenarda[1], E. Hrovatin[1], G. Sinagra[1], and F. Silvestri[2]

Departments of [1]Cardiology and [2]Pathology, Ospedale Maggiore and University of Trieste Trieste, Italy

Introduction

"Myocarditis" is defined "as a process characterized by an inflammatory infiltrate of the myocardium with necrosis and/or degeneration of adjacent myocytes not typical of the ischemic damage associated with coronary artery disease" [1]. A viral etiology is generally suspected in most cases observed in the western hemisphere, but it is very difficult to prove a viral pathogenesis in human myocarditis. The disease is, therefore, very often referred to as "idiopathic myocarditis."

Confusion has arisen in the past concerning the actual incidence of myocarditis in patients in whom the disease was clinically diagnosed. The development of the endomyocardial bioptome in 1962 [2] led to major improvement in the diagnosis, but it soon appeared that different histologic criteria were used by cardiac pathologists, as has been clearly demonstrated by Shanes et al. [3]. This may explain why the incidence of myocarditis in patients with idiopathic congestive heart failure, as reported in the literature, ranged from 6% [4] to 63% [5].

Quantitative criteria based on the mere presence of lymphocytes in the myocardium [6] are presently considered as not sufficient for a diagnosis of myocarditis since the need for coexistence of both an inflammatory infiltrate and necrosis of myocytes is unanimously recognized [1].

Therefore, the very high reported incidence of myocarditis in some patients with "idiopathic congestive cardiomyopathy" [5] is likely to reflect an overdiagnosis of myocarditis. Myocarditis was diagnosed by Zee-Cheng et al.[5] in 63% of cases with congestive heart failure of unknown cause, and this figure is well correlated with the 55% incidence of "foci of inflammation consisting solely of mononuclear cell" identified by Tazelaar and Billingham [7] in an unselected series of patients with dilated cardiomyopathy.

In order to overcome these discrepancies, the need for a uniform diagnostic and classification system was felt, which led to the definition of the histologic criteria suggested for the diagnosis of myocarditis, known as the Dallas criteria [1].

A viral implication in human myocarditis, although strongly suspected, has not been clearly demonstrated yet. The presence of viral particles in myocardial tissue of patients with myocarditis has only rarely been reported [8, 9]. However, Bowles et al. [10] were recently able to demonstrate the presence of coxsackievirus nucleic acid sequence in endomyocardial biopsy of patients with myocarditis. Such findings, although preliminary, clearly identify the need for contiued studies of

H.-P. Schultheiß (Ed.)
New Concepts in Viral Heart Disease

coxsackievirus-induced myocarditis in experimental animals to increase our knowledge of the temporal sequences of immunopathologic events that occur following virus infections.

A murine model of virus-induced myocarditis was developed by Woodruff using inbred BALB/c mice and a heart-adapted cardiotropic strain of coxsackievirus, group B, type 3 [11]. Virus can be isolated from the heart tissue within 24 h from intraperitoneal injection and is usually not detectable before the end of the 1st week.

At the time of the maximal amounts of virus in the heart no inflammatory cell infiltrate is evident. Inflammation begins 5 days after virus inoculation and is most prominent on the 9th and 10th days. Thereafter inflammation and cardiac necrosis persist for days to weeks after infecting virus is no longer detected in the heart. Woodruff [11] also demonstrated the dependency of cardiac injury on functional T lymphocytes in the mice.

Since these initial studies, other investigators have experienced varying degrees of success in demonstrating the importance of T cells in the pathogenesis of myocarditis [12]. Interestingly, in this experimental model cytolytic lymphocytes lyse both infected and uninfected myocytes in vitro. This could be explained either by a single cytolytic cell lysing both targets, or by two distinct cytolytic populations, one of which lyses infected and the other uninfected myocytes. The second hypothesis implies that autoimmunity is present in this experimental model.

From the data obtained in different experimental models of coxsackievirus B-infected mice, it may be suggested that the evolution of coxsackievirus B3 heart disease can be divided into three phases [12]. Very early in the infection, virus is present in the heart and may cause minimal myocyte damage in the absence of the inflammatory cell infiltrate. Later, an inflammatory cell infiltrate appears. This is predominantly of virus-specific cytolytic lymphocytes which lyse infected heart cell. Subsequently "autoimmune" T lymphocytes appear. This concept is based on an animal model in which virus is no longer detectable at the time of the appearance of autoimmune T cells.

However, there is no evidence that the same pathogenetic mechanism is present in human myocarditis, and, indeed, the recent observations of Bowles et al. [10], who detected viral nucleic acid in myocardial tissue of patients in the late stage of myocarditis, might suggest that the pathogenesis of human myocarditis is different from what has been observed in the inbred BALB/c mice infected with coxsackievirus B3.

The histologic aspect of allograft rejection after heart transplantation has been well defined [13]. The increasing use of the endomyocardial bioptome as a diagnostic tool, particulary in centers with an extensive experience in heart transplantation, led to a more frequent diagnosis of myocarditis in vivo and the similarities between the two histologic aspects were readily apparent.

The concept of an immune-mediated damage, derived from the animal models of myocarditis, and the well-defined autoimmune mechanism which is present in human allograft rejection were also considered. This led to the suggestion that the immunosuppressive treatment, used to control rejection, might be beneficial in human myocarditis too [14].

Immunosuppressive Therapy (Table 1)

Anti-inflammatory and immunosuppressive treatment has often been considered detrimental in the acute phase of myocarditis. Experimental data in animals have indeed suggested that corticosteroids given at an early stage of viral infection may increase both the severity of myocardial damage and the incidence of lethal disease in mice [15].

These data, associated with many others, led to the conclusion that corticosteroids are contraindicated in the early stages of viral myocarditis.

However, later studies [16] showed that steroids (in an experimental model of mice infected with encephalomyocarditis virus) may have different effects if given at different stages of viral myocarditis. They may in fact aggravate the course of the disease if administered when neutralizing antibody titers are scarce, but have no detrimental effects if given (on days 8–17) when the titer level of neutralizing antibodies is high. However, a positive beneficial effect is not obtained in this experimental model [16].

Similar results were obtained in mice infected with coxsackievirus B3 [17] or encephalomyocarditis virus [18] and treated with cyclosporine. In these animals, cyclosporine, when given in the early period of viral replication or shortly thereafter, may increase morbility and mortality, presumably by altering viral clearance and blunting the inflammatory response [17]. However, in the 1950s and 1960s some clinical observations suggested that anti-inflammatory treatment could be useful in patients with a clinical suspicion of acute myocarditis. In fact, Garrison and Swisher [19] treated a 7-year-old patient with a severe heart failure (considered as the result of an acute myocarditis) with adrenocorticotrophic hormone (ACTH). There was an early significant improvement, although relapse occurred after the end of the treatment. A new course of ACTH and digitalis caused a rapid improvement.

Only 11 years later did Ainger [20] describe eight infants and children with acute aseptic myocarditis associated with serologic evidence of recent infection with various types and strains of influenza virus. The clinical presentation was characterized by septic fever, myalgia, leukocytosis, rash, and congestive heart failure of sudden onset, which proved intractable to vigorous anticongestive therapy. All patients were treated with prednisolone (or prednisone or methylprednisolone) 1–2.2 mg/lb/day for 1–2 weeks. This dosage was then gradually reduced in a stepwise manner. The total duration of the therapy was 42–145 days. There was rapid (24–48 h) improvement with disappearance of "all signs of congestive heart failure." In no case there were signs of enhanced viral dissemination. Ainger [20] considered adrenocorticosteroid therapy a life-saving therapeutic adjunct in patients refractory to usual anticongestive treatment. The potent anti-inflammatory properties of corticosteroids were considered responsible for the improved myocardial function.

Other isolated cases of acute myocarditis in which a therapy with corticosteroids seemed to be useful or perhaps life saving are those reported by Voigt [21]. In this series of patients there was no biopsy demonstration of myocarditis, and the possibility of a pericardial effusion with tamponade could not be categorically ruled out.

Table 1

Reference	Patients (n)	Improved	Therapy	Biopsy	Specimens (n)	Duration of Treatment	Duration of Symptoms	Comments
[19]	1	1	ACTH	No				
[20]	8	8(100%)	Prednisone, prednisolone, methylpredni-solone	No		42–145 days		
[21]	1	1	"Steroid" therapy	No		?	?	
[14]	10	5 (50%) (but 1 died of pulmonary infection, 1 postoperatively)	Azathioprine + prednisone In 2 patients only prednisone	Yes	3–4	Up to 1 year	4 weeks – 6months	Two patients had a clinical and histologic worsening after discontinuation of therapy. Three patients had no inflammatory infiltrate but IgG myocardial infiltration (on immunofluorescence)
[24]	3	2 1 died	"Steroids"	Yes				Recent history of an acute "flu-like" disease
[38]	15	In 6 improvement of myocardial gallium 67 uptake and in ejection fraction In 9 gallium scans remained positive and diminution of ejection fraction	Azathioprine + prednisone	No		8 months or more	26.7 (1–180 months)	gallium 67 myocardial up-take considered as an in-dicator of "myocarditis"; biopsies were not made
[37]	3	3	Azathioprine + prednisone	Yes	3 or more	6 months (in 2 patients)	Few weeks (in 2 patients)	"peripartum" cardio-myopathy
[6]	5	3	4 prednisone + azathioprine 1 prednisone	Yes	4 (at least)	< 1 months – some months	Days – > 6 months	Histologic criteria of myocarditis no more accepted

Table 1. (Continued)

Reference	Patients (*n*)	Improved	Therapy	Biopsy	Specimens (*n*)	Duration of Treatment	Duration of Symptoms	Comments
[23]	4 Acute	0 (0%)	Prednisone or azathioprine + prednisone	Yes	3–4	?	1.9 ± 1.1 months	Three patients refused further studies. In one case after 55 months and in 1 after 60 months
	7 Rapidly progressive	0 (0%)	Immunosuppressive treatment				22.8 ± 12.3 months	deterioration of ejection fraction, persistence of inflammatory infiltrate and focal cell damage and increase of fibrosis
	7 Chronic	7 (100%)	Immunosuppressive treatment				39.2 ± 13.7 months	
[31]	6	6 Significant diminution of inflammation and significant increase of ejection fraction	Antithymocyte globulin; pulsed prednisolone; prednisone; azathioprine	Yes	?	1 month (?)	Less then 12 months	short-term follow-up
[22]	14	6 (43%) improved and resolved; 3 improved but died (2 of pneumonia, 1 postoperatively); 1 died (CHF) 4 little changes but no progression	Azathioprine + prednisone In 2 cases prednisone alone	Yes	3 or more	6 months or more (if well tolerated)	?	All held inflammatory infiltrate consistent with myocarditis
[5]	10	Chronic active myocarditis: in 4 improvement or stabilization in clinical function, hemodynamic status and biopsy grade; 1 died. Subacute myocarditis: 1 definite improvement, 1 symptomatic improvement	Azathioprine + prednisone	Yes	3–6	?	1 week-4 months	Four grades of myocarditis were considered: grade I (characterized by focal inflammatory infiltrates but without necrosis) and grade II were considered low grade active lymphocytic myocarditis

Table 1. (Continued)

Reference	Patients (n)	Improved	Therapy	Biopsy	Specimens (n)	Duration of Treatment	Duration of Symptoms	Comments
[27]	2	2 Case 1: regression of inflammation, progressive symptomatic improvement Case 2: regression of necrosis and of inflammatory edema, persistence of cellular infiltrate, clinical improvement and of ejection fraction	Azathioprine + prednisone	Yes	6 (in 1 patients)	6 months	3–6 weeks	In both cases atrial flutter with high ventricular rate
[32]	6	5 (at least 1 functional class) Histology: reduction of the inflammatory infiltrate and no further active necrosis 2 of these 5 died after 10 and 16 months after a sudden deterioration	Azathioprine +prednisone	Yes	?	6–16 months	1–9 years	In some cases long duration of the disease; clinical improvement (NYHA functional class) but echocardiographic left ventricular size and function did not improve
[33]	7	5 had functional and hemodynamic improvement no histologic improvement	Azathioprine + prednisone	Yes	?	7 ± 2 weeks	4 months – 4 years	Clinical and hemodynamic improvement may occur without histologic improvement on biopsy
[34]	1	1 (histologic and functional improvement)	Azathioprine + prednisone	Yes	?	More than 7 months	6 months	High dosage of prednisone (120 mg per day) and of azathioprine (200 mg per day)
[25]	9	7 (in 4 persistent improvement; in 1 reactivation of myocarditis; in 2 later progressive deterioration;	Azathioprine + prednisone or prednisone alone	Yes	3 (average)	6 months or more	1–16 weeks (mean 11 weeks)	Systematic clinical and hemodynamic study

Table 1. (Continued)

Reference	Patients (*n*)	Improved	Therapy	Biopsy	Specimens (*n*)	Duration of Treatment	Duration of Symptoms	Comments
[26]	6	6 Histology; absence or marked reduction of inflammation; absence of inflammation after discontinuation of therapy in 5 cases. In 4 of 5 patients ventricular tachycardia or ventricular fibrillation could not be provoked in laboratory. Clinical arrhythmias abolished in 5 patients	Azathioprine + prednisone	Yes	3–6	At least	? 6 months	Systematic clinical hemodynamic study; biopsy; histologic diagnosis of lymphocytic myocarditis; electrophysiological study in five patients
[28]	9	Histology disappearance of inflammation in 3, improvement in 3 Clinical and physiologic improvement in 3 Clinical recrudescence (in 4 patients) and histological worsening (in 2 patients) when immunosuppressive drugs were tapered or discontinued	In 7 patients azathioprine + prednisone In 2 patients prednisone alone	Yes	3–5 (usually)	6 months or less	6 months	Different dosage of prednisone in the first two patients; immunosuppressive drugs given also to one patient with negative biopsy
[29]	6	Histology: diminution of the degree of inflammation Significant diminution of mean heart rate and of pulmonary wedge pressure but no changes of	Azathioprine + prednisone	Yes	5 or more	3 months	1–55 months	

Table 1. (Continued)

Reference	Patients (*n*)	Improved	Therapy	Biopsy	Specimens (*n*)	Duration of Treatment	Duration of Symptoms	Comments
		cardiac index, end-diastolic diameter and fractional shortening						
[40]	17	9 "improved" 5 did not improve 3 died	Azathioprine + prednisone	Yes	?	6 months	Short in most patients (2–8 weeks)	Control biopsies performed in a minority of cases
[30]	12	Clinical improvement in 8	Prednisone or prednisolone + (in 2 patients) azathioprine cyclosporine (1 patient)	Yes	5 (average)	1–29 months	1–84 months	
[36]	2	1 improved 1 died after 2 months or treatment	Prednisone	Yes	5–6	4 months (case 1) 2 months (case 2)	1 week (case 1) "short" (case 2)	Granulomatous process with giant cells; necrosis; mild increase of collagen
[35]	1	Histologic and clinical improvement	Azathioprine + prednisone	Yes	?	About 6 months	More than 3 weeks	Giant cell myocarditis
present study	20	In 17 clinical and histologic improvement In 11/17 patients improvement of ejection fraction	Azathioprine + prednisone	Yes	Usually 4 or 5	4 months ore more	Less than 5 months (in 17 patients)	Two patients, with very low ejection fraction, slowly deteriorated and died. One died after the end of treatment (cerebral hemorrhage) two died after 1 and 3 years (sudden death)

The first report in which patients with biopsy–proven myocarditis were treated with immunosuppressive drugs is that by Mason et al. [14]. These authors used prednisone and azathioprine in nine cases and prednisone alone in one. Five patients improved, in four there was no further progression, while one worsened and died of congestive heart failure. However, it has to be noted that of the five "improved patients" one died of pneumonia, probably conditioned by immunosuppressor, and one "postoperatively." The authors concluded that their findings constitute "circumstantial evidence that immunosuppressive therapy is effective in eliminating myocardial cell inflammation and thereby improving myocardial performance."

It may be useful to compare some other relevant points. An endomyocardial biopsy (two to four separate myocardial specimens) was carried out in all ten patients, but lymphocytic infiltrate associated to myocytolysis with focal necrosis was absent in three in whom the diagnosis of myocarditis was based on "a striking IgG myocardial infiltration documented by immunofluorescent staining."

Finally, in two cases there was a characteristic relationship between immunosuppressive treatment, presence of inflammatory infiltrates, and severity of heart failure. Withdrawal of treatment was in fact associated with reappearance of inflammatory infiltrates and recrudescence of heart failure, while reinstitution of therapy apparently caused regression of histologic signs of inflammation and clinical improvement.

Billingham and Mason [22] presented more extensive experience in 1984. Fourteen patients were treated. All had an inflammatory infiltrate consistent with myocarditis while the authors felt that a positive immunofluorescence (for sarcolemmal IgG) is a nonspecific change seen in response to myocyte degeneration and not specific for immunologic injury.

In this group of fourteen patients treated with immunosuppressives, six (43%) improved and resolved, three (21%) improved initially but later died (Table 1), one had a downhill course and died of congestive heart failure, while in four (29%) there were slight changes without evident progression of the disease. According to the authors, in repeated endomyocardial biopsies the reduction of cellular infiltrate correlated well with clinical improvement and reduction of heart failure and heart size.

The paper by Fenoglio et al. [23] is oriented toward a classification of different forms of myocarditis which can be accomplished by endomyocardial biopsy. These authors studied 34 cases with unexplained heart failure, which, on the basis of both clinical and histologic findings, were divided into three groups:

1. Acute myocarditis (seven patients), characterized by an interstitial inflammatory infiltrate and extensive acute cell damage.
2. Rapidly progressive myocarditis (18 patients), characterized by patchy acute and healing cell damage and fibrosis.
3. Chronic myocarditis (nine patients) in which scattered foci of acute and healing, extremely focal cell damage were present, associated with an inflammatory infiltrate.

Four patients with acute myocarditis were treated with immunosuppressive therapy without clinical or hemodynamic improvement, while, on the contrary, two recovered spontaneously. It has to be noted that repeated biopsies after 1 week of immunosuppressive treatment showed a decrease in the inflammatory cell infiltrate and evidence of healing cell damage (although acute cell damage was still prominent).

Immunosuppressive treatment was instituted in seven patients with rapidly progressive myocarditis. There was no improvement of cardiovascular function, although in four patients, who had a repeated endomyocardial biopsy 1 week to 1 month after the beginning of therapy, a diminution of the inflammatory infiltrate and an apparent disappearance of foci of acute cell damage were observed. On the contrary, all the seven cases of chronic myocarditis treated with immunosuppressive therapy had symptomatic improvement, although only four could be clinically revaluated and rebiopsied 1 month after the beginning of therapy. In these four cases, the ejection fraction increased from 23% to 39% with the disappearance of cell damage and of inflammatory cell infiltrates. However, in two patients, who had repeated biopsies at early intervals and who were followed up for 55 and 60 months, respectively, there was a deterioration of ejection fraction with persistence of an inflammatory infiltrate, focal cell damage, and an increase of fibrosis.

In 1980 Sekiguchi et al. [24] reported on 12 patients with heart failure of acute onset (sometimes with shock and/or with heart block) in whom a diagnosis of myocarditis had been made after an endomyocardial biopsy. Three patients were treated with steroids and of these, two recovered. In most cases, however, the treatment consisted only of temporary cardiac pacing, various vasodilators and cardiotonic agents. After 7–11 years all patients were in NYHA class I–II.

In 1982 Edwards et al. [6] reported on ten patients with lymphocytic myocarditis, five of whom were treated with immunosuppressors (mainly prednisone). There was no difference between patients who had received steroids and those who had not (in both groups three improved although the criteria of improvement were not clearly defined).

In 1984 Daly et al. [25] published the results obtained in nine patients, presenting with an acute onset of cardiac failure, who were treated with prednisolone and azathioprine for 6 months. The patients were restudied 1 month after withdrawal of therapy. In the case of persistent clinical and histologic recovery, no further treatment was given. Seven of these patients showed an appreciable "clinical and hemodynamic improvement" after 2 months with the histologic feature of healed myocarditis, while two had evidence of persistent active myocarditis without clinical or hemodynamic improvement (one was treated with prednisone alone). However, only four of the seven improved patients remained asymptomatic off medical treatment, while one showed reactivated myocarditis (with a good response to a new course of immunosuppression). On the contrary, two deteriorated to a clinical picture of severe dilated cardiomyopathy, despite continuing treatment.

It has to be noted that the study of Daly et al. [25] is probably one of the first in which a systematic clinical and hemodynamic evaluation was carried out.

The six patients described by Vignola et al. [26] were clinically characterized by the presence of a history of aborted sudden death, or of ventricular tachycardia (resistant to treatment), or of high-grade ventricular arrhythmias. Signs of lymphocytic myocarditis were present in all cases in the endomyocardial biopsy. A complete cardiac catheterization and an electrophysiologic study were also carried out. After 6–8 weeks of therapy all six patients demonstrated histologic evidence of either a marked reduction or a disappearance of inflammation, frequently associated with an increase of interstitial fibrosis. Inflammation was also absent after interruption of treatment in five cases. In four of these five (in whom an electrophysiologic study was performed) ventricular tachycardia could not be provoked in the laboratory during repeat electrophysiologic testing, while "clinical arrhythmias" were abolished in five.

Also the two patients reported by Feneley et al. [27] were characterized by a recent history of progressive dyspnea and tachycardia determined by an atrial flutter. Both biopsies showed acute myocarditis of mild to moderate severity. Immunosuppressive treatment was followed by histologic and clinical improvement. It has to be noted that in both patients atrial flutter was abolished, in the first days after admission, with electrophysiologic or pharmacologic techniques. The authors concluded that "immunosuppressive therapy should be administered only to those patients with myocarditis confirmed by endomyocardial biopsy."

In the same year Zee-Cheng et al. [5] found that in a group of 35 patients with unexplained heart failure, a high percentage (63% = 22 patients) had signs of inflammation. Of these 22, however, 18 had low-grade (grade I and II) inflammatory changes which, by present histologic criteria, would not have been considered active myocarditis. Four other cases had grade III or IV myocarditis. Of the ten patients treated with immunosuppressors, six improved (or at least there was "no further progression of disease"), three died, while in one "follow up was in progress.".

In 1985 three papers were published with conflicting results. In the first one Dec et al. [28] analyzed nine patients, systematically studied from a clinical and also from a hemodynamic point of view, who received immunosuppressive treatment. Improvement occured in four of nine patients (44%) (one of these nine patients had negative biopsy) treated with immunosuppressors and also in six of 18 patients (33%) some with "focal inflammatory biopsy diagnosis and some with a negative (= no inflammation) biopsy," not receiving immunosuppressive therapy.

The biopsy was repeated 3 weeks after the beginning of treatment with disappearance of inflammation in three patients and a reduction in three. Clinical and functional improvement occurred in only three of the six patients.

Clinical recurrence was observed in four cases when the immunosuppressive drugs were tapered or discontinued. In a repeated biopsy in two of these four patients, the reappearance of a mononuclear cell infiltrate was demonstrated.

The authors concluded that many cases of dilated cardiomyopathy result from myocarditis, that a definite histologic diagnosis depends on the duration of the disease, and that "the efficacy of immunosuppressive treatment must still be established."

In the same year Hosenpud et al. [29] presented their data regarding six patients evaluated for an "unexplained dilated cardiomyopathy," who had histologic signs of myocarditis (but necrotic changes were sometimes absent) and who were treated with azathioprine and prednisone.

After 3 months, biopsy, right heart catheterization, and echocardiogram were repeated. There was a significant decrease of mean heart rate and of pulmonary wedge pressure, ascribed to increase of conventional therapy for heart failure, but no significant changes in cardiac index, end-diastolic dimension, and fractional shortening. Nonfatal complications occurred in three patients (acute psychosis, severe soft tissue infection, and adrenal insufficiency after surgery).

The authors concluded that in this group of cases, treated with immunosuppressors, there was lack of objective improvement, that these medications bring about significant side effects, and that a randomized study is necessary before this form of therapy becomes standard for patients with myocarditis.

Mortensen et al. [30] treated a group of 12 patients with prednisone or prednisolone associated with azathioprine in two, while one patient with insulin-dependent diabetes was treated with cyclosporine. After 1–2 months of therapy, eight patients were clinically improved. The clinical improvement was accompanied by a reduction or resolution of signs of inflammation in four. However, there were four deaths (two in patients with progressive deterioration, one in an improved, and one in a stabilized case). Moreover, in the recent literature there are other papers published as abstracts or as case reports. In 1983 Hess et al. [31] observed six cases of "inflammatory myocarditis" (predominantly lymphocytic), who were treated with antithymocyte globulin, azathioprine, and pulsed methylprednisolone, followed by prednisone. At follow-up, which was short (1 month), repeat biopsy showed a diminution of inflammation and a significant increase of left ventricular ejection fraction. According to the authors, selective (aimed at T lymphocytes) immunosuppressive therapy is safe and bears good short-term results.

Another short communication is that by Williams et al. [32] who evaluated six patients "with chronic cardiomyopathy" (duration: 1–9 years) who had, on endomyocardial biopsy, histologic signs of "chronic active myocarditis" (lymphocytic infiltration associated with cell destruction). These patients were treated with azathioprine/prednisone for 6–16 months. Five patients improved by at least one functional class, and the biopsy showed a marked reduction in the inflammatory infiltrates and no further active necrosis. However, two of these patients deteriorated after 10 and 16 months, respectively, and died. The sixth patient deteriorated progressively and treatment was stopped at 6 months. The authors' conclusion was that immunosuppression can improve functional status, but that the improvement may be transient; sequential biopsies and hemodynamics are useful "for monitoring of progress."

Also Strain et al. [33] studied seven patients with "chronic myocarditis," before and during immunosuppressive therapy with prednisone and azathioprine. After 7 $\pm$ 2 weeks of therapy, five of seven patients had a marked improvement in NYHA functional class and in left ventricular ejection fraction, but despite functional and hemodynamic improvement there was no corresponding evidence of reduced inflammation in the endomyocardial biopsy. The authors' conclusion was that in chronic myocarditis clinical and hemodynamic improvement may occur during immunosuppressive therapy without histologic improvement.

Finally, a few anedoctal cases have been published in which immunosuppressive treatment apparently corrected congestive heart failure and improved left ventricular function [34, 35] (see Table 1).

In the first paper [34] endomyocardial biopsy demonstrated mild to moderate patchy interstitial chronic inflammatory cell infiltration with scattered myocyte destruction. High-dosage prednisone and azathioprine treatment was finally associated with a histologic and clinical improvement.

The patient of Costanzo Nordin et al. [35] was one of the first cases of biopsy-documented isolated giant cell myocarditis (a disease usually considered to have a fulminant course unresponsive to treatment) in which immunosuppression determined a sustained hemodynamic and histologic improvement.

Two patients with "granulomatous myocarditis," in whom the diagnosis had been made by endomyocardial biopsy, have been also previously described by Mc Falls et al. [36]. In the first one, treated with oral prednisone, a repeat endomyocardial biopsy demonstrated resolution of the granulomatous disease; however, inducible ventricular tachycardia persisted. The second patient died suddenly after 2 months of treatment.

Other papers devoted to myocarditis in which some data on immunosuppressive treatment were also reported are those by Melvin et al. [37], O'Connell et al. [38, 39] and Ruzyllo et al. [40] (Table 1).

Personal Data

Patients. Twenty patients in whom myocarditis was histologically diagnosed during the years 1983–1987 entered a protocol to evaluate the effects of immunosuppressive treatment on histology and on left ventricular function. The group consisted of 13 males and 7 females with an age range of 16–66 years (mean 33.6). The clinical presentation was congestive heart failure in 16 patients, atrial and ventricular arrhyhtmias in two, and typical angina pectoris at rest in the other two. A history compatible with a viral illness in the preceding 4 months was found in nine patients. Seventeen patients showed a depressed (< 0.50) left ventricular ejection fraction at the time of diagnosis.

Endomyocardial Biopsy. Multiple endomyocardial biopsies were obtained during the period of study in each patient and immediately fixed in 10% buffered formalin. Sections were subsequently examined by light microscopy.

Histologic Classification. (Figs. 1 – 3) Both at the time of the diagnosis and during the whole period of study, the same histologic criteria were used [1].

Immunosuppressive Treatment. Immunosuppression consisted of oral azathioprine (75 mg/m^2 per day for at least 6 months) and oral prednisone (50 mg/m^2 per day for 2 weeks, then tapered until the dose of 0.3 mg/kg per day was reached and maintained for at least 4 months). The usual means of treatment of congestive heart failure, arrhythmia, and hypertension were kept constant during the period of study. In seven patients metoprolol 50–100 mg/day was added during the period of observation.

Patients' Assessment. Left ventricular ejection fraction values obtained at the time of diagnosis were compared to values obtained during the latest period of follow up in 17 of the 20 patients.

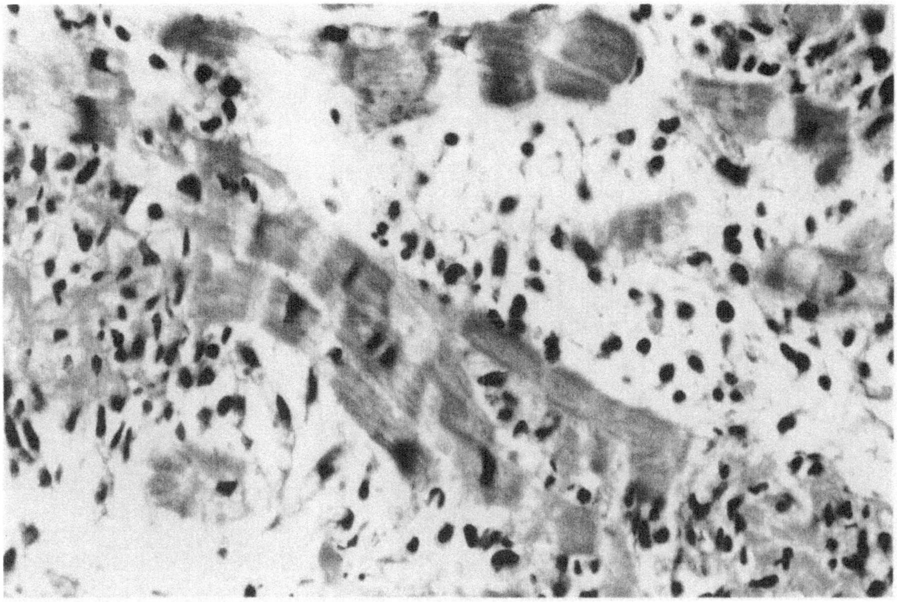

Fig. 1. Active myocarditis. Necrotic changes of the myofibers and infiltrations of lymphomonocytes and monocytes are visible. Hematoxylin and eosin, x 125

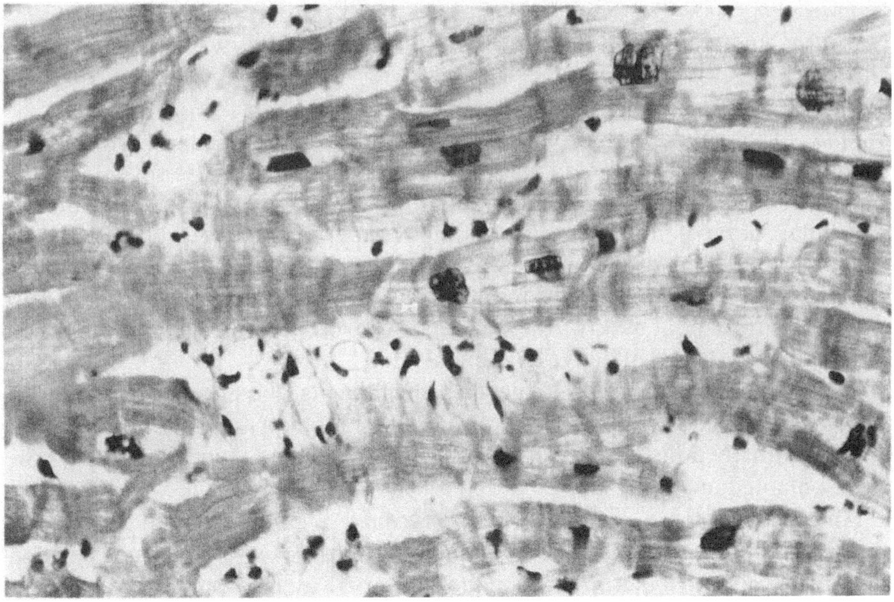

Fig. 2. Healing myocarditis. Few inflammatory infiltrates with minimal inflammatory activity still present. Hematoxylin and eosin, x 125

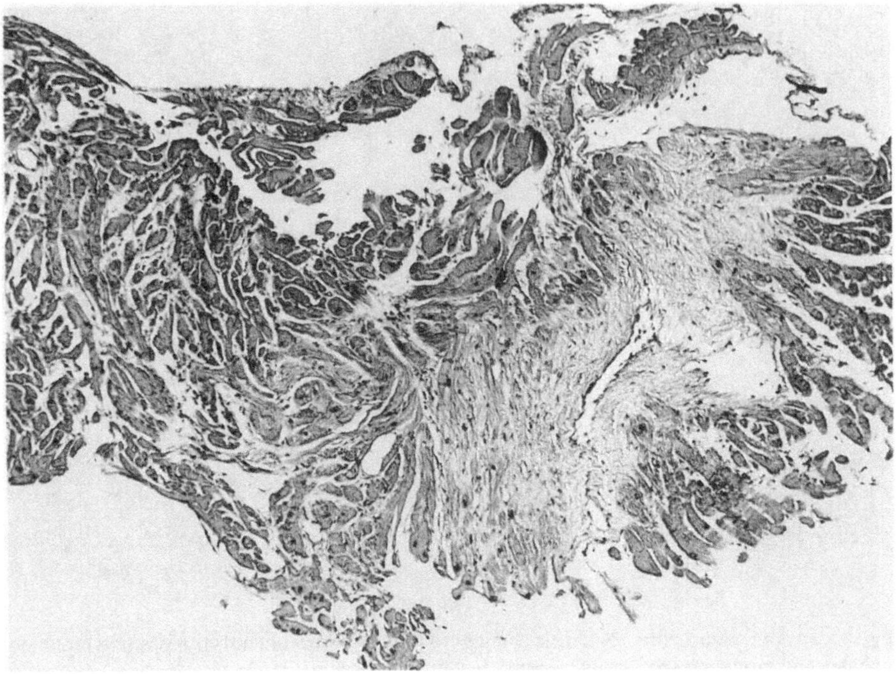

Fig. 3. Healed myocarditis. No signs of inflammation. Evident fibrotic changes. Hematoxylin and eosin, x 40

Statistical Analysis. Where appropriate, statistical analysis was performed using the test of Wilcoxon.

Results. (Table 2) After approximately 2 months of treatment and during a mean prednisone dose of 50 mg/day myocarditis was still persistent in seven patients. During continued treatment about 5 months later, at a mean prednisone dose of 25 mg/day, none of the patients showed signs of persistent myocarditis. When the immunosuppressive drug regimen was withdrawn, after approximately 9 months of treatment, and histology was reassessed, a worsening of the histologic picture was evident in about 50% of the patients. All were treated again at doses previously demonstrated effective or even higher.

The treatment was very prolonged in some patients who underwent several endomyocardial biopsies and several unsuccessful attempts at treatment withdrawal. Eventually a histologic picture of healing or healed myocarditis was, however, obtained in all (Fig. 5).

An overall increase of ejection fraction from 0.37 ± 0.14 to 0.46 ± 0.17, $P = 0.016$, was observed in the patients studied. This increase was not usually evident in the early period of therapy but, on the contrary, was more pronounced after several months of continuous treatment or after several unsuccessful attempts at treatment withdrawal (Figs. 4, 5).

Table 2. Summary of patients with active myocarditis treated with immunosuppression

Patient no.	Age (years)	Sex	Follow-up (months)	Worsening of histology after interruption	Initial LVEF (%)	Final LVEF (%)	Improved	Died
1	14	M	28	Yes	35	54	Yes	No
2	39	M	27	Yes	28	47	Yes	No
3	38	M	6	No	41	38	No	No
4	16	M	6	No	26	16	No	Yes
5	32	M	26	No	42	38	No	No
6	43	M	8	No	30	40	Yes	No
7	41	F	7	No	21	22	Yes	No
8	25	M	8	Yes	26	16	No	Yes
9	30	M	6	No	59	59	No	No
10	27	M	11	No	71	69	No	No
11	20	M	4	No	38	58	Yes	No
12	22	F	27	Yes	14	41	Yes	No
13	23	M	13	No	53	59	Yes	No
14	31	M	14	No	28	38	Yes	No
15	32	F	4	No	/	/	/	No
16	40	M	11	Yes	41	47	Yes	Yes[a]
17	43	F	12	Yes	/	/	/	No
18	60	F	27	Yes	42	75	Yes	No
19	33	F	7	No	41	57	Yes	No
20	36	F	7	No	/	/	/	No
20	32.2 ± ± 10.9	13 M 7 F	13 ± ± 8.8	7	37.4 ± ± 14.2	45.5 ± ± 17.1	11	2 (+1)

[a]This patient died of a cerebral hemorrhage
LVEF left ventricular ejection fraction

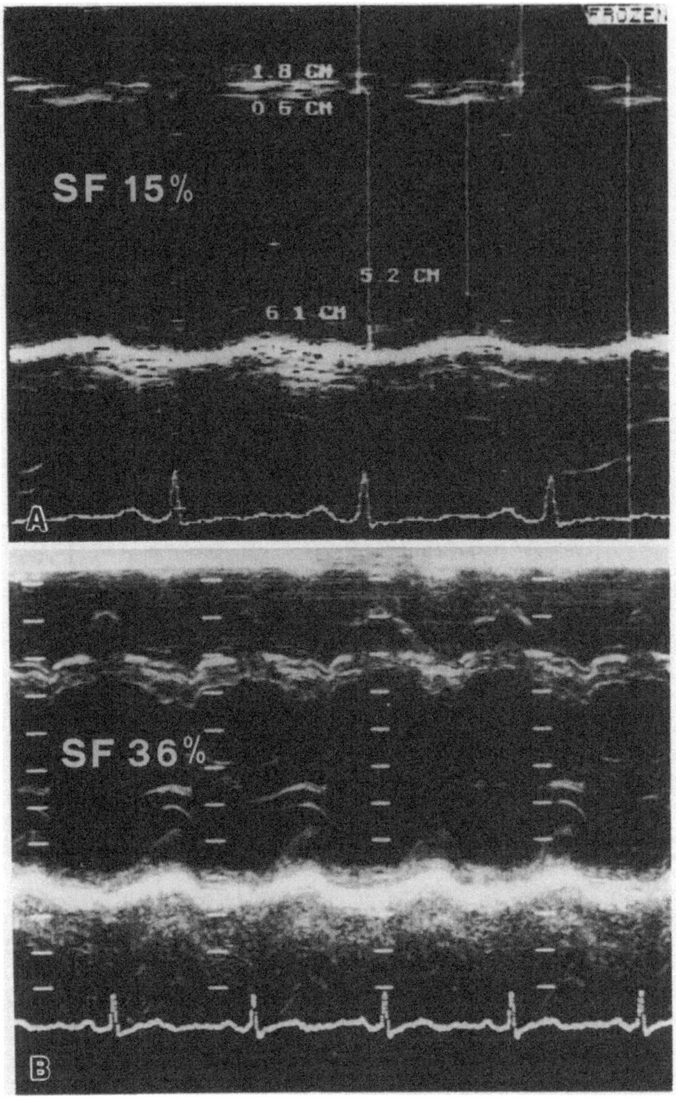

Fig. 4 A, B. Two M-mode echocardiographic tracings (at ventricular level) in a patient with myocarditis. **A** was performed a few days before the diagnosis of myocarditis was made: it shows an evident hypokinesis of left ventricular walls (shortening fraction 15%) and a mild increase of the end-diastolic diameter (6.1 cm). **B** was performed 14 months later: a normalization of left ventricular diameter (5.2 cm) and wall motion (shortening fraction 36%) is evident

Complications. One patient developed acute peritonitis as a complication of a previously known chronic appendicitis which perforated. He was treated surgically and made an uneventful recovery. In one patient the azathioprine dosage had to be reduced because of leukopenia, and he also developed oral candidiasis and epididymitis, which were readily controlled by mycostatin and antibiotics. All

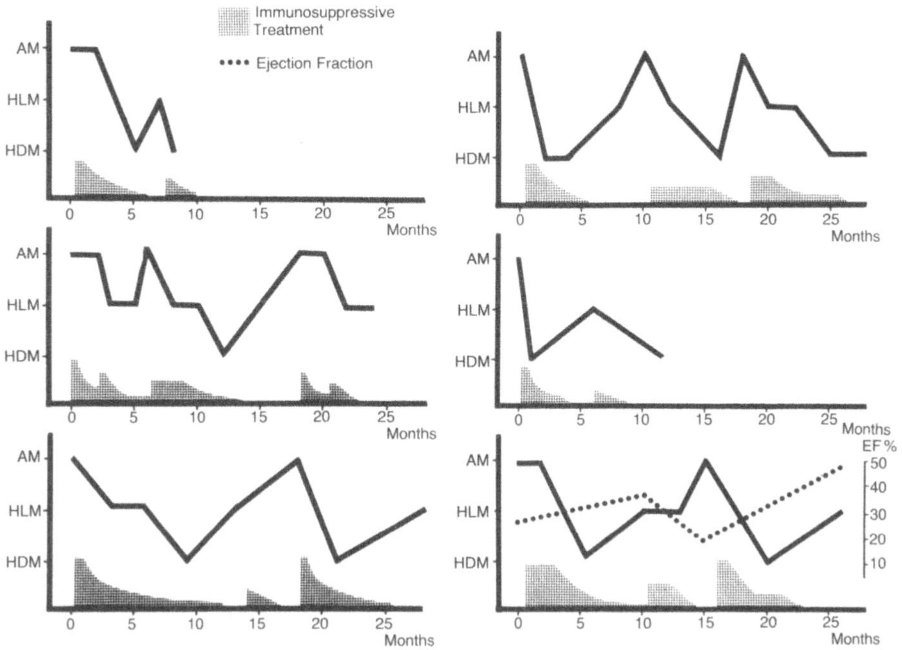

Fig. 5. Changes of histologic inflammatory activity in relationship to immunosuppressive treatment *(shaded areas).* A relationship between treatment reduction or withdrawal and worsening histology is evident. Vice versa, the reinstitution of a previous effective treatment is associated with an improvement of histology. In the last case improvement of histology seems to be associated with an improvement of the ventricular function. *AM,* acute myocarditis; *HLM,* healing myocarditis; *HLD,* healed myocarditis; *EF,* ejection fraction; *dotted line,* ejection fraction. The cases illustrated are (from left to right and from top to bottom): number 18, 8, 17, 12, 2, 1 (see table II).

patients became clinically cushingoid, and the oldest two also became hypertensive.

Deaths. Two patients died. Both had a low ejection fraction which deteriorated slowly during the observation period. Another patient died after the end of treatment for a cerebral hemorrhage.

Comment. In all patients included in this study, myocarditis was histologically diagnosed according to the Dallas criteria [1]. Treatment with azathioprine and prednisone was followed by a histologic picture of healing or healed myocarditis in all patients examined, although some needed very prolonged treatment periods. Attempts at treatment withdrawal were often associated with a worsening of the histologic picture which improved after reinstitution of treatment (Fig. 5). A direct effect of the treatment on the histologic status is therefore suggested. Improvement of ejection fraction was observed during treatment (Fig. 5), but it cannot be defined whether this improvement was related to the histologic changes.

Discussion

Myocarditis is a disease the causes of which are frequently obscure, the natural history various, the diagnostic criteria still discussed, and the indications and the results of therapy uncertain. Myocarditis not due to a definite etiology has been called by a variety of names such as "Fiedler's myocarditis," "idiopathic myocarditis," "pernicious myocarditis," etc. [5].

The possibility of immunomediated cardiac damage (as in animal models) and some similarities between the histologic picture in myocarditis and that seen in acute cardiac rejection suggested that immunosuppression might benefit both conditions. In the last 8 years a limited number of papers (Table 1) [5, 6 14, 19 – 26, 28 – 38, 40 – 42; personal data] regarding this problem have been published with uncertain or discordant results. Indeed, the problem of the results of treatment of myocarditis is very complex and difficult [14, 39] because the natural history of the disease is various and partially unknown, and histologic, clinical, and hemodynamic improvement may occur spontaneously. Moreover, in papers published on immunosuppression in myocarditis, histologic criteria, clinical and hemodynamic evaluation, type, dosage and duration of treatment, and duration of follow-up differed widely, and the total number of patients studied is limited.

Natural History of Myocarditis and the Possibility of Spontaneous Improvement

A good prognosis for cardiac function in uncomplicated myocarditis was observed in 1972 by Gerzen et al. [43] in 45 cases diagnosed on the basis of electrocardiographic alteration in the course of viral disease.

More recent studies, expecially by Japanese authors [24, 44, 45], have shown a favorable course of the disease in a great percentage of cases. Interestingly, many of these patients with spontaneous improvement and/or healing of the disease first presented with an acute disease probably of viral origin. A spontaneous improvement has also been observed in small subgroups of patients in other studies [6, 28, 46].

However, because of the different histologic criteria and short follow-up, at present it is very difficult to know the natural history and the spontaneous course of untreated myocarditis.

Previous Studies. Characteristics and Limitations

An analysis of the published studies in which immunosuppressive therapy was used shows many methodologic differences and many inconsistencies, which make it very difficult to draw conclusions on the results of this form of treatment. One of the first aspects deserving mention is the admission criteria. In the majority of the studies, congestive heart failure was the dominant symptomatology, while in others Adams-Stokes attacks [24] or high-grade unexplained ventricular arrhythmias, or aborted out-of-hospital sudden death [26], or an atrial flutter with

2:1 atrioventricular block [27] were the clinical aspects which prompted further investigation.

Data presented by Sekiguchi et al. [24] suggest that patients with atrioventricular block of sudden onset and Adams-Stokes attacks may present a spontaneous rapid improvement. Of the nine patients observed by these authors and treated in conventional manner, five had a third-degree atrioventricular block and Adams-Stokes attacks; one in shock and with acute renal failure died, but the other four rapidly recovered (within 1–2 months) and returned to their work.

Another aspect is the absence of standardization in timing of biopsies during treatment (from 1 week [23] to 2 or more months) and in the number of specimens; but probably more important is the absence of uniform criteria in the histologic diagnosis of active myocarditis.

In a study on myocarditis Edwards et al. [6] considered an increased number of interstitial lymphocytes (more than five per high-power field) "the best histologic indicator of active inflammatory disease," while active myocyte necrosis (with inflammatory cells within the sarcolemma) was not identified.

Zee-Cheng et al. [5] and Dec et al. [28] considered the presence of inflammatory infiltrates necessary for the diagnosis of myocarditis but also established a grading system for the pathologic changes present in the biopsy material, the most severe one being only characterized by the presence of myocytes necrosis. On the contrary, the criteria of Daly et al. [25] included the presence of a lymphocytic and inflammatory cell infiltrate but also of myocyte necrosis with fraying of adjacent myocardial fibers.

Billingham and Mason [22] also considered an inflammatory infiltrate the requisite for a diagnosis of myocarditis, while the presence of a "myocyte damage" was "helpful" but not "necessary." The same authors also believed that an essential criteria for the diagnosis of myocarditis was the absence of changes attributable to dilated cardiomyopathy such as myofiber hypertrophy, or the presence of bizarre-shaped nuclei. Identical criteria were used by Vignola et al. [26].

This variety of opinions, which preceded the establishment of more definite criteria for the diagnosis of myocarditis (Dallas criteria) [1] suggests that it is highly probable that some patients diagnosed in the past as "active myocarditis" (and consequently treated with immunosuppressors) would not be presently confirmed to be affected by the disease.

The different criteria of "improvement" used must also be considered as an important potential limitation. Indeed, while in some papers [25, 28, 29, 42] objective parameters such as hemodynamics, ejection fraction, and cardiothoracic ratio were considered, in others [6] less rigid and more subjective clinical criteria were used.

Immunosuppressive Treatment

Another factor which makes the analysis of the results difficult is the different duration of treatment and the various dosage of drugs. In fact, in many papers the results were presented after 6 months or more of treatment [14, 25–28, 32; personal data], but in others shorter periods were considered [6, 29, 31, 33]; finally, in a few others this information is not clearly provided. The most common form of treatment was the association of prednisone with azathioprine, but in a few prednisone or prednisolone alone [25, 28] or the association of prednisone with azathioprine and antithymocyte globulin [31] was used. Also the dosage used differed from author to author and sometimes reached high values (prednisone 120 mg and azathioprine 200 mg per day) [3].

Another factor which may potentially benefit the patients with myocarditis and the role of which is difficult to identify is the simultaneous use of other drugs as beta-blockers. This form of therapy may have a potentially favorable effect in patients with depressed left ventricular function secondary to dilated cardiomyopathy [47].

Another potential limitation of immunosuppressive treatment is that the autoimmune pathogenesis of myocarditis may appear less convincing after the experimental work of Bowles et al. [10] who detected coxsackievirus B-specific RNA in myocardial tissue of patients with myocarditis, even in the late healed stage. However, immunosuppression did not cause in any patients in our series an acute exacerbation of the disease, and detrimental effects, as described in the acute phases of myocarditis in experimental animals [8, 10, 11], were not observed. The absence of apparent deteriorations of the disease was also noted by Fenoglio et al. [23]. Also, the complications described, although sometimes relevant [29], were relatively few [42].

However, very frequently the patients become clinically cushingoid during treatment [42]. Some authors [47] stressed the risk of increased fluid retention or of decompensation during corticosteroid therapy in patients with left ventricular enlargement. This complication does not seem to be very important in published studies.

The most important limitation of all studies is, however, that it cannot be stated with certainty whether the observed histologic, clinical and hemodynamic changes are secondary to immunosuppression or spontaneous: indeed, the described "improvement" of symptomatology may be an aspect of the natural history of the disease and not a direct consequence of the immunosuppression. However, more authors [3, 5, 14, 25, 28] observed the important phenomenon of patients, previously improved during therapy, who presented a recrudescence of the symptomatology, with reappearance of histologic signs of activity after interruption of immunosuppressive treatment. In some of these patients reinstitution of immunosuppression was followed by improvement. Also in our group of twenty patients, reduction or interruption of treatment was accompanied by a worsening histology in seven. In six of them, treatment was reinstituted with subsequent improvement. This phenomenon was seen more than once and was also observed in individual patients in whom a reduction of the dosage of corticosteroids was associated with recurrence of signs of activity and deterioration of left ventricular function. The reinstitution of a previous effective dosage was again associated with improvement of histology and of left ventricular function (Fig. 5). The duration of treatment originally planned was 6 months, but the histologic response led to adjustment in dosage and duration of immunosuppression so that most patients had a period of follow-up longer than 6 months and some were treated for as long as 2 years.

All these data are highly suggestive of a direct beneficial effect of an immunosuppressive drug regimen in a subgroup of patients with active myocarditis. Moreover, treatment should probably not be carried out on the basis of a rigid protocol but rather modulated according to histology and continued for longer periods.

Finally, it has to be noted that a reduction of the signs of inflammatory activity is not always concomitant with a hemodynamic improvement. Indeed, while some authors noted a relationship between the two phenomena (although only in a few cases) [14, 25], others, such as Fenoglio et al. [23], observed apparent histologic improvement in patients without changes in left ventricular function. Vice versa in the study of Strain et al. [33], despite functional and hemodynamic improvement, there was no corresponding evidence of a reduction of inflammation in endomyocardial biopsy.

In conclusion, the role of immunosuppressive treatment continues to be uncertain, and many problems remain unsolved. For these reasons a controlled randomized trial was begun in 1984 and is presently in progress [48] and will be completed in 1991. Meanwhile some provisional conclusions may be drawn:

1. Immunosuppressive treatment may be associated with a reduction of histologic signs of active myocarditis and sometimes with an improvement of left ventricular function.
2. The reduction or interruption of treatment may be associated with a worsening histology (and sometimes with a deterioration of left ventricular function).
3. The improvement of left ventricular function may appear late.
4. The relationship between treatment and histologic and functional improvement and the possible deterioration after withdrawal of immunosuppression suggest that this form of therapy may be useful in a subset of patients with active myocarditis.

References

1. Aretz HT, Billingham ME, Edwards WD, Factor SM, Fallon JT, Fenoglio JJ, Olsen EGJ, Schoen FJ (1986) Myocarditis: a histopathologic definition and classification. Am J Cardiovasc Pathol 1:3–14
2. Sakakibara S, Konno S (1962) Endomyocardial biopsy. Jpn Heart J 3: 537–543
3. Shanes JG, Ghali J, Billingham ME, Ferrans VJ, Fenoglio JJ, Edwards WD, Tsai CC, Saffitz JE, Isner J, Furner S, Subramanian R (1987) Interobserver variability in the pathologic interpretation of endomyocardial biopsy results. Circulation 75: 401–405
4. Kunkel B, Lapp H, Kober G, Kaltenbach M (1978) Correlation between clinical and morphological findings and natural history in congestive cardiomyopathy. In: Kaltenbach M, Loogen F, Olsen ECJ (eds) Cardiomyopathy and myocardial biopsy. Springer, Berlin Heidelberg New York, pp 271–283
5. Zee-Cheng CS, Tsai CC, Palmer DC, Codd JE, Pennington DG, Williams GA (1984) High incidence of myocarditis by endomyocardial biopsy in patients with idiopathic congestive cardiomyopathy. J Am Coll Cardiol 3: 63–70
6. Edwards WD, Holmes DR, Reeder GS (1982) Diagnosis of active lymphocytic myocarditis by endomyocardial biopsy. Quantitative criteria for light microscopy. Mayo Clin Proc 57: 419–425
7. Tazelaar MD, Billingham ME (1986) Leukocytic infiltrates in idiopathic dilated cardiomyopathy. Am J Surg Path 10: 405–412
8. Lowry PJ, Edwards CW, Nagle RE (1982) Herpes-like virus particles in myocardium of patient progressing to congestive cardiomyopathy. Br Heart J 48: 501–503

9. Engblom E, Ekfors TO, Meurman OH, Toivanen A, Nikoskelainen J (1983) Fatal influenza A myocarditis with isolation of virus from the myocardium. Acta Med Scand 213: 75–78
10. Bowles NE, Olsen EGJ, Richardson PJ, Archard LC (1986) Detection of Coxsackie-B virus specific RNA sequences in myocardial biopsy samples from patients with myocarditis and dilated cardiomyopathy. Lancet 1: 1120–1122
11. Woodruff JF (1980) Viral myocarditis: a review. Am J Pathol 101: 427–479
12. Huber SA, Lyden DC, Lodge PA (1985) Immunopathogenesis of experimental coxsackievirus induced myocarditis; role of autoimmunity, Herz 10: 1–7
13. Billingham ME (1979) Some recent advances in cardiac pathology. Hum Pathol 10: 367–386
14. Mason JW, Billingham ME, Ricci DR (1980) Treatment of acute inflammatory myocarditis assisted by endomyocardial biopsy. Am J Cardiol 45: 1037–1044
15. Kilbourne ED, Wilson CB, Perrier D (1956) The induction of gross myocardial lesions by a Coxsackie virus and cortisone. J Clin Invest 35: 362–370
16. Tomioka N, Kishimoto C, Matsumori A, Kawai C (1986) Effects of prednisolone on acute viral myocarditis in mice. J Am Coll Cardiol 7: 868–872
17. O'Connell JB, Reap EA, Robinson JA (1986) The effects of cyclosporine on acute murine Coxsackie B3 myocarditis. Circulation 73: 353–359
18. Monrad ES, Matsumori A, Murphy JC, Fox JG, Crumpacker CS, Abelmann WH (1986) Therapy with cyclosporine in experimental murine myocarditis with encephalomyocarditis virus. Circulation 73: 1058–1064
19. Garrison RF, Swisher RC (1953) Myocarditis of unknown etiology (Fiedler's?) treated with ACTH. J Pediat 42: 591–599
20. Ainger LE (1964) Acute aseptic myocarditis: corticosteroid therapy. J Pediat 64: 716–723
21. Voigt GC (1968) Steroid therapy in viral myocarditis. Am Heart J 75: 575–576
22. Billingham ME, Mason JW (1984) Endomyocardial biopsy diagnosis of myocarditis and changes following immunosuppressive treatment. In: Bolte HD (ed) Virus heart diseases. Springer, Berlin Heidelberg New York, pp 200–216
23. Fenoglio JJ, Ursell PC, Kellogg CF, Phil M, Drusin RE, Weiss MB (1983) Diagnosis and classification of myocarditis by endomyocardial biopsy N Engl J Med 308: 12–18
24. Sekiguchi M, Hiroe M, Take M, Hirosawa K (1980) Clinical and histopathological profile of sarcoidosis of the heart and acute idiopathic myocarditis. Concepts through a study employing endomyocardial biopsy. II. Myocarditis Jpn Circ J 44: 264–273
25. Daly K, Richardson PJ, Olsen EGJ, Morgan Capner P, McSorley C, Jackson G, Jewitt DE (1984) Acute myocarditis. Role of histological and virological examination in the diagnosis and assessment of immunosuppressive treatment. Br Heart J 51: 30–35
26. Vignola PA, Kalutaka A, Swaye PS, Rozanski JJ, Blankstein RL, Benson J, Gosselin AJ, Lister JW (1984) Lymphocytic myocarditis presenting as unexplained ventricular arrhythmias: diagnosis with endomyocardial biopsy and response to immunosuppression. J Am Coll Cardiol 4: 812–819
27. Feneley MP, Hickie B, Gavaghan TP, Baron DW, Ralston M (1984) Diagnosis and management of acute myocarditis aided by serial myocardial biopsy. Aust NZ J Med 14: 826–830
28. Dec GW, Palacios IF, Fallon JT, Aretz HT, Mills J, Lee DCS, Johnson RA (1985) Active myocarditis in the sprectrum of acute dilated cardiomyopathies. Clinical features, histologic correlates, and clinical outcome. N Engl J Med 312: 885–890
29. Hosenpud JD, McAnulty JH, Niles NR (1985) Lack of objective improvement in ventricular systolic function in patients with myocarditis treated with azathioprine and prednisone. J Am Coll Cardiol 6: 797–801
30. Mortensen SA, Baandrup U, Buch J, Bendtzen K, Hvid-Jacobsen K (1985) Immunosuppressive therapy of biopsy proven myocarditis: experiences with corticosteroids and cyclosporin. Int J Immunother 1: 35–45
31. Hess ML, Hastillo A, Mohanty PK (1983) Inflammatory myocarditis: incidence and response to T-lymphocyte depletion. (Abstr). J Am Coll Cardiol 1: 584
32. Williams GA, Schwarze MW, Kennedy HL, Palmer D, Tsai CC (1984) Treatment of chronic active myocarditis with immunosuppressive therapy. (Abstr). J Am Coll Cardiol 3: 521

33. Strain JE, Grose RM, Hirsch CL, Kramer DC, Cho S, Factor SM (1984) Discordance of clinical and histological response to therapy in chronic myocarditis (Abstr). J Am Coll Cardiol 3: 520
34. Shanes JG, Krone RJ, Tsai CC, Fischer K, Williams GA (1984) Mild myocardial inflammation presenting as congestive cardiomyopathy responsive to immunosuppression. Am Heart J 107: 798–801
35. Costanzo-Nordin MR, Silver MA, O'Connell JB, Scanlon PJ, Robinson JA (1987) Giant cell myocarditis: dramatic hemodynamic and histologic improvement with immunosuppressive therapy. Eur Heart J, 8 [Suppl J]: 271–274
36. Mc Falls EO, Hosenpud JD, Mc Anulty JH, Kron J, Niles NR (1986) Granulomatous myocarditis. Diagnosis by endomyocardial biopsy and response to corticosteroids in two patients. Chest 89: 509–511
37. Melvin KR, Richardson PJ, Olsen EGJ, Daly K, Jackson G (1982) Peripartum cardiomyopathy due to myocarditis. N Engl J Med 307: 731–734
38. O'Connell JB, Robinson JA, Henkin RE, Gunnar RM (1981) Immunosuppressive therapy in patients with congestive cardiomyopathy and myocardial uptake of gallium-67. Circulation 64: 780–786
39. O'Connell JB, Mason JW (1987) The role of endomyocardial biopsy and gallium-67 scintigraphy in the assessment and treatment of active myocarditis. In: Kawai C, Abelmann WH (eds) Pathogenesis of myocarditis and cardiomyopathy. University of Tokyo Press, Tokyo pp 281–292
40. Ruzyllo W, Rosnowski A, Dabrowski M (1985) Profiles of biopsy proven cases with myocarditis, In: Sekiguchi M, Olsen EGJ, Goodwin JF (eds) Myocarditis and related disorders. Proceedings of the International Symposium on Cardiomyopathy and Myocarditis. Springer, Berlin Heidelberg New York, pp 107–110
41. Segal J, Harvey P, Gurel T (1965) Diagnosis and treatment of primary myocardial disease. Circulation, 23: 837–844
42. Salvi A, Hrovatin E, Dreas L, Silvestri F, Camerini F (1987) Changes in histology and left ventricular ejection fraction during immunosuppressive treatment in active myocarditis. Eur Heart J 8 [Suppl J]:267–269
43. Gerzen P, Granath A, Holmgren B, Zetterquist S (1972) Acute myocarditis. A follow up study. Br Heart J 34: 575–583
44. Kitaura Y, Morita H (1979) Virus myocarditis and cardiomyopathy. Jpn Circ J 43: 1017–1031
45. Hiroe M, Sekiguchi M, Take M, Kusakabe K, Shigeta A, Hirosawa K (1985) Long follow up study in patients with prior myocarditis by radionuclide methods. In: Sekiguchi M, Olsen EGJ, Goodwin JF (eds) Myocarditis and related disorders. Proceedings of the International Symposium on Cardiomyopathy and Myocarditis. Springer, Berlin Heidelberg New York, pp 199–203
46. Strain JE, Grose RM, Cho S, Factor SM (1985) Short term natural history of untreated myocarditis. Circulation 72 [Suppl III]: 439
47. Waagstein F, Hjalmarson A, Varnauskas E, Wallentin I (1975) Effect of chronic β-adrenergic receptor blockade in congestive cardiomyopathy. Br Heart J 37: 1022–1036
48. Anderson JL, Fowles RE, Unverferth DV, Mason JW (1987) Immunosuppressive therapy of myocardial inflammatory disease. Initial experience and future trials to define indications for therapy. Eur Heart J, 8 [Suppl J]: 263–266

The Role of Cardiac Transplantation in Patients with Active Myocarditis

J. B. O'CONNELL, G. W. DEC, and J. W. MASON

Utah Cardiac Transplant Program, University of Utah Medical Center, 50 North Medical Drive, Salt Lake City, UT 84132, USA

Introduction

Despite dramatic improvement in the success of cardiac transplantation, prediction of postoperative survival based on etiology of the pretransplant heart disease is not possible and cardiomyopathy remains the most common indication [4]. Of the patients presenting with dilated cardiomyopathy of recent onset without obvious etiologic factors, 10% have histologic evidence of active myocarditis [6]. Although preliminary studies suggest immunosuppression may result in clinical and hemodynamic improvement when myocarditis is biopsy-proven [7], a multicenter randomized controlled trial is required to determine the efficacy of immunosuppression in active myocarditis [5]. Until the results of this trial are reported, it can be stated that there is no known specific treatment for myocarditis.

Although only 0.85% of all patients who have undergone cardiac transplantation are identified as having myocarditis [4] and results have not been reported independently, this condition is being more widely recognized, and it is likely that some of the patients listed under "cardiomyopathy" in the Registry of the International Society for Heart Transplantation (ISHT) actually had myocarditis. Theoretically, pre-existent T lymphocytes and humoral antibodies committed to myocyte antigens may accentuate the immunologic response to the allograft after transplantation. A serious question, therefore, arises as to whether patients who have biopsy-proven myocarditis should be considered early for cardiac transplantation or whether they should be offered a trial of immunosuppression designed not only to improve symptoms and cardiac dysfunction but also to reduce the intensity of immunologic response prior to transplantation. The purpose of this study is to review retrospectively the pooled results of cardiac transplantation in patients with histologically proven myocarditis from cardiac transplant centers in the United States.

Materials and Methods

Patient Population

The data base of the Registry of the ISHT was reviewed and 26 patients from 15 transplant centers in the United States who had undergone transplantation between January 1, 1982 to May 1, 1988 were identified as having myocarditis as the

H.-P. Schultheiß (Ed.)
New Concepts in Viral Heart Disease

etiology of the pretransplant heart disease. Data from 21 (81%) of these patients were obtained through a survey questionnaire. Active myocarditis was defined histologically as the presence of an inflammatory infiltrate with associated myocyte necrosis in the absence of coronary artery disease (Dallas criteria [1]). Twelve of the 21 patients (57%) had no evidence of myocarditis or were classified incorrectly and hence deleted from data analysis. The Registry, therefore, provided nine patients with histologically verified myocarditis prior to transplantation from seven transplant centers. One of the transplant centers identified an additional patient with myocarditis and four patients from two centers not included in the Registry survey were identified. This report constitutes the analysis of the results in these 14 patients with active myocarditis who underwent cardiac transplantation and represents 1.7% of the 815 patients undergoing transplantation in the participating centers. Follow-up in survivors was complete by May 1, 1988 and averaged 506 ± 241 days.

Survey Questionnaire

A questionnaire was submitted to investigators at the transplant centers. This questionnaire requested demographic information, presenting signs and symptoms, description of the diagnostic technique used to confirm the diagnosis of myocarditis, histologic characteristics of the inflammatory heart disease, and the administration and duration of immunosuppressive therapy prior to cardiac transplantation. In addition, results of coronary angiography, hemodynamic and noninvasive cardiac studies, and severity of illness establishing priority status for donor allocation were tabulated. Immunologic reactivity based on panel reactive antibody (PRA), blood type, and donor-specific crossmatch were noted. The operative complications, immunosuppressive protocol, rejection incidence, and follow-up results constituted the remainder of the questionnaire.

Statistical Analysis

Survival in the patients with myocarditis was compared to 2033 age-matched (all patients between the ages of 22 and 50 years) and 364 sex-matched (all females between the ages of 22 and 50 years) controls transplanted from January 1, 1982 to May 1, 1988 provided by the Registry of the ISHT. Actuarial survival curves were compared using a Tarone-Ware test and a P value of <0.05 was considered statistically significant. $\varkappa^2$ analysis was used to compare survival in patients requiring mechanical assistance to those who did not. Individual data were expressed as mean ± standard error (SEM).

Results

Active myocarditis was documented histologically on pretransplant endomyocardial biopsy in the 14 patients and was present in the explanted heart in 11 (78%).

Table 1. Pretransplant characterics ($n = 14$)

Age (years)	38.3 ± 9.3 years
Sex	4 males 10 females
Duration of symptoms (days)	102 ± 152
Ejection fraction (%)	0.19 ± 0.06
Left ventricular end-diastolic dimension (cm)	5.8 ± 1.2
Right atrial pressure (mean)	14 ± 8 mmHg
Pulmonary artery pressure	40 ± 10/22 ± 6 mmHg
Cardiac index (liters per min per m^2)	1.9 ± 0.5
Hemodynamic support	
Oral therapy alone	3 (21%)
Intravenous inotropes	3 (21%)
Mechanical assistance	8 (57%)

The pretransplant characteristics are shown in Table 1. The patients averaged 38.3 ± 9.3 years of age with a female predominance (10/14 — 71%). Congestive heart failure was the presenting manifestation in 13 patients (93%) while arrhythmia requiring therapy was found in three (21%). One patient presented with angina pectoris despite normal coronary arteries. While the duration of symptoms prior to the diagnosis of myocarditis was 102 ± 152 days, eight (57%) patients presented within 30 days of onset of symptoms.

In three (27%) of the patients in whom the explanted heart showed active myocarditis, this diagnosis was unsuspected prior to cardiac transplantation, and therefore no pretransplant biopsy was performed. The signs and symptoms of myocarditis included fever in seven (50%), leukocytosis in five (36%), elevation of serum creatinine phosphokinase in five (36%), and pericarditis in two (14%). Six (43%) experienced no symptoms, suggesting inflammatory heart disease. Histologically, a lymphocytic infiltrate predominated in ten patients (71%), giant cells were noted in three (21%), and mixed infiltration in one (7%). Ten of 14 (72%) patients received immunosuppression including: prednisone (ten), azathioprine (eight), antithymocyte globulin (ATG) (one), or cyclosporine (one) prior to transplantation. Five of the patients with ongoing histologic myocarditis were receiving immunosuppression at the time of transplantation. Although the duration of immunosuppressive therapy averaged 87 ± 72 days, six (60%) of the patients received greater than a 4-month course of immunosuppressive therapy. Despite the presence of active myocarditis in the explanted heart, endomyocardial biopsy performed shortly before cardiac transplantation showed resolution of myocarditis in two (20%) patients.

Eight of the patients underwent coronary angiography prior to transplantation and all had normal results. The ejection fraction averaged 0.19 ± 0.06 and left ventricular end-diastolic dimension on echocardiogram was 5.8 ± 1.2 cm. Interestingly, five (50%) of those undergoing pretransplant echocardiography had a left ventricular end-diastolic diameter <5.0 cm, underscoring the acute nature of their illness. The mean right atrial pressure was 14 ± 8 mmHg, pulmonary artery pressure 40 ± 10 mmHg systolic, 22 ± 6 mmHg diastolic and the pulmonary capillary wedge presssure 23 ± 6 mmHg; cardiac index was depressed at 1.9 ± 0.5 liters per min per m^2, and pulmonary vascular resistance averaged 1.6 ± 1.3 Wood units. Only one patient had a pulmonary vascular resistance above four Wood units. Three patients were maintained solely on oral therapy of congestive heart failure, and three (21%) required intravenous inotropes with amrinone, dobutamine, dopamine, or norepinephrine alone or in combination. In addition, eight (57%) required mechanical assistance prior to transplantation, including intraaortic balloon counterpulsation in three, biventricular assist devices in two, and the total artificial heart as a bridge to transplant in three.

PRA averaged 3% ± 7%, and only one patient (8%) had a PRA greater than 10%. Five patients (38%) were blood group A, none blood group B, one blood group AB, and seven (54%) blood group 0. Blood group-compatible but not identical donor organs were transplanted in two (15%). The donor-specific crossmatch, one of which was performed before transplant, was positive in two patients (15%).

Ischemic time in the donor organ averaged 146 ± 39 min, and none exceeded 3 h and 30 min. All patients received corticosteroid therapy after transplantation, 13 of 14 received azathioprine, and 13 of 14 cyclosporine. In addition, seven patients (50%) received cytolytic immunoprophylaxis (induction): two–rabbit ATG, two–equine ATG, and three–antilymphoblast globulin (ALG).

In the first 4 months after cardiac transplantation, survivors averaged 1.9 ± 1.2 episodes of rejection (Table 2). The average duration from transplant surgery to first episode of rejection was 18 ± 23 days. However, six (46%) experienced their first episode of rejection within 1 week of cardiac transplantation. Three (21%) of the first rejection episodes were fatal, one of which occurred in a patient with a positive donor-specific crossmatch. Seven episodes of rejection (0.8/patient) were of such intensity that they required cytolytic therapy with ATG, Minnesota ALG, or OKT*3 monoclonal antibody, while six episodes did not respond to the first course of high-dose corticosteroid and cytolytic therapy. There were 15 infections (1.5/patient), and only two (20%) patients were infection free. Six (40%) of the infections were bacterial, six (40%) viral, and three (20%) fungal.

Actuarial survival (Fig. 1) was 56% at 1 year and 38% at 2 years, while in age-matched controls survival was 82% at 1 year and 79% at 2 years, and survival in an age-controlled female population had 78% at 1 year and 76% at 2 years ($P < 0.01$). Three of five patients requiring ventricular assistance or total artificial hearts (60%) died during the follow-up period, while four of the remaining patients (44%) died (P = NS). Two of the three patients with giant cell myocarditis and

Table 2. Characteristics of treated cardiac allograft rejection

Episodes per patient	
First 4 months *(n)*	1.9 ± 1.2
Total *(n)*	2.7 ± 1.9
Episodes requiring cytolytic therapy	7 (0.8/patient)
Interval to first rejection episodes (days)	18 ± 23
Death with first episode	3 (21%)

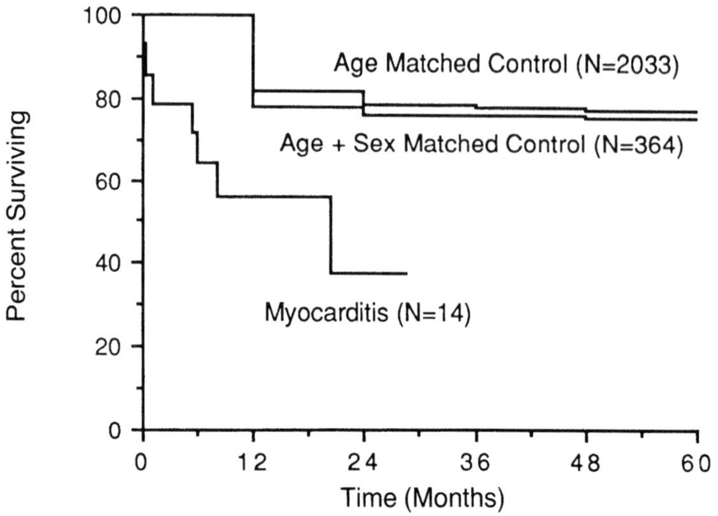

Fig. 1. Actuarial survival in patients with myocarditis compared to controls demonstrating a statistically significant difference ($P < 0.01$)

one of two patients with a positive donor-specific crossmatch survived. Of the patients who died, the cause of death was rejection in three (43%), infection in three (43%), and adult respiratory distress syndrome thought due to drug (amiodarone) toxicity in one. Five (71%) of the survivors are New York Heart Association functional class I and one each functional class II and III. One of four patients undergoing surveillance annual coronary angiography had developed arteriosclerosis of the graft by 2 years after transplantation, and left ventricular function on long-term follow-up is depressed in one (14%).

Discussion

Although cardiac transplantation is an accepted therapeutic modality for the treatment of end-stage heart disease, the results achieved in patients with dilated cardiomyopathy and coronary artery disease should not be extrapolated to patients with myocarditis. Because myocarditis is an immunologically mediated disease thought to result from autoimmune responses by cytotoxic T lymphocytes, patients with active myocarditis may be at high risk for rejection and death following cardiac transplantation. Furthermore, severely ill patients who require chronic invasive monitoring and receive immunosuppression may be predisposed to infections complications. In this retrospective analysis of pooled data from multiple cardiac transplant centers, patients who had undergone cardiac transplantation for biopsy-proven myocarditis showed differences from the general cardiac transplant population which may make definitive interpretation of these data difficult. There was a high percentage of females (71%), and more than half of the patients were symptomatic for less than 1 month. In addition, 36% of these patients were receiving immunosuppressive therapy at the time of cardiac transplantation. Despite the severity of their illness, these patients with myocarditis did not exhibit the degree of left ventricular dilatation and pulmonary artery pressure elevation that is commonly seen in other populations with chronic congestive heart failure. The degree of hemodynamic instability was significantly greater in this group compared to the general pretransplant population at most centers and necessitated mechanical assistance in 57% of patients with a ventricular assist device or total artifical heart in 36%. There was no indication of immunologic hyperreactivity as determined by PRA, and blood group-identical transplants represented the majority. Crossmatch positivity occurred in only two of the patients and the ischemic time was not excessively prolonged.

Rejection in this population could be characterized as early and severe with 21% of the first rejection episodes resulting in death. Furthermore, almost all patients required cytolytic therapy with a high proportion resistant to the first trial of therapy with both corticosteroids and cytolytic agents. Infection rate was high averaging 1.5 episodes per patient with a predominance of purulent fungal and bacterial infections (60%).

Actuarial survival was significantly lower in patients with myocarditis compared to age- and sex-matched controls. Because of unique characteristics of this patient population, however, firm conclusions and recommendations regarding the role played by myocarditis in the patients with poor outcome cannot be made. The Registry of ISHT reports that female sex provides a predictive risk for mortality [4]. In addition, females have a disproportionately high incidence of rejection when compared to males (2.4 ± 0.2 versus 1.7 ± 0.1 episodes in the first 4 months post transplant; [2]). Although use of mechanical assistance with the intra-aortic balloon pump or ventricular assist device is not a predictor of increased mortality after cardiac transplantation [8], the early experiences with the total artifical heart as a "bridge" document the high risk for death postoperatively [3]. In this population, however, the survival in patients requiring mechanical bridging versus those without was not significantly different. These data, therefore, should not be misconstrued to show conclusively that patients with active myocarditis are at high risk following cardiac transplantation but should serve as a stimulus for further cooperative efforts to examine these issues.

Acknowledgement. The authors wish to thank M. P. Kaye and the Registry of the International Society for Heart Transplantation for providing a preliminary patient list, and the following participating investigators:
G. Dec, Massachusetts General Hospital
G. Deeb, University of Michigan
I. Goldenberg, Minneapolis Heart Institute
M. Hess, Medical College of Virginia
J. Hosenpud, Oregon Health Sciences University
T. Icenogle, University of Arizona
M. R. Costanzo-Nordin, Loyola University
J. B. O'Connell, UTAH Cardiac Transplant Program
R. Starling, Ohio State University
This work was performed under the auspices of the Registry of the International Society for Heart Transplantation.

References

1. Aretz HT, Billingham ME, Edwards WD, Factor SM, Fallon JT, Fenoglio JJ Jr, Olsen EGJ, Schoen FJ (1986) Myocarditis: a histopathologic definition and classification. Am J Cardiovasc Pathol 1: 3–14
2. Crandall BG, Renlund DG, O'Connell JB, Burton NA, Jones KW, Gay WA Jr, Doty DB, Karwande SV, Lee HR, Holland C, Menlove RL, Hagan ME, Bristow MR (1988) Increased cardiac allograft rejection in female cardiac transplant recipients. J Heart Transplant (in press)
3. Joyce LD, Johnson KE, Pierce WS, DeVries WC, Semb BKH, Copeland JG, Griffith BP, Cooley DA, Frazier OH, Cabrol C, Keon WJ, Auger F, Bucherl ES, Wolner E (1986) Summary of the world experience with clinical use of the total artificial hearts as heart support devices. J Heart Transplant 5: 229–35
4. Fragomeni LS, Kaye MP (1988) Registry of the International Society for Heart Transplantation: fifth official report — 1988. J Heart Transplant 7: 249–253
5. Mason JW (1985) Endomyocardial biopsy: the balance of success and failure. Circulation 71: 185–8
6. O'Connell JB (1987) The role of myocarditis in end-stage dilated cardiomyopathy. Tex Heart Inst J 14: 268–75
7. O'Connell JB, Mason JW (1988) Immunosuppressive therapy in experimental and clinical myocarditis. Pathol Immunopathol Res (in press)
8. O'Connell JB, Renlund DG, Robinson JA, Fowler MB, Oyer PE, Pifarre R, Grady KL, Mullin AV, Menlove RL, Gay WA Jr, Bristow MR (1988) Effect of post-operative hemodynamic support on survival following cardiac transplantation. Circulation (in press)

Clinical and Experimental Aspects of Treatment for Viral Myocarditis

C. Kawai

Third Division, Department of Internal Medicine, Kyoto University Hospital,
54 Kawaracho Shogoin, Sakyo-Ku, Kyoto 606, Japan

Introduction

No reliable controlled study on immunosuppressive therapy, virus-specific vaccine, and antiviral treatment against viral or idiopathic (nonspecific) myocarditis has been carried out in a clinical setting. As a result, the treatment for viral or nonspecific myocarditis is still symptomatic and supportive in most instances. The purpose of the present study is to review the outcome of 31 patients, aged 11–68 years, with myocarditis of unknown etiology who had been treated symptomatically, and to assess the role of immunosuppressive treatment on the basis of experimental studies. Diagnosis of myocarditis was mainly based on endomyocardial biopsy or autopsy findings as well as clinical signs and symptoms. Viral etiology was highly suspected, but no confirmation of its definite evidence has ever been obtained.

Clinical Features of acute myocarditis [1]

All of the patients showed electrocardiographic abnormalities, in particular, disturbances in the conducting system. The ST segment elevation, flattening or inversion of T waves, abnormal Q waves, ventricular arrhythmias, and complete atrioventricular block were common. Stokes-Adams syndrome was frequently an initial symptom and found in more than half of the patients. A “flu-like” syndrome preceded the illness in 21 (68%) of the 31 patients. Mild to moderate fever was also common. Marked cardiac enlargement on the chest X-ray film was rather rare; however, slight but definite cardiac enlargement was usually recognized when serial films were carefully reviewed (Table 1).

Accelerated erythrocyte sedimentation rate, positive C-reactive protein (CRP), mild to moderate leucocytosis and a rise of serum enzymes such as glutamic oxalo-acetic transaminase (GOT), glutamic pyruvic transaminase (GPT), and lactate dehydrogenase (LDH) were usually, but not always, present in the acute phase of myocarditis. Elevation of antistreptolysin-O (ASO) titer was not common (Table 2).

Myocarditis may sometimes rapidly progress to circulatory collapse or it may lead to congestive heart failure as seen in Fiedler’s myocarditis. In our series of 31 patients with idiopathic myocarditis, however, 21 patients improved only with

H.-P. Schultheiß (Ed.)
New Concepts in Viral Heart Disease

Table 1. Clinical features in 31 patients with idiopathic myocarditis

	(*n*)	(%)
ECG changes (ST·T changes, Q waves, ventricular arrhythmias and atrioventricular block)	31	100
Stokes-Adams syndrome	18	58
Preceding "flu-like" syndrome	21	68
Mild to moderate fever	21	68
Cardiothoracic ratio $\geqq 0.55$	10	32

Table 2. Laboratory findings in 31 patients with idiopathic myocarditis

	(*n*)	(%)
Increased blood sedimentation rate (> 10 mm/h)	20/31	65
Positive C-reactive protein	18/31	58
Leucocytosis (> 8000/mm^3)	17/31	55
Glutamic oxalo-acetic transaminase (> 32 IU/liter)	21/31	68
Glutamic pyruvic transaminase (> 26 IU/liter)	22/29	76
Lactate dehydrogenase (> 475 IU/liter)	12/29	41
Antistreptolysin-O (> x 160)	5/29	17

symptomatic or supportive treatment. Two patients died of fulminent myocarditis. Three patients are alive with a permanent pacemaker, and two died despite implantation of a permanent pacemaker. Three patients with severe myocarditis and circulatory collapse survived the acute phase with the aid of an intra-aorta balloon pump (IABP) and temporary pacemaker and eventually completely recovered from the illness (Table 3).

The results indicate that 21 of 31 patients (68%) with acute myocarditis can recover with only symptomatic or supportive treatment. The use of an IABP may further increase the survival in more deteriorated cases.

Table 3. Treatment and outcome of 31 patients with idiopathic myocarditis

		(*n*)
Symptomatic of supportive treatment	Living	21
	Died	2
Permanent pacemaker implanted	Living	3
	Died	2
Intra-aorta balloon pump	Living	3

What is, then, the role of specific therapy with the use of immunosuppressive agents and/or antiviral drugs for acute myocarditis?

Kawamura et al. reported that at least eight of 218 patients with viral or idiopathic myocarditis collected from 62 institutions throughout Japan developed clinical pictures compatible with those of dilated cardiomyopathy within an average of 5.5 years after the onset of symptoms. We have no idea at all how many subclinical myocarditis may progress to dilated cardiomyopathy. If there is evidence to support the view that the patients with myocarditis can develop dilated cardiomyopathy, and that the immunological mechanisms are involved in the disease process, then immunosuppressive treatment should be considered in the treatment of patients with acute myocarditis.

Experimental Studies on Murine Viral Myocarditis

After many unsuccessful trials, we were able to produce severe myocarditis in an inbred strain of DBA/2 mice inoculated with the M variant of encephalomyocarditis (EMC) virus [2]. The mice which survived the acute stage of myocarditis developed cardiac lesions resembling dilated cardiomyopathy 3 months after viral inoculation. We believe that the development of cardiac dilatation and hypertrophy depends on the severity of myocardial lesions which have to be severe, but permit animals to survive. What then determines the severity of myocardial lesion in myocarditis?

Immunological Behavior of Lymphocytes in Experimental Viral Myocarditis

By using an immunofluorescence method, we examined the serial changes in the percentage of T and B lymphocytes in the heart,spleen, and peripheral blood of DBA/2 mice inoculated with EMC virus [3]. There was a marked decrease in T cells in the peripheral blood and a moderate decrease in the spleen on day 13. T cells accounted for approximately 80% of the cells in the myocarditium on days 7 and 14. There were no significant changes in B cells in the peripheral blood or the spleen throughout the entire experiment.

Pathological examination of the heart confirmed less prominent lesions in the nude mice (BALB/c-nu/nu) than in BALB/c-nu/+ mice or in BALB/c-nu/nu mice injected with spleen cells from BALB/c-nu/+ mice inoculated with EMC virus. Virus titrations of the heart and serum neutralizing antibody titers showed no significant differences among these three groups of mice [4]. The survival rate of the nude mice inoculated with EMC virus on days 9–15 was significantly better than that of nu/+ mice and nude mice injected with spleen cells from nu/+ mice. Mice died of congestive heart failure on days 9–15.

Recently, Kishimoto et al. [5] found that the percentage of pan-T cells (The1.2$^+$), particularly of the immature T cell (Lyt1$^+$, 23$^+$) subset decreased in peripheral blood of the mice on days 7 and 14 after virus inoculation. In the myocardium, the majority of infiltrating cells on days 7 and 14 after virus inoculation were pan-T cells, most of which were Lyt1-positive and Lyt2-positive cells confirmed immunohistologically. They concluded that myocardial Thy1.2$^+$ (pan-T) cells, especially Lyt1$^+$, 23$^+$ (immature and T cell subset precursors) subsets have some pathogenic role in the development of murine viral myocarditis.

Effects of Prednisolone, Cyclosporine, and Antithymocyte Serum on Acute Viral Myocarditis in Mice

The effect of prednisolone on viral myocarditis was studied in BALB/c mice inoculated with EMC virus [6]. When prednisolone was given early, i. e., on days 4–13 after inoculation, viral titers in the myocardium remained significantly high on day 10, at which time only one of five control mice had demonstrable, but very low, viral titers; on the other hand, the neutralizing antibody titers were significantly lower than those of the control group on days 8 and 10 ($P < 0.01$ for each). As a result, the survival rate of the prednisolone group was significantly lower ($P < 0.05$) than that of the control group on days 21–23. If prednisolone was started on day 8 when the neutralizing antibody titers were already high, the average titers of the prednisolone group were elevated to the same level as the control group on day 14, and the survival rate of the prednisolone group showed no difference from that of the control group throughout the experiment until day 27.

Similarly cyclosporine resulted in greater mortality when administered early in the illness; and there was no evident reduction in myocardial injury to suggest possible long-term benefit [7].

When the mice inoculated with Coxsachie B3 virus were treated with antithymocyte serum, histological changes of myocarditis were generally mild, but in a few mice it caused more severe and extensive myocardial lesions than in control mice [8]. The effect of antithymocyte serum on murine viral myocarditis needs further investigation.

Conclusion

From the series of animal experiments it is concluded that some immunological mechanisms are involved in the severity of viral myocarditis and thus in the development of dilated cardiomyopathy. The use of immunosuppressive agents with prednisolone or cyclosporine in murine viral myocarditis is associated with greater mortality if the agents are administered early in the illness. Although interpolation of the present studies on murine viral myocarditis to man should be cautious, it should be noted that immunosuppressive treatment in viral myocarditis may be detrimental in some cases of myocarditis in man. However, the possible use of antiviral agents associated with the immunosuppressive therapy may cause beneficial effects on the treatment of acute myocarditis. In addition, the role of immunosuppressive agents to prevent the transition of myocarditis to dilated cardiomyopathy should be elucidated with a reliable controlled study.

References

1. Kawai C, Matsumori A, Fujiwara H (1987) Myocarditis and dilated cardiomyopathy. Ann Rev Med 38: 221–239
2. Matsumori A, Kawai C (1982) An animal model of congestive (dilated) cardiomyopathy: dilatation and hypertrophy of the heart in the chronic stage in DBA/2 mice with myocarditis caused by encephalomyocarditis virus. Circulation 66: 355–360
3. Kishimoto C,Kuribayashi K, Masuda T, Tomioka N, Kawai C (1985) Immunological behavior of lymphocytes in experimental viral myocarditis: significance of T lymphocytes in the severity of myocarditis and silent myocarditis in BALB/c-nu/nu mice. Circulation 71: 1247–1254
4. Kishimoto C, Kuribayashi K, Masuda T, Tomioka N, Kawai C (1986) Animal model of silent myocarditis in athymic mice. Cardiovasc Res 20: 768–773
5. Kishimoto C, Kuribayashi K, Fukuma K, Masuda T, Tomioka N, Abelmann WH, Kawai C (1987) Immunologic identification of lymphocyte subsets in experimental murine myocarditis with encephalomyocarditis virus. Different kinetics of lymphocyte subsets between the heart and peripheral blood, and significance of Thy 1.2^+ (pan T) and Lyt 1^+23^+ (immature T) subsets in the development of myocarditis. Circ Res 61: 715–725
6. Tomioka N, Kishimoto C, Matsumori A, Kawai C (1986) Effects of prednisolone on acute viral myocarditis in mice. J Am Coll Cardiol 7: 868–872
7. Monrad ES, Matsumori A, Murphy JC, Fox JG, Crumpacker CS, Abelmann WH (1986) Therapy with cyclosporine in experimental murine myocarditis with encephalomyocarditis virus. Circulation 73: 1058–1064
8. Kawai C, Matsumori A, Kitaura Y, Takatsu T (1978) Viruses and the heart: viral myocarditis and cardiomyopathy. Prog Cardiol 7: 141–162

VII. New Aspects of Clinical Management

Alterations of Receptors and of Agonist Effectiveness in Myocardial Failure

E. ERDMANN, and M. BÖHM

Medizinische Klinik I der Universität München, Klinikum Großhadern, Marchioninistr. 15, D-8000 München 70, FRG

Introduction

In heart failure, besides the increased catecholamine levels [1], a downregulation of cardiac β-adrenoceptors is found, as well as a diminished positive inotropic effect of β-adrenoceptor agonists and of phosphodiesterase inhibitors [2–10]. The vasoconstrictive effects of noradrenaline, however, seem to be preserved, as can be judged by the increased peripheral resistance (impedance) in severe heart failure (Fig. 1).

β-Adrenoceptor agonists (dobutamine, dopamine, etc.) and phosphodiesterase inhibitors (amrinone, milrinone, enoximone, etc.) play a role in the treatment of heart failure [11, 12]. Therefore, it is important to find out:

- To what extent β-adrenoceptors are downregulated in the various states of heart failure
- Whether only the positive inotropic effects of β-adrenoceptor agonists are diminished or whether this applies to other drugs as well (for instance, digitalis, calcium channel agonists, histamine, etc.)

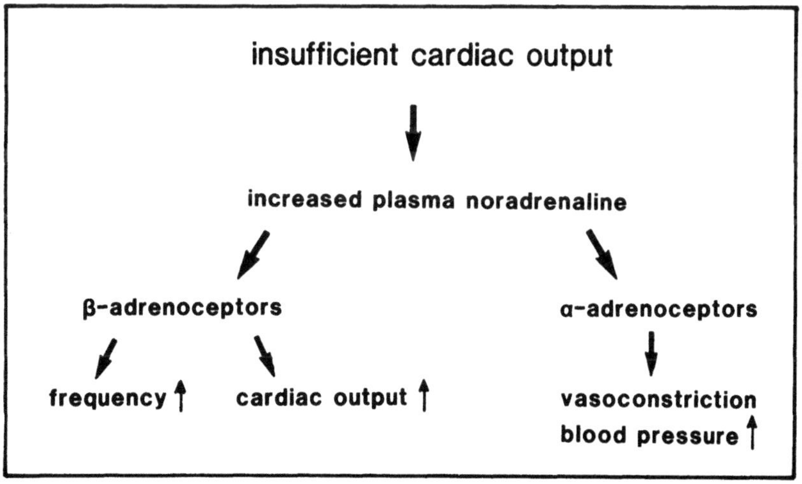

Fig. 1. In acute heart failure, increased levels of plasma noradrenaline stimulate β- and α-adrenoceptors to increase cardiac output and blood pressure. In chronic heart failure, increased plasma noradrenaline is less effective in respect to cardiac output due to a downregulation of β-adrenoceptors. The unaffected α-adrenoceptors, however, are responsible for an increased afterload in this situation

H.-P. Schultheiß (Ed.)
New Concepts in Viral Heart Disease

- Whether changes of cardiac cell membranes or the type of heart disease is related to this functional state of the heart

We have tried to answer these questions by measuring α-adrenoceptors by ^{3}H-prazosin binding, β-adrenoceptors by ^{3}H-CGP 12177 binding to human cardiac cell membrane preparations and by quantitating the positive inotropic effects of the various drugs on papillary muscle strips from human hearts, either after explantation (before cardiac transplantation), during mitral valve replacement surgery, or from donor hearts (normals) that could not be transplanted because of technical reasons. Detailed methodological information on these experiments has been published elsewhere [8–10].

Results and Discussion

Adrenoceptors

In accordance with other groups [2–7], we found a significantly decreased number of β-adrenoceptors in human myocardial tissue from left ventricle in heart failure (Fig. 2). It is interesting that in severe heart failure there are only about 25% as

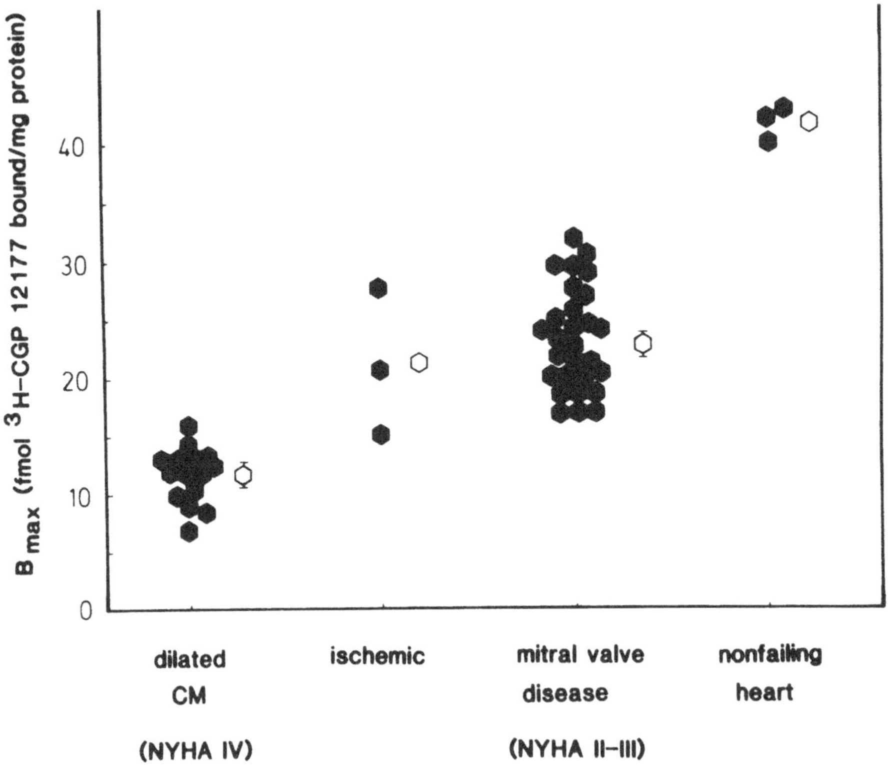

Fig. 2. β-Adrenoceptors in the human heart. The number of myocardial β-adrenoceptors in cardiac membranes from patients with dilated cardiomyopathy *(CM)*, ischemic heart diseases, mitral valve disease (NYHA II-III) and nonfailing hearts. (Modified from [10])

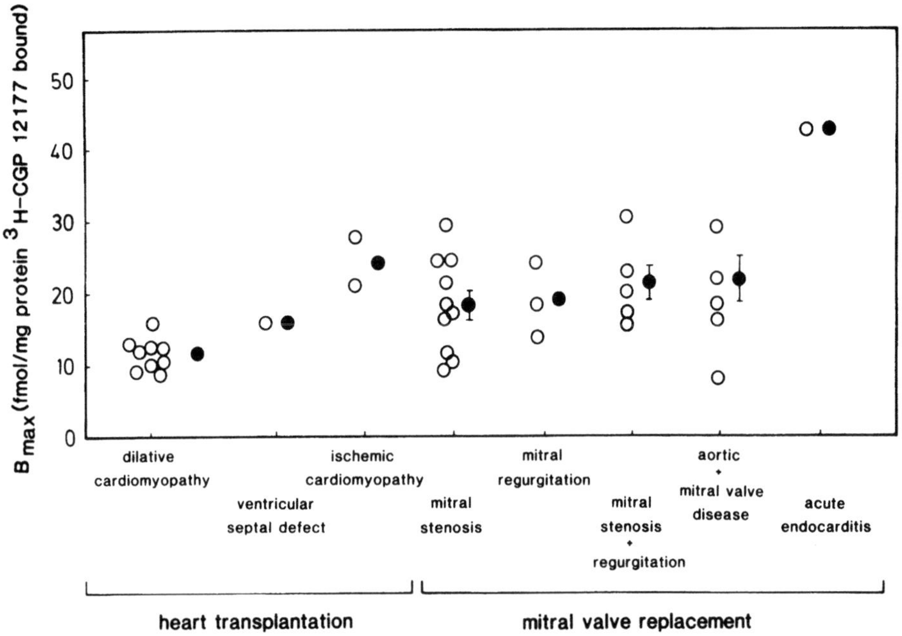

Fig. 3. The number of cardiac β-adrenoceptors in cardiac membranes from patients with various heart diseases. The worse the state of heart failure, the less β-adrenoceptors are measurable in cardiac tissue. The patient with acute endocarditis was operated on because of uncontrolled bacterial infection, not because of heart failure.

many ^{3}H-CGP 12177 binding sites as in the healthy explanted human myocardium, whereas in moderate heart failure (mitral valve diseases, NYHA II-III) there are about 50% fewer β-adrenoceptors. This concurs with the findings of Cohn et al. [12] of extremely high noradrenaline levels in severe heart failure and moderately increased serum noradrenaline concentrations in less severe heart failure. The downregulation of β-adrenoceptors is apparently dependent on the noradrenaline serum levels which might lead to permanent occupation of β-adrenoceptors and stimulation of adenylate cyclase. A detailed analysis of the various diseases leading to downregulation of β-adrenoceptors (Fig. 3) supports this idea. Independently of the disease, the number of receptors is apparently decreased by the state of heart failure. The myocardium of the patient without heart failure but with uncontrolled endocarditis contained a similar number of β-adrenoceptors as the nonfailing myocardium (Fig. 3).

The affinity of the receptors for ^{3}H-CGP 12177 was the same in all conditions tested. α-Adrenoceptors were the same in number and in affinity for ^{3}H-prazosin in terminal (NYHA IV) and in moderate heart failure (NYHA II-III). This may be the reason why the detrimental effects of noradrenaline in severe heart failure, i.e. raised afterload (Fig. 1), are maintained and finally lead to a vicious circle.

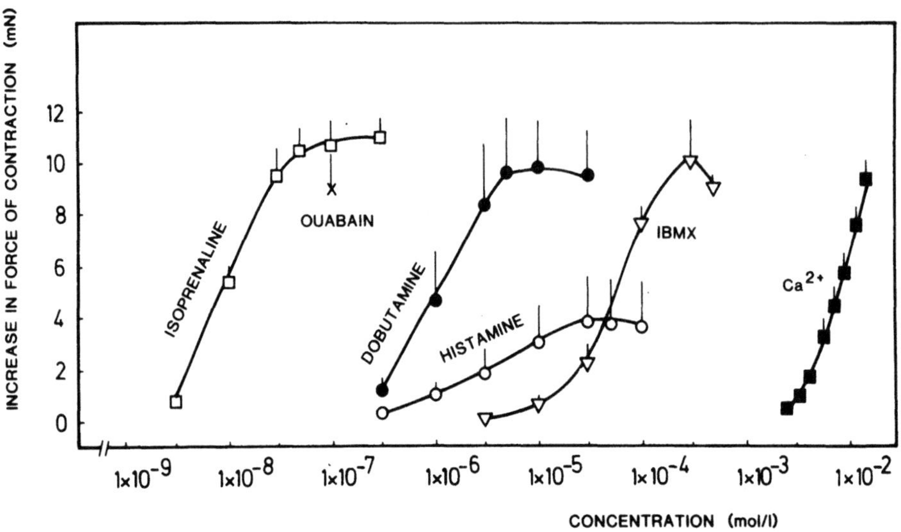

Fig. 4. Increase in isometric contractile force in the healthy human papillary muscle strips. As can be seen clearly, the drugs used stimulated contractile force to about the same maximum as Ca^{2+} did — with the exception of histamine which was less effective. Any higher concentration of the drugs used caused arrhythmia or contracture

Contractile Force

In papillary muscle strips from healthy human explanted hearts, maximal contractile force is about the same for isoprenaline, ouabain, dobutamine, isobutylmethylxanthine (IBMX) and Ca^{2+} (Fig. 4). Histamine was found to be of significantly less effectiveness. In severe heart failure, isoprenaline, dobutamine, IBMX, and the calcium channel agonist Bay-k-8644 [13] were significantly less effective than Ca^{2+}, digitalis, forskolin and dibutyryl-cyclic adenosine monophosphate (db-cAMP) (Fig. 5). This indicates that the β-adrenoceptor is defect or nonfunctioning, whereas contractile force can be stimulated almost up to the same maximum as in the healthy heart by drugs acting independently of the β-adrenoceptor [14]. However, phosphodiesterase inhibitors such as IBMX, amrinone, and milrinone also cannot increase contractile force significantly in heart failure [4, 8–10]. Therefore, one must consider the possibility of decreased intracellular cAMP-levels in heart failure [4, 15]. This hypothesis is supported by the finding of increased potency and effectiveness of milrinone after prestimulation of the failing papillary muscle with isoprenaline, which by itself has only minimal positive inotropic effects in these hearts (for details see [15]). Phenylephrine as an α-adrenoceptor agonist in contrast to isoprenaline has a similar effectiveness in moderate and severe heart failure (Fig. 6). This finding agrees with the small but unchanged number of α-adrenoceptors in severe heart failure [16]. However, due to the overall small positive inotropic effect, let alone the potentially detrimental effects on the coronary arteries, we do not suggest the use of α-adrenoceptor agonists as therapeutic agents in myocardial failure.

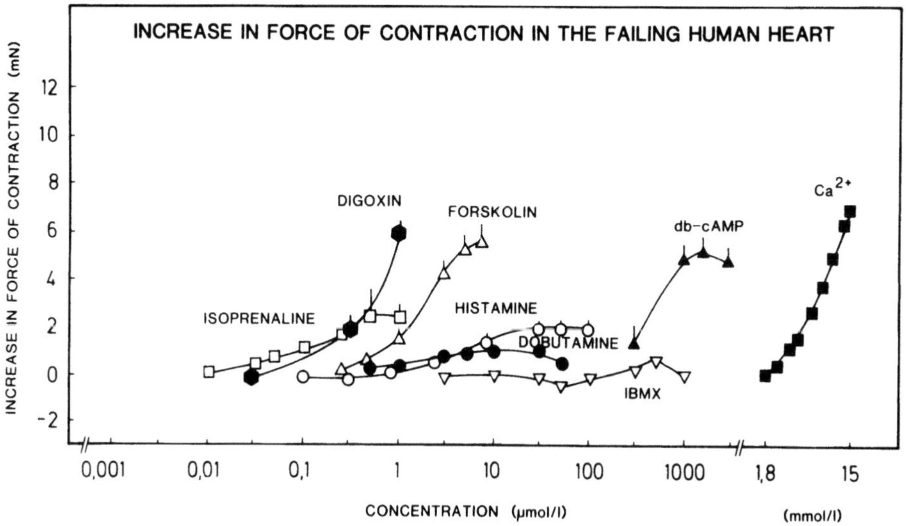

Fig. 5. In papillary muscle strips from explanted human hearts (dilated cardiomyopathy, end-stage heart failure), isoprenaline, histamine, dobutamine, and IBMX were much less effective than digoxin, forskolin, dibutyryl-cAMP or Ca^{2+} with respect to contractile force

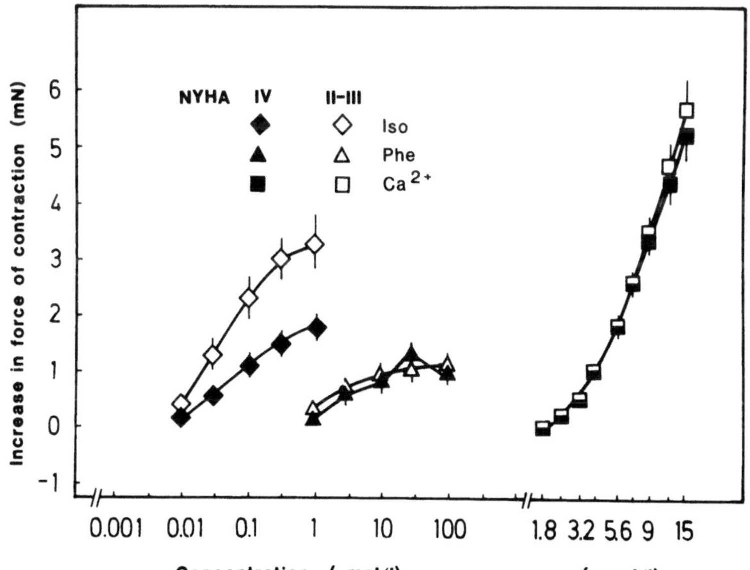

Fig. 6. Concentration-response curves for the effects of isoprenaline *(Iso)*, phenylephrine *(Phe)* or Ca^{2+} in electrically driven papillary muscle strips from patients with moderate (NYHA II-III) or severe (NYHA IV) heart failure. The effectiveness of Phe is the same in moderate and terminal heart failure in contrast to that of isoprenaline. (Modified from [16])

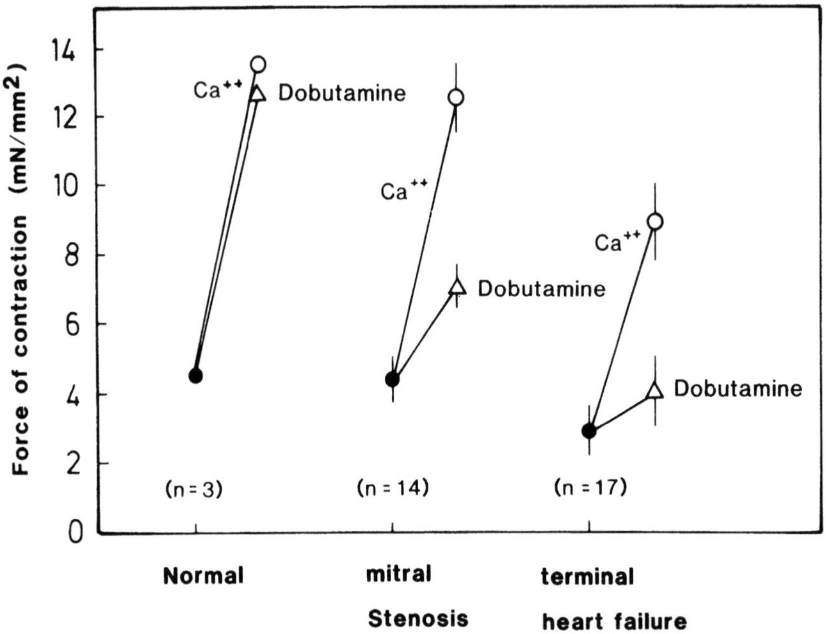

Fig. 7. Contractile force in the "normal" and in the failing human myocardium. It can be seen clearly that the positive inotropic effect of dobutamine ($5x10^{-5}$ *M)* is less the more severe the heart failure. Also, basal contractile force (in the presence of tyrode solution after optimal preload has been achieved) and maximal contractile force (in the presence of 15 m*M* Ca^{2+}) declines with progression of heart failure

We wanted to find out whether the more or less unaffected papillary muscle in mitral stenosis shows any differences as compared with terminal heart failure in dilated cardiomyopathy. Therefore, we measured basal contractile force (after optimal preload in tyrode solution) and maximal increase in contractile force (after 15 m*M* Ca^{2+} or $5x10^{-5}$ *M* dobutamine) (Fig. 7). The results show that maximal effects of Ca^{2+} in heart failure NYHA II-III are slightly less pronounced than in the normal heart but that dobutamine has only about 50% of the Ca^{2+} effects. In terminal heart failure due to dilated cardiomyopathy, basal as well as maximal (Ca^{2+}-induced) contractile forces are reduced, and dobutamine is almost ineffective.

Thus, apart from the well-known downregulation of β-adrenoceptors, there must be a further defect in one or several of the chain links that connect the cell membrane with the contractile proteins in severe heart failure. Nevertheless, it is noteworthy that some drugs are ineffective in end-stage myocardial failure or cannot make full use of the "contractile reserve" of the diseased muscle. Thus, these drugs probably should not be used for therapy. If positive inotropic effects are wanted in this situation, one should make use of effective agents.

Summary

The number of α- and β-adrenoceptors, as well as the effect of positive inotropic drugs on isometric contractile force was investigated in human papillary muscle strips from healthy heart donors, patients with mitral valve disease, and patients with terminal heart failure due to dilated cardiomyopathy.

A decreased number of β-adrenoceptors was found corresponding inversely to the degree of heart failure (NYHA II-III-IV). The number of α-adrenoceptors was very low but stayed constant in severe myocardial failure. Maximal increases of contractile force of the papillary muscle strips were about the same for Ca^{2+} (12 m*M*), dobutamine, isoprenaline, IBMX, milrinone, ouabain, and digoxin in the healthy hearts. In heart failure all drugs acting through β-adrenoceptors and phosphodiesterase inhibition were less effective than agents with other modes of action (digitalis, forskolin, db-cAMP, etc.). The effectiveness of phenylephrine was unchanged but low in heart failure.

Thus, the search for drugs with positive inotropic effects through systems other than the β-adrenoceptors seems worthwhile, as most of the presently known new positive inotropic drugs have reduced effects just when they are needed most, in severe heart failure.

Acknowledgements. The experimental studies were supported by the Deutsche Forschungsgemeinschaft (Er 65/4–5) and Friedrich-Baur-Stiftung, Munich, FRG.

References

1. Francis GS (1985) Plasma catecholamine levels in patients with congestive heart failure. Cardiovasc Rev Rep 6: 444–454
2. Bristow MR, Kantrowitz NE, Ginsburg R, Fowler MB (1985) β-Adrenergic function in heart muscle disease and heart failure. J Mol Cell Cardiol 17 [Suppl 2]: 41–52
3. Bristow MR, Ginsburg R, Minobe W, Cubicciotti RS, Sageman WS, Lurie K, Billingham ME, Harrison DC, Stinson EB (1982) Decreased catecholamine sensitivity and β-adrenergic-receptor density in failing human hearts. N Engl J Med 307: 205–211
4. Feldman MD, Copelas L, Gwathmey JK, Phillips P, Warren SE, Schoen FJ, Grossman W, Morgan JP (1987) Deficient production of cyclic AMP: pharmacologic evidence of an important cause of contractile dysfunction in patients with end-stage heart failure. Circulation 75: 331–339
5. Baumann G, Riess G, Erhardt WD, Felix SB, Ludwig L, Blümel G, Blömer H (1981) Impaired beta-adrenergic stimulation in the uninvolved ventricle post-acute myocardial infarction: reversible defect due to excessive circulating catecholamine-induced decline in number and affinity of beta-receptors. Am Heart J 101: 569–581
6. Brodde O-E (1987) Cardiac beta-adrenergic receptors. ISI Atlas of Science: Pharmacol 1: 107–112
7. Fowler MB, Laser JA, Hopkins GL, Minobe W, Bristow MR (1986) Assessment of the β-adrenergic receptor pathway in the intact failing human heart: progressive receptor down-regulation and subsensitivity to agonist response. Circulation 74: 1290–1302
8. Böhm M, Beuckelmann D, Brown L, Feiler G, Lorenz B, Näbauer M, Kemkes B, Erdmann E (1988) Reduction of beta-adrenoceptor density and evaluation of positive inotropic responses in isolated, diseased human myocardium. Eur Heart J 9: 844–852
9. Brown L, Näbauer M, Erdmann E (1986) The positive inotropic response to milrinone in isolated human and giunea pig myocardium. Naunyn Schmiedebergs Arch Pharmacol 334: 196–201

10. Böhm M, Diet F, Feiler G, Kemkes B, Kreuzer E, Weinhold C, Erdmann E (1988) Subsensitivity of the failing human heart to isoprenaline and milrinone is related to β-adrenceptor downregulation. J Cardiovasc Pharmacol 12: 726–732
11. Brown L, Lorenz B, Erdmann E (1986) Reduced positive inotropic effects in deseased human ventricular myocardium. Cardiovasc Res 20: 516–520
12. Cohn JN, Levine TB, Olivari MT, Garberg V, Lura D, Francis GS, Simon AB, Rector T (1984) Plasma norepinephrine as a guide to prognosis in patients with chronic congestive heart failure. N Engl J Med 311: 819–823
13. Näbauer M, Brown L, Erdmann E (1988) Positive inotropic effects of the calcium channel activator Bay K 8644 on guinea-pig and human isolated myocardium. Naunyn Schmiedebergs Arch Pharmacol 337: 85–92
14. Erdmann E (1988) The effectiveness of inotropic agents in isolated cardiac preparations from the human heart. Klin Wochenschr 66: 1–6
15. Böhm M, Diet F, Kemkes B, Erdmann E (1988) Enhancement of the effectiveness of milrinone to increase force of contraction by stimulation of cardiac beta-adrenoceptors in the failing human heart. Klin Wochenschr 66: 1051–1052
16. Böhm M, Diet F, Feiler G, Kemkes B, Erdmann E (1988) α-Adrenoceptor and α-adrenoceptor-mediated positive inotropic effects in failing human myocardium. J Cardiovasc Pharmacol 12: 357–364

Positive Inotropic Drugs in the Treatment of Chronic Congestive Heart Failure: A Critical Update

W. Motz, and B. E. Strauer

Universität Düsseldorf, Medizinische Klinik und Poliklinik B, Abteilung für Kardiologie, Pneumologie und Angiologie, Moorenstraße 5, D-4000 Düsseldorf 1, FRG

There are many patients with severe forms of congestive heart failure who cannot be hemodynamically recompensated even through the combined use of digitalis glycosides, diuretics, vasodilators and angiotensin-converting enzyme inhibitors [30]. In this particular group of patients, sustained improvement of myocardial contractility through positive inotropic stimulation of the myocardium is a critical therapeutic goal [29].

Catecholamines such as dobutamine, dopamine, and epinephrine are potent inotropic agents which are highly effective in enhancing left ventricular pump function through positive inotropic stimulation. However, their use is limited to the intravenous route of application and they cannot be used in the long-term management of congestive heart failure [30, 34]. Recently, the pharmaceutical industry has developed new oral cardiotonic agents to improve symptoms and functional capacity of patients with chronic cardiac failure. Because of the differences in their mechanism of action [6] (Table 1), these new positive inotropic substances can be principally divided into

a) derivatives of catecholamines; and
b) phosphodiesterase inhibitors.

Table 1. Sympathomimetics

Reference	Substance	Short-term effects	Long-term effects
[17]	Dobutamine	++	
[34]	Dobutamine	+	+
[21]	Salbutamol	++	+ (4/20 patients VT)
[5]	Pirbuterol	++	+ Tolerance
[3]	Pirbuterol	++	+
[13]	Prenalterol	++	(+) Tolerance
[31]	Prenalterol	++	(+) Tolerance
[27]	Corwin	++	+
[26]	Levodopa	++	++

VT, ventricular tachycardia

H.-P. Schultheiß (Ed.)
New Concepts in Viral Heart Disease

Derivatives of Catecholamines

Stimulation of beta$_1$-receptors causes an enhancement of myocardial contractility as well as increasing heart rate and facilitating atrioventricular conductance, whereas stimulation of beta$_2$-receptors leads to relaxation of smooth muscle. The relation of beta$_1$ and beta$_2$-effects determines and characterizes the hemodanymic profile of catecholamine derivatives. Salbutamol, terbutaline, and pirbuterol were originally developed for asthma therapy and have strong beta$_2$-effects [3, 5, 7, 21]. Accordingly, these substances have a pronounced vasodilating capacity which causes arterial hypotension.

Pirbuterol

Awan [3] and Dawsib [7] described, in uncontrolled studies, a favorable effect of pirbuterol in the treatment of chronic cardiac failure, whereas Colucci [5] observed the occurrence of tolerance and could not find a long-term efficacy. In the latter study, the density of beta-receptors on the lymphocyte surface was found to be decreased.

Prenalterol

Prenalterol is a stronger beta$_1$-stimulator than pirbuterol and additionally posseses some beta-receptor-blocking effects. The intrinsic sympathetic-mimetic activity is about 56% of the mere beta$_1$-agonist isoprenaline. After an acute intravenous application of prenalterol, cardiac output, stroke volume, and maximal dp/dt increase along with a drop in left ventricular filling pressure. As a consequence of afterload reduction through the decrease in end-diastolic volume, myocardial energy demand does not increase despite the increment in isovolumetric contractility indexes. However, two studies have clearly indicated the development of tolerance in the long-term management of heart failure [13, 31].

Corwin

Under resting conditions corwin (ICI 118587) stimulates beta$_1$-receptors, whereas during physical exercise beta$_1$-receptors are blocked by corwin. In patients with coronary artery disease, corwin has an antianginal effect [22]. Due to its pharmacodynamic profile, corwin seems to have a potential in the management of less severe forms of heart failure. Two studies demonstrate a long-term efficacy of corwin [22, 27]. The risk of development of tolerance might be less as a consequence of its beta-receptor-blocking properties than under therapy with Beta-Stimulators.

Salbutamol, Dobutamine, Levodopa

Salbutamol, an antiasthmatic substance, has been found to be effective in the improvement of symptoms of chronic cardiac failure [21]. On the other hand, in 20% of the patients treated, ventricular tachycardias occurred [21]. Regarding dobutamine, there is still a lack of evidence for a long-term efficacy of long-term intermittent dobutamine infusions. The efficacy of levodopa, which is decarboxylated to form dopamine, can also not be considered as an established alternative to digitalis, because a sustained improvement in cardiac function was observed only in five patients enrolled in a trial to evaluate the effectiveness of long-term therapy with levodopa [26].

Phosphodiesterase Inhibitors

Phosphodiesterase inhibitors are characterized by their pharmacodynamic ability to inhibit type 3 phosphodiesterase at very high doses and thus increase the intracellular cyclic adenosine monophosphate (AMP) concentration [10]. Myocardial calcium influx is increased simultaneously. Some substances such as UDCG 115 BS are considered as also having a direct effect on the actomyosin system.

Amrinone

In the isolated papillary muscle amrinone increases myocardial contractility by 50% when administered at the high dose of 3×10^{-3} *M* per liter. Vasodilation can already be achieved at lower doses of 5×10^{-3} *M* per liter. Clinically therapeutically effective plasma levels of amrinone are between 5×10^{-6} and 2×10^{-5} *M* per liter.

Doses in vitro and in vivo conditions are markedly different. Accordingly, experimental results based on papillary muscle experiments in respect of amrinone should be very carefully tested for their clinical value. Furthermore, it should be considered that the healthy rat papillary muscle does not reflect the disabled papillary muscle of a patient with dilative cardiomyopathy.

After short-term treatment with amrinone, cardiac output and stroke volume increase, whereas pulmonary artery pressure drops markedly. Myocardial energy

Table 2. Positive inotropic substances

Sympathomimetics		Phosphodiesterase inhibitors	Others
Epinephrine	Prenalterol	Theophyllin	Impromidine
Norepinephrine	Pirbuterol	Amrinone	Forskulin
Dopamine	Salbutamol	Milrinone	
Dobutamine	Terbutaline	ARL 115 BS	
Dobexamine	Corwin	UDCD 11555	
	Levodopa	MDL 17043 (enoximon)	
	MDL 19205		

demand remains unchanged, most likely as a consequence of the marked afterload reduction through the decrease in heart size. Amrinone seems to be more effective than dobutamine regarding its effectiveness on cardiac index and preload reduction. As far as tolerance is concerned, no study has yet been published which clearly denies the development of tolerance to the effects of amrinone [1, 2, 11, 12].

Early studies [1] revealed substantial improvement of congestive heart failure symptoms and of functional capacity. These preliminary findings could not be confirmed by controlled randomized studies [8, 20] which found a progressive improvement in exercise tolerance both in the placebo group and the amrinone group. Along with the lack of evidence for long-term effectiveness, a high incidence rate of severe side effects such as thrombocytopenia, jaundice, and drug-related fever was observed. Accordingly, oral amrinone did not become a drug of choice in the long-term therapy of heart failure [11, 12, 20, 25].

Milrinone

Milrinone was developed from amrinone. At a comparable dose it is about five times more potent than amrinone as regards its hemodynamic effects, whereas the incidence rate of side effects is about four times lower [15]. Administered directly into the coronary artery, milrinone, at a dose of 50 μg/min, causes an increase in dp/dt_{max} without any indication of peripheral arteriolar vasodilation [16]. It seems that during milrinone treatment an increased positive inotropy contributes toward about one-third of the improved left ventricular function apart from vasodilation. Long-term studies [2, 14, 18] suggest that milrinone is an effective substance in the treatment of heart failure. However, the results of controlled randomized studies will decide on the clinical role of this substance in the future. Furthermore its arrythmiogenic effects should be considered [9].

ARL 115, UDCG 115 BS

ARL 115 was withdrawn from clinical studies because this substance was suspected to cause cancer at very high doses in animal experiments. UDCG 115 BS is a derivative of ARL 115 and is characterized by a long half-life (15 h).

Enoximone (MDL 17043), Piroximone (MDL 19205)

Both enoxinone and piroxinone are recently developed phosphodiesterase inhibitors with both positive inotropic and vasodilatory properties. Short-term enoximone or piroximone treatment improves cardiac output and stroke volume along with a decrease in left ventricular filling pressure and in total peripheral arterial resistance. Accordingly, both these substances are both positive inotropic agents and arterial vasodilatiors. Consequently, these substances have a very promising hemodynamic profile. During long-term therapy with enoximone, improvement in symptoms and maximal oxygen uptake was observed in many patients with moderate-to-severe heart failure. Inspite of symptomatic improvement, the 12-month survival rate was as high as 44% in Maskin's study [19]. Diarrhea in particular is the most common side effect of both of these drugs. Controlled trials can be expected to establish the safety and efficacy of enoximone [19, 28, 33, 35].

Conclusions

In the treatment of acute forms of heart failure, diuretics, vasodilators, and catecholamines are the agents of choice and have completely replaced digitalis [30]. However, digitalis remains the only oral positive inotropic drug in the management of chronic cardiac failure [14, 23, 24] because all the oral catecholamines show a development of tolerance to their effects. There is still lack of evidence for a high long-term efficacy of phosphodiesterase inhibitors as far as improvement in left ventricular pump function is concerned. Until now there has been no indication that positive inotropic drugs might prolong life and improve prognosis of patients with heart failure. Angiotensin-converting inhibitors are currently the only agents which might prolong life in severe heart failure [32].

This speaks in favor of the hypothesis that unloading the disabled myocardium is more reasonable than long-term positive inotropic stimulation which might exhaust the myocardial reserve and may even accelerate the natural course of disease.

References

1. Alousi AA, Dobreck HP (1983) Amrinone. In: Scriabine A (ed) New drugs annual, cardiovascular drugs. Raven, New York
2. Anderson JF, Jack M, Askins C, Gilbert EM, Menlove RL, Lutz JR (1986) Occurrence of ventricular arrhythmias in patients receiving acute and chronic infusions of milrinone. Am Heart J 111: 466–473
3. Awan MA, Jeedham K, Everson MK, Hermanovich J, Joyce JA, Mason DT (1981) Therapeutic efficacy of oral pirbuterol in severe chronic congestive heart failure: acute hemodynamic and long-term ambulatory evaluation. Am Heart J 102: 555–564
4. Baim DS, McDowell AV, Cherniles J, Monrad ES, Parker JA, Edelson J, Braunwald E, Grossman W (1983) Evaluation of a new bipyridine inotropic agent — milrinone — in patients with severe congestive heart failure. N Engl J Med 309: 748–756
5. Colucci WS, Alexander RW, Williams GH, Rude RE, Holman BL, Konstam MA, Wynne J, Mudge GH, Braunwald E (1981) Decreased lymphocyte beta-adrenergic-receptor density in patients with heart failure and tolerance to the beta-adrenergic agonist pributerol. N Engl J Med 305: 185–190
6. Colucci WS, Wright RF, Braunwald E (1986) New positive inotropic agents in the treatment of congestive heart failure. N Engl J Med 314: 290–299
7. Dawsib JR, Canepa-Anson R, Kuan P, Reuben S et al. (1983) Symptoms, hemodynamics, and exercise capacity during long-term treatment of chronic heart failure. Experience with pirbuterol. Br Heart J 50: 282–290
8. Dibianco R, Shabetai R, Siverman BD, v Leier C, Benotti JR (1984) Oral amrinone for the treatment of chronic congestive heart failure: results of multicenter randomized double-blind and placebo-controlled withdrawal study. J Am Coll Cardiol 4: 855–866
9. Holmes JR, Kubo SH, Cody RJ, Kligfield P (1985) Milrinone in congestive heart failure: observation on ambulatory ventricular arrhythmias. Am Heart J 110: 800–806

10. Honerjäger P, Schäfer-Korting M, Reiter M (1981) Involvement of cyclic AMP in the direct inotropic action of amrinone. Naunyn Schmiedebergs Arch Pharmacol 318: 112–120
11. Klepzig M, Baur X, Hauser F, Mernitz G, Fruhmann G, Strauer BE (1984) Rechtsventrikuläre Hämodynamik und Lungenfunktion nach Amrinone-Injektion. Z Kardiol 73: 623–627
12. Klepzig M, Kleinhans E, Büll U, Strauer BE (1985) Amrinone in Akut- und Langzeittherapie. Z Kardiol 74: 85–90
13. Lambertz H, Meyer J, Erbel R (1984) Long-term hemodynamic effects of prenalterol in patients with severe congestive heart failure. Circulation 69: 298–305
14. Lee DCH, Johnson RA, Bingham JB, Leathy M, Dinsmore RE, Goroll AH, Newell JB, Strauss HW, Haber E (1982) A randomized trial of digoxin versus placebo. N Engl J Med 306: 699–705
15. Likoff MJ, Weger KT, Andrews V, Janicki JS, Wilson H, Rocci ML (1985) Milrinone in the treatment of chronic cardiac failure: a controlled trial. Am Heart J 110: 1035–1042
16. Ludmer PL, Wright RF, Malcolm J, Arnold O, Ganz P, Braunwald E, Colucci WS (1986) Separation of the direct myocardial and vasodilator actions of milrinone administered by an intracoronary infusion technique. Circulation 73: 130–137
17. Maskin CS, Forman R, Sonnenblick EH, Frishman WH, Le Jemtel TH (1983) Failure of dobutamine to increase exercise capacity despite hemodynamic improvement in severe chronic heart failure. Am J Cardiol 51: 177–182
18. Maskin CS, Sinoway L, Chadwick B, Sonnenblick EH, Le Jemtel TH (1983) Sustained hemodynamic and clinical effects of a new cardiotonic agent, WIN 47203, in patients with severe congestive heart failure. Am Heart J 67: 1065–1070
19. Maskin CS, Weber KT, Janicki JS (1987) Long-term oral enoximone therapy in chronic cardiac failure. Am J Cardiol 60: 63C–67C
20. Massie B, Bourassa M, DiBianco R, Hess M, Konstam M, Likoff M, Packer M (1985) Long-term oral administration of amrinone for congestive heart failure: lack of efficacy in a multicenter controlled trial. Circulation 71: 963–971
21. Mettauer B, Rouleau J-L, Burgess JH (1985) Detrimental arrhythmogenic and sustained ventricular beneficial hemodynamic effects of oral salbutamol in patients with chronic congestive heart failure. Am Heart J 109: 840–847
22. Molajo A, Bennett DH (1985) Effect of xamoterol, a new $beta_1$ adrenoceptor partial agonist, on resting haemodynamic variables and exercise tolerance in patients with left ventricular dysfunction. Br Heart J 54: 17–21
23. Moss AJ, Davis HT, Conrad DL, DeCamilla JJ, Odoroff CL (1981) Digitalis-associated cardiac mortality after myocardial infarction. Circulation 64: 1150–1155
24. Muller JE, Turi ZG, Stone PH, Rude RE, Raabe DS, Jaffe AS, Gold HK, Gustafson N, Poole WK, Passamani E, Smith TW, Braunwald E, Milis Study Group (1986) Digoxin therapy and mortality after myocardial infarction. N Engl J Med 314: 265–271
25. Packer M, Medina N, Yushak M (1984) Hemodynamic and clinical limitations of long-term inotropic therapy with amrinone in patients with severe chronic heart failure. Circulation 70: 1038–1047
26. Rajfer I, Anton AH, Rossen JD, Goldberg LI (1984) Beneficial hemodynamic effects of oral levodopa in heart failure. N Engl J Med 310: 1357–1362
27. Rousseau MF, Pouleur H, Nannau M, Defry JMR, Brasseur LA, Lavenne F (1984). Improved physical capacity, during ICI 118, 587 administration in moderate heart failure: Preliminary results after 1 year of follow-up, European Heart J 5 Suppl 1
28. Rubin SA, Tabak L (1985) MDL 17043: short- and long-term cardiopulmonary and clinical effects in patients with heart failure. J Am Coll Cardiol 5: 1422–1427
29. Strauer BE (1981) Pathophysiologie der Herzinsuffizienz. Therapiewoche 31: 477–490
30. Strauer BE (1981) Differentialdiagnose und -therapie digitalisresistenter Formen der Myokardinsuffizienz. Therapiewoche 31: 2895–2924
31. Strauer BE, Bohn I, Hahn B, Kment A, Motz U (1984) Contractile, coronary and metabolic effects of the acute and long-term treatment of cardiac failure with prenalterol. J Cardiovasc Pharmacol 6: 491–498
32. Swedberg (1987): Effect of Enalapril on mortality in severe congestive heart failure. Results of the Cooperative North Scandinavian Enalapril Survival Study (consensus). N Engl J Med 316: 1429–1435
33. Treese N, Erbel R, Rhein S, Diefenbach C, Alken RG, Mohr-Kahaly S, Meyer J (1987) Langzeittherapie der chronischen Herzinsuffizienz mit Enoximon. Dtsch Med Wochenschr 112: 1163–1169

34. Unverferth DV, Magorien RD, Lewis RP, Leier CV (1980) Long-term benefit of dobutamine in patients with congestive cardiomyopathy. Am Heart J 100: 622–630
35. Uretzky BF, Generalovich T, Verbalis JG, Valdes AM, Reddy S (1985) MDK 17043 therapy in severe congestive heart failure: characterization of the early and late hemodynamic, pharmacokinetic, hormonal and clinical response. J Am Coll Cardiol 5: 1414–1421

Effect of Angiotensin-Converting Enzyme Inhibitors on Myocardial Perfusion

G. Ertl

Medizinische Klinik, Julius-Maximilians-Universität, Josef-Schneider-Straße 2, D-8700 Würzburg, FRG

Inhibition of the angiotensin I-converting enzyme (ACE) results in diminished production of angiotensin II which is the active vasoconstrictor peptide of the renin-angiotensin system. The effect of ACE inhibitors is, however, complex. Apart from their interference with the plasmatic or vascular renin-angiotensin system, they may affect the central [5] and, in particular, the sympathetic nervous system [3, 34] as well as the kallikrein-kinin system [28] and prostaglandins [43]. Finally, angiotensin II acts in a directly positively inotropic way [32], increases vascular permeability [26], and may attract neutrophils [15, 21]. Some converting enzyme inhibitors have been claimed to have free radical scavenger potency as do other drugs possessing an (SH) sulfhydryl-group [37]. Thus, there are multiple ways how ACE inhibitors may influence the heart and coronary circulation. Coronary circulation, however, is a peculiar topic for the pharmacologist since it is influenced by many factors including metabolic demand of the heart and mechanical compression ot the coronary vasculature by cardiac contraction.

In order to allowACE inhibitors to evoke a vasodilatory effect by inhibition of angiotensin II production, most investigators have considered an activation of the renin-angiotensin system to be essential. Several states of an activated renin-angiotensin system have been reported to be associated with the coronary dilatory effect of ACE inhibitors: dogs in sodium deprivation [30] or hypoxia [29], man in dependence on plasma renin activity [16] and after diuretic therapy [33], patients with dilative cardiomyopathy [17], and several models of myocardial ischemia [9, 12, 13].

We have studied the effect of ACE inhibitors and the role of the renin-angiotensin system in mediating this effect in several models of myocardial ischemia. Dogs were anesthesized and the chest was opened. Myocardial ischemia was induced either by simple coronary occlusion or by a snare around a coronary artery and atrial pacing. To study the role of the renin-angiotensin system in the control of coronary vascular tone more precisely, a coronary artery was cannulated and perfused at constant pressure. Regional myocardial function was assessed by ultrasonic crystals implanted into the subendocardial layer of myocardium. In this model, which has previously been described in detail [14], lowering coronary perfusion pressure mimics a progressive coronary stenosis [9].

Figure 1 shows the reaction of the coronary vascular system to the lowering of the coronary perfusion pressure in analogy to a progressive coronary stenosis. An increase in coronary conductance suggests coronary dilatation, a decrease sug-

H.-P. Schultheiß (Ed.)
New Concepts in Viral Heart Disease

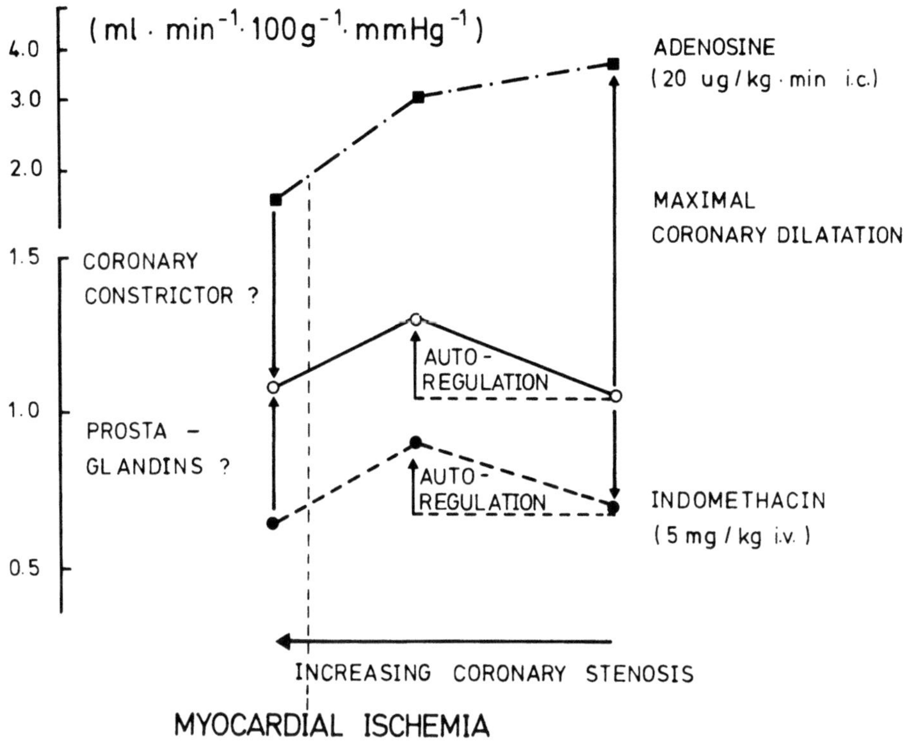

Fig. 1. Effect of lowering perfusion pressure of a cannulated coronary artery *(from right to left)* in analogy to progressive coronary stenosis on coronary conductance as determined distal to "coronary stenosis." *Solid line* represents control situation with an initial autoregulatory increase in coronary conductance (coronary vasodilatation). Further lowering of coronary perfusion pressure to induce myocardial ischemia *(left of broken vertical line)*, as indicated by lactate production and depression of myocardial function, results in a decrease rather than a further increase in coronary conductance indicating exhaustion of coronary autoregulation. Intracoronary infusion of adenosine increases coronary conductance, suggesting a vasodilatory reserve or a constrictive tone despite myocardial ischemia. The cyclooxygenase inhibitor indomethacin reduces coronary conductance, suggesting a vasodilatory tone of cyclooxygenase products of arachidonic acid in this model of myocardial ischemia

gests coronary constriction. As the degree of stenosis increases, coronary conductance increases. This vasodilatory response represents autoregulation which tends to maintain coronary blood flow. When, however, perfusion pressure is further lowered to a level which represents high-grade coronary stenosis and induces myocardial ischemia, as evidenced by myocardial lactate production and depression of myocardial function, coronary conductance falls, indicating not only exhaustion of autoregulation but even coronary constriction. Surprisingly, the unspecific vasodilator adenosine may almost double coronary conductance and thus unmask the coronary constrictor tone present during myocardial ischemia. Other groups have shown, by the use of radioactive microspheres, that this holds true for perfusion of subendocardial myocardial layers with the most critically reduced coronary flow [1, 2]. The cyclooxygenase inhibitor indomethacin does not change the pattern of this response of the coronary vascular system to mechanical restriction of coronary flow and myocardial ischemia. The fact that indomethacin may reduce coronary conductance even in the presence of myocardial ischemia supports previous reports on a role of products of the cyclooxygenase enzyme as mediators of a vasodilatory tone on coronary vasculature [38]. Thus, during myocardial ischemia coronary vascular tone is subject to vasoconstrictory as well as vasodilatory influences.

In addition to lowering perfusion pressure, we have paced the hearts in this model to intensify myocardial ischemia. When coronary perfusion pressure is maintained at the level of mean aortic pressure (considered "normal" coronary perfusion pressure), atrial pacing results in an increase in coronary blood flow proportional to myocardial metabolic demand. When, however, coronary perfusion pressure is lowered to 35 mmHg, a level at which myocardial ischemia is invariably present, atrial pacing results in a depression of coronary blood flow. The time course of this fall in coronary flow suggests that it results not only from mechanical inhibition (shortening of diastole) but, in addition, from any humoral or reflex mechanism [9]. Numerous other studies by our group and others have indicated that a coronary constrictive component may take place despite the presence of myocardial ischemia [1, 2, 9, 11, 18, 22, 24]; the mediator, however, has not yet been identified.

If we consider angiotensin II as a potential mediator of coronary constriction during myocardial ischemia, an activation of the renin-angiotensin system by myocardial ischemia has to be assumed. Figures 2a and b show the effect of brief (30-s) coronary occlusions on plasma renin activity and angiotensin II concentration in anesthetized dogs. Both plasma renin activity and angiotensin II concentration substantially increase as early as 30 s after coronary occlusion. This activation of the plasma renin-angiotensin system is maintained if coronary occlusion is maintained for up to 6 h [12]. Figure 2c shows an increase in plasma renin activity when myocardial ischemia is induced by coronary stenosis and atrial pacing, while pacing in the absence of coronary stenosis has no effect.

Finally, we measured plasmy renin activity in patients with coronary heart disease as compared to a control group (Fig. 3). Plasma renin activity was higher at rest and increased significantly during exercise only in patients with coronary heart disease. It must be noted that there was no difference between medication, blood pressure, and heart rate between the two groups. Patients with heart failure, hypertension, or other severe diseases were excluded from the study, and neither patients nor controls had received diuretics, ACE inhibitors, or beta-receptor blocking agents. In addition, other groups have reported elevated plasma renin activity and angiotensin II concentration in patients with acute myocardial infarction [4, 35].

The question is whether this activation of the plasma renin-angiotensin system has consequences for coronary vascular tone. Figure 4 shows the effect of the angiotensin II receptor blocker saralasin and the ACE inhibitor captopril on coronary vascular resistance in dogs following coronary artery occlusion as compared to sham-operated control dogs. While both captopril (0.25 mg/kg as initial bolus, followed by 0.25 mg/kg per hour) and saralasin (1 μg/kg per minute i. v.) reduced

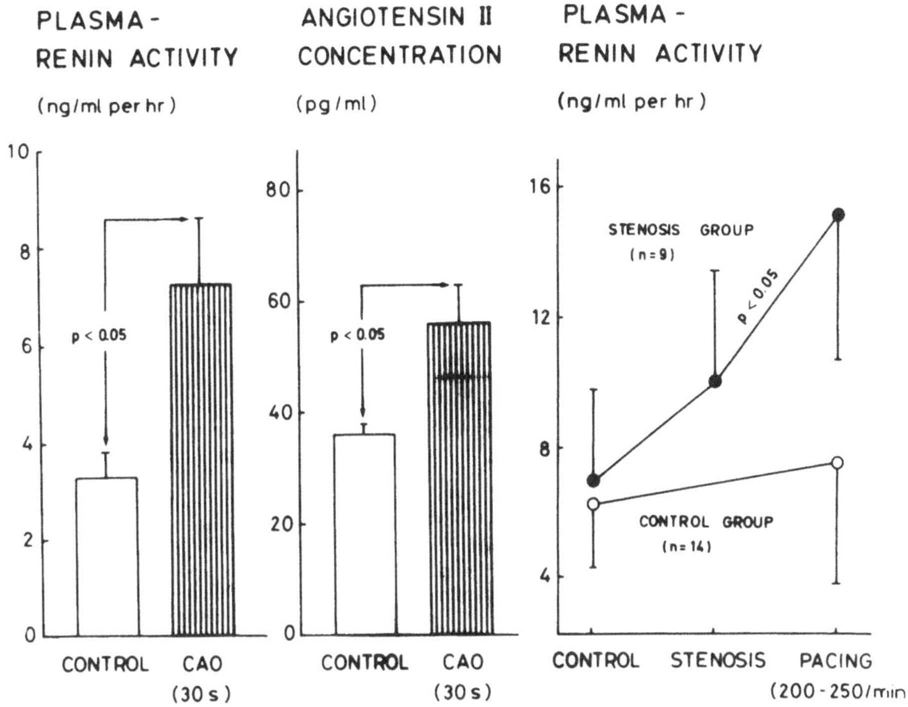

Fig. 2 a–c. Coronary artery occlusion *(COA)* for a period of 30 s **(a, b,)** induces an immediate increase in plasma renin activity and angiotensin II concentration (both measured by radioimmune assay). Atrial pacing, when induced in face of coronary stenosis to induce myocardial ischemia, results in elevated plasma renin activity; a control group showed no change of plasma renin activity during atrial pacing **(c)**. All experiments were performed in anesthetized dogs

coronary vascular resistance in dogs after 30 min and 6 h of coronary occlusion, respectively, suggesting coronary dilatation, they had no effect on coronary vascular resistance in the control group.

To sum up the results reported so far, Fig. 5 shows a schematic diagram of endogenous plasma levels and the effect of exogenous agonists and antagonists of the renin-angiotensin system. Endogenous plasma levels of renin and angiotensin I and II are elevated [4, 12, 13, 35]. The effect of exogenous angiotensin I and II is diminished, suggesting that receptors are occupied by endogenous angiotensin II, and the effect of converting enzyme inhibitor and angiotensin II receptor blocker is enhanced by myocardial ischemia [9, 13]. Thus, the renin-angiotensin system appears to be activated, may mediate coronary constriction, and thus offers ACE inhibitors the possibility of acting as coronary dilators.

One fact, however, was still puzzling in this hypothesis. During coronary occlusion in dogs, the angiotensin II receptor blocker saralasin reduced only coronary vascular resistance while systemic vascular resistance remained unchanged [13]. Our hypothesis was that the coronary effect of saralasin was mediated by a local coronary vascular renin-angiotensin system [13]. In fact, some more evidence has been accumulated for the functional importance of the system [8].

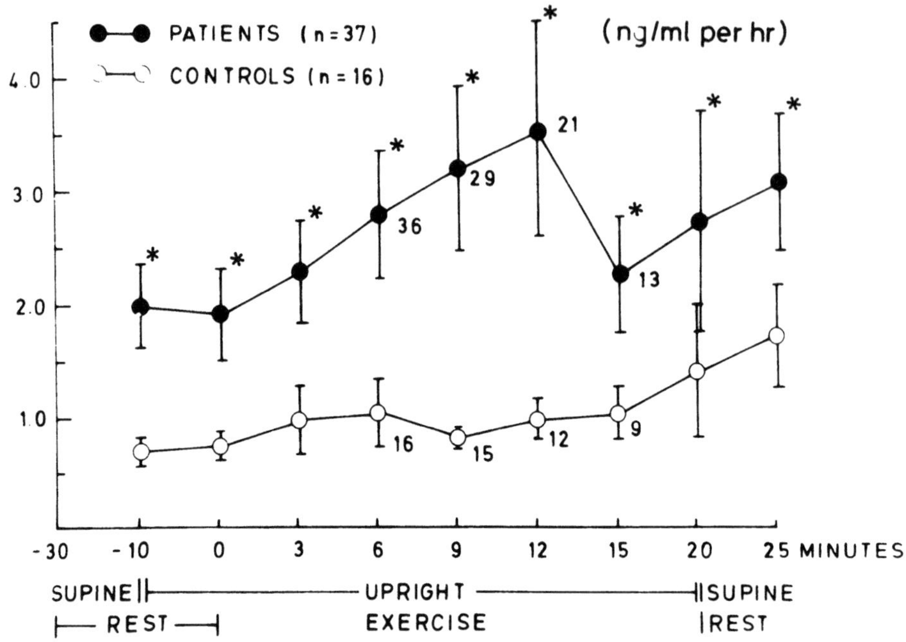

Fig. 3. Plasma renin activity is elevated in patients with proven coronary heart disease (*solid circles; n* = 37; no heart failure, hypertension of diuretic therapy) versus a control group (*open circles;n* = 16) at rest and during bicycle exercise

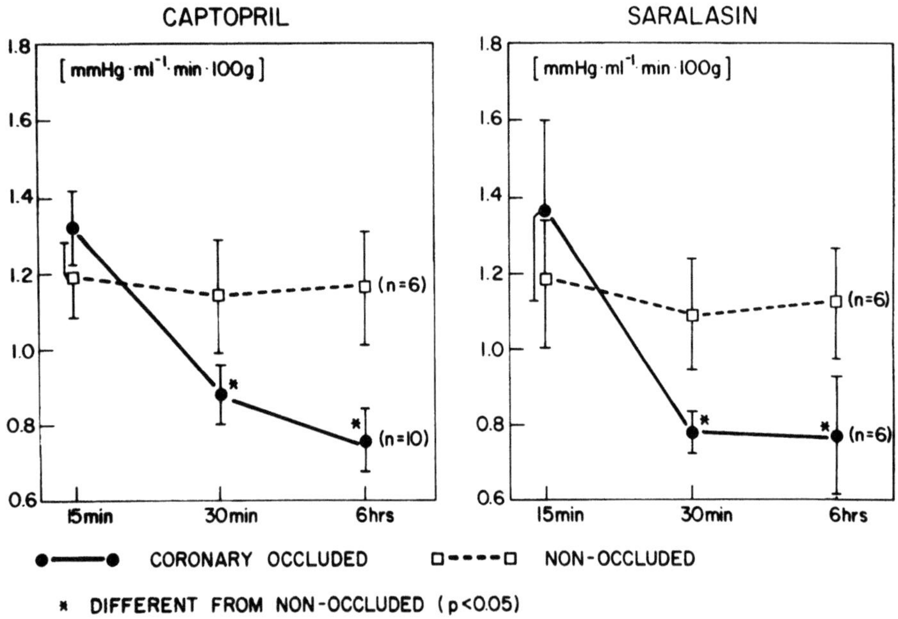

Fig. 4. The converting enzyme inhibitor captopril (0.25 mg/kg per hour, i. v.) and the angiotensin II receptor blocker saralasin (1 μg/kg per minute, i. v.) reduce coronary resistance (coronary dilatation) in dogs with coronary artery occlusion *(circles)* but not in sham-operated control dogs *(squares)*. Drug infusions were started between 15 min and 30 min post coronary occlusion. *Asterisks*, different from non-occluded controls ($P < 0.05$)

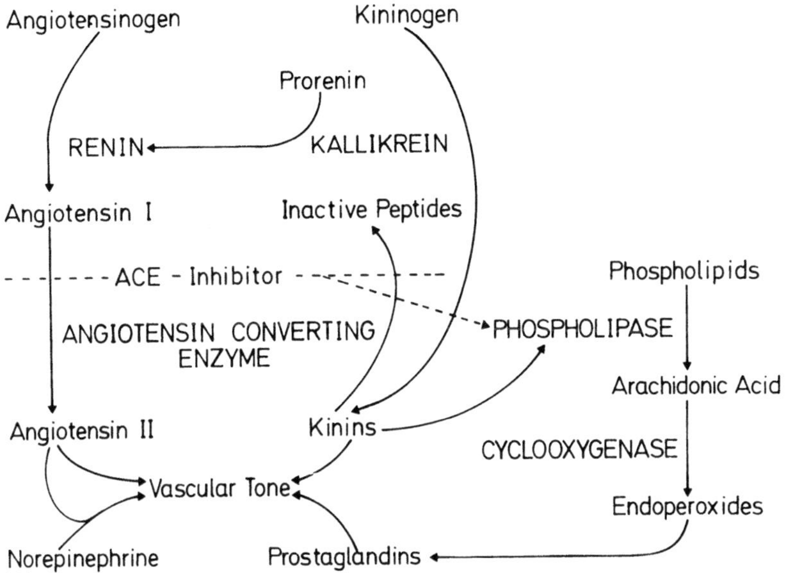

Fig. 5. Schematic diagram on endogenous hormone plasma levels and the effect of exogenous agonists and antagonists of the renin-angiotensin system on coronary circulation. *AI* and *AII*, angiotensin I and II. See text for details

1. Vasodilatory effects of ACE inhibitors and angiotensin II receptor blockers habe been observed in nephrectomized animals [25, 44].
2. The effect of ACE inhibitors in patients with "low renin hypertension" might be explained by a vascular renin-angiotensin system [20].
3. ACE inhibitors may be effective without changing plasma ACE concentrations, and it has been shown that ACE operates in the vascular endothelium [41, 42, 46].
4. Renin and angiotensin II have been demonstrated in the vascular wall and myocardial cells [6, 7, 19, 23, 27, 31, 39, 40].
5. Isolated hearts perfused with blood-free solutions possess the complete renin-angiotensin system [47].
6. Finally, regional differences in the effects of antagonists, as demonstrated for coronary and systemic circulation, might be explained by a regional renin-angiotensin system [13].

A local cardiac renin-angiotensin system might participate directly in the control of coronary smooth muscle tone or myocardial function, or might modulate the sympathetic nervous system [34]. Angiotensin II potentiates the sympathetic nervous system, an effect that is blocked by the angiotensin II receptor blocker saralasin [34]. ACE inhibitors inhibit the sympathetic nervous system at a pre- and postsynaptic site [3]. ACE inhibitors induce systemic vasodilation and a fall of blood pressure without reflex tachycardia and without increasing plasma nore-

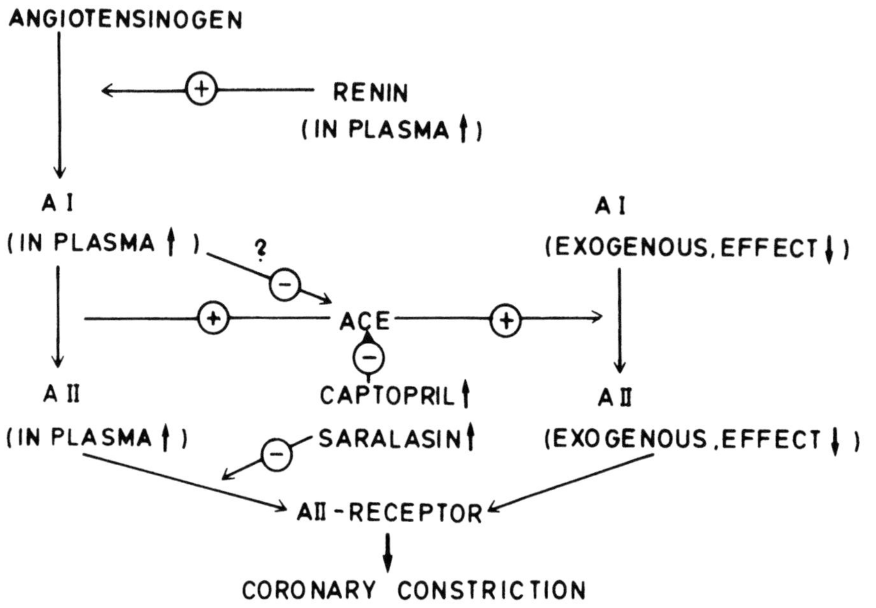

Fig. 6. Schematic diagram of the interrelations among the renin-angiotensin system, kallikrein-kinin system, and prostaglandins, and their potential relevance for the effect of ACE inhibitors on vascular tone

pinephrine [36]. Finally, the ACE inhibitors may have effects on the sympathetic nervous system which are independent of the renin-angiotensin systems. Prostaglandins might play a role in this context [45].

Figure 6 shows potential links between the effect of ACE inhibitors and prostaglandins. ACE is identical with kininase II which degrades bradykinin to inactive peptides. ACE inhibitors, therefore, potentiate the availability of bradykinin [28]. Some ACE inhibitors may have direct effects on phospholipase. Thus, ACE inhibitors may act on vascular tone directly by bradykinin or by prostaglandin production.

In conclusion, coronary vascular tone is controlled by local and systemic mediators. Apart from systemic and local angiotensin II production, ACE inhibitors may interfere with the sympathetic nervous system or may potentiate prostaglandin production, perhaps by inhibition of bradykinin degradation, and may thus increase coronary blood flow in certain groups of patients [16, 17, 33] and animals including models of myocardial ischemia. Improvement of myocardial perfusion may participate in the beneficial effect of ACE inhibitors in heart failure. In addition, experimental data indicate a potential indication of ACE inhibitors in ischemic heart disease, and several clinical trials are currently testing this possibility.

References

1. Aversano T, Becker LC (1985) Persistence of coronary vasodilator reserve despite functionally significant flow reduction. Am J Physiol 248: H 403–411
2. Canty JM, Klocke FJ (1985) Reduced regional myocardial perfusion in the presence of pharmacologic vasodilator reserve. Circulation 71: 370–377
3. Clough DP, Collis MG, Conway J, Hatton R, Keddie JR (1982) Interaction of angiotensin-converting enzyme inhibitors with the function of the sympathetic nervous system. Am J Cardiol 49: 1410–1414
4. Dargie HJ, McAlpine HM, Morton JJ (1987) Neuroendocrine activation in acute myocardial infarction. J Cardiovasc Pharmacol 9 [Suppl 2]: 21–24
5. Deboben A, Inagami T, Ganten D (1983) Tissue renin. In: Genest J, Kichel O (eds) Hypertension: physiophathology and treatment. McGraw, New York pp 194–209
6. Dengler H (1956) Über einen reninartigen Wirkstoff in Arterienextrakten. Arch Exp Pathol Pharmacol 227: 481–487
7. Desjardins-Giasson S, Gutkowska J, Garcia R, Genest J (1981) Renin substrate in rat mesenteric artery. J Physiol Pharmacol 59: 528–532
8. Dzau VJ (1984) Vascular wall renin-angiotensin pathway in control of the circulation. A hypothesis. Am J Med 77: 31–36
9. Ertl G (1987) Coronary vasoconstriction in experimental myocardial ischemia. J Cardiovasc Pharmacol 9 [Suppl 2]: 9–17
10. Ertl G (1988) Angiotensin converting enzyme inhibitors and ischemic heart disease. Eur Heart J 9: 716–727
11. Ertl G, Fuchs M, Oswald S, Wichmann J, Lochner W (1982) Influence of the ischemic coronary bed on collateral blood flow. Basic Res Cardial 77: 520–535
12. Ertl G, Kloner RA, Alexander RW, Braunwald E (1982) Limitation of experimental infarct size by an angiotensin-converting enzyme inhibitor. Circulation 65: 40–48
13. Ertl G, Alexander RW, Kloner RA (1983) Interactions between coronary occlusion and the renin-angiotensin system in the dog. Basic Res Cardiol 78: 518–533
14. Ertl G, Fiedler VB, Bauer B, Schwarzenberger P, Kochsiek K (1986) Effects of nifedipine and indomethacin on leukotriene C_4 and D_4-induced coronary constriction at normal and reduced coronary perfusion in dogs. J Cardiovasc Pharmacol 8: 1078–1085
15. Farber HW, Center DM, Rounds S (1985) Bovine and human endothelial cell production of neutrophil chemoattractant activity in response to components of the angiotensin system. Circ Res 57: 898–902
16. Faxon DP, Creager MA, Halprin JL, Sussmann HA, Gavras H, Ryan TJ (1982) The effect of angiotensin converting enzyme inhibition on coronary blood flow and hemodynamics in patients without coronary artery disease. Int J Cardiol 2: 251–262
17. Foult J-M, Tavolaro O, Antony J, Nitenberg A (1988) Direct myocardial and coronary effects of enlaprilat in patients with dilated cardiomyopathy: assessment by a bilateral intracoronary infusion technique. Circulation 77: 337–344
18. Frame LH, Powell WJ (1976) Progressive perfusion impairment during prolonged low flow myocardial ischemia in dogs. Circ Res 39: 269–276
19. Garst JB, Koletsky S, Wisenbaugh PE, Hadady M, Matthews D (1979) Arterial wall renin and renal venous renin in the hypertensive rat. Clin Sci 56: 41–46
20. Gavras H, Brunner HR, Turini GA, Kershaw GR, Tifft CP, Cuttelod S, Gavras J, Vukovich RA, McKinstry DN (1978) Antihypertensive effect of the oral angiotensin converting-enzyme inhibitor SQ 14225 in man. N Engl J Med 298: 991–995
21. Goetzl EJ, Klickstein LB, Wass KWK, Wintroub BU (1980) The preferential human mononuclear leukocyte chemotactic activity of the substituent tetrapeptides of angiotensin II. Biochem Biophys Res Commun 97: 1097–1102
22. Gorman MW, Sparks HV (1982) Progressive coronary vasoconstriction during relative ischemia in canine myocardium. Circ Res 51: 411–420
23. Gould AB, Skeggs LT, Kahn JR (1964) The presence of renin activity in blood vessel walls. J Exp Med 119: 389–399
24. Guyton RA, McClenathan JH, Michaelis LL (1977) Evolution of regional ischemia distal to a proximal coronary stenosis: self-propagation of ischemia. Am J Cardiol 40: 381–392

25. Hutchinson JS, Mendelsohn FAO, Doyle AE (1980) Hypotensive action of captopril and saralasin in intact and anephric spontaneously hypertensive rats. Hypertension 2: 119–124
26. Jährhult J (1971) Comparative effects of angiotensin and noradrenaline on resistance, capacitance, and precapillary sphincter vessels in cat skeletal muscle. Acta Physiol Scand 81: 315–324
27. Jimenez-Diaz C (1949) Le rôle de la paroi artérielle dans la régulation neurochimique de la pression artérielle. Schweiz Med Wochensch 38: 920–923
28. Johnston CI, Clappison BH, Anderson WP, Yasujima M (1982) Effect of angiotensin-converting enzyme inhibition on circulating and local kinin levels. Am J Cardiol 49: 1401–1404
29. Liang C-S, Gavras H (1978) Renin-angiotensin system inhibition in conscious dogs during acute hypoxemia. Effects on systemic hemodynamics, regional blood flows and tissue metabolism. J Clin Invest 62: 961–970
30. Liang C-S, Gavras H, Hood WB (1978) Renin-angiotensin system inhibition in conscious sodium depleted dogs. Effects on systemic and coronary hemodynamics. J Clin Invest 62: 874–883
31. Lilly LS, Pratt RE, Alexander RW, Gimbrone MA, Dzau VJ (1983) Cultured vascular endothelial cells contain the complete renin-angiotensin system. Clin Res 31: 332 A
32. Lorber V (1942) The action of angiotonin on the completely isolated mammalian heart. Am Heart J 23: 37–42
33. Magrini F, Shimizu M, Roberts N, Fouad FM, Tarazi RC, Zanchetti A (1987) Converting-enzyme inhibition and coronary blood flow. Circulation 75 [Suppl I]: 168–174
34. Malik KU, Nasjletti A (1976) Facilitation of adrenergic transmission by locally generated angiotensin II in rat mesenteric arteries. Circ Res 38: 26–30
35. Michorowski B, Ceremuzynski L (1983) The renin-angiotensin-aldosterone system and the clinical course of acute myocardial infarction. Eur Heart J 4: 259–264
36. Muiesan G, Alicandri CL, Agabiti-Rosei E (1982) Angiotensin-converting enzyme inhibition, catecholamines and hemodynamics in essential hypertension. Am J Cardiol 49: 1420–1424
37. Myers ML, Bolli R, Lekich RF, Hartley CJ, Robert T (1986) *N*-2-mercaptopropionylglycine improves recovery of myocardial function after reversible regional ischemia. J Am Coll Cardiol 8: 1161–1168
38. Needleman P, Kaley G (1978) Cardiac and coronary prostaglandin synthesis and function. N Engl J Med 298: 1122–1128
39. Re R, Fallon JT, Dzau VJ, Quay SC, Haber E (1982) Renin synthesis by canine aortic smooth muscle cells in culture. Life Sci 30: 99–106
40. Rosenthal JH, Pfeifler B, Michailov ML, Pschorr J, Jacob ICM, Dahlheim H (1984) Investigations of components of the renin-angiotensin system in rat vascular tissue. Hypertension 6: 383–390
41. Saye JA, Singer HA, Peach MJ (1984) Role of endothelium in conversion of angiotensin I to angiotensin II in rabbit aorta. Hypertension 6: 216–221
42. Stalcup SA, Lipset JS, Woan J-M, Leuenberger P, Mellins RB (1979) Inhibition of angiotensin converting enzyme activity in cultured endothelial cells by hypoxia. J Clin Invest 63: 966–976
43. Swartz SL, Williams GH (1982) Angiotensin-converting enzyme inhibition and prostaglandins. Am J Cardiol 49: 1405–1409
44. Thurston H, Swales JD (1977) Blood pressure response of nephrectomized hypertensive rats to converting enzyme inhibition: evidence for persistent vascular renin activity. Clin Sci 52: 299–304
45. Van Gilst WH, De Graeff PA, Wesseling H, De Langen CDJ (1986) Reduction of reperfusion arrhythmias in the ischemic isolated rat heart by angiotensin converting enzyme inhibitors: a comparison of captopril, enalapril, and HOE 498. J Cardiovasc Pharmacol 8: 722–728
46. Waeber B, Brunner HR, Brunner DB, Curtet A-L, Turini GA, Gavrs H (1980) Discrepancy between antihypertensive effect and angiotensin converting enzyme inhibition by captopril. Hypertension 2: 236–242
47. Xiang J-Z, Linz W, Becker H, Ganten D, Lang RE, Schölkens B, Unger T (1985) Effects of converting enzyme inhibitors: ramipril and enalapril on peptide action and sympathetic neurotransmission in the isolated heart. Eur J Pharmacol 113: 215–223

Experimental Evidence for Cardioprotection Afforded by Ramipril, an Inhibitor of Angiotensin Converting Enzyme

B. F. BECKER, M. HEIER, and E. GERLACH

Physiologisches Institut der Universität München, Pettenkoferstraße 12, D-8000 München 2, FRG

Introduction

Inhibitors of angiotensin-converting enzyme (ACE) are being increasingly applied in the therapy of high blood pressure and in the treatment of congestive heart failure (see also Motz et al., this volume). A dual mode of action is generally invoked: inhibition of the formation of the vasoconstrictor angiotensin II, and attenuation of the degradative inactivation of the vasodilator bradykinin. This is because ACE is identical to kininase II, the chief enzyme involved in the breakdown of bradykinin in the pulmonary vascular bed [13]. Alterations of the levels of the two peptides in plasma and tissue can cause secondary changes, such as a reduction of sympathetic tone [6, 20], an increase in prostaglandin production [1, 18], and natriuresis [8].

Several in vitro studies, based on coronary ligation and subsequent reperfusion of the myocardium in isolated rat heart preparations, have recently demonstrated cardioprotection by the ACE inhibitors captopril and ramipril [11, 12, 17]. Greater myocardial resistance to hypoxia incurred by coronary embolization with microbeads has also been reported for ramipril in an in situ dog heart model [15]. These effects may be a direct consequence of the ability of the drugs to inhibit ACE since the exogenous application of angiotensin II increased myocardial sensitivity towards hypoxia, whereas infusions of bradykinin proved highly protective [11]. However, another ACE inhibitor, enalapril, proved ineffective [17]. Furthermore, the beneficial actions of both captopril and ramipril were abolished in the presence of indomethacin [17], which suggests that stimulation (direct or indirect) of arachidonic acid metabolism via cyclooxygenase is causally involved.

To gain further insight into the potential of inhibitors of ACE to render the myocardium more resistant towards hypoxia under in vivo conditions, guinea pigs and spontaneously hypertensive rats (SHR) were pretreated with ramipril or captopril. After a period of hypoxia combined with sympathetic stimulation of the heart and an additional phase of in situ ischaemia, myocardial function was assessed ex vivo in the isolated hearts, prepared so as to perform pressure-volume work under strictly defined conditions of perfusion [2, 5]. This latter experimental approach precludes complications arising from inadvertent and generally uncontrollable changes in systemic blood pressure, autonomous nervous stimulation or in humoral parameters (pH, PO_2, the concentrations of Ca^{2+}, K^+, etc.) and humoral cardioactive substances subsequent to cardiovascular shock of any genesis.

H.-P. Schultheiß (Ed.)
New Concepts in Viral Heart Disease

To determine further the mode of action of the substance ramipril, the effect of pretreatment in combination with indomethacin was investigated, and the effect was also compared to that of pretreatment with metoprolol, a cardioselective β_1-adrenoceptor antagonist.

Methods and Materials

The experiments were performed on guinea pigs (male, 250–350 g body weight) and SHR (male, 220–250 g body weight). The standard experimental protocol is outlined in Fig. 1. Pharmacological pretreatment of the animals was generally begun 60 min prior to the initiation of cardiac hypoxia. The drugs used were ramipril (1 mg/kg body weight, given p. o. by gavage in isotonic saline), captopril (50 or 100 mg/kg, given by gavage), metoprolol (2 mg/kg, s. c.) and indomethacin (30 mg/kg, i. p. in Tris buffer, pH 8). Control animals received just the respective vehicle.

Cardiac sympathetic stimulation was initiated by neck dislocation and bleeding of the animals (rats were additionally subjected to mild anaesthesia with ether), and hypoxia was induced by immediately opening the thorax. After 60 s the hearts were arrested by cooling to about 20 ° C (brief, intermittent superfusion with cool saline). The hearts remained in this state of in situ ischemia for 5 min before coronary perfusion with modified Krebs-Henseleit buffer was commenced via a canula which had meanwhile been inserted into the aorta. Retrograde coronary perfusion of hearts in a non-working Langendorff mode was maintained for 25 min to allow completion of other preparatory steps and instrumentation. In some

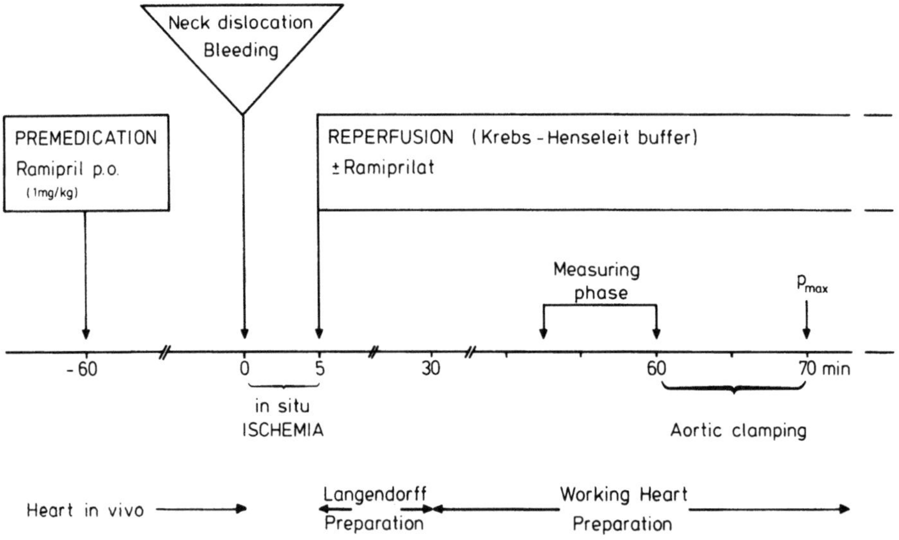

Fig. 1. Outline of the experimental protocol generally applied illustrating the in vivo pretreatment phase, the in vivo/in situ cardiac stress phase and the ex vivo reperfusion steps. The application of ramipril in vivo and of ramiprilat in vitro is exemplified. P_{max} represents the maximal pressure development, i. e. the mean aortic pressure during aortic clamping minus the filling pressure

cases, ECG electrodes had been affixed during the pretreatment phase to record changes in heart rate subsequent to stunning and bleeding. The high K^+ concentration in the perfusate (5.9 m*M*) prevented reperfusion arrhythmias.

To assess postischaemic myocardial function ex vivo, perfusion of the isolated hearts was switched to a mode enabling the performance of left ventricular pressure-volume work. All pertinent features of this preparation have been described previously [2, 5] (see also Schultheiss et al., this volume). The left ventricular filling pressure was maintained at 8.8 mmHg throughout, the mean aortic pressure was set at 70 mmHg for hearts of guinea pigs, and at 100 mmHg for hearts of SHR. The haemodynamic parameters determined after 25–30 min in the working mode (steady-state values) were aortic flow, coronary flow, spontaneous heart rate and ejection time. Cardiac output and external heart work were calculated from these parameters (cf. [2]). Furthermore, the arterial and coronary effluent oxygen partial pressures were monitored (Clark electrodes) to determine myocardial oxygen consumption. Samples of coronary effluent were taken to determine myocardial lactate release and the liberation of adenosine monophosphate (AMP), adenosine, inosine, hypoxanthine, xanthine and uric acid, as described by Becker and Gerlach [3]. After taking these measures, the aorta was clamped, and the hearts were allowed to develop as much pressure as they were capable of, preload remaining fixed at 8.8 mmHg. This maximal value of developed pressure (i.e., mean aortic pressure minus preload), termed "P_{max}", was determined 10 min later.

In several experiments, the pharmacologically active moiety of ramipril, the dianion ramiprilat [4], had been added to the Krebs-Henseleit perfusion buffer (10 mg per litre). This served to establish whether the ACE inhibitor had a significant effect on cardiac performance if present only in the in vitro reperfusion phase of the experiment.

To analyse tissue levels of creatine phosphate, adenosine tri- (ATP) and diphosphate (ADP), AMP, inosine monophosphate (IMP) and of oxidized nicotinamide adenine dinucleotide (NAD^+), immediately after the ischaemic stress, hearts were perfused with Krebs-Henseleit buffer for only 3 min. This removed contaminating blood constituents before the tissue was clamped between aluminium tongs precooled in liquid nitrogen. The tissue was then lyophilised. The nucleotides were extracted and quantified by HPLC, as described in detail elsewhere [3].

The drugs employed were kind donations of the companies Höchst, Frankfurt, FRG (ramipril and ramiprilat); Heyden, Munich, FRG (captopril); and Ciba-Geigy, Wehr, FRG (metoprolol). Indomethacin was purchased from Sigma, Munich, FRG.

Results

Hearts of both guinea pigs and SHR pretreated with ramipril in vivo were able to perform greater external work than control hearts (Fig. 2). Such an effect could not be elicited by merely adding ramiprilat (the pharmacologically active catabolite of the prodrug ramipril) to the Krebs-Henseleit perfusate from the onset of in vitro perfusion, and ramiprilat in vitro did not improve upon the response obtained with

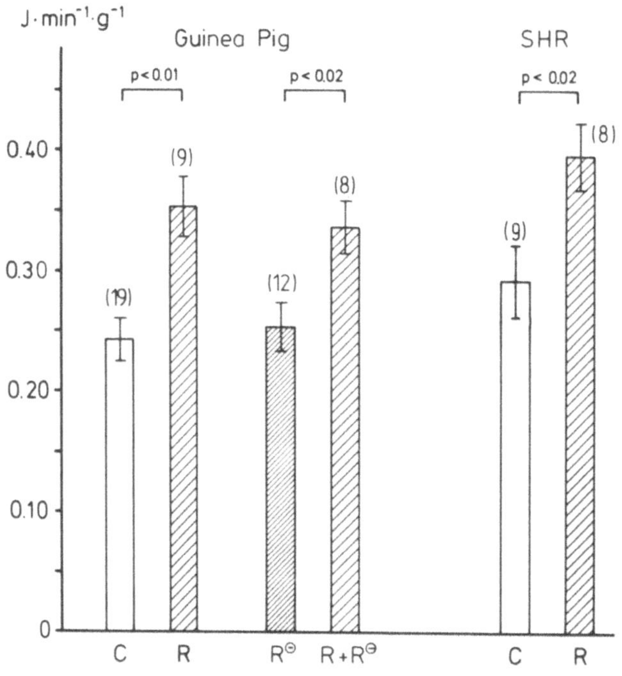

Fig. 2. The effects of in vivo pretreatment with ramipril and/or of applying ramiprilat ex vivo with the perfusate on the performance of external work by hearts isolated from guinea pigs and spontaneously hypertensive rats *(SHR)*. The values are means ± SEM from the number of hearts given in *brackets*. Statistical significance was calculated by Student's *t* test for unpaired data. *C*, control; *R*, ramipril (1 mg/kg); $R^{\ominus}$, ramiprilat (10 mg per litre)

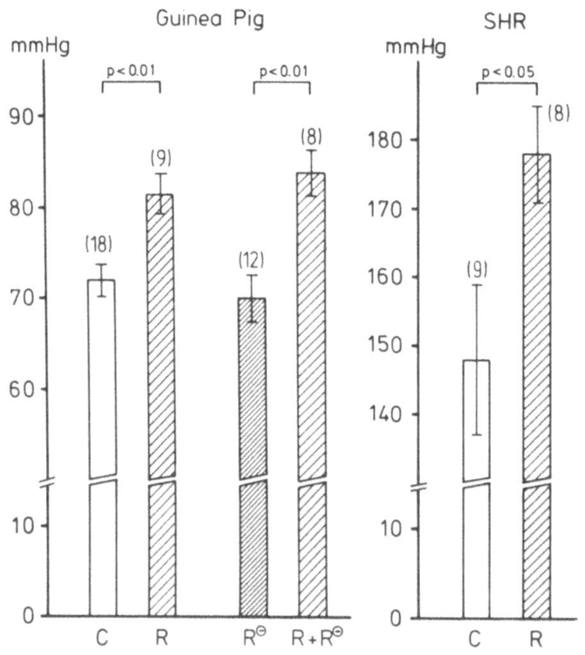

Fig. 3. The effects of in vivo pretreatment with ramipril and/or of applying ramiprilat ex vivo with the perfusate on the maximal development of arterial pressure (P_{max}) by hearts isolated from guinea pigs and spontaneously hypertensive rats *(SHR)*. Values are means ± SEM from the number of hearts given in *brackets*. For further details, see the legend to Fig. 2

Table 1. Effect of pretreatment with ramipril on functional and metabolic parameters of isolated perfused guinea pig hearts performing pressure-volume work

Parameter	Control (Means ± SEM)	(*n*)	Ramipril (Means ± SEM)	(*n*)
Heart rate (beats per minute)	233 ± 5	19	233 ± 5	17
Coronary flow ($ml\,min^{-1}\,g^{-1}$)	10.6 ± 0.5	19	11.2 ± 0.3	17
MVO_2 ($\mu mol\,min^{-1}\,g^{-1}$)	5.8 ± 0.3	14	6.2 ± 0.3	15
Lactate release ($\mu mol\,min^{-1}\,g^{-1}$)	0.40 ± 0.03	19	0.36 ± 0.02	16
Purine release ($\mu mol\,min^{-1}\,g^{-1}$)	3.06 ± 0.38	8	2.92 ± 0.28	12

MVO_2, myocardial oxygen consumption; Purine release, sum of adenosine, inosine, hypoxanthine, xanthine, uric acid and AMP release rates; *n*, number of hearts

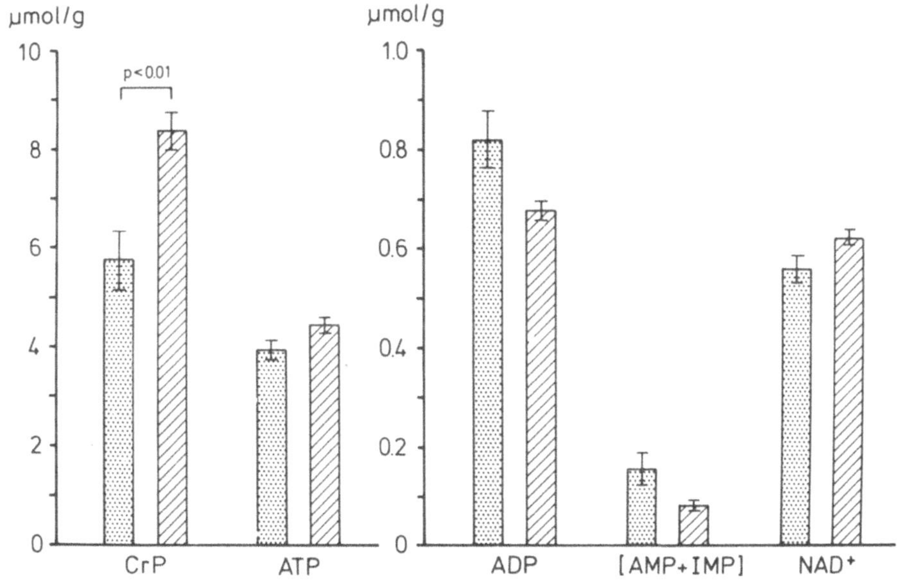

Fig. 4. Tissue levels of myocardial high-energy phosphates and of oxidized nicotinamide adenine dinucleotide (NAD^+) in hearts isolated from guinea pigs subjected to in vivo ischaemia (5 min) and flushed with Krebs-Henseleit buffer (3 min). The values are means ± SEM for seven control hearts *(dotted columns)* and seven hearts of animals pretreated with ramipril (1 mg/kg, *hatched* columns). Statistical significance was calculated according to Student's *t* test for unpaired samples. *CrP,* creatine phosphate; *ATP, ADP, AMP,* adenosine 5'-tri-, -di-, -monophosphate; *IMP,* inosine 5'-monophosphate

ramipril pretreatment in vivo (Fig. 2). The same effects were found with respect to cardiac output (not shown) and the maximal development of pressure (P_{max}, Fig. 3).

The intrinsic ability of hearts of SHR to develop far greater pressures than the non-hypertrophied hearts of the normotensive guinea pigs was readily apparent also in the isolated perfused working heart preparation (Fig. 3) and validates the use of this experimental model for studying myocardial function.

There were no untoward changes in any of the metabolic parameters measured. Myocardial oxygen consumption and coronary flow were slightly elevated in hearts after ramipril, in accordance with the increased external heart work; lactate release and total purine release were essentially unaltered (Table 1).

The levels of creatine phosphate and of nucleotides found in the myocardium of guinea pigs immediately after the ischaemic insult are illustrated in Fig. 4. It is noteworthy that hearts from animals pretreated with ramipril had retained a significantly higher level of creatine phosphate; also the global tissue content of ATP tended to be greater. Correspondingly, these hearts contained less ADP, AMP and IMP than the control hearts. Moreover, the level of oxidized NAD^+ of pretreated hearts was higher, though statistical significance was not quite reached ($0.1 > P\,0.05$).

Neither the ACE inhibitor captopril (given at doses of 50 and 100 mg/kg) nor the cardioselective β_1-adrenoceptor antagonist metoprolol were able to mimic the effect of ramipril in guinea pigs. This is exemplified in Fig. 5 with cardiac out-

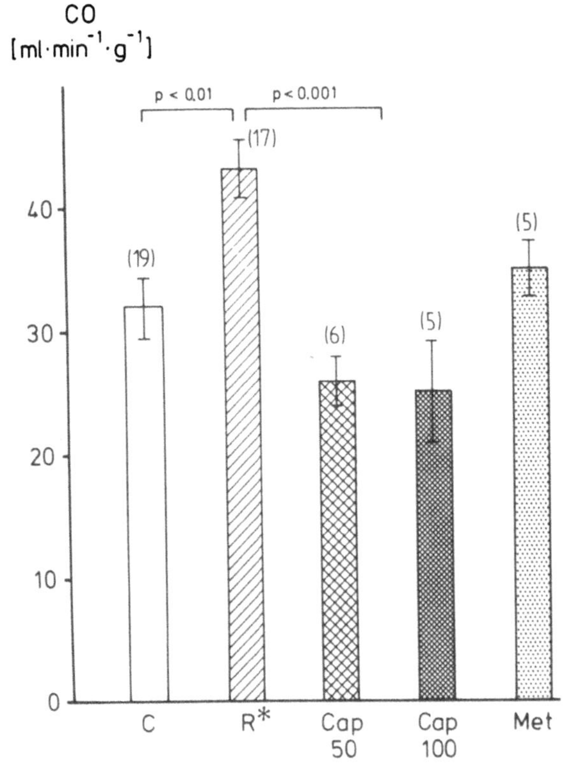

Fig. 5. Influence of in vivo pretreatment of guinea pigs with ramipril (*R**, 1 mg/kg), captopril (*Cap,* 50 and 100 mg/kg) or metoprolol (*Met,* 2 mg/kg) on ex vivo cardiac output *(CO).* Values are means ± SEM from the number of hearts shown *above the respective columns.* Statistical significance was determined with Student's *t* test for unpaired data. *C,* control; *R**, combined mean of the experiments with ramipril alone, and those with ramipril plus ramiprilat

Table 2. Haemodynamic performance of working heart preparations subsequent to pretreatment of guinea pigs in vivo with indomethacin or indomethacin + ramipril, 1 h prior to isolation of the hearts

Parameter	Control ($n = 19$) (Means ± SEM)	Indomethacin (30 mg/kg, i.p., $n = 7$) (Means ± SEM)	Indomethacin + ramipril (1 mg/kg, p.o., $n = 7$) (Means ± SEM)
EHW ($J\ min^{-1}\ g^{-1}$)	0.24 ± .02	0.28 ± .04	0.23 ± .03
CO ($ml\ min^{-1}\ g^{-1}$)	32.1 ± 2.3	35.8 ± 5.5	25.7 ± 5.1
P_{max} (mm Hg)	72 ± 2	75 ± 2	70 ± 4

EHW, external heart work; CO, cardiac output; P_{max}, maximal aortic pressure development

put as the functional parameter. Captopril was also given to SHR (50 mg/kg), and in this species, too, failed to improve myocardial function, in clear contrast to ramipril (result not shown).

Pretreatment of guinea pigs with ramipril did not influence the in vivo heart rate at rest, the mean value (± SD) being 339 (± 24) beats per minute for both controls ($n = 8$) and pretreated animals ($n = 10$). In the 60 s following neck dislocation intermittent tachycardia (up to 480 beats per minute) and atrioventricular block (bradycardia down to 60 beats per minute) could be observed in almost all animals, but the total number of beats did not differ significantly (300 ± 70 for controls, 270 ± 95 with ramipril).

The simultaneous application of indomethacin with ramipril fully prevented the enhancement of postischaemic heart function seen in all other experiments with the ACE inhibitor. Indomethacin itself was without significant effect on the performance of external heart work (Table 2).

Discussion

It is scarcely realized that the very process of isolating hearts for ex vivo perfusion can present, in itself, an acute phase of combined hypoxic and ischaemic hypoxia, associated with cardiac sympathetic stimulation. The latter arises from stunning the animals by neck dislocation and/or applying ether as an anaesthetic and is intensified by bleeding. Indeed, with respect to the heart and circulatory system, the evoked situation strongly resembles that of haemorrhagic shock. Hypoxia develops rapidly if the thorax is opened without the lungs being mechanically ventilated and ischaemia ensues when the heart is arrested, either spontaneously or by cooling. This intervening state of in situ ischaemia can be easily aggravated by prolonging the time until reperfusion which, in the present studies, was set at 5 minutes.

Under the chosen conditions, in vivo pretreatment with ramipril clearly afforded cardioprotection in guinea pigs and SHR. This was evidenced ex vivo by the larger cardiac output, the greater ability to perform external heart work and the higher pressure development of the isolated working heart preparations of both species.

The improved performance in comparison with controls obviously resulted from better-maintained myocardial levels of energy-rich phosphates immediately after the ischaemic stress. Furthermore, the tendency for a higher state of tissue oxygenation was reflected by the level of NAD^+. These observations, together with the lack of any substantial improvement in heart function when the drug was applied only during the in vitro perfusion (in the form of its active moiety ramiprilat; cf. [4]), clearly indicate that the effect of ramipril was largely dependent on its being present at least prior to reperfusion. Far more, an initiation of the mechanism already in vivo is very likely.

Both the measurements of heart rate prior to and after initiation of the circulatory shock situation and the inability of the cardioselective β-blocker metoprolol to mimic the action of ramipril, speak against a decisive reduction in sympathetic tone subsequent to pretreatment with the ACE inhibitor. It was, therefore, particularly significant to find that the cyclooxygenase inhibitor indomethacin was able to prevent cardioprotection by ramipril. This strongly suggets an involvement of prostaglandins (PGI_2, PGE_2).

PGI_2 is well known to be cardioprotective [9, 16]. Furthermore, PGI_2 formation is enhanced by ramipril in the heart [18] and vascular tissue [14], possibly via elevated levels of bradykinin [1]. However, similar properties have been reported for the ACE inhibitor captopril [7, 10, 18] which was definitely ineffictive in the ischaemic stress model employed here, a most unexpected result.

Since ramipril and captopril at the dosages applied produced equal inhibition of plasma ACE activity (80%–100% each, unpublished observations), the causal distinction between the two drugs must be sought elsewhere. Linz and Schölkens [11] suggested that the beneficial actions of ramipril seen in hearts subjected to in vitro coronary ligation and reperfusion arise from interference with converting enzyme in the coronary vascular wall or heart tissue. The tissue enzyme(s) could respond quite differently to plasma ACE with respect to various substrates and ACE inhibitors (cf. [17]). Indeed, ramipril is apparently able to induce a greater and longer-lasting synthesis of PGI_2 in the rat heart than captopril [18].

Though no general extrapolation of the results presented here is at present possible (the experiments were performed with animals and involved only brief ischaemia and acute pretreatment), it can be concluded that treatment of patients with ACE inhibitors may afford cardioprotection, independently of the inhibition of plasma ACE. However, this ability may not be equally pronounced for all ACE inhibitors. Furthermore, interactions with inhibitors of cyclooxygenase should be taken into consideration [19].

Summary

Hearts of guinea pigs and SHR were subjected in vivo to hypoxia combined with sympathetic stimulation, followed by global ischaemia in situ. Myocardial function was then assessed ex vivo in the isolated hearts by determining the capacity to perform pressure-volume work and the ability to develop aortic pressure. It was found that in vivo pretreatment with the ACE inhibitor ramipril (1 mg/kg, given p.o. 1 h prior to heart isolation) improved external heart work by about 40% over control in both species. Cardiac output and maximal pressure development were likewise increased. The improved heart function was directly ascribable to better preserved tissue oxygenation and higher levels of energy-rich phosphates immediately after ischaemia. Since the addition of the pharmacologically active moiety ramiprilat to the heart perfusate only from the onset of reperfusion was without any appreciable beneficial effect, cardioprotection by ramipril clearly depended on an in vivo action. However, inhibition of plasma ACE activity cannot be the causal mechanism because pretreatment with another ACE inhibitor, captopril, did not enhance myocardial function in either guinea pigs or SHR. Furthermore, the effect of ramipril was abolished by simultaneous treatment with the cyclooxygenase inhibitor indomethacin. It is thus proposed that ramipril affords cardioprotection by inducing synthesis of prostacyclin, a known cardioprotective agent, possibly via a specific elevation of tissue bradykinin levels.

References

1. Barrow SE, Dollery CT, Heavey DJ, Hickling NE, Ritter JM, Vial J (1986) Effect of vasoactive peptides on prostacyclin synthesis in man. Br J Pharmacol 87: 243–247
2. Becker BF, Gerlach E (1984) Acute effects of nicotine on hemodynamic and metabolic parameters of isolated, perfused hearts of guinea pigs and rats. Klin Wochenschr 62 [Suppl II]: 58–66
3. Becker BF, Gerlach E (1987) Uric acid, the major catabolite of cardiac adenine nucleotides and adenosine, originates in the coronary endothelium. In: Gerlach E, Becker BF (eds) Topics and perspectives in adenosine research. Springer, Berlin Heidelberg New York, pp 209–222
4. Becker RHA, Schölkens B (1987) Ramipril: review of pharmacology. Am J Cardiol 59: 3D–11D
5. Bünger R, Sommer O, Walter G, Stiegler H, Gerlach E (1979) Functional and metabolic features of an isolated perfused guinea pig heart performing pressure-volume work. Pflugers Arch 380: 259–266
6. Clough P, Collis MG, Conway J, Hatton R, Reddie JR (1982) Interaction of angiotensin converting enzyme inhibitors with the function of the sympathetic nervous system. Am J Cardiol 49: 1410–1414
7. Düsing R, Scherag R, Landberg G, Glänzer K, Kramer HJ (1983) The converting enzyme inhibitor captopril stimulates prostacyclin synthesis by isolated rat aorta. Eur J Pharmacol 91: 501–504
8. Düsing R, Moritz J, Glänzer K, Kramer HJ (1985) Effect of angiotensin II and captopril on renal tubular function in man. Br J Clin Pharmacol 19: 29–35
9. Faber NE, Pieper GM, Thomas JP, Gross GJ (1988) Beneficial effects of iloprost in the stunned canine myocardium. Circ Res 62: 204–215
10. Kudo K, Abe K, Chiba S, Sato M, Yasujima M, Kohzuki M, Omata K, Tanno M, Tsunoda K, Yoshinaga K (1988) Role of thromboxane A_2 in the hypotensive effect of captopril in essential hypertension. Hypertension 11: 147–152
11. Linz W, Schölkens BA (1987) Influence of local converting enzyme inhibition on angiotensin and bradykinin effects in ischemic rat hearts. J Cardiovasc Pharmacol 10 [Suppl 7]: S75–S82
12. Linz W, Schölkens BA, Han YF (1986) Beneficial effects of the converting enzyme inhibitor, ramipril, in ischemic rat hearts. J Cardiovasc Pharmacol 8 [Suppl 10]: S91–S99

13. Regoli D, Barabe J (1980) Pharmocology of bradykinin and related kinins. Pharmacol Rev 32: 1–46
14. Scherf H, Pietsch R, Landsberg G, Kramer HJ, Düsing R (1986) Converting enzyme inhibitor ramipril stimulates prostacyclin synthesis by isolated rat aorta: evidence for a kinin-dependent mechanism. Klin Wochenschr 64: 742–745
15. Schölkens BA, Martorana PA, Göbel H, Gehring D (1986) Cardiovascular effects of the converting enzyme inhibitor ramipril (HOE 498) in anesthetized dogs with acute ischemic left ventricular failure. Clin Exp Hypertens A 8: 1033–1048
16. Smith EF III, Gallenkämper W, Beckmann R, Thomsen T, Mannesmann G, Schrör K (1984) Early and late administration of a PGI_2-analogue, ZK 36374 (iloprost): effects on myocardial preservation, collateral blood flow and infarct size. Cardiovasc Res 18: 163–173
17. van Gilst WH, de Graeff PA, Wesseling H, de Langen CDJ (1986) Reduction of reperfusion arrhythmias in the ischemic isolated rat heart by angiotensin converting enzyme inhibitors. A comparison of captopril, enalapril and HOE 498. J Cardiovasc Pharmacol 8: 722–728
18. van Gilst WH, van Wijngaarden J, Scholtens E, de Graeff PA, de Langen CDJ, Wesseling H (1987) Captopril-induced increase in coronary flow: an SH-dependent effect on arachidonic acid metabolism? J Cardiovasc Pharmacol 9 [Suppl 2]: S31–S36
19. Witzgall H, Hirsch F, Scherer B, Weber PC (1982) Acute haemodynamic and hormonal effects of captopril are diminished by indomethacin. Clin Sci 62: 611–615
20. Xiang J, Linz W, Becker H, Ganten D, Lang RE, Schölkens B, Unger T (1985) Effects of converting enzyme inhibitors: ramipril and enalapril on peptide action and sympathetic neurotransmission in the isolated heart. Eur J Pharmacol 113: 215–223

Energetic Consequences of Pharmacological Therapy in Patients with Myocardial Failure

G. Hasenfuss, C. Holubarsch, H. W. Heiss, and H. Just

Department of Internal Medicine — Cardiology, University of Freiburg, Hugstetter Str. 55, D-7800 Freiburg, FRG

Introduction

Reducing ventricular load and increasing the contractile force of the myocardium are fundamental principles in the pharmacological therapy of heart failure. Therapeutic goals are the relief of symptoms and the improvement of the prognosis of the disease. Although most vasodilators and inotropic agents have yielded short-term hemodynamic improvement in congestive heart failure, beneficial effects on prognosis have only been shown for vasodilators [5, 7, 8]. The long-term benefit of inotropic agents is controversial. Moreover, detrimental effects of increased myocardial oxygen consumption due to the inotropic stimulation of the myocardium have recently been discussed [14, 16, 25, 28].

Vasodilators may have beneficial effects on myocardial energetics by reducing left ventricular load and thus the hemodynamic determinants of myocardial oxygen consumption. Inotropic drugs may influence myocardial energetics by altering the hemodynamic determinants of myocardial energy demand. In addition, myocardial stimulation may influence the coupling between hemodynamics and myocardial energy consumption. Therefore, myocardial energy consumption must be analyzed in relation to its hemodynamic determinants in order to characterize the energetic effects of pharmacological therapy on the myocardium.

No data analyzing the influence of pharmacological therapy on myocardial energy consumption relative to the hemodynamics changes in patients with heart failure are yet available. Therefore, in the present study the following questions were investigated:

- How is myocardial oxygen consumption related to hemodynamics in patients with heart failure?
- How does vasodilation by nitroprusside influence the relationship between myocardial oxygen consumption and its determinants?
- How do inotropic agents influence the relationship between myocardial oxygen consumption and hemodynamics?

H.-P. Schultheiß (Ed.)
New Concepts in Viral Heart Disease

Methods

Patients and Study Protocol

The study was performed in patients with idiopathic dilative cardiomyopathy (NYHA II–III). Idiopathic dilative cardiomyopathy was defined by increased left ventricular volume (> 220 ml) and reduced left ventricular ejection fraction (< 55%) in the absence of coronary or valvular heart disease, arterial hypertension, or systemic diseases known to be associated with cardiomyopathy. All patients were in sinus rhythm. The investigations were performed during routine diagnostic cardiac catheterization. Previous medications were withheld at least 48 h before catheterization. After coronary heart disease had been excluded by coronary angiography, left ventriculography with simultaneous pressure measurement was done using a Millar microtip catheter pressure transducer. Thereafter myocardial blood flow was measured by the argon method. Blood samples were taken from the aorta and the coronary sinus for oxygen saturation measurements. Upon completion of basal measurements, the following intervention were performed:

- Nitroprusside was given in ten patients, at an average dosage of 71 ± 58 μg/min, in order to analyze the influence of pure vasodilation on hemodynamics and myocardial oxygen consumption.
- The phosphodiesterase inhibitor enoximone was given intravenously in 12 patients at an average dose of 1.42 ± 0.52 mg/kg at 12.5 mg/min.
- UDCG-115 was infused intravenously in nine patients at 1.25 mg/h over 3–5 h. UDCG-115 is a new cardiotonic agent which was shown to increase the affinity of troponin C for calcium ions in skinned cardiac muscle fibers in addition to moderate phosphodiesterase inhibiting properties [21, 24].

Hemodynamic and myocardial oxygen consumption measurements were repeated during steady state conditions following the respective interventions.

Hemodynamic Measurements

Aortic and left ventricular pressure were measured with pigtail Millar microtip catheter pressure transducers (PC-485 A). Left ventricular angiography was performed by power injections of 40 ml nonionic contrast solution at 50 frames per second. The projection was a 10 ° caudally angulated 45 ° right anterior oblique view. Left ventricular pressure was measured simultaneously. Maximal rate of left ventricular pressure rise was determined from continuous differentiation of left ventricular pressure tracings. Heart rate was taken from the electrocardiogram which was recorded continuously. During one cardiac cycle, left ventricular volumes were calculated from each cine frame at intervals of 20 ms using the Sandler and Dodge method [22]. Left ventricular muscle mass was calculated according to a modification of the method of Rackley et al. [17]. Instantaneous wall thickness was determined at intervals of 20 ms using a special computer programe (Herrath II) which calculates instantaneous wall thickness from wall mass, instantaneous volume, major and minor hemiaxes. Instantaneous circumferential wall stress values were calculated using the ellipsoid model of Mirsky [15]. The systolic stresstime integral was calculated by integrating instantaneous stress values from end-diastole to end-systole. End-diastolic wall stress was defined as the wall stress value at end-diastolic pressure, end-systolic wall stress was defined as the stress value derived from the cine frame before left ventricular volume increased. Pressure-volume work was calculated as the area of the pressure-volume loop obtained by relating instantaneous pressure to volume every 20 ms and normalizing for 100 g wall mass.

Myocardial Energetics

Myocardial blood flow was measured by the argon method [4, 18]. Myocardial oxygen consumption was determined as the product of myocardial blood flow times arterial-coronary-sinus oxygen content difference. Myocardial oxygen consumption per beat was calculated from myocardial oxygen consumption per minute and heart rate.

The relationship between hemodynamic parameters and myocardial oxygen consumption was investigated by multiple regression analyses using the SPSS PC+ computer programe. For each drug the variables were compared with control values by the paired *t*-test. *P* values of less than 0.05 were accepted as significant.

Results

Coupling Between Hemodynamics and Myocardial Oxygen Consumption

The coupling between measured myocardial oxygen consumption and the hemodynamic parameters left ventricular peak systolic wall stress, systolic stress-time integral, pressure-volume work, maximal rate of left ventricular pressure rise, and mean velocity of circumferential fiber shortening was evaluated by multiple regression analyses in 29 patients. Differences in heart rate were taken into account by calculating both myocardial oxygen consumption and the hemodynamic parameters for one single cardiac cycle.

The best correlation was obtained between myocardial oxygen consumption per beat and left ventricular systolic stress-time integral ($r = 0.71$; $P < 0.001$) (Fig. 1). The correlation between myocardial oxygen consumption and peak systolic wall stress was less relevant ($r = 0.55$; $P < 0.005$). There was no significant correlation between myocardial oxygen consumption and pressure-volume work, or the velocity parameters maximal rate of left ventricular pressure rise and mean velocity of circumferential fiber shortening. Consideration of pressure-volume work, maximal rate of left ventricular pressure rise, or mean velocity of circumferential fiber shortening did not significantly improve the relationship between myocardial oxygen consumption per beat and systolic stress-time integral.

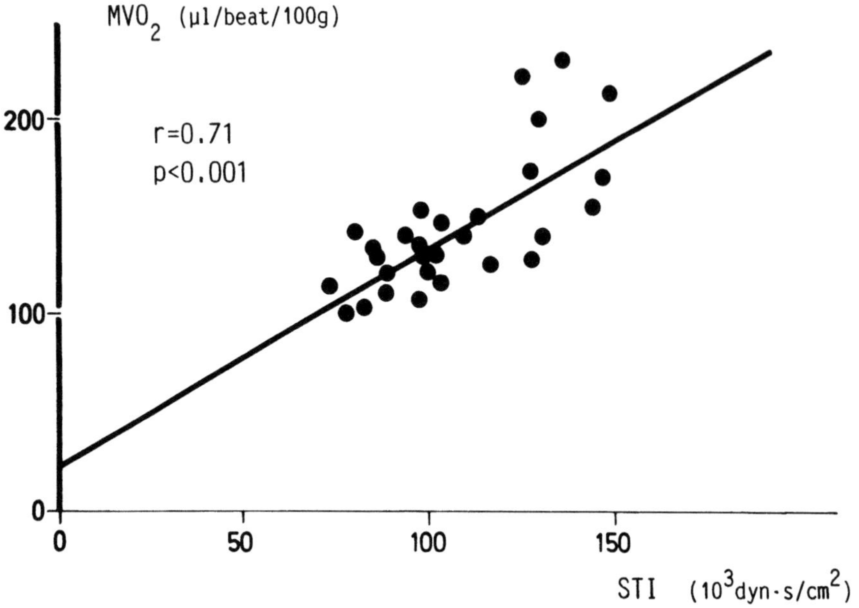

Fig. 1. Linear correlation between myocardial oxygen consumption per beat *(MVO_2/beat)* and left ventricular systolic stress-time integral *(STI)* in 29 patients with idiopathic dilative cardiomyopathy

Influence of Vasodilation and Inotropic Interventions on Hemodynamics and Myocardial Energetics (Table 1)

Following vasodilation with nitroprusside, systolic stress-time integral decreased by 35% due to a decrease of left ventricular pressure and volume. Myocardial oxygen consumption per minute decreased by 26%, and myocardial oxygen consumption per beat decreased by 30%. As shown in Fig. 2, there was a significant linear correlation between the decrease of myocardial oxygen consumption per beat and the corresponding reduction of systolic stress-time integral ($r = 0.82$; $P < 0.005$).

With enoximone, systolic stress-time integral decreased by 60%, whereas myocardial oxygen consumption per minute did not change significantly and myocardial oxygen consumption per beat decreased by only 19%. The decrease of myocardial oxygen consumption per beat correlated significantly with the decrease of systolic stress-time integral ($r = 0.65$; $P < 0.05$). The slope of the calculated regression line was significantly lower compared to nitroprusside conditions (Fig. 3).

Following UDCG-115, the decrease of stress-time integral and myocardial oxygen consumption per beat were comparable to the changes of these parameters with nitroprusside. Stress-time integral decreased by 36%, myocardial oxygen consumption per minute by 26%, and myocardial oxygen consumption per beat by 33%. The slope of the calculated regression line was not significantly different compared to nitroprusside conditions (Fig. 3).

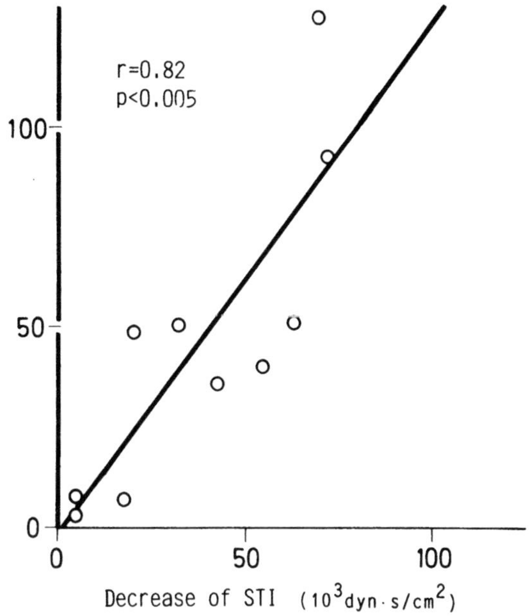

Fig. 2. Linear correlation between the decrease of myocardial oxygen consumption per beat *(MVO_2/beat)* and the decrease of the corresponding left ventricular systolic stress-time integral *(STI)* following the administration of nitroprusside

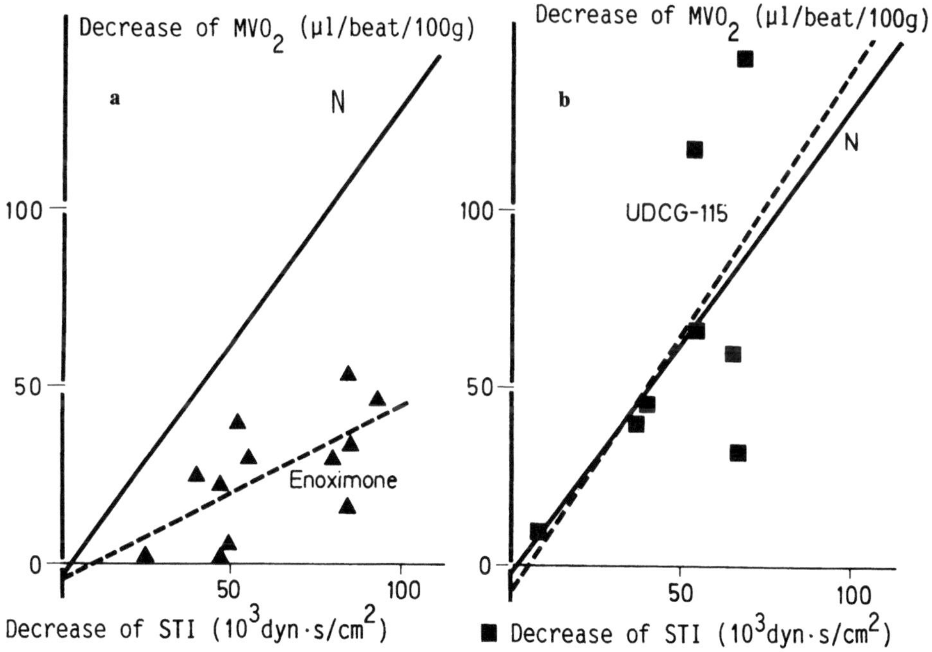

Fig. 3 a, b. Linear correlation between the decrease of myocardial oxygen consumption per beat *(MVO_2/beat)* and the decrease of the corresponding left ventricular systolic stress-time integral *(STI)*. **a)** With enoximone *(triangles; broken line)* the slope of the regression line is significantly lower compared to nitroprusside *(N; solid line)*. **b)** With UDCG-115 *(squares; broken line)* the slope of the regression line is not statistically different compared to nitroprusside *(N; solid line)*. The following regression lines were calculated: nitroprusside: $y = 1.27\,x - 1.69$; $r = 0.82$; $P < 0.005$; enoximone: $y = 0.51\,x - 5.6$; $r = 0.65$; $P < 0.05$; UDCG-115: $y = 1.48\,x - 9.3$; $r = 0.69$; $P < 0.05$

Table 1. Influence of nitroprusside, enoximone, and UDCG-115 on hemodynamics and myocardial energetics in patients with dilative cardiomyopathy

	C	N	C	E	C	UDCG-115
HR (min^{-1})	82 ± 13	88 ± 14 NS	86 ± 18	100 ± 20 **	87 ± 15	95 ± 16 NS
LVSP (mmHg)	122 ± 15	96 ± 6 **	128 ± 18	96 ± 16 **	122 ± 15	103 ± 10 **
LVEDP (mmHg)	17 ± 8	9 ± 5 **	16 ± 8	5 ± 3 **	21 ± 10	12 ± 8 *
LVEDV (ml)	338 ± 96	280 ± 108 *	299 ± 48	208 ± 58 **	396 ± 133	331 ± 112 **
dp/dt_{max} (mmHg/s)	1046 ± 202	947 ± 245 NS	1134 ± 186	1144 ± 225 NS	935 ± 248	1056 ± 284 *
STI (10^3 dyn s/cm^2)	109 ± 22	71 ± 21 **	104 ± 23	42 ± 10 **	117 ± 25	75 ± 19 **
MVO_2/min (ml/min/100 g)	13.3 ± 5.0	9.8 ± 2.2 *	11.6 ± 2.4	10.7 ± 3.2 NS	14.4 ± 5.2	10.6 ± 3.8 *
MVO_2/beat (μl/beat/100 g)	159 ± 44	112 ± 23 **	134 ± 28	109 ± 21 **	165 ± 47	111 ± 32 *

* $P < 0.05$ versus control; ** $P < 0.01$ versus control. C, control; N, nitroprusside; E, enoximone; HR, heart rate; LVSP, left ventricular systolic pressure; LVEDP, left ventricular end-diastolic pressure; LVEDV, left ventricular end-diastolic volume; dp/dt_{max}, maximum rate of left ventricular pressure rise; STI, left ventricular systolic stress-time integral; MVO_2/min, myocardial oxygen consumption per minute; MVO_2/beat, myocardial oxygen consumption per beat; NS, non significant versus control

Discussion

The concept of the present study relies on the fact that cardiotonic drugs exert inotropic and vasodilating effects. The inotropic effect may increase myocardial energy demand by altered activation and contraction processes. Vasodilation — resulting from direct effects on the peripheral vasculature or from reflective changes — is supposed to reduce energy demand due to altered myocardial working conditions. Measured myocardial oxygen consumpting reflects the sum of both effects. In order to separate the energetic consequences of vasodilation from inotropism, myocardial oxygen consumption must be related to its hemodynamic determinants.

The coupling of myocardial oxygen consumption with hemodynamic parameters was investigated in 29 patients with dilative cardiomyopathy. The rationale for the multiple regression analyses performed were the experimental data showing that myocardial oxygen consumption is closely related to heart rate, wall stress, and velocity of myocardial contraction [2]; the influence of external myocardial work on energy consumption is controversial [6, 27].

Heart rate can be taken into account by either calculating both myocardial oxygen consumption and the hemodynamic parameters per minute [3, 20] or per beat [23, 26]. In the present study, myocardial oxygen consumption and hemodynamic parameters were calculated for one single cardiac cycle. A close and significant linear correlation was found between myocardial oxygen consumption per beat and systolic stress-time integral. Additional considerations of pressure-volume work or indices of contraction velocity did not statistically improve this correlation in this group of patients.

To evaluate the influence of pure vasodilation on the relationship between myocardial oxygen consumption per beat and stress-time integral, nitroprusside was administered in ten patients with dilative cardiomyopathy. A close and significant correlation was found between the decrease of myocardial oxygen consumption per beat and the decrease of the corresponding stress-time integral.

Compared to nitroprusside, enoximone induced a more pronounced decrease of stress-time integral and a less pronounced decrease of myocardial oxygen consumption per beat. Accordingly, the slope of the respective regression line was significantly lower, as calculated under nitroprusside conditions. In other words, a certain reduction of left ventricular load lowers myocardial energy demand less with enoximone than with nitroprusside. The different slopes may represent the additional myocardial action of enoximone although a direct inotropic effect was not obvious by hemodynamic measurements in these patients.

Following the administration of UDCG-115, both myocardial oxygen consumption and stress-time integral decreased comparable to the changes with nitroprusside. The slopes of the regression lines with UDCG-115 and nitroprusside were not statistically different. The energetic effects of the interventions performed may be explained by the respective pharmacological mode of action. On a cellular level, myocardial energy consumption of the working heart muscle is mainly determined by excitation contraction coupling, i.e., energy consumption for calcium transport and actomyosin interaction, i.e., energy consumption for cross-bridge cycling [9].

Nitroprusside has no direct myocardial effects. It reduces left ventricular load by peripheral vasodilation. Thus less wall stress must be developed and maintained by the myocardium. Consequently less energy is consumed for activation and contraction processes.

Enoximone has both vasodilating effects on the peripheral vasculature and direct effects on the myocardium by elevating myoplasmic cyclic adenosine monophosphate (AMP) concentration [19]. Cyclic AMP increases the number of calcium ions delivered into the cytoplasma on the one hand and reduces the sensitivity of troponin C for calcium ions on the other [13]. Thus the amount of energy-dependent calcium transport is increased relative to the degree of activation when cyclic AMP levels are elevated.

In addition, cyclic AMP may decrease the energy transduction efficiency of the contractile proteins. Hoh et al. [11] have recently demonstrated that catecholamines increase the cycling rate of cross-bridges. Since one molecule of adenosine triphosphate (ATP) is hydrolyzed for each cross-bridge cycle, increased cycling rate may result in augmented energy consumption of the contractile machinery [1]. Accordingly, increased energy turnover of both the calcium transport system and the contractile proteins under catecholamines and phosphodiesterase inhibitors was shown in myothermal studies [10, 12].

UDCG-115 influences myocardial energetics comparable to the vasodilator nitroprusside. This implies that UDCG-115 either acts as a pure vasodilator without a direct effect on the myocardium or that the myocardial effect of this drug is different from enoximone with respect to the energetic consequences. Since UDCG-115 significantly increased the maximal rate of pressure rise, despite a reduction of pre- and afterload, a direct effect on the myocardium must have been present in these patients. Inotropic stimulation of the myocardium with energetic consequences different from pure phosphodiesterase inhibition might be explained by the pharmacological properties of this drug as determined in vitro experiments. UDCG-115 was shown to increase the calcium sensitivity of troponin C whereas the phosphodiesterase-inhibiting properties were only moderate compared to current phosphodiesterase inhibitors such as IBMX, theophylline, milrinone, and amrinone [21, 24]. Accordingly, in myothermal experiments the energy turnover for calcium transport did not change significantly and the energy demand of the contractile proteins was influenced in proportion to developed tension or tension-time integral [10]. Whether the calcium-sensitizing effect is also relevant in vivo or whether phosphodiesterase inhibition is dominant under in vivo conditions is still controversial. The energetic effects of UDCG-115 observed in these patients may underline that the calcium-sensitizing effect contributed significantly to the inotropic action of UDCG-115.

The present data implicate that pharmacological interventions may influence myocardial energetics

a) by altering the hemodynamic determinants of myocardial energy consumption; and
b) by influencing the coupling between hemodynamics and myocardial energy demand. The pharmacological mode of action determines the proportions of both effects on myocardial energetics.

Summary

The relationship between hemodynamics and myocardial oxygen consumption was investigated in 29 patients with idiopathic dilative cardiomyopathy (NYHA II–III). Multiple regression analyses yielded that left ventricular systolic stress-time integral is the major determinant of myocardial oxygen consumption per beat in these patients ($r = 0.71$; $P < 0.001$).

Intravenous application of the vasodilator nitroprusside reduced myocardial oxygen consumption by 30% and systolic stress-time integral by 35%. In contrast, the phosphodiesterase inhibitor enoximone decreased stress-time integral more (60%) and myocardial oxygen consumption less (19%).

From this data we conclude that pure vasodilators have energy-saving effects by reducing left ventricular load. The phosphodiesterase inhibitor enoximone exhibits both energy-saving (vasodilation) and energy-consuming effects (myocardial stimulation). The influence an global myocardial oxygen consumption may depend on the balance of both pharmacological properties.

The new inotropic and vasodilating agent UDCG-115, which acts by phosphodiesterase inhibition as well as by increasing the sensitivity of troponin C for calcium ions, influenced myocardial energetics comparable to the vasodilator nitroprusside: myocardial oxygen consumption decreased by 35% and systolic stress-time integral by 36%.

It seems to be obvious that the mode of action of positive inotropic substances determines their energetic consequences.

References

1. Alpert NR, Mulieri LA (1984) Hypertrophic adaptation of the heart to stress: a myothermal analysis. In: Zak R (ed) Growth of the heart in health and disease. Raven, New York, p 363
2. Braunwald E (1971) Control of myocardial oxygen consumption. Physiologic and clinical considerations. Am J Cardiol 27: 416–432
3. Bretschneider HJ, Hellige G (1976) Pathophysiologie der Ventrikelkontraktion — Kontraktilität, Inotropie, Suffizienzgrad und Arbeitsökonomie des Herzens. Verh Dtsch Ges Herzkreislaufforschg 42: 14–30
4. Bretschneider HJ, Cott L, Hilgert G, Probst R, Rau G (1966) Gaschromatographische Trennung und Analyse von Argon als Basis einer neuen Fremdgasmethode zur Durchblutungsmessung von Organen. Verh Dtsch Ges Herzkreislaufforsch 32: 267–273
5. Cohn JN, Archibald DG, Ziesche S, Franciosa JA, Harston WE, Tristani FE, Dunkman B, Jacobs W, Francis GS, Flohr KH, Goldman S, Cobb FR, Shah PM, Saunders R, Fletcher RD, Loeb HS, Hughes VC, Baker B (1986) Effect of vasodilator therapy on mortality in chronic congestive heart failure. Results of a Veterans Administration cooperative study. N Engl J Med 314: 1547–1552
6. Coleman HN, Sonnenblick EH, Braunwald E (1969) Myocardial oxygen consumption associated with external work: the Fenn effect. Am J Physiol 217: 291–296
7. Colucci WS, Wright RF, Braunwald E (1986) New positive inotropic agents in the treatment of congestive heart failure. Mechanisms of action and recent developments. N Engl J Med 314: 290–299
8. Consensus Trial Study Group (1987) Effects of enalapril on mortality in severe congestive heart failure. N Engl J Med 316: 1429–1439
9. Gibbs CL (1978) Cardiac energetics. Physiol Rev 58: 174–254

10. Hasenfuss G, Holubarsch Ch, Just H, Blanchard E, Mulieri LA, Alpert NR (1987) Energetic aspects of inotropic interventions in rat myocardium. Basic Res Cardiol 82 [Suppl 2]: 252–259
11. Hoh JFY, Rossmanith GH, Kwan LJ, Hamilton AM (1988) Adrenaline increases the rate of cycling of cross-bridges in rat cardiac muscle as measured by pseudo-random binary noise-modulated perturbation analysis. Circ Res 62: 452–461
12. Holubarsch Ch, Hasenfuss G, Just H, Mulieri LA, Alpert NR (1986) Energetic aspects of inotropic interventions in guinea pig papillary muscles. Circulation 74 [Suppl III]: 326
13. Katz A (1983) Cyclic adenosine monophosphate effects on the myocardium: a man who blows hot and cold with one breath. J Am Coll Cardiol 2: 143–149
14. Katz AM (1986) Potential deleterious effects of inotropic agents in the therapy of chronic heart failure. Circulation 73 [Suppl III]: 184–190
15. Misky I (1979) Elastic properties of the myocardium: a quantitative approach with physiological and clinical applications. In: Berne RM (ed) Handbook of physiology. The cardiovascular system. American Physiological Society, Washington DC, p 497
16. Packer M, Medina N, Yushak M (1984) Hemodynamic and clinical limitations of long-term inotropic therapy with amrinone in patients with severe chronic heart failure. Circulation 70: 1038–1047
17. Rackley CE, Dodge HT, Coble YD, Hay RE (1965) A method for determining left ventricular mass in man. Circulation 29: 666–671
18. Rau G (1969) Messung der Koronardurchblutung mit der Argon-Fremdgasmethode. Arch Kreislaufforsch 58: 322–398
19. Roebel LE, Dage RC, Cheng HC, Woodward JK (1982) Characterization of the cardiovascular activities of a new cardiotonic agent, MDL 17.043 (1,3-dihydro-4-methyl-5 (4-methylthio-benzoyl)-2*H*-imidazol-2-one). J Cardiovasc Pharmacol 4: 721–729
20. Rooke A, Feigl EO (1982) Work as a correlate of canine left ventricular oxygen consumption, and the problem of catecholamine oxygen wasting. Circ Res 50: 273–286
21. Rüegg JC (1986) Effects of new inotropic agents on Ca^{++} sensitivity of contractile proteins. Circulation 73 [Suppl III]: 78–84
22. Sandler H, Dodge HT (1968) The use of single plane angiocardiograms for the calculation of left ventricular volume in man. Am Heart J 75: 325–334
23. Sarnoff SJ, Braunwald E, Welch GH, Case RB, Stainsly WN, Marcuz R (1958) Hemodynamic determinants of oxygen consumption of the heart with special reference to the tension-time-index. Am J Physiol 192: 148–156
24. Scholz H, Meyer W (1986) Phosphodiesterase-inhibiting properties of newer inotropic agents. Circulation 73 [Suppl III]: 99–108
25. Shah PK, Amin DK, Hulse S, Shellock F, Swan HJC (1985) Inotropic therapy for refractory congestive heart failure with oral fenoximone (MDL-17,043): poor long-term results despite early hemodynamic and clinical improvement. Circulation 71: 326–331
26. Suga H, Hisano R, Gota Y, Yamada O, Igarashi Y (1983) Effect of positive inotropic agents on the relation between oxygen consumption and systolic pressure volume area in canine left ventricle. Circ Res 53: 306–318
27. Weber KT, Janicki, JS (1977) Myocardial oxygen consumption: the role of wall force and shortening. Am J Physiol 233: H421–H430
28. Webster MWI, Sharpe DN (1986) Adverse effects associated with newer inotropic agents. Med Toxicol 1: 335–342

Effect of Angiotensin-Converting Enzyme Inhibitors on Energy Metabolism in Chronic Heart Failure – Enzymatic Studies of Myocardial Biopsies

K. SCHULZE, M. SCHINDLER, and H.-P. SCHULTHEISS

Department of Internal Medicine, Klinikum Großhadern, University of Munich, Marchioninistr. 15, D-8000 München 70, FRG

Introduction

Severe chronic heart failure is characterized by poor systolic function with low cardiac output, increased systolic and diastolic volumes, augmented diastolic pressures and ventricular hypertrophy secondary to cavity dilatation [10, 22, 23]. The myocardial oxygen demand is heightened due to wall stress, myocardial contractility and heart rate [21, 24]. However, the oxygen supply, at least to the subendocardium, is inadequate, which is caused by decreased coronary blood flow, an increased intramyocardial vascular resistance by elevated filling pressures and a shortened diastolic perfusion time [17, 24]. There is evidence that progressive deterioration of the heart is due to the inadequate coronary blood flow to the subendocardium, leading to decreased oxygen delivery to this area with a resultant loss of its ability to contribute to the work of the heart. Thus the imbalance between myocardial oxygen supply and demand might initiate a vicious cycle that worsens heart failure and further decreases coronary blood flow [1, 22].

The evidence for an impairment of cardiac energy metabolism in severe heart failure is supported by former results of our group showing an increase in lactate dehydrogenase (LDH) 1 and a simultaneous decrease in LDH 5 in patients with dilative cardiomyopathy correlating with the degree of hemodynamic deterioration [18].

Angiotensin converting enzyme (ACE) inhibitors decrease left ventricular end-diastolic pressure, ventricular size and heart rate thus improving myocardial efficiency [11]. We determined the LDH isoenzyme pattern in myocardial biopsies of patients with severe chronic heart failure before and after therapy with enalapril in order to prove whether ACE inhibition is able to correct the myocardial imbalance between energy delivery and demand and can thus interrupt the self-perpetuation of the worsening of heart failure.

The structure of LDH is well known. Each isoenzyme is formed by four monomeres. LDH 1 consits of "H subunits" only, LDH 5 of "M subunits" only. The other three existing isoenzymes consist of H and M subunits in random association. The LDH isoenzyme pattern is closely related to the redox state of the pyridinenucleotides, which is a sensitive marker for the energy state of the cell [16]. In aerobically metabolizing tissues, LDH 1 is active serving as lactate dehydrogenase. In this condition the heart is a lactate consumer with the NADH quickly reoxidated by oxidative phosphorylation via mitochondrial shuttle systems. In

H.-P. Schultheiß (Ed.)
New Concepts in Viral Heart Disease

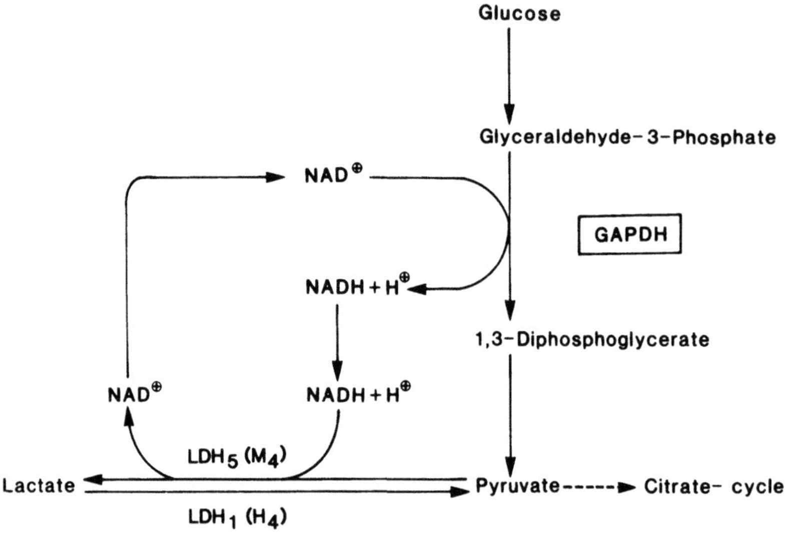

Fig. 1. Relation of LDH isoenzymes and NAD/NADH ratio in energy metabolism. LDH_5 (M-type LDH) serves as a pyruvate reductase in anaerobically metabolizing tissues, whereas LDH_1 (H-type LDH) serves as lactate dehydrogenase in aerobically metabolizing tissues

anaerobically metabolizing tissues pyruvate formation by glycolysis in enhanced. NADH reoxidation occurs by LDH 5 serving as pyruvate reductase. In this condition the heart is a lactate producer [6]. This coherence of the LDH isoenzymes and the NAD/NADH ratio is illustrated in Fig. 1. Therefore it is obvious that any change in the myocardial energy state should alter the LDH isoenzyme pattern.

Material and Methods

Thirty-three patients with severe chronic heart failure, NYHA classes III and IV, were given enalapril 5 mg twice daily for a period of 6 months. Pretreatment with digitalis and diuretics was not changed during the time of the study. Before and after treatment haemodynamic data and myocardial biopsies were obtained by right and left heart catheterization. LDH isoenzymes were measured by microisoelectric focusing as described elsewhere [19]. In brief, the biopsy material was put into phosphate buffer without delay. After homogenization and centrifugation the supernatant was diluted and mixed with glycerine in a 25% end concentration. Electrophoresis was run in 10-μl capillaries filled with 7% acrylamide. After electrophoresis, gels were stained by the tetrazolium method [19] and measured densitometrically in a photometer. For statistical analysis the t test for paired samples was applied.

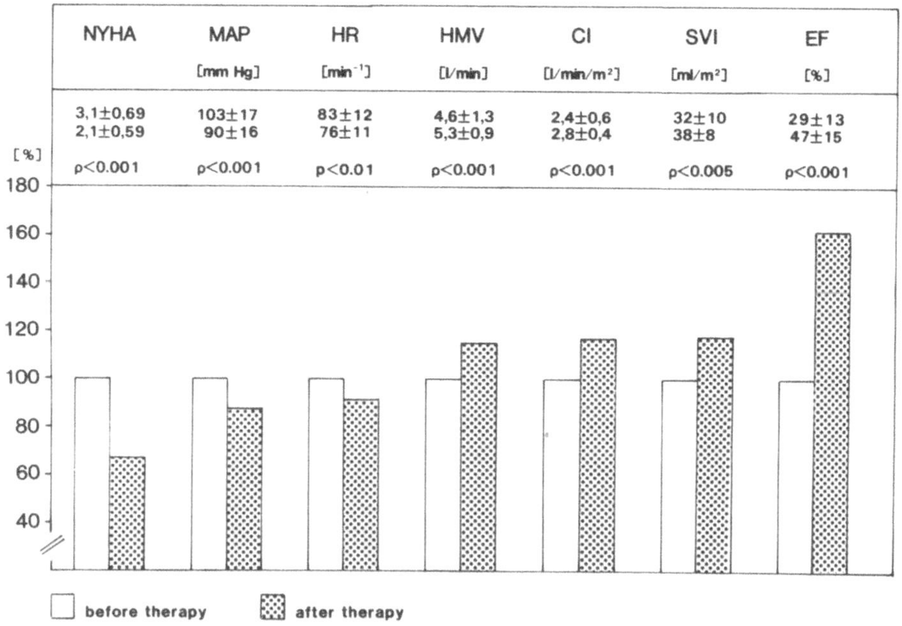

Fig. 2. Haemodynamic parameters before *(unshaded bars)* and after *(shaded bars)* 6 months of enalapril treatment. Values before treatment are given as 100%

Results

After 6 months of enalapril treatment we found, in full agreement with the results of numerous studies, a clinical improvement (NYHA classification decreased from 3.1 ± 0.7 to 2.1 ± 0.6), a reduction of mean aortal pressure (MAP) from 104 ± 17 to 93 ± 18 mmHg and heart frequency (HF) from 80 ± 16 to 76 ± 12 beats per minute, as demonstrated in Fig. 2. The parameters reflecting cardiac function increased: heart minute volume (HMV) from 4.6 ± 1.3 to 5.3 ± 0.9 litres per minute, cardiac index (CI) from 2.4 ± 0.6 to 2.8 ± 0.4 litres/min/m^2, stroke volume index (SVI) from 32.1 ± 10.5 to 37.6 ± 8.2 ml/m^2, ejection fraction (EF) from 29.7% ± 13.2% to 47.9% ± 15.3%. All these changes were significant ($P < 0.01$).

Preload determining parameters were reduced after the period of study: end-diastolic volume index (EDVI) from 126 ± 59 to 106 ± 33 ml/m^2 ($P < 0.01$), left ventricular end-diastolic pressure (LVEDP) from 24.9 ± 9.3 to 15.6 ± 7.8 mmHg ($P < 0.05$). The calculated diastolic wall stress decreased about 15% from 58 ± 34 to 51 ± 46 10^3 dyn/cm^2 (not significant).

The afterload-determining parameters were found diminished, too: the systemic vascular resistance fell from 1760 ± 594 to 1346 ± 401 dyn x s/cm^{-5} ($P <$ 0.001) and the systolic wall stress from 173 ± 56 to 147 ± 61 10^3 dyn/cm^2 (not significant).

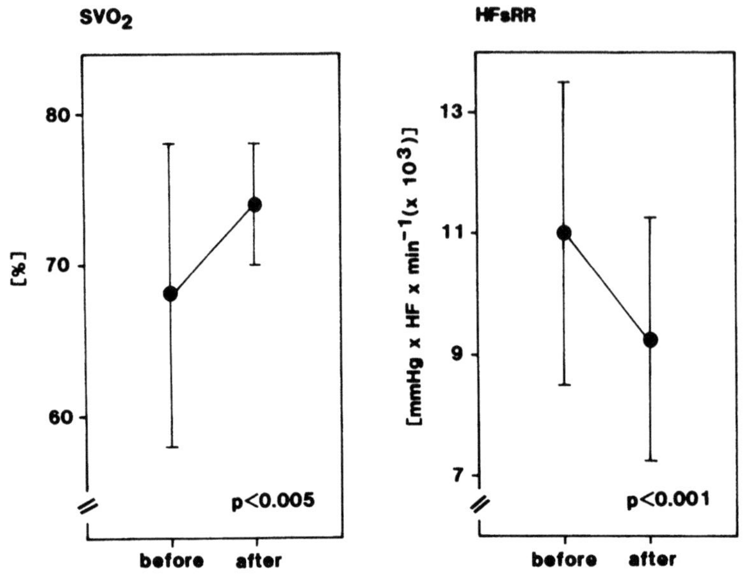

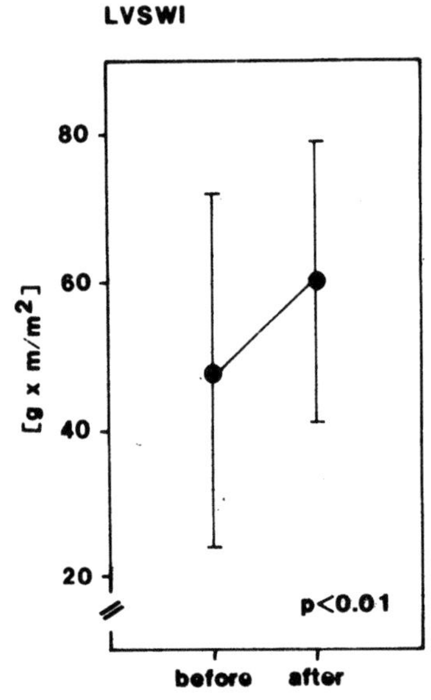

Fig. 3. Parameters of myocardial efficiency before and after 6 months of enalapril treatment. *SVO*$_2$, mixed venous oxygen saturation; *HFsRR,* rate pressure product; *LVSWI,* left ventricular stroke work index

As illustrated in Fig. 3, the mixed venous oxygen saturation increased from 68% ± 10% to 74% ± 5% ($P < 0.005$), and the rate pressure product decreased from 11055 ± 2629 to 9358 ± 1763 mmHg · beats/min ($P < 0.001$). The left ventricular stroke work index was found heightened from 53 ± 26 to 59 ± 20 g · m/m^2 ($P < 0.01$).

Figure 4 demonstrates the changes in LDH isoenzyme pattern before and after ACE inhibitor treatment: LDH 1 rose from 38.7% ± 6.7% to 42.3% ± 5.5% ($P < 0.005$) and the H subunits from 59.7% ± 7.7% to 64.8% ± 4.4% ($P < 0.001$)

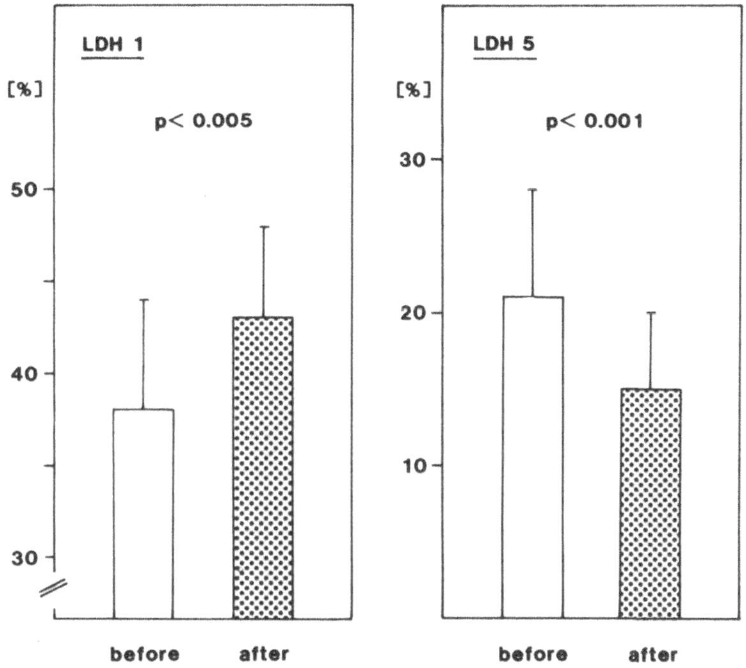

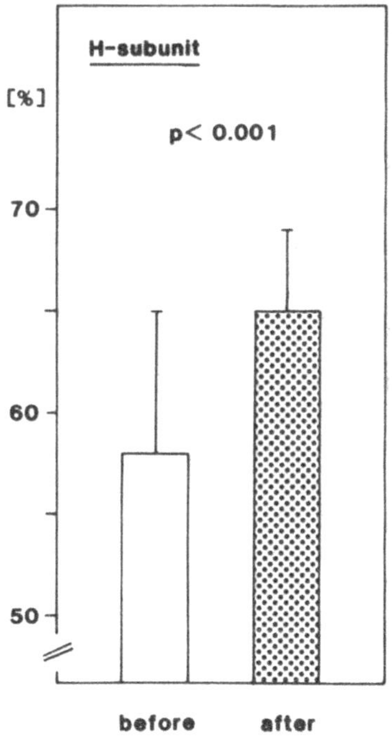

Fig. 4. LDH isoenzymes before and after 6 months of enalapril treatment, measured by microisoelectric focusing as described in the text

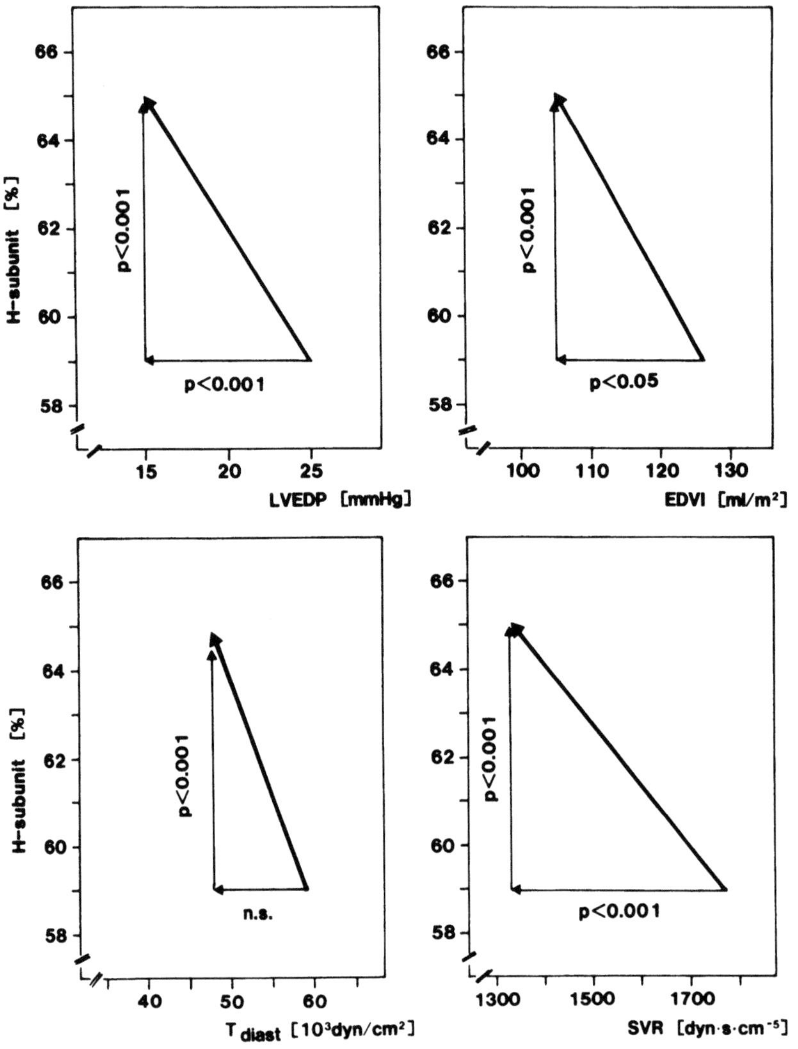

Fig. 5. Haemodynamic and metabolic changes after 6 months of enalapril treatment. *LVEDP*, left ventricular end-diastolic pressure; *EDVI*, end-diastolic volume index; T_{diast}, diastolic wall stress; *SVR*, systemic vascular resistance

paralleled by a decrease of LDH 5 from 20.8% ± 7.0% to 15.8% ± 4.7% ($P < 0.001$) and the M subunits (not shown in the figure) from 40.3% ± 7.7% to 34,9% ± 5.0% ($P < 0.001$). The simultaneous changes of the haemodynamic data and the LDH isoenzymes are shown in Fig. 5.

Discussion

Treatment with ACE inhibitors decreases cardiac pre- and afterload and wall stress [2, 3, 15], thus lowering myocardial oxygen demand as shown by the decrease in the rate pressure product. In addition, the diminished sympathetic nervous activity further reduces oxygen demand [12]. However, left ventricular function is improved, indicating a better efficiency of the myocardium.

On the other hand, myocardial oxygen supply is augmented due to a diminished coronary resistance by decreased filling pressures and wall stress (extravascular component) and by direct dilation of the coronary arteries (vascular component) [7, 8, 13, 14, 20]. The existence of a local coronary renin angiotensin system with converting enzyme present in the coronary circulation and consequently an increase in coronary flow after converting enzyme inhibition could be shown in isolated hearts perfused with blood-free solution [5, 9, 25].

The supposed correction of the imbalance between oxygen delivery and demand, which is caused by ACE inhibitor treatment, is confirmed by an improved myocardial energy metabolism. After 6 months of enalapril therapy we could demonstrate a shift in the LDH isoenzyme pattern towards a higher LDH 1 and a lower LDH 5 activity. This indicates a change from an anaerobic to an aerobic cardiac metabolism.

Thus our results show an improvement of cardiac function paralleled by an amelioration of the myocardial energy state by ACE inhibitor treatment. The efficiency of the heart is elevated. The observed correction of the impairment of energy supply and demand might terminate, or at least influence, the self-perpetuating process of the progressive subendocardial perfusion deficit in chronic heart failure. ACE inhibitor treatment could be decisive with regard to an improvement of the poor prognosis of severe chronic heart failure [4].

Summary

The LDH isoenzyme pattern was determined in 33 patients with severe chronic heart failure (NYHA classes III and IV) in right ventricular myocardial biopsies before and after 6 months of ACE inhibitor (enalapril) treatment. Coincidentally with the marked improvement of the haemodynamic parametes (systemic vascular resistance from 1760 $\pm$ 594 to 1346 $\pm$ 401 dyn $\cdot$ s cm^{-5}, systolic wall stress from 173 $\pm$ 56 to 147 $\pm$ 61 10^3 dyn/cm^2, left ventricular end-diastolic pressure from 24.9 $\pm$9.3 to 15.6 $\pm$ 7.8 mmHg, end-diastolic volume index from 126 $\pm$ 59 to 106 $\pm$ 33 ml/m^2, diastolic wall stress from 58 $\pm$ 34 to 51 $\pm$ 46 10^3 dyn/cm^2) and a slight decrease of heart rate from 80 to 76 beats per minute, the LDH 1 rose from 38.7% $\pm$ 6.7% to 42.3% $\pm$ 5.5% ($P < 0.001$) paralleled by a reduction of the LDH 5 from 20.8% $\pm$ 7.0% to 15.8% $\pm$ 4.7% ($P < 0.001$). This shift of the LDH isoenzyme pattern indicates an improvement of the myocardial energy state caused by an improved efficiency of the myocardium during ACE inhibitor treatment.

References

1. Bishop SP, Altschuld RA (1971) Evidence for increased glycolytic metabolism in cardiac hypertrophy and congestive heart failure. In: Alpert NA (ed) Cardiac hypertrophy. Academic, New York, pp 567–585

2. Captopril Multicenter Research Group (1983) A placebo-controlled trial of captopril in refractory chronic congestive heart failure. J Am Coll Cardiol 2: 755–763
3. Cleland JGF, Dargie HJ, Ball SG, Gillen G, Hodsman GP, Morton JJ, East BW, Robertson I, Ford I, Robertson JIS (1985) Effects of enalapril in heart failure: a double blind study of effects on exercise performance, renal function, hormones, and metabolic state. Br Heart J 54: 305–312
4. CONSENSUS Trial Study Group (1987) Effects of enalapril on mortality on severe congestive heart failure. Results of the Cooperative North Scandinavian Enalapril Survival Study (CONSENSUS). N Engl J Med 316: 1429–1435
5. Ertl G (1986) The effect of angiotensin I and II at normal and reduced coronary perfusion. Pflugers Arch 406 [Suppl]: R49
6. Everse J, Kaplan NO (1975) Mechanisms of action and biological functions of various dehydrogenase isoenzymes. In: Markert CL (ed) Isoenzymes. II. Physiological functions. Academic, New York, pp 29–43
7. Faxon DP, Creager MA, Halperin JL, Sussmann HA, Gavras H, Ryan TJ (1982) The effect of angiotensin converting enzyme inhibition on coronary blood flow and hemodynamics in patients without coronary artery disease. Int J Cardiol 2: 251–262
8. Foult JM, Nitenberg A, Tovolaro O, Antony I (1987) Selective coronary vasodilator effect of enalaprilat in patients with dilated cardiomyopathy: demonstration by a bilateral intracoronary infusion technique. J Am Coll Cardiol 9: 192 A
9. Gerlings ED, Gilmore JP (1974) Evidence for myocardial conversion of angiotensin, part 1. Basic Res Cardiol 69: 222–227
10. Goodwin JF (1974) Prospects and predictions for the cardiomyopathies. Circulation 50: 210–219
11. Halperin JL, Faxon DP, Creager MA, Bass TA, Melidossian CD, Gavras H, Ryan TJ (1982) Coronary hemodynamic effects of angiotensine inhibition by captopril and teprotide in patients with congestive heart failure. Am J Cardiol 50: 967–972
12. Hasking GJ, Esler, MD, Jennings GL, Burton D, Johns JA, Korner PI (1986) Norepinephrine spillover to plasma in patients with congestive heart failure: evidence of increased overall and cardiorenal sympathetic nervous activity. Circulation 73: 615–621
13. Holtz J, Busse R, Sommer O, Bassenge E (1987) Dilation of epicardial arteries in conscious dogs induced by angiotensin converting enzyme inhibition with enalaprilat. J Cardiovasc Pharmacol 9: 348–355
14. Magrini F, Shimizu M, Roberts N, Fouad FM, Tarazi RC, Zanchetti A (1987) Converting-enzyme inhibition and coronary blood flow. Circulation 75 [Suppl I]: 168–174
15. McGrath BP, Arnolda L, Matthews PG et al. (1985) Controlled trial of enalapril in congestive cardiac failure. Br Heart J 54: 405–414
16. Nadal-Ginard B, Markert CL (1975) Use of affinity chromatography for purification of lactate dehydrogenase and for assessing the homology and function of the A and B subunits. In: Markert CL (ed) Isoenzymes. II. Physiological functions. Academic, New York, pp 45–67
17. Nitenberg A, Foult JM, Blanchet F, Zouioueche S (1985) Multifactorial determinants of reduced coronary flow reserve after dipyridamole in dilated cardiomyopathy. Am J Cardiol 55: 748–754
18. Schultheiss H-P, Bolte HD, Fischer S, Cyran J (1980) Enzymatic analysis and collagen content in endomyocardial biopsy samples of patients with congestive cardiomyopathy of unknown etiology. Clin Cardiol 3: 329–334
19. Schultheiss H-P, Bispink G, Neuhoff V, Bolte HD (1981) Myocardial lactate dehydrogenase isoenzyme distribution in the normal heart. Basic Res Cardiol 76: 681–689
20. Schultheiss H-P, Zähringer J, v. Scheidt W, Ulrich G (1986) Myocardial lactate dehydrogenase (LDH) isoenzyme distribution in chronic heart failure (CHF) before and after treatment with enalapril. Circulation 74 [Suppl II]: 508
21. Strauer BE (1987) Cardiac energetics in clinical heart disease. In: Jacob R, Just HJ, Holubarsch C (eds) Cardiac energetics. Steinkopff, Darmstadt; Springer, New York
22. Unverferth DV, Magorien RD, Lewis RP, Leier CV (1983) The role of subendocardial ischemia in perpetuating myocardial failure in patients with nonischemic congestive cardiomyopathy. Am Heart J 105: 176–179

23. Weber KT, Janicki JS, Campbell C, Replogle R (1987) Pathophysiology of acute and chronic heart failure. Am J Cardiol 60: 3c–9c
24. Weiss MB, Ellis K, Sciacca RR, Johnson LL, Schmidt DH, Cannon PJ (1976) Myocardial blood flow in congestive and hypertrophic cardiomyopathy. Relationship to peak wall stress and mean velocity of circumferential fiber shortening. Circulation 54: 484–494
25. Xiang J-Z, Linz W, Becker H, Ganten D, Lang RE, Schölkens B, Unger TH (1985) Effects of converting enzyme inhibitors: ramipril and enalapril on peptide action and sympathetic neurotransmission in the isolated heart. Eur J Pharmacol 113: 215–223

Subject Index